Oil and Gas Pipelines and Piping Systems

Oil and Gas Pipelines and Piping Systems

Design, Construction, Management, and Inspection

Alireza Bahadori, PhD, CEng, MIChemE, CPEng, MIEAust, RPEQ
School of Environment, Science & Engineering,
Southern Cross University, Lismore, NSW, Australia
Australian Oil and Gas Services, Pty Ltd, NSW, Australia

ELSEVIER

AMSTERDAM • BOSTON • HEIDELBERG • LONDON
NEW YORK • OXFORD • PARIS • SAN DIEGO
SAN FRANCISCO • SINGAPORE • SYDNEY • TOKYO
Gulf Professional Publishing is an imprint of Elsevier

G P P

Gulf Professional Publishing is an imprint of Elsevier
50 Hampshire Street, 5th Floor, Cambridge, MA 02139, United States
The Boulevard, Langford Lane, Kidlington, Oxford, OX5 1GB, United Kingdom

British Library Cataloguing-in-Publication Data
A catalogue record for this book is available from the British Library

Library of Congress Cataloging-in-Publication Data
A catalog record for this book is available from the Library of Congress

ISBN: 978-0-12-803777-5

For Information on all Gulf Professional publications
visit our website at https://www.elsevier.com

Publisher: Joe Hayton
Senior Acquisition Editor: Katie Hammon
Senior Editorial Project Manager: Kattie Washington
Production Project Manager: Kiruthika Govindaraju
Cover Designer: Victoria Pearson Esser

Typeset by MPS Limited, Chennai, India

Dedication

Dedicated to the loving memory of my Parents, grandparents, and to all who contributed so much to my work over the years

Contents

Biography

Alireza Bahadori, PhD, CEng, MIChemE, CPEng, MIEAust, RPEQ, is an academic staff member in the School of Environment, Science, and Engineering at Southern Cross University, Lismore, NSW, Australia and Managing Director of Australian Oil and Gas Services, Pty Ltd, www.AustralianOilGas.com.au. He received his PhD from Curtin University, Perth, Western Australia. During the past 20 years, he has held various process- and petroleum-engineering positions and was involved in many large-scale projects at National Iranian Oil Co. (NIOC), Petroleum Development Oman (PDO), and Clough AMEC Pty Ltd.

He is the author of approximately 300 articles and 14 books. His books have been published by several major publishers, including Elsevier. He is the recipient of the highly competitive and prestigious Australian Government's Endeavour International Postgraduate Research Award as part of his research in the oil and gas area. He also received a Top-Up Award from the State Government of Western Australia through Western Australia Energy Research Alliance (WA:ERA) in 2009. He serves as a member of the editorial board and reviewer for a large number of journals. He is a chartered member of the Institution of Chemical Engineers (IChemE, London, United Kingdom), chartered professional engineer (CPEng), and member of The Institution of Engineers Australia.

Preface

In designing the pipeline and its associated piping systems, due account should be given to the operation, inspection, and maintenance requirements for the predicted life cycle and the planned conditions and criteria for the operation and maintenance of the pipeline. Due regard should also be given to manning levels, pipeline condition monitoring, and maintenance system, remote operations, communications, means of access to the right-of-way, bypass requirements for components needing regular maintenance without interruption of the pipeline operation, etc.

Moreover, the materials engineer shall establish the preferred materials selection based on the process requirements such as medium, pressure, temperature, flow, and the environment of the process facility. Design life and cost considerations shall also be taken into account in this respect.

This book covers essential requirements for all aspects to be considered in design of piping for oil, gas, and petrochemical plants to be designed, which includes, but is not limited to, the following:

1. Loading and unloading terminals.
2. Crude oil and gas gathering central facilities.
3. Process units.
4. Package equipment.
5. Pump house and compressor stations (booster stations).
6. Tank farms and oil/gas depots.

This book outlines requirements for the procedure in handing over, loading, hauling, unloading, keeping an inventory list, and the storing of materials required for installation of pipelines.

This book also covers requirements for trenching operations for buried pipelines. It explains conditions of trench before pipe is laid, conditions under which blasting can be performed, staking, and the marking and use of excavated materials.

This book also covers requirements to be observed in clearing a right-of-way and trenching operation in terrains containing natural rock, large boulders, or other materials that have to be removed by a blasting operation.

This book also deals with manners to be considered in taking pipes from stockpile and stringing along a right-of-way for subsequent welding, coating, lowering, and backfilling.

Moreover, this book covers checking to be made in respect of cleaning of pipe ends and interior of pipes as well as pipe condition. It also emphasizes the necessity of repair before lining up pipe and welding into a string.

It covers requirements for cold field bending. It describes limits to be observed in field bending in terms of minimum bend radius as well as maximum stretching or thinning of the pipe wall thickness.

In addition, this book covers technical requirements for welding of pipe over the ditch and double-jointing yard as well as laying pipe in trench, exposed, and casing sections. The requirements include complementary trenching works, considerations to be given in laying pipe when trench contains water or mud, and test of pipe coating just before laying.

This book also covers the requirements for backfilling of buried lines that are carried out immediately after a lowering operation and before hydrostatic testing. It explains how to check important features before performing backfilling operation, the quality of backfill material, and considerations to be given at crossings with waterways, roads, and irrigated lands. This book covers clean up and restoration activities that need to be performed immediately after the pipe is backfilled.

This book outlines specific considerations to be given in the construction of pipelines at crossings. It deals with surveys required to be carried out before giving schedules and timing for work execution at crossings.

This book also covers requirements for installation of casings for correct insertion of carrier pipe where pipeline crosses major public roads, freeways, and railways. It also explains boring methods for installation of casings in order to obviate interruption of traffic, test and installation of fabricated assemblies, such as launchers and receivers, line break valves, isolated valves, vent, and drain valves.

This book gives technical specifications and minimum requirements for welding of transportation pipeline and related facilities for use in oil, gas, and petrochemical industries, the arc welding of butt, fillet, and socket welds in carbon and low alloy steel for liquid and gas transmission pipelines, and related facilities, including pig traps. The welds may be produced by position or roll welding or by a combination of position and roll welding. Roll welding is only acceptable when using a fully automatic welding process. Oxyacetylene welding (otherwise known as gas welding) and flash butt welding processes shall not be used.

The standard also covers the acceptance standards to be applied to production welds tested to destruction or inspected by radiographic, ultrasonic, or magnetic particle techniques. It includes the inspection procedures for using these techniques.

This book covers the minimum requirements for welding work to be carried out for installation of on-plot piping in oil, gas, and petrochemical industries. The book relates to the requirements pertaining to welding techniques to be used, the qualification of welder/welding operators, and welding procedures together with testing and recording involved. It also deals with inspection, testing, and the limits of acceptability and heat treatment of production welds, if required.

This work covers the minimum requirements of pressure test to be carried out on plant piping systems. Upon completion of piping systems and before commissioning, it shall be pressure-tested in order to prove the strength of the system, its tightness (absence of leaks), and the integrity of weldments and materials.

This book covers the essential requirements of hydrostatic pressure test to be carried out on off-plot piping. Upon completion of a pipeline and before purging and commissioning operations, the line shall be high-pressure tested in order to prove the strength of the pipeline, its tightness (absence of leaks), and the integrity of weldments and materials. The test is also intended to confirm acceptability of pipeline for the service.

It covers the requirements for the internal chemical cleaning of piping systems on upstream sides of process machineries and lube oil/seal oil systems for which removal of rust, mill scale, grease, and foreign matter is essential.

This book covers minimum technical requirements for design, manufacture, quality control, testing, and finishing of pig launching and receiving traps that are installed in oil, gas, and petrochemical industries under the service conditions.

This book also provides technical specifications and general requirements for the purchase of steel line pipe for use in oil, gas, and petrochemical industries under nonsour as well as sour conditions.

Dr. Alireza Bahadori
Southern Cross University, Lismore, NSW, Australia
May 7, 2016

Acknowledgments

I would like to thank the Elsevier editorial and production team, Ms. Katie Hammon and Ms. Kattie Washington of Gulf Professional Publishing, for their editorial assistance.

Chapter 1

Transportation Pipelines

This chapter provides a baseline for minimum technical requirements and recommended engineering practices for design of pipelines used for transportation of hydrocarbons in Oil, Gas and Petrochemical Industries.

1.1 FLUID CATEGORIES

Based on the hazard potential of a fluid transported in the pipeline, it should be categorized in one of the following four groups (Table 1.1).

1.2 DESIGN

The relevant sections of ANSI/ASME and API codes and standards referred to and supplemented by this chapter should be used for design of the pipeline in which the operating conditions and requirements, ease of inspection and maintenance, environmental conditions, safety requirements, geographic location, climatic, geotechnic, and seismic conditions as well as future changes and expansions should be taken into account over the pipeline entire projected life cycle, including its final abandonment.

1.2.1 Operational Requirements

In designing the pipeline and its associated piping systems, due account should be given to the operation, inspection, and maintenance requirements for the predicted life cycle and the planned conditions and criteria as set by and/or agreed to in advance with the personnel responsible for the operation and maintenance of the pipeline. Due regard should also be given to manning levels, pipeline condition monitoring and maintenance system, remote operations, communications, means of access to the right-of-way, by-pass requirements for components needing regular maintenance without interruption of the pipeline operation, etc.

1.2.2 Economic Considerations (Optimization)

When there are alternatives for designing and constructing a pipeline, an economic analysis should be carried out to determine the optimum design

Oil and Gas Pipelines and Piping Systems. DOI: http://dx.doi.org/10.1016/B978-0-12-803777-5.00001-0

TABLE 1.1 Fluid Categories

Category	Description	Example
A	Nonflammable, stable and nontoxic fluids that are in liquid form at ambient temperature and 50 kPa (0.5 bar) above atmospheric pressure, i.e., having vapor pressure lower than 150 kPa (1.5 bar) (abs) at ambient temperature	Water, slurries
B	Flammable, or unstable or toxic fluids that are in liquid form at ambient temperature and 50 kPa (0.5 bar) above atmospheric pressure, i.e., having vapor pressure lower than 150 kPa (1.5 bar) (abs) at ambient temperature	Stabilized crude, gas oil
C	Nonflammable, stable, nontoxic fluids that are in gaseous form or a mixture of gas and liquid at ambient temperature and 50 kPa (0.5 bar) above atmospheric pressure, i.e., having vapor pressure higher than 150 kPa (1.5 bar) (abs) at ambient temperature	Nitrogen, carbon dioxide
D	Flammable, unstable, or toxic fluids that are in gaseous form or a mixture of gas and liquid at ambient temperature and 50 kPa (0.5 bar) above atmospheric pressure, i.e., having vapor pressure higher than 150 kPa (1.5 bar) (abs) at ambient temperature	Natural gas, LPG, ammonia

specifications to meet the specified operating requirements with the highest technical integrity in the best possible way at the lowest possible cost. The analysis should consider the following parameters as well as other factors that could have significant cost implications on the one hand and safety risks and environmental impacts on the other:

1. Different pipe diameters, operating pressures, flow velocities, materials, etc.
2. Distances between booster stations, with due consideration to other facilities required for operation and maintenance of booster stations.
3. Alternative routes with their problems, peculiarities, impacts and risks with due consideration to the interaction between the pipeline and the environment during each stage of the pipeline life cycle.
4. Various construction methods particularly at different crossings, difficult terrains, marshy areas, etc.

For fluid category B and C in location classes 3 and 4 and for category D in all location classes, risk assessment should be carried out to confirm that the selected design factors (refer to Section 1.4.2) and proximity distances (refer to Section 1.6.2) are adequate. Consideration should be given to the potential causes of failure such as internal corrosion and Hydrogen Induced

Cracking (HIC), internal erosion, external corrosion and bicarbonate and sulfide stress corrosion cracking, mechanical impacts and external interferences, fatigue, hydrodynamic forces, geotechnical forces, material defects, thermal expansion forces, etc. and their frequencies and the factors critical to public safety. The environment protection should be analyzed for the life span of the pipeline. While the risk of failures should be reduced to as low as reasonably practicable, the economic-risk aspects including cost of repairs, liabilities to the public and environment as well as loss of revenue, should be evaluated for various alternatives and for each phase of the pipeline-operating life, all of which should be compatible with the overall objectives of the company.

1.2.3 Hydraulic Design

Flow rate and/or pressure drop calculations may be made for the pipelines in various services using the formulas and methods set out and/or referred to in this subsection. Although the equations and methods for calculating the pressure drops quoted or referred to in this subsection have proved to be generally consistent with the actual experienced results during operation, nevertheless, more accurate methods of calculation should be considered for particular cases and where the fluid characteristics are fully known.

For a given pipe size, fluid characteristics, and flow rate, a hydraulic analysis should be carried out to establish the possible range of operational parameters that should provide the pressure and temperature profiles along the pipeline for steady state and transient conditions by taking full account of the possible changes in flow rates and operational modes over the life span of the pipeline.

The analysis should provide data to address the following:

- Surge pressure during sudden shut-down of the liquid lines.
- Turn-down limitation and inhibition or insulation requirements to avoid wax or hydrates or other impurities to deposit.
- Effect of flow rates on the efficiency of the corrosion inhibitors.
- Liquid catching and slug control requirement especially at the downstream end of two-phase lines or at the low pressure points.
- Effect of higher velocity ranges on impingement, cavitation and erosion on pipe wall, fittings and valves.
- Cleaning requirements for water and other corrosive substances which may deposit in the line. (Refer to API 1160)

1.2.4 Velocity Limitations

For liquid lines the normal average flow velocities should be selected between 1 and 2 m/s. Operations above 4 m/s should be avoided and lines containing a separate water phase (even in small quantity such as 1% water cut) should not operate at velocities below 1 m/s (to prevent water dropout which may create corrosive situations).

For gas lines the normal average flow velocities should be selected between 5 and 10 m/s. In special cases, continuous operations up to 20 m/s. Velocities lower than 5 m/s may have to be used for fluids containing solid particles where maximum velocity will be dictated by the occurrence of erosion.

Note:

The maximum velocity that can be attained by a compressible fluid is the critical or sonic velocity. In no case should the operating velocity exceed one half of the critical velocity.

$$V_c = \sqrt{kgRT}$$

where:

V_c critical velocity (m/s)
$k = C_p/C_v$ specific heat ratio
g gravity acceleration, 9.81 m/s^2
$R = R_0/M$ gas constant
$R_0 = 8314$ universal gas constant (J/kg mol/K)
M mole weight (kg)
T gas absolute temperature (K)

where a mixture of gas and liquid is being transported, the erosional velocity may be determined by

$$V_e = \frac{1.22C}{\sqrt{\rho_m}}$$

where:

V_e erosional velocity (m/s)
C empirical constant = 125 for noncontinuous operation and 100 for continuous operation
ρ_m density of the gas/liquid mixture in kg/m^3 at operating pressure and temperature
T flowing temperature (K)
M mass(kg/mol)
V_c critical velocity
V_e erosional velocity

If sand or other erosive solids are expected to be present, the fluid velocity should be reduced and/or special materials selected to avoid or reduce erosion.

However in two-phase lines (especially for long lines with elevation changes) the velocity should be selected to have a suitable flow regime with minimum pressure drop across the lines.

1.2.5 Pressure Drop Calculations

1.2.5.1 Pressure Drop in Crude Oil Pipelines

1. For sizes up to DN 750 (NPS 30) and fully turbulent flow, the Service Pipeline Co. formula may be used.
2. For sizes above DN 750 (NPS 30), the Shell/MIT (Massachusetts Institute of Technology) formula may be used for both laminar and turbulent flow.

Notes:

1. As for diameter sizes of DN 750 (NPS 30) and below, the MIT equation also gives acceptable results; it may be used for laminar flow in all pipe sizes.
2. The above formulas have given accurate results for the crude oils presently produced from most of the fields in south of (with API No. ranging between 30 and 34). However, for crude oil properties that are substantially different, these formulas may not be accurate enough and therefore basic hydraulic principles should be applied to determine the friction factor.
3. Flow lines should be sized primarily on the basis of flow velocity which should be kept at least below fluid erosional velocity (see Note in Section 1.2.4).

The pressure drop in the flow line as well as other design parameters should be such that gas separation from the oil cannot occur in the pipeline.

For estimating the pressure drop, the use of a simplified Darcy equation is recommended. However, for appropriate hydraulic design of flow lines, B.T Yocum computer program should be used with due consideration to flow regime and well characteristics.

1.2.5.2 Pressure Drop in Refinery Products Pipelines

1. Gas oil and fuel oil

   Service pipeline formula may be used.
2. Kerosene, aviation turbine kerosene, motor spirit, jet petroleum naphtha, and condensate

The T.R Aude or Hazen–William's formulas may be used.

Note:

Hazen–William's formula also gives very accurate result for water pipelines.

1.2.5.3 *Pressure Drop in NGL Pipelines*

Pressure drop calculations may be made by using one of the appropriate methods given for liquid transmission. Due consideration should be given to the thermal expansion and contraction of the liquid due to temperature variations. Also, pressure loss should not create vaporization and hence or otherwise two-phase flow in the pipeline.

1.2.5.4 *Pressure Drop in Natural Gas Pipelines*

For line sizes up to DN 300 (NPS 12) and operating pressures below 450 kPa, Weymouth formula may be used.

For line sizes up to DN 300 (NPS 12) and operating pressures above 450 kPa, the Panhandle revised (or B) equation may be used. For pipelines with diameters greater than DN 300 (NPS 12) the IGT/AGA formula may be used.

Notes:

1. Gas gathering lines between wellhead separators and production units or NGL plants may contain liquids and so the effect of two-phase flow should be taken into account in pressure drop calculations. Also, the effect of liquid accumulation at low sections of the pipelines with provision of liquid knock-out traps, if necessary and where permitted, should be considered in the design.
2. If periodical cleaning of the pipeline from liquids and other deposits is considered necessary by running pigs during operation, due regard should be given to the additional pressure requirements for pigging.
3. Increased system availability by having the possibility of line packing should be considered by excluding sections of decreasing design pressure in the pipeline.

1.3 MECHANICAL DESIGN

1.3.1 Application of Codes (Category B Fluids)

Pipelines carrying Category B fluids should be designed and constructed in accordance with ANSI/ASME B 31.4 and the additional requirements of this chapter.

1.3.2 Application of Codes (Category C and D Fluids)

Pipelines carrying category C or D fluids should be designed and constructed in accordance with ANSI/ASME B 31.8 and the additional requirements of this chapter.

Notes:

1. Although LPG and anhydrous ammonia are covered by ANSI/ASME B 31.4 but according to this chapter they fall under category D and therefore pipelines carrying these products should be designed to ANSI/ ASME B 31.8.
2. Flow lines should also meet the requirements of Standard Production Facilities Manual except for what is specified in this chapter especially for inhabited areas 50% of SMYS should be used.

1.3.3 Pigging Requirements

All pipelines should be designed to have the capability of passing suitable types of pigs through them as and when required.

Permanent pigging facilities should be considered for those pipelines that require frequent pigging and/or have operational constraints. The distance between pigging stations should be determined on the basis of anticipated pig wear and amount of collected solids that can be pushed through as well as time required for traveling of pig between launcher and receiver. Bends should have a sufficient radius to allow passage of those types of pigs that are anticipated to pass through them. The minimum radius of bend should be 7D.

Permanent pig signalers should only be considered when frequent pigging operations are anticipated. Flush mounted ancillary equipment, barred tees, and sphere tees with suitable drainage facilities, should be considered where appropriate. Pig launcher and receiver systems for pipelines should be designed in accordance with standards.

Valves to be used in the pipeline which will be pigged should be full bore through-conduit gate valves or full bore ball valves.

Reduced bore wedge gate or ball valves may be used in piping that is not to be pigged.

Check valves should not normally be installed in pipelines which will be pigged unless they have special design to make them capable of passing pigs.

1.3.4 Block Valves

Block valves should be provided at each end of all pipelines, at all connections and branches of the pipeline, and where necessary for safety and maintenance reasons to isolate long pipelines into sections as to limit the release of line content in case of leaks or line raptures.

The appropriate method of operating block valves (i.e., locally or automatically) should be determined from the likely effects of a leak or line rupture and its acceptable released volume based on the total time in which a leak can be detected, located, and isolated.

Automatic valves can be activated by detection of low pressure, increased flow, rate of loss of pressure or a combination of these, or a signal from a leak detection system. Automatic valves should be fail-safe. The closure time of the valves should not cause unacceptably high surge pressures.

The emergency shutdown valves should be automatically actuated when an emergency shutdown condition occurs at the plant or facility.

1.3.5 Thermal Relief Valves

Thermal relief valves should be considered for each section of liquid filled pipeline (including pig traps) that could be isolated by or between valves.

1.3.6 Vents and Drains

Vent and drain connections should be provided where necessary for satisfactory testing, commissioning, and operation.

1.3.7 Valves and Flanges

The rating of valves should be adequate for MAIP and test pressures of the pipeline subject to ANSI/ASME B 31.4 and B 31.8 pressure and temperature limitations.

The number of flanges in the pipeline and piping systems should be kept to a minimum and should be installed only to facilitate maintenance and inspection and where construction conditions or process requirements dictate. Tie-in welds are preferred.

1.3.8 Double Block and Bleed System

Double block and bleed system should be used in the situations where isolation of the main stream from the ancillary equipment is needed for safe operation and maintenance without depressurizing the pipeline.

1.3.9 Emergency Depressurization Facilities

Emergency depressurization facilities should be considered at one end of all pipelines and for category C and D fluids, at each sectionalizing valve location. The material specified for the blowdown system should be suitable for low temperatures encountered during blowdown of category C and D fluids. The capacity of the blowdown system should be such that the pipeline can be depressurized as rapidly as practicable. Due regards should be given to the control of excessive movements and vibration of the system due to forces created by sudden blowdown.

1.3.10 Overpressure Protection System

Any type of pressure control system should not be considered as an overpressure protection system. An overpressure protection system (consisting of mechanical safety/relief valves) should be fitted between the pipeline and the upstream facilities which can generate pressures in excess of MAIP of the pipeline. MAOP should not be exceeded at any point along the pipeline during normal continuous operations and MAIP should not be exceeded at any point along the pipeline during upset conditions of limited frequency and duration.

The pipeline system should be designed such that surge pressure cannot exceed MAIP at any point along the pipeline and will not trigger the overpressure protection system if fitted for protection from upstream facilities.

The occurrence of pressure surges should be determined for fluids with high density and low compressibility (such as liquid fluids) by transient pressure analysis, using a specialized simulation computer program. The location of the highest pressure points along the pipeline should be recognized especially in hilly terrain.

Unacceptably high surge pressures should be prevented by one or a combination of the following methods:

- Valve closure speed reduction.
- Special fast-response pressure relief systems close to the point of surge initiation.
- Strict adherence to well-formulated operating procedures (especially when other methods are insufficient).

1.3.11 Pipeline Stability

Sections of the pipelines in swamps, floodable areas, high water table areas, river crossings, etc., should be stable under the combined action of hydrostatic and hydrodynamic forces. The negative buoyancy should be sufficient to prevent unacceptable lateral and vertical movements and displacement of the pipeline. One or a combination of the following methods can be employed to achieve on-bottom stability:

- Increasing the pipe wall thickness.
- Applying concrete weight coating.
- Installing spaced anchor points, set-on weights, or bolt-on weights.
- Burying the pipeline.

The pipeline should be stable while empty or filled with water (for test) or with fluid for which it is designed. When calculating the negative buoyancy the density of water-logged backfill mud should be taken into account.

Special consideration should be given to possible differential settlements in weak soils which may cause damage to the pipeline.

1.4 PIPELINE WALL THICKNESS CALCULATING BASIS

1.4.1 Minimum Wall Thickness

The nominal pipe wall thickness should be calculated according to ANSI B 31.4 for category B service and ANSI B 31.8 for categories C and D services.

Special attention should be paid to the requirements given in the above-mentioned standards for the least wall thickness of the pipe when the ratio of pipe nominal diameter to wall thickness exceeds 96.

1.4.2 Design Factors (for Hoop Stress Limitation)

The recommended design factors for the calculation of the nominal wall thickness (excluding any corrosion allowance) are given in Table 1.2, derived from ANSI/ASME B 31.8 Table 841.114 B but expanded.

1.4.3 Strain-Based Design for Hot Products Pipelines

For hot products pipelines (above 80°C), a strain-based approach may be used. In this case, a maximum permanent deformation strain of 2% is acceptable.

1.4.4 Derating Factors

Derating factors for carbon steel materials operating at above 120°C should be used in accordance with Table 841.116 A of ANSI/ASME B 31.8. For duplex stainless steel, derating is required at lower temperatures (above 50°C).

1.5 MATERIALS

Depending mainly on the type of the fluid to be transported, especially its corrosivity, flow regime, temperature and pressure, the selection of pipeline material type can become a fundamental issue which should be decided at the conceptual design stage of a pipeline project. The most frequently used pipeline materials are metallic, especially carbon steel. Since the protection of internal corrosion and erosion of the pipe wall are governed by a variety of process conditions such as corrosivity of the fluid (particularly due to presence of water combined with hydrogen sulfide, carbon dioxide, or oxygen), temperature, pressure and velocity of the fluid as well as deposition of solids, etc., cannot be easily achieved in the same manner as for the protection of external corrosion, the selection of pipeline material should be made after careful consideration of all conditions to ensure that the pipeline can remain fit-for-purpose throughout its life time.

TABLE 1.2 Design Factors for Onshore Steel Pipelines

Fluid Category	B	C and D			
Applicable ANSI/ASME Code	**B 31.4 (Note 1)**	**B 31.8**			
Location classes	1, 2, 3, and 4	1	2	3	4
Pipelines	0.72	0.72	0.60	0.50	0.40
Crossings (Note 2)					
Private roads	0.72	0.72	0.60	0.50	0.40
Unimproved public roads	0.60	0.60	0.60	0.50	0.40
Roads, highways, streets, and railways	0.60	0.60	0.60	0.50	0.40
Rivers, dunes, and beaches	0.60	0.60	0.60	0.50	0.40
Parallel encroachments (Note 3)					
Private roads	0.72	0.72	0.60	0.50	0.40
Unimproved public roads	0.72	0.60	0.60	0.50	0.40
Roads, highways, streets, and railways	0.72	0.60	0.60	0.50	0.40
Fabricated assemblies (Note 4)	0.60	0.60	0.60	0.50	0.40
Pipelines on bridges	0.60	0.60	0.60	0.50	0.40
Near concentration of people	0.72	0.50 (Note 5)	0.50 (Note 5)	0.50	0.40
Pipelines, block valve stations, and pig trap stations (Note 6)	0.60	0.60	0.60	0.50	0.40
Compressor station piping	–	0.50	0.50	0.50	0.40

Notes:

1. ANSI/ASME B 31.4 does not use design factors other than 0.72, which is considered inappropriate at critical locations (e.g., crossings, within plant fences), and for fabricated assemblies. In these situations, design factors in line with ANSI/ASME B 31.8 location Class 1 are recommended.
2. ANSI/ASME B 31.8 differentiates crossings with casings and without casings. Because of the poor experience of cased crossings (i.e., annular corrosion), the same design factor is recommended, whether a casing is used or not. Design factors for crossings of rivers, dunes and beaches, not included in ANSI/ASME B 31.8, are provided.
3. Parallel encroachments are defined as those sections of a pipeline running parallel to existing roads or railways, at a distance less than 50 m.
4. Fabricated assemblies include pig traps, valve stations, headers, finger type slug-catchers, etc.
5. Concentrations of people are defined in ANSI/ASME B 31.8 Article 840.3.
6. This category, not specifically covered in ANSI/ASME B 31.8, is added for increased safety.

When sour service conditions are foreseen (as specified in NACE MR 0175) the line pipe material and other materials should be specified to resist sour services, regardless of whether or not the fluid is to be dehydrated and inhibitors are to be used.

Carbon steel linepipe material may be used in "light" sour corrosive conditions (typically where rate of corrosion is less than 0.5 mm/year without inhibition) but with sufficient corrosion allowance, inhibitor injection, appropriate inspection and controlled operation. Corrosion allowances in excess of 3 mm should not be considered without detailed analysis by corrosion specialists.

If conditions that may cause erosion cannot be avoided, special materials with improved designs to reduce or eliminate erosion should be used.

When selecting higher grades of steel linepipe (×60 and higher), special attention should be given to weldability and welding procedure (especially requirement for preheating to 300°C the unfinished welds before re-welding, and required yield-to-tensile ratio. Use of grades higher than ×70 is not recommended at present.

When low temperatures are expected (e.g., at downstream of gas pressure reducing stations), attention should be given to the fracture toughness properties of pipe material (for possibility of long running fractures).

1.5.1 Material Procurement

All materials should comply with relevant codes, standards, specifications, and technical requirements set and/or approved by the company and should be procured from company-approved vendors/ manufacturers/suppliers.

Depending on the criticality of pipeline, type of material, past performance, and the quality control system of the manufacturer, the company should specify the level and extent of inspection that the company intends to perform (if any).

For each pipe size, sufficient spare materials for possible route deviations, transportation, and construction damages, testing, and set-up of contingency stock should be estimated and ordered with the actual quantities required for the project.

1.5.2 Line Pipe Materials

Carbon steel line pipe should be in accordance with API Spec. 5L

Line pipe materials other than carbon steel should comply with ANSI/ASME B 31.4 and B 31.8 and this supplement as well as other specific relevant supplements and codes specified by the company.

1.5.3 Valves

The valve inlet and outlet passages should be specified to match the pipe internal diameter.

Check valves should preferably be swing type to API-6D. Other types may be considered subject to prior approval of the company.

1.5.4 Branch Connections, Fittings, Etc.

Threaded connections (pipe to pipe, fittings, etc.) and slip-on flanges should not be used in any part of the pipeline system.

"Pup" pieces should not be less than 0.15 m or 2½ pipe diameter, whichever is greater.

Flanges should preferably be of welding neck type, and the neck should match the internal diameter of the line pipe for welding.

Flanged connections should conform to the followings

- Raised face flanges for classes 900 and below.
- Ring type joint flanges for classes above 900 and flow lines.

Note:

Gaskets should be raised face spiral wound for raised face flanges.

Branch or instrument connections smaller than DN 50 (NPS 2) should not be used on pipeline for mechanical strength reasons. For pipelines smaller than DN 50 (NPS 2), the branch connections should be of the same diameter as the pipeline.

Weldolets larger than DN 75 (NPS 3) should not be used.

1.6 PIPELINE ROUTE SELECTION

In selecting the route, full account should be taken of the associated risks (particularly safety and environmental risks based on location classes, fluid categories, expected frequency of failure, etc.), the accessibility for maintenance and inspection, as well as economic factors (length of line, difficult terrains and crossings, etc.).

Site checks of alternative routes should be made, and available maps and geotechnical/geological information should be studied before selecting a suitable route for detailed survey.

1.6.1 Route and Soil Surveys

Detailed survey data should be made available before finalizing the pipeline route and carrying out detailed design. These data should comply with standards. Additional plan and profile drawings at enlarged scales should be provided for difficult sections such as crossings at rivers, roads, railways, etc. Full topographic surveys may be required for certain areas.

The profile drawings should also indicate areas in which major excavation or elevated pipeline supports may be required.

The radius of curvature of the pipeline foundation along the route should not be less than 500× the pipeline diameter (bends should be used when lower values are necessary). Additional data to be furnished as follows:

1. Geotechnical and other environmental data (such as landslides, faults, earthquakes, floods, currents at river crossings, climatic data, vegetation, fauna, etc.).
2. Soil investigation for type and consolidation of ground for assessing the degree of excavation difficulties.
3. Soil investigation for foundation design (burial and/or support design), subsidence areas (e.g., underground erosion and cavitation by acidic water or mining activities).
4. Water table levels at mid-spring and winter along the route of the pipeline where it is to be buried.
5. Soil resistivity along the pipeline route for coating selection and cathodic protection design. Areas where soil properties may change due to causes such as sulfide-reducing bacteria, which increases the current required for cathodic protection systems, should be identified.

1.6.2 Proximity to Occupied Buildings

For minimum distance of pipeline from occupied buildings, reference should be made to the safety regulations enforced by the related company.

1.6.3 Proximity to Other Facilities

- For categories B, C, and D, the separation requirements of the pipeline to other facilities within plant fences should be in accordance with standards.
- For separation requirement at crossings see Section 1.8 of this chapter.
- Refer to the Institute of Petroleum Model Code of Safe Practice Part 15 for area classifications around the pipeline.

1.6.4 Right-of-Way

Every pipeline should have a permanent right-of-way with sufficient width to enable the line to be constructed (including future additional lines) and to allow access for pipeline inspection and maintenance.

Land acquisition drawings should be prepared, and necessary coordination with related authorities should be made.

1.6.4.1 Right-of-Way Width

For every pipeline project, the width of the right-of-way should be decided based on the following criteria:

- Pipeline being buried or above ground.
- Diameter of the pipeline.

- Method of construction.
- Zig-zag configuration of above-ground pipeline.
- Pipeline being in flat areas or in mountainous or hilly areas, etc.
- Future pipelines along the same route (particularly in hilly and mountainous areas where blasting and/or excavation for widening the existing right-of-way may create problem).
- Type of fluid and pressure of the pipeline and the consequential risks of pipeline failure. Buried pipeline widths of right-of-way should conform to standard drawing.

The following figures can be considered as minimum widths of right-of-way and may be increased where necessary to suit the particular requirements of a specific project or may be reduced, subject to prior approval of the company, if certain restrictions do not permit widening of the right-of-way to the required ideal widths:

a. For above-ground pipelines in flat areas:

For DN 150 (NPS 6) and below	25 m
For DN 200 (NPS 8) up to and including DN 650 (NPS 26)	40 m
For above DN 650 (NPS 26) and based on 1–3 lines per track	60 m

b. For above-ground pipelines in hilly and mountainous areas:

For DN 400 (NPS 16) and below	21 m
For above DN 400 (NPS 16)	24 m

c. For buried pipelines, Table 1.3 should be used for one line per track:

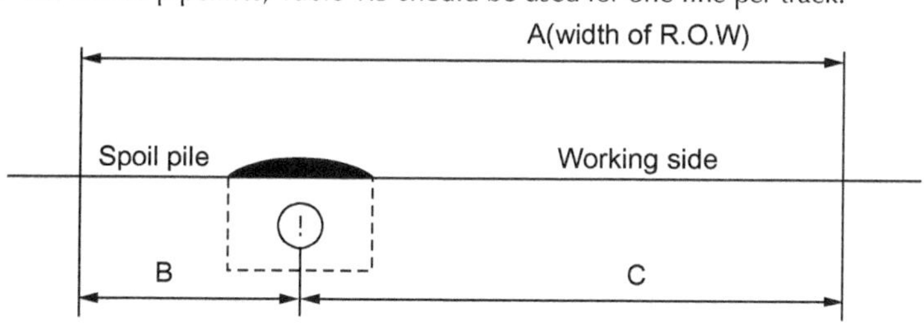

1.6.4.2 *Other Considerations*

The longitudinal slope of right-of-way should not exceed 22%. However, for short distances (less than 1 km), the longitudinal slope of the right-of-way may be up to 30% in which case the service roads with maximum longitudinal slope of 22% should be considered for access to these sections. In high longitudinal slope and depending on depth of trench coverage and type of soil and seasonal inundation where pipeline may lose its full restraint, it should be ensured that the equivalent stresses in the pipe wall are within acceptable limits or else remedial provisions are considered to reduce or

TABLE 1.3 Right-of-Way Width

Pipeline Diameter (mm)		Widths of R.O.W in Soft Soil (m)			Widths of R.O.W in Rocky Ground (m)		
		A	B	C	A	B	C
Up to DN 300	(NPS 12)	16	4	12	11	3	8
DN 350–DN 400	(NPS 14–16)	18	5	13	11	3	8
DN 450–DN 550	(NPS 18–22)	20	5	15	12	3	9
DN 600–DN 900	(NPS 24–36)	23	5	18	15	3.5	11.5
DN 950–DN 1050	(NPS 36–42)	25	7	18	17	4	13
Above DN 1050	(NPS 42)	28	8	20	19	5	14

Notes:
1. For additional lines the width should be increased by one dimension B for each new line (when two lines are not of the same diameter, the larger diameter should be assumed for both lines).
2. When several pipelines have to be installed in the same trench, the minimum separation between two adjacent pipelines should be 0.3 m.
3. The crossing of existing pipelines, cables, power lines, roads, railways and waterways should be at an angle between 90 and 60 degrees.
4. When installing a pipeline along power lines, the horizontal distance from any of the power cables and posts should be at least 4 m for power lines up to 63 kV and 10 m for power lines above 63 kV.

eliminate longitudinal forces due to effective component of the dead weight of the pipeline and its content.

The design of right-of-way should comply with line bending specification and also civil construction standard specification (Earthworks).

1.7 PIPELINE PROTECTION AND MARKING

1.7.1 Burial Philosophy

Pipelines are normally buried to protect them from mechanical damage, unusual environmental and climatical conditions, fires, tampering, etc., and to assure that they are fully restrained. As a general rule, pipelines of DN 400 (NPS 16) and larger should normally be buried unless the terrain would make burial impracticable or the length is too short to justify burial advantages. Pipelines of DN 300 (NPS 12) and smaller and short life pipelines of all sizes (such as flow lines) may be laid above ground unless there are good reasons for burial; e.g., process requirements or where protection from diurnal temperature variation is necessary or where the line passes through populated areas, etc.

1.7.2 Trench Dimensions

The recommended minimum covers are given in Table 1.4, based on ANSI/ ASME B 31.8 Article 841.142 but with some modifications for increased safety margins.

Additional depth may be required in certain locations such as agricultural areas where depth of ploughing and of drain systems should be taken into account. A cover of 1.2 m would be adequate in most cases. The trench width should be not less than 400 mm wider than the pipeline outside diameter in all ground conditions including rock.

When pipelines are coated and/or insulated, the outside diameter of coated or insulated pipe should be assumed as outside diameter for minimum coverage.

1.7.3 Thermal Expansion and Other Forces

Buried pipelines operating at very high temperatures may be prone to upheaval buckling caused by high compressive loads due to expansion. In such cases, the depth of burial cover should be increased to prevent the upheaval buckling. In general, the recommended cover depth should be enough to make the pipeline fully restrained and to contain thermal expansion and contraction of the pipeline as well as other forces due to internal pressure and pipeline weight in slopes.

Pipeline anchors should be installed at end points of buried pipelines and at other locations where the pipeline rises above ground level for connections to facilities, etc.

Pipeline anchors should be designed for the particular application to withstand forces due to MAIP and temperature variations and to suit the ground conditions especially where subject to seasonal inundation or in dry water courses in high slopes where pipeline dead weight creates longitudinal stresses.

TABLE 1.4 Recommended Minimum Cover for Buried Pipelines

Local	Minimum Cover (m)	Minimum Cover (m)
	In Normal Ground	In Rock Requiring Blasting
Class 1	0.9	0.6
Class 2	1.0	0.8
Classes 3 & 4	1.2	1.0
Public road and railway crossings	2	2

Notes:
1. The cover refers to the undisturbed ground level to the top of the pipe.
2. A minimum vertical clearance of 0.3 m should be kept between the pipeline and other buried structures (see also Section 1.8.3 for crossing other pipelines).

1.7.4 Nonburied Pipelines

Any nonburied pipeline sections should be justified on an individual basis and hence should be installed in such a way that stay clear of the ground all the time to avoid external corrosion.

The height of supports should be chosen to suit local conditions but should be sufficient to keep the bottom of the pipeline at least 300 mm above the highest recorded flood level.

Nonburied pipelines should normally be laid in a zig-zag configuration to cater for the effect of thermal expansion and contraction. The zig-zag configuration may be in accordance with Fig. 1.1. However, for specific cases, the correct configuration should be determined by appropriate design (Table 1.5).

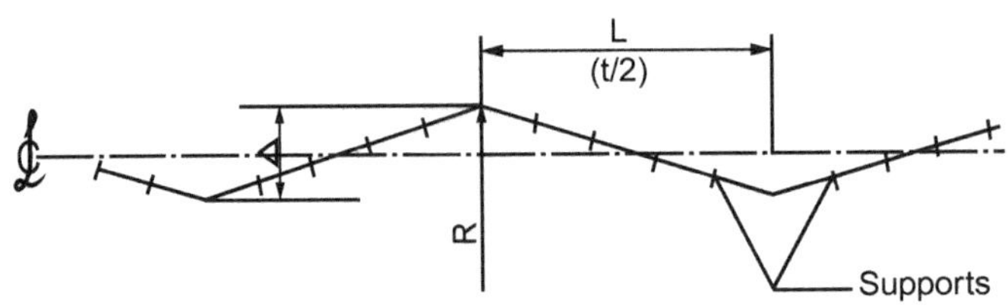

FIGURE 1.1 Plan view of zig-zag configuration for above-ground pipeline.

TABLE 1.5 Zig-Zag Configuration Dimensions

Pipe Size DN (NPS)		Pipe Material Grade Per API 5L	Straight Length L (m)	Offset Δ(m) (Minimum)	Bend Radius R (m) (Minimum)
Up to DN 300	(NPS 12)	GR B	60	4	25 × Dia.
DN 400	(NPS 16)	GR B/×42	116	9.1	17
		×52/×60	100	6.5	17
DN 500	(NPS 20)	GR B/×42	116	9.1	22
		×52/×60	100	6.5	22
DN 550	(NPS 22)	GR B/ ×42	116	9.5	23
		×52/×60	100	6.5	23
DN 600	(NPS 24)	GR B/×42	116	9.5	25
		×52/×60	100	6.5	25
DN 650	(NPS 26)	GR B/×42	116	7.1	28
		×52/×60	100	6.5	28
DN 750	(NPS 30)	GR B/×42	116	7.1	31
		×52/×60	100	6.5	31

Where zig-zag configuration is not or cannot be employed, alternative means, such as fully restraining the pipeline from movements (e.g., by adequate anchoring at appropriate intervals), should be provided to contain thermal expansion and contraction as well as other prevailing forces.

Pipeline anchors should be considered for nonburied pipelines at all tie-in connections to other facilities and at other positions where restraint may be necessary.

Hillside anchors should be designed, as and where required and should be installed on steep hills to restrain pipeline movement and to keep the combined stresses in the pipeline wall within the acceptable limits. The effect of the weight of the pipeline and its contents on the longitudinal stress in the pipeline wall should be considered in calculating the combined stresses.

1.7.5 Corrosion Protection

As a general rule, in normally dry climates, no external anticorrosion coating is required for above-ground pipelines that are supported clear of the ground. However, where the climatic conditions or the ground are such that external corrosion may occur, either a corrosion allowance on the pipe wall thickness may be required or, alternatively, a suitable anticorrosion coating should be considered.

Where sections of above-ground pipelines are to be buried (e.g., road, railway, or river crossings), the buried sections should be suitably coated, cathodically protected, and electrically isolated from the rest of the pipeline.

Those sections of pipeline that pass above waterways and rivers should be externally coated for protection against corrosion caused by condensation of water vapor on the pipeline exterior.

Where above-ground pipelines pass through culverts or below bridges (which are normally for pipelines crossing the main roads and/or for surface water passages), these sections of the lines should be suitably coated for protection against splashing water and blown sand and dirt.

All metallic buried pipelines including duplex material pipelines, should be coated externally by a suitable anticorrosion coating, supplemented by cathodic protection, and electrically isolated from the plants and facilities to which they are connected.

The design of cathodic protection systems should be carried out in accordance with standards.

Protective coatings should be selected to suit the soil and other environmental conditions and should comply with standards.

1.7.6 Pipeline Markers

The location of buried pipelines should be clearly identified by markers. In areas where the risk of interference or disturbance by mechanical excavators or boat anchors (at river crossings) is high, additional warning signs should be installed to lower the risk. Pipeline markers should be installed at the following locations along buried pipelines:

1. At 1-km interval.
2. At all major changes in direction of the pipeline.
3. At both sides of every road, railway, and under-water crossings.
4. At changes in wall thickness or material.
5. At branches.
6. At buried valves and fittings such as check valves, vents, drains, slug-catchers, etc.

Fabrication and installation details should be as per Standard Drawing.

1.8 CROSSINGS

1.8.1 River Crossings

Where pipeline has to cross a major river, careful studies should be carried out to determine the most suitable way of crossing that will ensure maximum reliability during the pipeline operating life with minimum maintenance problems. The selection of the most suitable location and type of crossing should be based on the survey results and information on geotechnical and hydroclimatological conditions and other prevailing environmental issues. The migration of the river course should also receive particular attention.

Elevated pipe supports should be high enough to carry the line at least 300 mm clear of highest flood level (oldest available return conditions). This clearance should be increased if there is likelihood of large floating objects being carried by flood water and where the river is navigable. Elevated pipe supports should be designed to suit the particular circumstances and be strong to withstand the forces imposed on them by flood water and the objects that are carried by the flood and may be caught by the supports. In wide rivers and where there is the possibility of torrential flood, pipe bridges are preferred to single pipe supports. If pipeline is to be cathodically protected, means of isolating the pipeline from the supports should be considered.

The sections of pipelines laid under the river bed should be coated and wrapped in accordance with standards.

The sections of pipeline laid in trenches in the river bed should be weight-coated to give the necessary negative buoyancy to the pipeline to fully restrain the pipeline in position at all times during construction, operation, and while shut down for maintenance or inspection. The weight coating should normally be designed to maintain pipeline stability in mud of specific gravity 1.2. In any case, the nature of the river bed should be taken into account in determination of required weight.

Depth of cover and the curvature of the pipeline during laying and henceforth as well as method of laying the pipeline should be selected for the particular application to avoid damage to the pipeline especially when it is being installed.

Isolating block valves fitted with automatic line-break-operators should be installed in fenced areas on either side of the major river crossings. If valves are installed in valve pits, the top of the pits should be above the maximum recorded high-water level, and if there is a possibility of water ingressing into the pits, facilities should be considered for emptying the water.

The automatic line-break-operators should be designed to close the valve in the event of pipeline failure and subsequent rapid rate of change of pressure in the pipeline but should not be affected by normal operational pressure fluctuations. The design should ensure that changes of the water course and/or collapse of the river side walls will not endanger the integrity of the valve support.

1.8.2 Road and Railway Crossings

Pipelines crossing roads and railways should preferably be through culverts or concrete box and bridges (new or existing). The use of casing pipe should be discouraged (due to external corrosion problems and electrical contact between casing pipe and carrier pipe). (See API RP 1102 for recommendations in this respect.) Suitable protection should be provided on both sides of the road to prevent damage to the pipeline by vehicles leaving the road.

If the right-of-way is intended for more than one pipeline, culverts or bridges should be wide enough to accommodate future pipeline(s). In this case the horizontal space between two adjacent pipelines should not be less than 400 mm.

1.8.3 Crossing Other Pipelines

Where above-ground pipelines cross each other, a minimum clearance of 300 mm should be maintained between adjacent lines.

Where a buried pipeline is to cross an existing above-ground pipeline, an increased depth of cover should be specified for the whole width of the right-of-way.

Where an above-ground pipeline is to cross an existing buried pipeline means should be provided to allow continued use of the buried pipeline right-of-way.

Where a buried pipeline is to cross an existing buried pipeline, the new line should pass under the existing line with at least 900 mm clearance between the two lines.

Potential test points, current test points, and bonding points (direct or resistance) should be installed on both lines at the crossing to enable the cathodic protection systems to be interconnected, if required.

For a minimum distance of 15 m on either side of the pipeline crossing, the new pipeline should be double wrapped.

Where a pipeline crosses an existing pipeline owned by an outside company, the design of the crossing and cathodic protection should satisfy the requirements of the outside company.

1.8.4 Crossing Land Faults

When pipeline has to cross a passive fault, the necessity of provision of any protection system should be decided after geotechnical survey results are studied by the company's geological department or by a company-appointed geologist or otherwise their recommendations as given. Crossing an active fault should be avoided if feasible. When, however, pipeline has to cross an active fault or a passive fault which is expected to become active, the following considerations should be given at the crossing for the protection of the pipeline.

Design factors similar to those indicated for rivers, dunes, and beaches should be used.

Crossing angle should be selected for minimum bending and shear stresses in the pipeline wall due to movement of the fault banks.

There should be no horizontal bends, flanges, tees, valves, or similar constraints such as concrete weights in at least 200 m of the pipeline either sides of the fault center.

The trench dimensions and the backfill material around the pipeline at 200 m on either side of the fault center should be selected in such a way that the pipeline is subjected to minimum restraint.

Line break valves with automatic shut-down operators should be installed at 250 m either side of the fault center. These valves should be secured against movements of the section of the pipeline which crosses the fault by means of adequately designed anchors.

1.8.4.1 Land Slides

Passing near the areas where there is evidence of a landslide should be avoided by using alternative routes or going around the suspected areas.

1.9 RECORDS

A comprehensive set of design documents should be produced and retained for the life of the pipeline. These documents should include all the design criteria, calculations, and assessments that led to the technical choices during conception and design of the pipeline. They should also include a pipeline-operating and -maintenance manual that covers the range of key operating conditions that can be envisaged for the entire life span, major features, parameters, contingency plans, etc.

APPENDIX

Service Pipeline co. Formula

For turbulent flow up to $Re = 170{,}000$

$$P = \frac{31.92 \times 10^3 \times Q^{1.748} \times \nu^{0.2518} \times S}{D^{4.748}}$$

or

$$Q = \frac{2.6527 \times 10^{-3} \times D^{2.716} \times P^{0.572}}{\nu^{0.144} \times S^{0.572}}$$

where:

- P pressure drop (bar/km)
- Q flow rate (m^3/day)
- D pipe inside diameter (mm)
- ν kinematic viscosity (cSt)
- S specific gravity

Shell/MIT Formula

Calculations are made in accordance with the method recommended by Wilson and McAdams and reported in "Contribution No. 19 from the Department of Chemical Engineering, Massachusetts Institute of Technology." After conversion to the metric system, the formulas are as follows:

$$P = \frac{4.4191 \times 10^6 \times f \times S \times Q^2}{D^5}$$

For viscous flow

$$f = 0.16025\left(\frac{\nu}{DV}\right)$$

For turbulent flow

$$f = 0.0018 + 0.013685\left(\frac{\nu}{DV}\right)^{0.355}$$

where:

- P pressure drop (bar/km)
- Q flow rate (m^3/day)
- D pipe inside diameter (mm)
- V kinematic viscosity (cS)
- S specific gravity
- F friction factor
- V average velocity of fluid (m/s)

Simplified Darcy Equation

$$P = 6.254 \times 10^5 \times \frac{f_m \times W_T^2}{d^5 \rho_m}$$

where:

P pressure drop (bar/km)
f_m Darcy or moody friction factor
W_T total liquid plus vapor flow rate (kg/h)
D inside diameter of pipe (mm)
ρ_m gas/liquid mixture density at operating pressure and temperature (kg/m^3)

Notes:
ρ_m may be calculated from the following derived equation:

$$\rho_m = \frac{28,829.6 S_L \times P + 35.22 R \times G \times P}{28.82 P + 10.12 R \times T \times Z}$$

where:

S_L relative density of oil (water = 1)
P operating pressure (kPa Absolute)
R gas/oil ratio (m^3 of gas/m^3 of oil at metric standard conditions)
G gas relative density = (MW/28.9) at standard conditions
MW molecular weight of the gas at 20°C and 760 mm Hg
T operating temperature (K)
Z gas compressibility factor

The above Darcy equation loses its accuracy when the pressure drop is more than 10% (due to changes in density of gas). For pressure drops higher than 10%, calculations should be carried out for smaller segments of the pipeline and the values added together.

T.R. Aude and Hazen–William's Formulas

1. *T.R. Aude Formula*

$$Q = \frac{3.4657 \times 10^{-3} \times D^{2.66} \times P^{0.552} \times C}{S^{0.448} \times \mu^{0.104}}$$

or

$$P = \frac{28,635Q^{1.812} \times \mu^{0.188} \times S^{0.812}}{C^{1.812} \times D^{4.819}}$$

2. *Hazen–William's Formula*

$$Q = \frac{2.6 \times 10^{-3} \times D^{2.63} \times P^{0.54} \times C}{S^{0.54}}$$

or

$$P = \frac{61.07 \times 10^{3} \times Q^{1.852} \times S}{C^{1.852} \times D^{4.87}}$$

where:

Q rate of flow (m^3/day)
D inside diameter of pipe (mm)
P pressure drop (bar/km)
C T.R. Audi's friction factor (pipe efficiency)
= 1.2 for new steel pipe
= 1.0 for old steel pipe
S specific gravity of liquid
μ absolute viscosity (cP)

Panhandle Revised or B Formula

$$Q = 10.024 \times 10^{-3} \times E \times \left(\frac{P_1^2 - P_2^2}{S^{0.961} \times T \times L \times Z}\right)^{0.51} \times D^{2.53} \times \left(\frac{T_0}{P_0}\right)^{1.02}$$

where:

Q flow rate (m_3/day)
S specific gravity of gas (Air = 1)
L length of line (km)
T_0 temperature basis for defining gas (K)
T mean flowing temperature (K)
P_1 inlet pressure (bar (abs))
P_2 final pressure (bar (abs))
P_0 pressure basis for defining gas (bar (abs))
D pipe inside diameter (mm)
E efficiency factor (taken as 0.9)
Z compressibility factor

IGT/AGA Formula

$$Q = 2.298 \times 10^{-3} \times \frac{T_0}{P_0} \left(\frac{P_1^2 - P_2^2 - \left(\frac{0.06834(h_2 - h_1)p^2}{TXZ} \right)}{S \times T \times Z \times L} \right)^{0.5} D^{2.5} \times \text{Log} \frac{3.7D}{\text{Ke}}$$

where:

Q flow rate (m^3/day)
S specific gravity (Air = 1)
T_0 temperature basis for defining gas (K)
T mean flowing temperature (K)
P_1 inlet pressure (bar (abs))
P_2 final pressure (bar (abs))
P_0 pressure basis for defining gas (bar (abs))
P average pressure (bar (abs))
L length of line (km)
D pipe inside diameter (mm)
Z average compressibility factor at P and T
Ke effective roughness of pipe wall (mm)
h_1 initial elevation of line (m)
h_2 final elevation of line (m)

Weymouth Formula

$$Q = 0.003749 \times \frac{T_0}{P_0} \left(\frac{(P_1^2 - P_2^2) D^{5.333}}{S \times T \times L} \right)^{0.5}$$

where:

Q flow rate (m^3/day)
S specific gravity of gas (Air = 1)
L length of line (km)
T_0 temperature basis for defining gas (K)
T mean flowing temperature (K)
P_1 inlet pressure (bar (abs))
P_2 final pressure (bar (abs))
P_0 pressure basis for defining gas (bar (abs))
D pipe inside diameter (mm)

Moody Friction Factor Chart

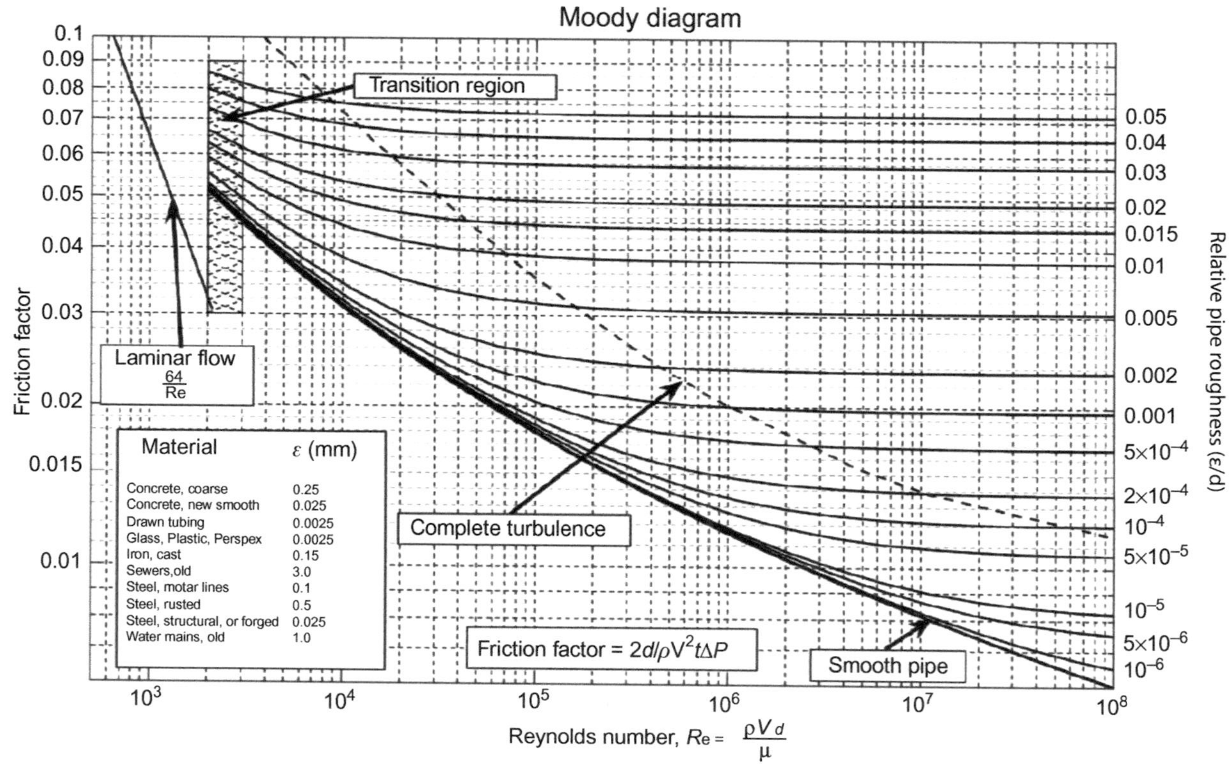

Chapter 2

Construction Guidelines for Onshore Transportation Pipelines Applications

This chapter covers the requirements and works to be carried out for construction of gathering, flow lines, and transportation pipelines within oil, gas, and petrochemical industries.

Every part of this chapter should be deemed supplementary and complementary to every other part and should be read with it.

2.1 MATERIAL HANDLING

This part outlines requirements for a procedure to specify responsibility of Company and Executor in handing over, loading, hauling, unloading, keeping list of inventory, and storing of materials required for installation of pipelines except for explosive materials.

The land necessary for sites suitable as storage centers should be leased by Executor as early as possible. A procedure for handover of materials should be agreed upon between the Company and the Executor. The procedure, to be specified in the particular conditions of contract agreement, should give details for handing over of pipe, valves, fittings, coating, and other materials and equipment by the Company to the Executor.

The procedure should include making a list which indicates the extent of all repairs found necessary when the pipe is received by the Executor. The list should be made by the Executor and must be approved by the Engineer.

After taking delivery by the Executor, if any defect is found in materials, specified as sound in the list, it should be made good or replaced at the Executor's expense. Unless otherwise specified, site preparation of storage areas, loading, transporting of materials to storage areas and from there to the installation site is the responsibility of the Executor. The Executor should provide equipment required for loading, unloading, and hauling of pipes and all other materials necessary for installation of pipeline.

Oil and Gas Pipelines and Piping Systems. DOI: http://dx.doi.org/10.1016/B978-0-12-803777-5.00002-2

2.1.1 Storing Materials

The methods of storage and the location of storage facilities should be approved by the Engineer prior to use.

Materials that are liable to deterioration or damage should be suitably stored and protected. Any loss of material caused by inadequate storage or protection should be removed and replaced at the Executor's expense.

Special attention should be paid to the storage of rolls of wrapping materials intended for the protective coating system. This material should be raised off the ground in layers to a height not exceeding 1.50 m in a dry and protected building. The rolls should be stored with their core standing upright.

Tools or equipment should not be piled on top of the rolls of wrapping materials.

The Executor should strictly conform to the supplier's instruction regarding storage temperatures and conditions of all materials. The Executor should provide air conditioned storage, where it is deemed necessary.

Materials with machined faces (such as flanges, fittings, equipment, and valves) should be handled and stored in such a manner to ensure that contact of the machined faces with the ground or other materials that may damage them will be prevented. Such materials should be stored under temporary shelters or covered with plastic or tarpaulin sheets and raised above the ground on wooden planks or other isolating devices approved by the Engineer. Both ends of valves and insulating flanges should be temporarily sealed by wooden plugs or plastic. Machined surfaces should be protected from shocks by maintaining their proper packaging.

Primer barrels may be stored in stacks. However, the barrels should be supported clear from the ground by use of timbers and the maximum number of row should not exceed 3.

Pipe coated with P-E (polyethylene) can be kept at ambient temperature range of $-45°C$ to $80°C$. P-E-coated pipes can be exposed to sun for a maximum period of 6 months provided that temperature does not exceed $60°C$. Therefore, for longer storage P-E-coated pipes should be kept under roof or shed.

All lengths and sections of coated and wrapped pipe should be picked up clear of ground. Walking on coated pipe should not be permitted.

2.1.2 Loading and Unloading Pipes

Loading and unloading of pipes must be done carefully. Pipes should not be dropped. Pipes larger than DN 150 (NPS 6) must be handled by crane or other suitable lifting equipment complete with proper slings, belts, or approved end hooks.

Lifting devices to be used for handling pipe should be so designed and used that no damage to the pipe and coating will result.

Belts used for pipe handling should have a minimum width equal to the diameter of the pipe being handled and should not consist of abrasive material.

The Executor should replace any lifting equipment or apparatus considered unsafe or unsuitable by the Engineer.

The use of nonferrous hooks is not allowed for lifting pipe.

Lifting shoes may be made of malleable iron or other approved material and should be faced with belting materials, brake lining, plastic, or other material approved by the Engineer. Each shoe should be properly shaped and sized to engage at least one-tenth of the pipe's inside circumference and 100 mm wide for single joint and 150 mm for double joints.

Rivets or other metallic fasteners used to hold the protective facing should be suitably countersunk to prevent contact with the pipe.

During hoisting, wire slings should be of such length that the angle formed between each leg of the sling and the pipe is not less than 45 degrees.

Yard concrete coated pipe should be raised or lowered to or from the stockpile, ground, or truck by means of end hooks engaging the ends of the pipe and should be carried by a wire rope or chain sling with "Spreader Bar" between lifting lines.

2.1.3 Stockpiling of Pipes

Pipes should be placed on level-graded and adequately compacted areas. Pipes should be stockpiled at approved locations by grade of steel, diameter, thickness, and also by pipe manufacturer and clearly identifiable as required by the Engineer. In addition, the pipes should be secured against collapse or movement. The Executor should limit the tiers of pipe so that distorting or flattening of pipe will not occur. The maximum number of tiers is given in the table below.

NPS	DN	Maximum Number of Tiers
6	150	12
8	200	9
10	250	7
12	300	6
16	400	5
20	500	4
24	600	4
30	750	3
36	900	3
42	1050	3
48	1200	3
52	1300	3
56	1400	3
60	1500	3
72	1800	3

Pipes should be stocked in such a manner that at intervals of 30 m two stacks of pipe be stored leaving adequate space approximately 6 m for passage of trucks and crane between successive stacks.

Storage area in locations consisting of soft ground should be compacted to the satisfaction of the Engineer using sand, gravel, and roller. Compaction should be extended with the same degree 6 m on either side of stack boundaries.

The pipes should be kept clear from the ground using skids or timbers approved by the Engineer.

Every length of percolated pipes should be separated from next tier by use of plastic belt.

2.1.4 Transportation of Pipes

During transportation, the pipe should be wedged and protected in such a way as to prevent any distortion, flattening, and damage and to have longitudinal welds located at the point of minimum possible stress and to have no contact with the adjacent pipe.

Metal bolsters should not be used. Approved bolsters supporting the lower tier of pipes should be shaped to support at least 30 degrees of the circumference of the pipe. Padded bolsters should be used.

Each load should be tied down using cables of suitable proportions and arranged in such a manner that tie-down is achieved over the bolsters. Tie-down cables should be covered by rubber or similar material to protect the pipe and to prevent metal-to-metal contact. During transportation of pipe, tie-down cables should be inspected and retightened by the Executor during and after hauling.

Extreme care should be exercised for hauling precoated pipes using suitable and approved material for padding.

In transportation of yard concrete coated pipes, the bolsters of truck and trailer should have adequate bearing surface (as directed by the Engineer), to give necessary support to padding which protects the coating.

2.1.5 Hauling of Coating Materials

Hauling of coating materials to right-of-way should be limited only to quantities necessary for the daily coating application.

Coating materials should be transferred directly to vehicles supplying the coat and wrap machine.

During unloading, transport and application of wrapping materials, any contact with water, mud, earth, dust, crushed stone, and foreign materials should be avoided.

2.1.6 Damaged Pipes

Pipes that have been found damaged, especially those showing ovality, dents, flattening, or other permanent deformation, should be stored and stockpiled separately. These pipes can be taken to site for incorporation in the works only when defects have been repaired or eliminated by cutting and beveling provided that such repairs have been approved by the Engineer.

2.1.7 Final Storage

In line with completion of construction activities all excess company supplied materials (e.g., pipe, valve, fitting, usable crop-ends) should be collected, transported, and stored by the Executor and kept under his care. Location for storing such materials should be approved by the Engineer and materials should be classified according to their characteristics (dia, thickness, grade, type, etc.). This operation must be finished at the date of "work completion."

2.2 RIGHT-OF-WAY

This part of the Standard describes requirements and responsibilities of Executor in connection with verification of the staked route with that shown in the working drawings and setting out the right-of-way in width shown in approved working drawings.

The pipeline route should be in accordance with the relevant working drawings and the route as staked on the site which also indicate the working width. In the event of any discrepancy between the route as staked on site and the route on the working drawings, the route on the drawings should prevail.

2.2.1 Route

The Executor should restake any alterations necessary to comply with the route shown in the working drawings. The Executor is responsible for the accuracy of survey line established by him and should incur expenses to perform survey work.

The Executor may suggest changes in the original route in some sections resulting from observation to avoid presence of water wells, qanats, rocks, inaccessible locations, and other conditions which may affect the construction, maintenance, and safe operation of pipeline. In such cases the following procedure should be adhered to:

1. The Executor should prepare the relevant drawings and submit them to the Engineer for his review and written approval.
2. The Executor should mark the proposed route for the Engineer to inspect.

3. The Executor should give in writing the reasons for route alteration and explain advantages.
4. All relevant information should be provided to the Engineer by the Executor sufficiently in advance of construction to enable the Engineer to examine the Executor proposals.

The Executor should never deviate from the original route before obtaining written approval of the Engineer. The Engineer's approval does not relieve the Executor from any of his responsibilities or obligations in respect of timely and orderly completion of the project.

The Engineer may instruct the Executor to make any deviations from the original route deemed necessary.

Where surface features or any foreign underground installations are indicated on the working drawings, the Executor should verify that the locations so indicated are correct.

The Executor should be held responsible for any claim arising from incomplete verifications. The Executor should locate all installations that are to be crossed particularly underground installations.

The Executor should explore and precisely locate existing pipelines and all other installations which are parallel or adjacent to the pipeline being constructed, especially when the clearance between the pipelines is less than required by the working drawings.

The Engineer should be informed by the Executor of any foreign installations encountered that have not been indicated on the working drawings; he will advise the Executor of action to be taken.

2.2.2 Drainage and Temporary Staking

Drainage channel should be made along the route in hilly terrains and also any other place where there are the possibility of erosion of right-of-way by surface water. The channel should be properly sloped.

2.2.3 Clearing and Grading

The right-of-way may be adjacent to an existing pipeline or the Company may have existing line on the right-of-way of the pipeline under construction. In such cases, the clearing and grading operations should be carried out in a manner that:

1. The cover of existing buried pipelines should be strictly maintained.
2. The passage of the Executor's vehicles or equipment along or across existing pipelines is forbidden except at crossings located and approved by the Engineer. The number of such crossings should be kept to a minimum and they should be constructed and maintained so that the cover height to the existing pipeline is sufficient to prevent damage to the existing pipeline and is a minimum of 2.5 m over a width of at least 8.5 m on either side of the center line of the pipeline, or as instructed by the Engineer.

2.2.4 Clearing

Before commencing work in any new section of right-of-way the Executor should get a clearance certificate from the Engineer for that section.

The Executor should clear the entire width of right-of-way.

The Executor should observe the forest protection law and obtain necessary permits before cutting the trees. The Engineer should provide the Executor with such permits if requested by the Executor. The ecological nature of the terrain that the right-of-way traverses must be preserved as far as possible. When trees have to be removed, any resulting saleable timber will remain the property of the Company.

All cleared trees removed from the right-of-way should be cut into 2.5 m lengths and neatly piled along the edge of the right-of-way.

Existing fences should not be removed without giving previous notice to owners and/or users through the Engineer, where it is necessary to remove permanent fences, walls, or gates.

Before cutting any fence, the Executor should furnish material and adequately brace the existing fence to prevent damage and, if the fence is in the vicinity of a power transmission line, should adequately earth the fence on both sides of the cut.

Where a fence paralleling the right-of-way must be removed for construction purposes, Executor should install a temporary fence or, with the Engineer's and landowner's approval, temporarily relocate the existing fence.

Access points through fences which cross the right-of-way should be securely locked and kept closed by the Executor and to be opened only for the passage of vehicles and authorized personnel.

The Executor should not dump any kind of material resulting from clearing and grading on roads, railroads, streams, ditches, drains, and any other place where it might obstruct the flow of water or passage of traffic.

2.2.5 Grading

After clearing has been carried out the right-of-way must be reasonably leveled and sharp changes in contour along the right-of-way should be graded down. The maximum gradient should be 12 and if it is continuous, a 15-m stretch of level track should be provided at 150-m intervals.

Should any damage occurs to any existing installation in clearing and grading operation, the Executor is responsible for repair to owner(s) satisfaction.

The Executor is responsible to maintain right-of-way until completion of construction of the project.

2.2.6 Access/Service Roads

Access/service roads (on temporary bases-during construction).

Access roads should be provided by the Executor to connect existing roads to the following locations:

1. Right-of-way
2. Valve sites
3. Site for water source
4. Either side of rail road and river crossings
5. Cathodic protection and telecommunication stations
6. Camp site
7. Any other locations as instructed by the Engineer.

Service roads along the right-of-way should be provided by the executor to make passage of vehicles and construction plants possible.

The temporary access and service roads during construction period should have conditions satisfactory to the Engineer and be suitable for driving four-wheel-drive vehicles along them.

2.2.7 Access and Service Roads (Permanent)

The Executor should upgrade the temporary roads and also construct permanent roads at the following locations:

1. On either side of marshy lands.
2. At locations where the slope of right-of-way is more than 16 degrees.
3. At any other locations where it is deemed necessary and instructed by the Engineer.

Requirements for construction of the above-mentioned permanent roads (e.g., compaction of subgrade, fill and backfill, finish grading, base course, road crossings, drainage and culverts, and surface conditions) should comply with specification set forth in *standards*.

2.2.8 Permits and Authorization

The Company should undertake the acquisition of land for permanent access service roads, and will secure at his own expense the necessary permits for roads, railways, stream/river crossing, and also crossing of existing structures. The Executor should comply with all conditions required in the permits. Prints of drawings showing details of such crossing, should be provided to the Engineer sufficiently in advance of construction to enable the Engineer to obtain the necessary permits in time to comply with the construction schedule.

The Executor should obtain necessary easement and permit through the Engineer for construction of the above-mentioned access/service roads.

When transit of the vehicles and construction equipment or, in particular, the construction of R.O.W. interferes or requires the relocation of the electric poles, telephone lines, electric cables, underground structure, and/or installation, the Executor should notify the Engineer in advance of his requirements. The Engineer will obtain the necessary approval and/or permits for the Executor within a reasonable time after notification. The Executor should not commence the work in that section of R.O.W. without the Engineer's written approval.

The Executor should give the Engineer advance notice of his needs for authorization to use public and private roads for access to the work by all vehicles, equipment and personnel. The Engineer will obtain the necessary authorization for the Executor.

2.2.9 Care of Services and Properties

The Executor should promptly repair any damage caused by his operation to bridges, private roads, fences, buildings or other property on or off right-of-way to the satisfaction of the Engineer. All relevant costs should be borne by the Executor.

Where right-of-way operations run parallel to or crossing installations such as pipelines, cables, etc., the Executor should take care to prevent any damage to these installations and should conform to the requirements of the relevant authorities.

When telephone or utility lines are damaged due to the ingress and egress of the Executor's tools and equipment for the construction of the pipeline, the Executor should, at his cost, undertake the necessary repairs of such telephone and utility lines to the satisfaction of the Engineer. The Executor should pay for such repairs if the repairs are accomplished by the concerned authority.

The Executor should, where possible, avoid causing damage to irrigation canals, qanats, ditches, or other irrigation installation and should also avoid interruption in flow of water.

During the progress of right-of-way operations, the Executor should endeavor to minimize disruption to any existing cultivated areas. The Executor may be requested by the Engineer to amend his right-of-way operation program from time to time, particularly during harvesting, to ensure that such disruption is minimized by the Executor.

The Executor should perform the works in such a way so as to minimize the disruption of activities to authorities, owners, users of adjacent roads, bridges, and the owner of buildings or lands. The Executor should endeavor to use his/her best judgment to avoid unnecessary damage to crops, trees, and properties adjacent to the right-of-way.

The Executor should make every endeavor to keep traditional right-of-way of tribes that are open during their migratory periods.

2.3 DITCHING

This part covers requirements for trenching operations for buried pipeline.

It explains conditions of a trench before pipe is laid, conditions under which blasting can be performed as well as the staking, marking, and use of excavated materials.

Trench should be located in the cut/graded section (working side of right-of-way).

The bottom of the trench should be uniformly graded and free from loose rock, large gravel, and other objects which may damage a pipe's coating. Wherever additional depth of ditch is required, such as for roads, railways, canal, underground pipelines, or utility crossings, it should be provided at no additional expense to the Company.

When the excavated material does not meet the specified requirements for backfilling, in the opinion of the Engineer, the Executor should remove such material from the site as the digging of the trench progresses and should supply suitable material approved by the Engineer.

2.3.1 Method of Ditching and Padding of Trench Bed

The Executor should use such system of trenching, equipment, and methods as may be required to excavate the ditch to satisfaction of the Engineer, regardless of the type of soil or rock encountered.

The Executor should familiarize himself with the location of the pipeline route in relation to the existing roads, bridges, railways, houses, etc., and should be deemed to have made provisions for excavating by hand or machine (but not explosives) in such locations where the use of explosives would result in damage, injury, or disturbance.

When blasting is permitted, the Executor should strictly observe and adhere to all requirements specified in Section 2.4.

The sides of the trench should be free of rock, loose stones, blasted debris, or other spoil likely to fall or blown around or on top of the pipe.

To prevent the excavated material from falling into the trench, a strip with a minimum width of 40 cm should be left clear between the edge of the trench and the edge of the pile of excavated earth. In steep areas of hillside, before starting the work and where required, the Executor should provide suitable barricades or other similar protection in order to prevent the material from falling down the hill particularly when lines of communication, houses, services, water courses, and cultivations are found on the downslope side of the ditch.

The pipe should be laid in the water-free trench bottom on a soft layer (padding). Padding material should be stone-free dry sand or soft and sieved earth. Minimum depth of padding for trench bed should be 20 cm.

In steep rocky areas where padding cannot be laid and contained due to steep slope, prior to lowering-in, it is necessary to put sacks filled with soft material at the bottom of the trench.

2.3.2 Setting Stakes

The Executor should supply and set temporary stakes along the right-of-way as follows:

1. Stakes should be placed on each extremity of the entire length of the right-of-way at intervals not exceeding 100 m and close enough to be visible one from the other. Stakes should be of such heights as to be visible above any growing crops.
2. Along the working side of the right-of-way 1.80-m high stakes should be set to indicate each pipeline change in horizontal direction.
3. At each kilometer a post 1.80 m high should be set with an indicating plate.
4. Auxiliary markers with 1.80-m high stakes (with indicating plates) should be placed at the following locations:
   a. At each point where the pipes to be laid changes its characteristics in wall thickness, diameter, pipe manufacturer, type of coating as required by the Engineer.
   b. At each intersection with underground installations such as water, electricity, telephone, etc.
   c. The Executor should avoid causing damage to drainage, irrigation, and waterway systems. The exact location of such installations should be staked out in an easy and visible manner immediately next to the trench to be dug.
5. The Executor should indicate on the stakes the trench depth within the zone.

2.3.3 Care of Services

Prior to performing any machine excavation closer than 5 m to such underground installations, the Executor should expose the existing installation by hand excavation. Spoil should be placed so that no heavy equipment operates over the existing installations.

Any structure that may be threatened during the trench excavation such as electric poles, foundations, etc., should be braced and adequately supported.

Bracing should not be removed until the backfilling progress allows its removal without any risk of damage to structure or injury to personnel. All excavated material should be placed in such a way as to avoid any inconvenience to property owners, or interference with the access of pedestrians or vehicles as well as to operation of adjacent installations.

The excavated material should, when crossing cultivated areas, be placed so that the top soil can be replaced on the surface after backfilling. General backfilling material should be stored separately from the top soil and care should be taken to ensure that these materials are not intermixed.

Wherever it is deemed necessary by the Engineer to gain access to houses, buildings, etc., or where live stock is confined and a passage way across the trench is desired, the Executor should provide safe temporary bridges or suitable plate cover for crossing the trench or temporary fencing to prevent animals from falling in the trench.

2.3.4 Rock Trench and Padding

Rock trench should be defined as a trench which cannot be excavated by use of normal hand excavation or machine excavation techniques. Such trenches would require drilling or blasting to enable formation of the trench. In such rocky areas blasting operations should be carried out as specified in Section 2.4.

2.3.5 Ditch Dimensions

For pipeline to be laid without damage to protective coating, trench dimensions should not be less than the following:

	Trench in Rocky Terrain (mm)	Trench in Uncultivated Terrain Other Than Rocky (mm)	Trench in Cultivated Terrain (mm)
Minimum depth of cover	600	900	1200
Width of trench excess of pipe dia	400	400	400

Notes:

1. The minimum depth of cover is the vertical distance measured from the graded ground level (excluding crown) to the top of the pipe. In swampy areas, the minimum depth is the distance from the grade level to the top of the weighting. In rocky areas, to the top of rockshield over the pipe. In case of transverse slope the depth of cover should be measured from the level of the downslope side of the trench.
2. Minimum depth of cover, when crossing with roads, rivers, and seismic faults should be as indicated in alignment and working drawings.
3. Width of trench, when crossing seismic faults should be as per related working drawings.

2.3.6 Bell Holes for Tie-Ins

The Executor should dig bell holes where tie-in welds are to be performed. The dimensions of bell holes should be as follow to permit ease of maneuvering for welding and coating of pipe:

- Length: 3 pipe O.D. as minimum
- Depth: Such that the minimum clearance under the pipe should be 1 m.

2.3.7 Damages and Claims

The Executor should be fully responsible for any claims resulting from his interference with surface and subterranean water flow (qanats) and for any accident resulting therefrom.

Any damage to a drainage or irrigation system should be temporarily repaired by the Executor in order to allow its proper usage during construction works. Permanent repair should be made by the Executor prior to backfilling the trench.

In such final repair, the Executor should restore the former gradient alignment of the damaged system, and should use materials of a quality and size at least equal to that of the materials to be replaced.

Minimum depth of cover	600 mm 900 mm 1200 mm
Width of trench in excess of pipe	dia 400 mm 400 mm 400 mm

2.4 SUPPLY, STORING, HANDLING, AND USE OF EXPLOSIVE MATERIALS

This part covers requirements to be observed by the Executor in clearing right-of-way and trenching operations in terrains containing natural rock, large boulders, or other materials that have to be removed by blasting operations.

2.4.1 Supply, Handling, and Storage of Explosive Materials

The use of explosive materials may be permitted provided the approval is obtained from the authorities concerned. Taking delivery of explosive materials and its use should be the responsibility of the Executor.

The Executor should notify the Engineer in writing, 4 months in advance of its need for explosives. This notification should include all necessary documentation to satisfy the relevant authorities with the type and amount of explosives requested by the Executor. The notification should also include a consumption schedule for explosive materials throughout execution of the project. The authorities concerned may supply the explosives or grant permit for importing explosive materials by the Executor.

The Engineer assists the Executor to obtain explosives or permit for its importation if it is found necessary.

For delivery, transportation, storage, handling, and use of explosives, the Executor should strictly conform to the laws and regulations.

The Executor should deliver to the blasting sites all necessary explosives, detonators, and all materials associated with blasting operations.

2.4.2 Engineer's Approval Prior to the Use of Explosives

The Executor should obtain the Engineer's written approval prior to the use of explosives. Each approval should be limited to a specific length of the line. The Engineer will refuse his approval in a particular case if, in his opinion, blasting cannot be carried out with complete safety. The Engineer's approval does not obligate him for any of the Executor's responsibilities concerning the use of explosives.

The Executor should obtain all necessary local permits.

2.4.3 Blasting Operations

Blasting operation will be permitted upon approval of the Engineer.

The Executor should use only experienced workmen to supervise, handle, haul, load, and shoot explosives.

The Engineer or his representative (nominated in writing) should be present at all times during the use of any explosive, unless the Engineer has provided written notice that blasting may proceed without his attendance.

The Executor should provide all necessary warnings and control procedures required for blasting operations.

2.4.4 Care of Life, Services, and Property

The Executor should exercise extreme care to safeguard against loss of life, livestock, injury, and accident and damage to other installations, springs, irrigation facilities, or water courses.

Where the pipeline runs parallel to an existing pipeline or adjacent to any petroleum installations or underground structures and in those areas, where the Engineer deems it necessary to take particular safety measures, the Executor should, prior to blasting operations, present to the Engineer for his approval a detailed blasting procedure and plan. The procedure and plan should indicate the type of explosives, exact location, and size of each charge in addition to blasting schedule. The procedure should include, but not limited to, minimum distance to existing installations.

Blasting operations will be restricted near the existing installations. The minimum distance where blasting will be permitted is subject to agreement of the Engineer. The agreement should be obtained prior to proceeding with any blasting, taking into account the blasting procedure proposed by the Executor and approved by the Engineer.

The Executor is fully responsible for damages affected to any person, animal, adjacent property, springs, irrigation facilities, or water courses as result of blasting operations. The Engineer's approval does not relieve the Executor of his responsibility for correctness in blasting.

When telephone or utility lines are damaged by or require relocation due to the blasting operations, the Executor should repair or relocate such services to the satisfaction of the Engineer at the Executors expense.

Any damage caused to existing installations should be immediately repaired to the full satisfaction of the Engineer and the owner. All direct or indirect losses and costs should be borne by the Executor.

2.4.5 Care for Loss of Explosive and Its Dispositions

Proper disposition must be made of any and all refuse from explosive containers and cartridges, and in no case should they be disposed of in the backfill of the trench.

If loose rock is scattered over cultivated fields, the Executor should pick up such rock and dispose of it to the satisfaction of the Engineer.

The loss or partial loss of explosives or detonators during the blasting operations will render the spoil material from that area unacceptable for backfilling. Such materials should be removed from the right-of-way.

2.5 STRINGING

This part deals with manners to be considered in taking pipes from stockpile and stringing along right-of-way for subsequent welding, coating, lowering, and backfilling.

Pipes should be transferred from stockpile to the installation site. Care must be taken to ensure that pipes are taken from the correct stock.

The Executor should submit his stringing plan of at least 15 km for approval of the engineer in advance of stringing operations. The engineer retains the right to review such distance during the progress of works.

In order to keep and maintain the work within reasonable lengths, the pipe stringing should precede welding operation by a distance of not greater than 5 km. The Engineer reserves the right to review such distance during the progress of the works and, if necessary, to instruct the Executor to limit the daily pipe stringing activity accordingly.

In the stringing operation, the Executor should observe the characteristics of pipes as indicated in the working drawings and that auxiliary stakes are placed in the ground along the right-of-way.

In the part of the right-of-way directly subject to flood, swamps, and in close vicinity of corrosive soil, no pipe should be strung until the time of assembly.

Pipe should not be in direct contact with ground. The Executor should provide appropriate wooden skids in such cases.

Stringing should be made after completion of explosion operations.

2.5.1 Dimensions to Be Observed

The minimum distance between the pipe and the edge of the trench should be 1 m.

2.5.2 Percentage of Pipe Allowed for Rejection

The total maximum tolerated length for the nonusable crop-ends and for the pipes rejected for reasons attributable to the Executor activities including pipes used to qualify the welding procedures and the welders, is fixed at 1% maximum of the total length of the Company supplied pipes. The lengths exceeding this amount should be charged to the Executor.

2.5.3 Care for Services

During pipe stringing care should be taken to ensure that water courses, water supplies, electricity, and telecommunication and traffic services are not interrupted or affected in any way by the operation.

2.6 PREPARATION OF PIPES

This part of the Standard covers checking to be made in respect of the cleaning of pipe ends and interior of pipes as well as pipe condition. It also emphasizes the necessity of repair before lining up pipe and welding it into a string.

2.6.1 Care of Line Pipe

During loading, transportation, unloading, and stringing, the Executor should take all necessary precautions to keep the inside of the pipe free from dirt, waste, and other foreign matter. When lifting pipe care must be taken to prevent buckling and any damage.

2.6.2 Checking Pipe Condition

The Executor should inspect all line pipe to ensure that no pipe with buckles, gouges, grooves, nicks, notches, or other defects is incorporated in the finished pipeline. Bevels on the ends of pipe are to be checked immediately before lining up and production welding.

2.6.3 Swabbing of Pipes

The interior of each length of pipe or fitting should be visually and thoroughly examined to make sure that it is clean and empty.

Each single length of pipe which is contaminated in any way and if necessary, as determined by the Engineer, should be thoroughly swabbed before lining up and being welded into a string. Swabbing should be made to remove all dirt or foreign matter from the inside of the pipe. Any solvent, if required to be used, should be approved by the Engineer.

The swab should be constructed of two separate discs of 1-cm thick rubber or similar material sized to fit the inside diameter of the pipe and rigid backup plate. The type of SWAB, used by the Executor, should be approved by the Engineer.

2.6.4 Cleaning Pipe Ends

The ends of the pipe including bevel and root face should be thoroughly cleaned of rust, scale, dirt, grease, protective coating, burrs, or other foreign matter on both the inside and outside edges which may adversely affect the weld ability of the beveled ends.

2.6.5 Cutting and Beveling

If necessary, as determined by the Engineer, bevels on the ends of pipe should be made in the field with a beveling machine. Normally these bevels should be made to an angle of $30°$, $+5°$, $-0°$, measured from a line drawn perpendicular to the axis of the pipe, and with a root face (1/16 in $\pm$ 1/32 in) 1.6 ± 0.8 mm. Manual oxygen cutting should not be used for beveling the ends of pipe for welding unless it is impracticable, to the Engineer's judgment, to use machining or machine flame cutting equipment. In such cases permission should be obtained from the Engineer, and the cut face ground back to clean metal before welding.

2.6.6 Repair Work on Line Pipe

Defective longitudinal welds should be repaired by cutting out as a cylinder the section of pipe containing the defective weld and rewelding new section with normal circumferential welds. Patching is prohibited and minimum length of the new section should be according to standard.

A list indicating the extent of all repairs found necessary, when the pipe is received, is to be made by the Executor and approved by the Engineer before repair work commences.

2.7 CHANGE OF DIRECTION

This part of the Standard covers requirements for cold field bending. It describes limits to be observed in field bending in terms of minimum bend radius as well as maximum stretching or thinning of the pipe wall thickness.

Performed bends should be used when changes of slope or direction are required.

No bends will be permitted on pipe which has been or is to be concrete coated. For river crossings the pipe may only be laid in natural curves that cannot overstress the pipe or the reinforced concrete sheath.

Maximum permissible angle of cold field bending should conform to ANSI/ASME B 31.8.

No bend should be made within 2 m of a circumferential butt weld or beveled end of pipe unless the pipe is double jointed, in which case this dimension may be reduced to a minimum of 1 m.

No wrinkle bends or hot bends are permissible.

If bend is made from pipe lengths containing longitudinal seam, the seam should be positioned on the neutral axis of the bend and length should be arranged so that successive lengths have the seam properly staggered as per standards.

All bends should be made cold and uniform by the use of an approved bending machine with bending shoes of proper size and skilled operator.

2.7.1 Minimum Bend Radius to Be Observed

Minimum radius of field bends should be based on 1.5 degrees per length of pipe equal to nominal diameter.

Where physical requirements, obstructions, etc., dictate that a bend be made of tighter radius than the limits given in relevant documents and standards, this should be done with the Engineer's prior approval. However, the limit of thinning and deformation in Section 2.7.2 should be fulfilled.

2.7.2 Limit of Thinning and Deformation

Stretching or thinning of the pipe wall thickness should not exceed 1.5% at any point along a performed bend.

The pipe diameter should not be reduced by more than 2% of nominal pipe diameter. Care must be taken to avoid wrinkling and not to exceed this limit of deformation. Wrinkled pipe or pipe exceeding the said deformation should be rejected.

2.8 WELDING AND LAYING OF PIPE

This part of the Standard covers technical requirements for welding of pipe over the ditch and double jointing yard as well as laying pipe in trench, exposed, and casing sections. The requirements include complementary trenching works, considerations to be given in laying pipe when trench contains water or mud, and test of pipe coating just before laying.

2.8.1 Welding of Pipe at Double Jointing Yard

For expediency efficiency, line pipe may be double jointed by the Executor before transportation to site for stringing.

When double jointing of line pipe is agreed upon between the Company and the Executor, the Executor should weld single joints of pipe into double joints at the pipe storage site, wherever the double jointing is considered efficient, practical and suitable, prior to transportation and stringing the pipe along the right-of-way.

Prior to commencement of the work, the Executor should submit his proposal regarding the welding process in addition to welding procedure for the Engineer's review and approval.

All welding work, welding materials, qualification tests for welding procedure and welders, inspection and testing, acceptance standard to be applied to production weld, repair or removal of defects should conform to requirements cited in widely accepted standards.

Nondestructive testing should be carried out at frequency of 100% by radiographic or ultrasonic method as agreed by the Engineer.

Destructive test, when instructed by the Engineer, should be carried out as per requirements of standards.

2.8.2 Capping Pipe Ends

At the end of each day's work, the Executor should cap all open ends of welded sections of pipe with a night cap.

Failure to cap ends of sections that cannot be swabbed should be cause for requiring the Executor to run a scraper through that section before tie it in pipeline.

Night caps should be placed using suitable technique to prevent damage to the bevels at the ends.

Unattended sections of the pipeline and strings with open ends should be effectively capped when work is discontinued for any reason to prevent entrance of animals or foreign matter into the pipeline. These night caps should remain in position where work is interrupted at crossings, under railways, roads, rivers etc.

The Executor should be responsible for any obstruction inside the pipeline.

2.8.3 Laying Operation

2.8.3.1 Verification of Trench Conditions Before Laying Pipe

The Executor should verify the trench conditions and dimensions prior to lowering-in. The trench should also be visually examined for proper finishing of its walls, bottom, and padding material.

When necessary, complementary trenching works should be carried out in order to ensure that the pipeline rests uniformly on the bottom of the trench. In such cases, the padding operation should be completed before lowering-in is done to the satisfaction of the Engineer.

All hard objects, such as rocks, large clods, sticks should be removed from the bottom of the trench into which the coated and wrapped pipeline is to be lowered so that the protective coating is not punctured or abraded.

2.8.4 Laying Pipe Into Trench

The Executor should, when the trench contains water or mud, dry, and clean out the trench before lowering in the pipe. For certain locations of limited length, the Executor may propose, with the approval of the Engineer, a process that, without drying out the trench, enables the pipeline to be laid as normal in the bottom of the trench.

Pipe should normally be lowered into the trench immediately after the coating and wrapping test has been passed by the Engineer.

The Executor should test the pipeline for coating integrity with a holiday detector just prior to lowering in. The testing method and instruments should be approved by the Engineer.

The line should be lowered into the ditch in such a manner as to obtain required amount of slack. The Executor may employ any acceptable means of lowering for laying, provided that such means secure the necessary amount of slack uniformly in the bottom of the ditch and does not injure the pipe or its protective coatings.

Where coated and wrapped pipe is supported on padded skids, their number should be sufficient to ensure that no damage will be caused to the coating and wrapping. Wide nonabrasive belts or canvas padded sling should be used at all times in handling the pipeline.

Protective shields, plywood (or equivalent material) should be placed along the side walls of trench containing rock or hard objects, if the Engineer deems it necessary. The shields should be taken out when pipe is not subject to further movement.

If during the laying operation the coating is unreasonably damaged by equipment or rough handling or rough trench, the Engineer may require the equipment be replaced or that the procedure be altered to eliminate the defect. Damaged coating should be repaired and tested with holiday detector before lowering pipe into trench. It is most important to ensure that pipe coating is sound and undamaged before pipe reaches the bottom of trench.

If after inspection and repair of coating the pipeline has to be again placed on skids, the skids should be suitably padded. Upon removal of skids, the Executor should check the coating with a holiday detector and make repairs as required.

Walking on coated pipe is absolutely prohibited.

Except for water crossings, sections of coated pipe being tied into the line should not be dragged or pulled into position and the length of such sections should be regulated to allow handling without damaging the protective coating.

At crossings or locations where it may be necessary to pull or drag sections of pipe into place, the coated pipe should be properly protected and handled in a manner to prevent damage to the coating or to the pipe.

2.8.5 Laying Pipe in Exposed Sections

Assembling and laying pipe in exposed sections, e.g., in scrapper traps or line break valve stations, should be performed only after the construction of pipe supporting structures are completed. The Executor should erect suitable support structures ensuring their correct alignment in plan and elevation.

Care should be taken to ensure that the pipe should have an additional protection; when this comes into contact with the support structure, such methods of protection must be approved by the Engineer.

The methods and techniques employed for lifting and moving and laying of exposed pipe work and the control procedures to be followed should be the same as specified for buried pipeline.

The pipeline should under no circumstances rest directly on the ground and at no point should the clearance between the bottom of the pipe and the ground be less than 40 cm.

When the supports consist of steel structures, the Executor should avoid setting the supports coinciding with the pipe field welds.

2.8.6 Laying Pipe in Casing Sections

Laying pipe in casing is not recommended. However, in case where using of other means is impractical, casing should be laid with the following requirements.

Sections of carrier pipe to be installed in casing should be totally straight and at least 2 m longer than casing on each side.

The welded joints of the carrier pipe should be 100% radiographed and then the carrier pipe should be externally coated, as specified in Section 2.9. After the completion of coating, the insulators should be fitted on the coated pipe as per working drawings.

All debris must be removed from inside surface of casing before the carrier pipe is pulled into it.

The pipe should then be pulled into place in such a manner that the pipe is centered in the casing, the pipe insulation is undamaged and the rate of travel within the casing is uniform.

Centering cradles and end seals should be installed immediately after the pipe is in place. After installation of pipe section in the casing, the end seals should be fitted and secured in position by stainless steel straps.

Before tying in the pipeline-cased section into the remainder of the pipeline on both sides of the crossing, the Executor should verify the electrical insulation between the casing and the pipeline-cased section using an ohmmeter. During the resistance measurement, the pipeline cased section should have no contact with the earth or water.

Under no circumstance the measured value should be less than 100 ohms. If this minimum value is not obtained, the Executor should locate and remove the cause of improper insulation to the satisfaction of the Engineer.

2.8.7 Lowering Bend Sections

The length of pipe between bends should be adjusted, if required, by the Executor to obtain an adequate amount of slack.

All bends should generally fit the trench in the following manner.

Sag-bends should fit into the bottom of the trench. They should be lowered in, such that the legs of the bends are firmly supported.

Over-bends should clear the high point of the bottom of the ditch. They should be lowered in, such that the crutch of the bend is firmly supported so that backfill will tend to close rather than open the bend.

Side-bends should be laid, such that the crutch of the bend is at a minimum of 200 mm distance from the side of the trench.

2.9 BACKFILLING

This part of the Standard covers the requirements for backfilling of buried lines that should be carried out immediately after the lowering operation and before hydrostatic testing. It explains the procedure of checking some important features before performing backfilling operation, the quality of backfill material, and considerations to be given at crossings with waterways, roads, and irrigated lands.

Backfilling should be carried out immediately after the pipeline is lowered into the trench, but the Executor should first obtain approval of the Engineer prior to backfilling any section of the trench.

If any backfilling is carried out without approval of the Engineer, he will have the right to require the Executor to remove the backfill for examination of coat and wrap.

Before any backfilling operation is undertaken, some important features of the work should be checked by the Executor. These include the following:

1. The pipe is laid in the trench bottom on a soft layer with minimum thickness of 20 cm.
2. The quality of backfill materials conforms to the requirements and meets the satisfaction of the Engineer.
3. Correct positioning of accessories such as weights, anchors, etc.
4. Removal of loose rocks or debris from the trench.
5. Any repair to coating or welding is made.

2.9.1 Initial Backfilling (With Soft Materials)

The lowered-in pipe should be surrounded with soft materials around the sides and the top. Minimum thickness of soft material on top of pipe should be 20 cm.

The soft material should be rock and stone-free sand or soft and sieved spoil or earth. The Engineer should approve the source and quality of the sand or soft earth.

For this soft layer the Executor may use, with the approval of the Engineer, the softest material from sieved spoil earth cleared of all debris.

If the softest material cleared of all debris is unavailable from the excavated spoil earth, the Executor may use sieved earth containing some loose gravel provided that in the opinion of the Engineer no damage to the pipe coating would result from the inclusion of such gravel.

The soft material should be immediately furnished and placed around and at top of the pipe by the Executor to protect pipe coating from excessive temperatures or inclement weather.

Where rock shield wrap is used, the initial backfill may contain loose gravel and small-size rock fragments provided that in the opinion of the Engineer no damage to the coating would result from the inclusion of such materials.

2.9.2 Normal Backfilling

After covering coated pipe with at least 20 cm of soft materials, the backfill should be placed in the ditch in the same order of soil structure as removed provided the Executor has obtained approval of the Engineer.

The first 50 cm of the normal backfill may contain 30% by volume the rocky debris, with a maximum size of 10 cm, provided that they are thoroughly mixed with 70% loose earth. Other layers of backfill should not contain rocks with size of more than 30 cm.

The backfill material should be free from nonearthy debris. Large clods of dirt or clays, stumps or foreign material that could cause voids in the backfill should not be allowed in the trench.

The use of highly alkaline, acid, or sulfurous material is strictly prohibited for backfilling.

The Engineer should reject backfilling material at his own discretion.

The backfill should normally be extended and crowned to a height of not less than 20 cm, and not more than 30 cm, above the adjacent ground level or as otherwise directed by the Engineer.

Backfilling should be such that to allow natural surface drainage. Surplus spoil should be removed from the premises and disposed of by the Executor to the satisfaction of the Engineer.

Backfilling operation should never be carried out when trench contains ice, water or when earth or backfill material is frozen.

2.9.3 Backfilling in Special Locations

2.9.3.1 Backfilling in Areas Subject to Erosion

At water crossing and similar areas that are subject to erosion (as indicated by the Engineer), protection measures should be taken by the Executor against the same. In such locations the Executor should reinforce backfill with earth-filled bags or sand mixed with cement bags after the trench has been filled and the backfill has been solidly compacted to the surrounding ground level.

2.9.3.2 Backfilling at Road Crossings

Backfilling of trench through roads should be carried out immediately after the pipe has been laid. The backfill should be compacted and finished level with road surface. Degree of tamping and compaction should be to the satisfaction of the Engineer.

Such sections should be maintained by the Executor until the works are completed. The road surface should be finally restored by the Executor to the original condition.

2.9.3.3 Use of Stoppage and Drainage Ditch

On steep slopes or similar areas that are subject to severe erosion and in the opinion of the Engineer there is danger of trench backfill being washed out, the Executor should take the following preventive measures:

The Executor should furnish and place bag breakers filled with sand, soil, or sand mixed with cement, as directed by the Engineer. In addition, ditches should be dug, as necessary, to channel water diverted by breakers away from the pipe trench.

2.9.3.4 Backfilling in Irrigated Areas

In irrigated areas where water packing of the backfill is necessary, the following procedure should be adopted:

1. Soil should be placed in the ditch to a level not exceeding one half the depth from top of pipe to ground surface and then saturated with water.
2. Soil filling should be made in 50 cm maximum increments, and water packing should continue as above until trench is full.
3. Trench should be windrowed approximately 15 cm high and then compacted to the satisfaction of the Engineer.

Such reinstatement of the backfilling in irrigated areas should be carried out in a way to ensure that there is no disruption or disturbance to the overall irrigation system.

2.10 CROSSINGS

This part of the Standard outlines specific considerations to be given by the Executor in construction of pipeline at crossings. It deals with surveys required to be carried out by Executor before giving schedules and timing for work execution at crossings.

Construction should be organized so that traffic interruption and interference with activities of adjacent property and utility owners are kept at a minimum.

In addition to information given in route and profile drawings, the Executor should endeavor to determine the locations of buried lines, utilities, and other underground structures that may exist across the proposed route of pipeline and have to be crossed.

Owners of any possible affected structures should be given notice well in advance of construction work so that they may make necessary preparations and assign a representative to witness the construction activities at the crossing concerned. To submit the notice, the Executor is required to perform following.

Before starting crossing work, the Executor should carry out an adequate geological, hydrographic, and meteorological survey in order to determine the necessary precautions and the period most favorable for the execution of the works.

The survey results and detailed work schedule on crossings should be submitted to the Engineer for approval. Work should not be started without this approval.

- Regardless of the method of crossing the Executor elects to use, he should submit to the Engineer for approval details of the method and equipment to be used together with timing of operation.

2.10.1 Overhead Structures Crossings

Pipeline routes may cross existing overhead structures, e.g., power or telecommunication lines. In such cases, the distance of ditch from poles must be observed as shown in the drawing.

If blasting is required for ditching, the Executor should take necessary precautions to avoid any damage to poles and overhead lines.

2.10.2 Road Crossings Without Casings

Road crossings should be performed in accordance with relevant drawings, taking the following into consideration.

Crossings without casing should be made by trenching. The Executor should open the trench only after the pipeline section relevant to the crossing is prepared. Before opening the trench, the Executor should have welded the crossing pipeline section to be laid.

The crossing section should have a minimum length extending about 1.5 m on either side overall width of the road.

The radiographic test should be 100% of all welded joints along crossing section.

All backfilling should be accomplished by placing suitable and compactable material (in accordance with the relevant standard) in layers of 15 cm, in thickness.

Each layer should be mechanically tamped with a pneumatic tamping device (or equal) until the degree of compaction is equal to or more than the density of the road. Whenever the material removed from the open trench is not suitable for backfilling and tamping, the Executor should obtain and use suitable material from other sources.

After backfilling, the surface of the road should be replaced with material having at least equal quality with the surrounding surface and in a manner satisfactory to both the Engineer and the concerned authority.

2.10.3 Major River and Lake Crossings

In major river crossings and lakes continuity of operation and the safety of the general public should be controlling factor in construction.

Depending on the crossing width, depth, and velocity of flow, the crossing may be laid either by launching and pulling the pipe across or by laying pipe by conventional land method.

When major river crossings are long and require particular laying methods, the Executor should prepare and submit, for the Engineer's approval, a detailed method and work plan.

If underwater pipeline is pulled into place, the Executor should follow the following procedures.

The pulling line should be attached to a pulling head welded to the pipeline such that no bending stresses of any magnitude are introduced into the pipe as a result of the pulling operation.

The trench should be graded to give the maximum amount of support to the pipeline when being lowered or pulled into place immediately after the pipeline is in the trench. The Executor should take the necessary measurements to determine the location and length of underwater spans. Any spans found to be in excess of 2 mshould be remedied in a manner acceptable to the Engineer.

For crossings to be pulled, no cold bends or fabricated bends should be permitted in that section of the crossing between the sag-bends at each bank.

Pipe for underwater crossing should be cleaned coated and wrapped as for underground sections in accordance with the specification and drawings, unless otherwise specified.

Precautions should be taken during construction to limit stress below the level that would produce buckling or collapse due to out-of-roundness of the completed pipeline.

All circumferential welds in river crossings should be tested 100% by radiography. All river crossings should be pretested hydrostatically prior to being tied into the line.

All water and river crossings should be constructed to provide adequate cover throughout the flood plain to prevent exposure of the pipe due to bottom scour or erosion of the banks.

The river bed should be restored as near as possible to its former elevations, and obstructions resulting from construction of the pipeline should be removed and disposed of by the Executor to the satisfaction of the Engineer.

On completion of the lowering in of the pipe, the trench should be backfielded such that the level of the water course is restored to its original level. Banks should be restored to their original lines and levels. Bank protection works should be provided in accordance with the working drawings and to the satisfaction of the Engineer.

2.10.4 Waterway Crossings

Waterways with a minimum rate of flow or small water way where a peak seasonal flow occurs should be treated as normal pipelaying except that the top cover should in no case be less than 1.2 m.

All the water way crossings should be carried out in accordance with the Standard or working drawings.

Welded joints should be inspected by radiography at the rate agreed by the Engineer.

Pretesting before tie-in is not required for water way crossing.

2.10.5 Qanat Crossings

The Executor should establish from local information the size and depth of these water carriers at the actual point of crossing and construct the crossing in accordance with the drawings. Should the construction of the qanat be disturbed, the Executor should make repairs at his own cost to the satisfaction of the controlling authority. The Executor should prepare all drawings if they are not provided by the company or if they require modifications.

2.10.6 Land Drains

The pipeline should be laid underneath all land drains exposed during trenching unless the Engineer directs otherwise.

The course of any land drain cut by the trench should be clearly marked and it should be permanently reinstated after the pipelaying is completed.

If so directed by the Engineer, land drains should be temporarily reinstated during construction to minimize inconvenience to land owners and occupiers or to prevent water from gaining access to the trench.

2.10.7 Crossing of Existing Buried Installations

Crossing of existing buried installations such as pipes and cables should be carried out in compliance with working drawings.

The uncovering of the buried installations should be carried out by hand excavation.

As long as the buried installations remain uncovered, the executor should install and maintain any necessary and/or required protection in order to avoid damage or breakage occurring to existing installations.

At all crossings of buried steel installation a reinforced concrete slab should be applied to the pipeline under construction for a distance of 5 m on each side of the crossed installation.

The Executor should be responsible for any damage that occurs to the buried installations during the works.

Should the Executor for his own convenience request the concerned authority to move or put out of service an electrical or telephone cable, all such requests should be arranged with the knowledge of the Engineer.

2.10.8 Bridge Attachments

Special requirements are involved in this type of crossing. The use of higher strength lightweight steel pipe, proper design and installation of hangers, and special protection to prevent damage by the elements of bridge and approach traffic should be considered.

2.10.9 Other Crossings

The overhead crossings should be carried out in compliance with specific working drawings and the laying of the pipeline should be carried out as per specification. All welded joints should be 100% radiographed. For all such overhead crossings, the pipe should be externally protected by painting.

At the points where the pipe enters the ground, the external coating of buried pipe should be extended for a minimum of 1 m on the riser above-ground level or for a distance equal to a one-time pipe diameter, whichever is greater.

The overhead portion of the buried pipe coating overlapping the overhead section should be adequately secured and reinforced to avoid possible water infiltrations beneath the coating joint.

2.11 CASING INSTALLATIONS

This part of the Standard covers requirements for installation of casings for correct insertion of carrier pipe where pipeline crosses major public roads, freeways, and railways. The Standard also explains boring methods for installation of casings in order to obviate interruption of traffic.

Crossing of major public roads, freeways, railways, and such other facilities shown on working drawings should be made by laying pipeline in steel casings. Crossing of railways and freeways should be made without any interruption of railway traffic.

2.11.1 Permits and Authorizations

Vehicle traffic should not be interrupted unless prior permission has been obtained by the Engineer from concerned authority.

The Executor is required to advise the Engineer of the time that work will commence on any crossing at least 1 month in advance. In addition to a notice of at least 1 week before commencement of work begins at a specific crossing, the Executor should notify the Engineer in writing the need to obtain permit from the relevant authorities.

Upon receipt of the written notification from the Executor, the Engineer should obtain necessary permits from the relevant authorities and the Executor should comply with all requirements of the concerned authorities, as instructed by the Engineer.

2.11.2 Methods of Casing Installation

The Executor should install steel casing by thrust boring. If this method is not practicable because of the nature of the ground, with permission of the concerned authority and the approval of the Engineer, casings may be laid by other approved methods. Where conventional boring methods become impracticable because of rock formation, the Executor should complete the crossing using a tunneling method and employing suitable means for this purpose.

In order to obviate settlement of pipe, the pipe should have firm bearing on the bottom for a distance of not less than 8 m from each end of the casing. This should be accomplished by placing sand filled bags under the pipe at 1-m intervals. The Executor may suggest an alternative means for firm bearing for the bottom, where pipe rests, and this should be approved by the Engineer.

The Executor should install vent and drain pipes as per working drawings.

The Executor should install insulators and cradles onto carrier pipe at intervals given in the working drawing.

The Executor should install end seal at either end of the casing.

2.11.3 Inspection of Casing Pipe

Prior to welding, each length of casing pipe should be thoroughly checked by the Executor to ensure that no out-of-roundness or dents are existing. These defects, if found, should be repaired before lining up the casing pipes. The pipe length containing such defects should be rejected if the defects cannot be repaired.

Welding of casing pipes should be in accordance with standards without needing destructive or nondestructive tests.

The casing should be considered ready for insertion of the carrier pipe after inspection and removal of earth, mud, and other foreign matters to the Engineer's satisfaction and all internal welds on the bottom 90 degrees have been ground smooth. The casing and the trench on either side of the casing ends should be free of water.

Vent pipes should be coated externally up to an elevation of 30 cm above the ground level.

2.11.4 Backfilling and Tampting

When permission is obtained to cross road bed by open trenching, backfilling and tampting should be performed in accordance with requirements given in standards. The Executor should plan and organize trenching, welding, laying, and backfill works so as to minimize delay and traffic interruption.

2.11.5 Safety of Traffic and Pipeline

During the entire period of the casing installation, the Executor should furnish and install, to the satisfaction of the Engineer, adequate and proper traffic aids such as warning signs, guards, and other safeguards necessary for the safety of the public at all crossings. The traffic aids should be in accordance with the regulations in force concerning traffic safety. Suitable protection should be provided on both sides of the road to prevent damage to the pipeline by vehicles leaving the road. Where any ditch during the casing operation remains open across public or private roads, the Executor should construct bypass roads, temporarily backfill the ditch, or install substantial temporary bridgework of adequate strength and width to ensure safety of traffic.

2.12 PREFABRICATED ASSEMBLIES

This part of the Standard covers test and installation of fabricated assemblies such as pig launcher and receiver, line break valves, isolated valves, vent, and drain valves.

2.12.1 Checks

The Executor should carry out preinstallation checks on all block valves, isolated valves, and scraper trap assembly to ascertain they are operational. The checks should be made in presence of the Engineer at storage sites and prior to transportation of the assemblies to the right-of-way. If the Executor feels needs of assistance or service from supplier or manufacturer, he can ask the Engineer for this.

2.12.2 Installations

The fabricated assemblies should be installed at the positions indicated on the working drawings.

The Executor should submit his method of installation to the Engineer for his approval and should exercise every care to ensure that damage does not occur to valves and assemblies as a result of his method of installation.

Piping, pipe fittings, and special components connected to the valves or to scraper traps should be assembled by the Executor in the ditch or at field workshop, if possible for some parts of assembly. Such parts of assembly which might be fabricated in workshop should receive corrosion protective coating before transportation to installation site.

Maximum torque used for stud bolts in flanged end valves should be agreed by the Engineer.

All valves should be fitted over an insulated pad on the surface of their concrete support, where concrete valve support is shown on working drawing.

Actuators for valves should be installed by the Executor. All tubing and wiring should be installed by the Executor in accordance with the working drawings or the valve manufacturer's instruction as directed by the Engineer.

2.12.3 Coating and Painting

Surface of all buried parts of fabricated assemblies should be cleaned, coated, and inspected as specified in standards.

Wherever buried parts or apparatus exit from the ground, they should be coated with the same specification as applied to buried parts for a distance corresponding to at least one pipe diameter above finished ground level. This distance should never be less than 1000 mm.

All above-ground parts of fabricated assemblies should be cleaned and painted.

2.13 CLEANUP OPERATION

This part of the Standard covers cleanup and restoration activities that should be performed immediately after the pipe is backfielded. The activities are aimed at leaving right-of-way to its original conditions.

As soon as possible after the pipe is laid and backfielded, the Executor should perform such work as required to leave the right-of-way as nearly like its original condition as possible. The work should include removing surplus and defective materials, disposing waste materials, removing temporary bridges, restoring the land traversed, and restoring any neighboring land and utilities affected by the Executor operations to their original condition.

The Executor should use a special "cleanup crew" to work systematically down the line behind the pipelaying and backfilling operations until the work is complete to the satisfaction of the Engineer. At any time during progress of the work, completed cleanup should not lag more than 5 km behind completed backfilling unless written approval of the Engineer is obtained.

2.13.1 Stages of Cleanup and Restoration Activities

Major cleanup and restoration activities should get started immediately after the pipe is backfielded. Remaining cleanup and restoration activities should be carried out upon completion of the final tie-in corresponding to hydrostatic test section.

Final cleanup should be carried out after the completion of final dewatering and before commissioning.

2.13.2 Surplus, Defective, and Waste Materials

All surplus or defective construction materials that are the property of the company should be transported and delivered by the Executor to the company's storage depot as directed by the Engineer.

Waste materials such as damaged coat and wrap, damaged drums of primer, etc., should be disposed of by the Executor. Outside of built-up areas, it is permissible, subject to approval by the Engineer, to bury such waste outside of the right-of-way to a depth giving a minimum of 500 mm of backfill cover.

2.13.3 Restorations of Public and Private Facilities

The Executor may use public and private roads, bridges and other structures, and utilities for access or haulage. Assessment of any damage to these facilities caused by his operation is the sole judgment of the Engineer and the concerned authorities. Repair of damage should be made as cited in relevant standards.

In carrying out the restorations and repairs, the Executor should comply with all the clauses or regulations issued by the relevant authorities for the purpose.

All drainage and irrigation systems, masonry works, and other items listed in the reports concerning site conditions, prepared before opening the right-of-way, should be brought back to a Standard equal to that existing before the start of the works and accepted by the Engineer.

The Executor should ensure that any water which was collected, as a result of the work, is evacuated by constructing, at his own cost, a drainage system without causing any environmental damage.

All creeks, water courses, wells, qanats, and ditches should be restored to their original condition and their banks should be pitched with stone to prevent from washing out or erosion.

2.13.4 Activities in Cleanup Operations

The service road section of a right-of-way and also the edge of the backfill crown should be cleaned and graded smoothly.

Temporary roadways provided for construction should be removed unless required by the company for future operational use.

The right-of-way should be continually maintained by the Executor against wash-outs or erosion until final acceptance of the work. Therefore, if the Executor attempts to perform cleanup operations during adverse weather or a wet-ground condition, he should assume full risk of acceptance and the Executor may be required to perform again such cleanup activities.

2.14 RECORDS AND AS-BUILT SURVEY

This part of the Standard describes the Executor responsibility in maintaining record books and locating repairs. It also covers requirements and information in respect of the as-built surveys.

Upon completion of the final cleanup and restoration operation, a complete record of the condition of the right-of-way and access ways should be submitted to the Engineer by the Executor.

The Executor should maintain during the work record books describing and locating repairs, of whatever nature, to the pipe and pipeline and should make three copies of them available to the Engineer.

The Executor should provide and submit to the Engineer all reports, documentations cited in individual parts of this chapter, as well as result of inspection and tests performed in accordance with Standards.

Together, with the final survey drawings, the Executor should submit a technical report on the as-built survey that follows construction work as well as copies of all field books used on the survey and copies of all computations made during the survey.

The report on the survey and its computations should give full details of all basic data used, methods employed, instruments used, adjustments made, Standards of accuracy observed, reference and bench marks established, special problems or difficulties experienced, index of all drawings, and data used and associated with the survey. The Executor should submit final survey drawings in five copies.

Unless otherwise agreed upon with the Engineer, as-built data to be prepared by the Executor should be in accordance with those shown on route and profile drawings.

2.14.1 As-Built Survey

On completion of the pipeline construction, or of a section as approved by the Engineer, the Executor should carry out the "as-built" survey in accordance with the procedures approved by the Engineer. The survey report and drawing should provide the primary information given below:

1. The total length of the pipeline.
2. The continuous, contour-line profile of the pipeline along the route in relation to existing topographical features and Company or other development.
3. The positions and relevant lengths of above-ground sections of the pipeline and facilities.
4. Accurate connections to existing control points (NCC) to enable the new works to be shown on Company maps and drawings.
5. Data for showing the relation of the new works in relation to properties and land boundaries and rights-of-way of other properties and as a basis for assessing any claims for compensation from other parties arising from the construction of the works.

2.14.2 Detail Information

The as-built survey and final survey drawings should include the following information:

The elevations in meters, above mean sea level, of the ground level and at the top of the pipe at intervals not exceeding 150 m and also at all changes in grade, river stream and road crossings, above-ground sections along the pipeline, and its terminal and crossing with other pipelines. Elevations at all kilometer posts, bench marks should be given.

The horizontal positions, in metric coordinates, of the pipeline and all topography features specified in Paragraph 3.1e and 3.1b above within the following distance on each side of the pipeline.

150 m on each side of liquid pipeline.
250 m on each side of gas pipeline.

The slope changes in meters at intervals not exceeding 150 m for the complete length of the pipeline.

Changes should also be shown at river-stream crossings, major road crossings, changes in pipe diameters, pipe type, grade, and wall thickness, valves, scraper traps, start and end of all above-ground sections of the pipeline and reference markings, adjacent building, survey control points, reference marks, bench marks, and other identifiable features.

2.14.3 Topographical Feature Survey

Other information cited in route and profile drawings, given to the Executor as part of tender documents, should also be reflected and relevant spaces should be filled in. The survey report and survey drawings should be accepted and approved in writing by the Engineer before final certificate is issued.

2.14.4 Survey Procedure

The Executor should submit details of his proposed survey personnel for approval of the Engineer. The Executor should replace any of his survey staff who are considered by the Engineer to be incompetent, inexperienced, negligent, or incapable of their duties.

The survey is to be carried out using theodolite and chain (steel survey tape), and any other apparatus necessary, to an accuracy of not less than 1/7500. Electronic distance measuring equipment, if approved by the Engineer, may be used instead of steel tape.

Horizontal angles should be observed to 10 seconds of arc and individual chain measurements measured to 0.025 m.

Elevation of the ground surface at the top of pipe should be observed and recorded to 0.025 m.

The slope change is to be used to show the actual distance along the pipeline of any point from the start of the line. The Executor should make the measurement over pipeline entire length by a continuous change commencing with chain at the start of the line and continuing to the end of same.

On completion of the survey, or before any drafting commences, the Executor should produce all field books and computations for examination by the Engineer. The Executor may be required to clarify such field notes or computations as required by, and to the satisfaction of, the Engineer. If the survey drawings and survey report are not acceptable in part or whole, the final certificate will only be issued after the Executor has carried out such additional survey work or prepared revised drawings or survey report that are acceptable to the Engineer.

2.15 ABOVE-GROUND PIPELINES

Requirements specified for buried lines in preceding parts of this chapter should be observed in construction of above-ground pipelines where applicable.

Considerations which apply specifically to above-ground pipelines include the following which should be taken into account during construction.

The pipe should be laid to follow the ground contours of the cleared, right-of-way surface on sleepers and supports.

In grading operation, the gradients should be uniform between changes in vertical direction.

Crash barriers should be provided to protect the pipeline from traffic accidents, if found necessary, using suitable barriers to prevent the pipeline from becoming an unauthorized footpath.

Chapter 3

Welding of Transportation Pipeline

3.1 INTRODUCTION

This chapter gives technical specifications and minimum requirements for welding of transportation pipeline and related facilities for use in oil, gas, and petrochemical industries and is based on API Standard 1104 19th edition Sep. 1999 and should be read in conjunction with that document.

This chapter covers the arc welding of butt, fillet, and socket welds in carbon and low-alloy steel for liquid and gas transmission pipelines and related facilities including pig traps. The welding may be carried out by a shielded metal arc welding (SMAW), submerged arc welding, gas tungsten arc welding, gas metal arc welding, or flux cored arc welding process, or a combination of these processes using a manual, semiautomatic, or automatic welding technique or a combination of these techniques. The welds may be produced by position or roll welding or by a combination of position and roll welding. Roll welding is only acceptable when using a fully automatic welding process.

Oxyacetylene welding (otherwise known as gas welding) and flash butt welding processes should not be used.

The use of gas metal arc, gas tungsten arc, and flux cored arc welding (except the self-shielding type) processes should be restricted to construction areas protected against wind and draught.

The standard also covers the acceptance standards to be applied to production welds tested to destruction or inspected by radiographic, ultrasonic, or magnetic particle techniques. It includes the procedures for inspection using these techniques.

3.2 MATERIALS

3.2.1 Filler Metal

3.2.1.1 Type and Size

Consumables should conform to one of the following specification:

- AWS A5.1
- AWS A5.5

Oil and Gas Pipelines and Piping Systems. DOI: http://dx.doi.org/10.1016/B978-0-12-803777-5.00003-4

- AWS A5.17
- AWS A5.18
- AWS A5.20
- AWS A5.28
- AWS A5.29

All welding consumables should be selected to produce welds with yield strength exceeding that specified for the parent material. Where steels with different specified properties are joined the weld metal yield strength should match or exceed that of the higher strength grade.

For sour service applications, the deposited weld metal should comply with the requirements of NACE MR0175.

If low-hydrogen electrodes are selected, the diffusible hydrogen content should not exceed 10 mL/100 g in the resulting deposited weld metal.

3.2.1.2 Storage and Handling of Filler Metals and Fluxes

Electrodes should be supplied with fully sealed packages and stored in a dry storage room with a maximum relative humidity of 50%. Manual types of electrodes should be properly identifiable up to the time of usage, with each electrode being distinguishable by proper coding. If the coding is destroyed by baking, handling, or other causes, the electrodes should not be used.

Low-hydrogen electrodes should not be stored in heated cabinets containing electrodes of other types, such as rutile or organic type electrodes.

Wire spools for automatic and semiautomatic processes should be stored in cabinets with supplier wrapping not removed and remain clearly identifiable up to the time of use. Unidentifiable wire should not be used.

Filler metals and fluxes should be handled and stored in accordance with the manufacturer's recommendations.

Each batch of flux and wire should be labeled with the information from the supply container.

Unidentifiable, damaged, wet, rusty, or otherwise contaminated or deteriorated consumables should not be used.

3.2.2 Shielding Gases

3.2.2.1 Types

Unless otherwise indicated by the company the maximum variation of specified gas additions, e.g., 5% carbon dioxide, should be +10% of the value stated. The moisture content should correspond to a dew point not exceeding −30°C.

3.2.2.2 Storage and Handling

Shielding gas containers should have clear identification labels that include the gas type.

3.3 QUALIFICATION OF WELDING PROCEDURES FOR WELDS CONTAINING FILLER METAL ADDITIVES

3.3.1 Procedure Qualification

Before production welding is started, a detailed WPS should be prepared. Prior to carrying out qualification testing, this WPS should be submitted to the company for review.

Welding procedures should be tested to demonstrate that acceptable welds can be made by the procedure. The quality of the welds should be determined by both nondestructive and destructive testing. The welding procedure qualification testing (WPQT) should be witnessed by the company. Only qualified and approved welding procedures should be used for production welding.

Unless otherwise stated in the contract documents, for each contract all existing welding procedures should be requalified by the contractor, and submitted for approval by the company.

Repair welding procedures should be prepared and approved in the same manner as production welding procedures.

3.3.2 Record

Qualified procedures should be recorded by the contractor and submitted to the company.

3.3.3 Procedure Specification

3.3.3.1 Pipe and Fitting Materials

The contract materials to which the procedure applies should be identified on the WPS. Grouping of materials of different pipe manufacturers, supply condition, diameter, wall thickness, or steel specification/grade should not be done unless agreed by the company.

Where contract materials have been supplied in the same dimensions by more than one manufacturer, a qualification test may be performed using two pipes from different manufacturers. This may be used to qualify the procedure for use on the pipes from each manufacturer provided the specified range and number of tests in the heat-affected zone are taken from both sides of the weld.

3.3.3.2 Diameters and Wall Thicknesses

Procedures should be qualified for each combination of diameter and nominal wall thickness of contract materials.

3.3.3.3 Joint Design

The specification should indicate the allowable tolerances on each of the joint design details. Permanent backing should not be used.

3.3.3.4 *Filler Metal and Number of Beads*

Details of the filler metal sizes, classification, and manufacturer/brand identity should be given together with a sketch showing the location, minimum number, deposition sequence, and characteristics (stringer or weave) of each weld bead.

3.3.3.5 *Electrical Characteristics*

Additionally, the aim voltage and amperage for each bead should be stated. The ranges of voltage and amperage should not vary from the aim values by more than $+10\%$.

3.3.3.6 *Time Between Passes*

For pipeline butt welds, this time should be 5 minutes or fewer.

3.3.3.7 *Cleaning and/or Grinding*

The WPS should state the methods to be used for interrun cleaning, final weld surface preparation, and treatment to backside of the weld, if any. The type of tools (power, manual, or both) should be specified.

3.3.3.8 *Pre- and Postheat Treatment*

Similar measures should be taken to specify and monitor weld interpass temperatures.

3.3.3.9 *Shielding Flux*

The type of shielding flux, the name of the flux manufacturer, and the flux identity and/or brand name should be designated.

3.3.3.10 *Speed of Travel*

Alternatively, for SMAW the range of electrode run-out length for each pass in each in each electrode size should be clearly specified. Speed of travel or electrode run-out length should be within a range of $+10\%$ of the nominal value for the specified electrode type and size as stipulated by the manufacturer.

3.3.3.11 *Heat Input Range*

The allowable range of heat input rates to be applied by the welding processes for each weld bead should be clearly specified. The units to be used should be kilojoules per millimeter (kJ/mm) based on the following formula:

$$\text{Heat input(kJ/mm)}\frac{\text{Volts}^*\text{Amps}}{1000^*\text{Welding speed(mm/s)}}$$

3.3.3.12 *Number of Welders*

For pipeline girth welds the WPS, the WPS should designate:

1. the number of root pass welders.
2. the number of hot pass welders.

3.3.3.13 *Partially Complete Joint*

The minimum number of passes before the joint is allowed to cool to ambient temperature and action required for partially completed welds.

3.3.3.14 *Removal of Line-Up Clamp*

The stage at which the line-up clamp is removed (see Section 3.8.2).

3.4 ESSENTIAL VARIABLES

3.4.1 Changes Requiring Requalification

3.4.1.1 *Base Material*

Unless agreed to otherwise by the company for pipeline welding, a change of pipe manufacturer, manufacturing process, or steel specification/grade should constitute an essential variable. In addition, requalification should be required if the carbon content or carbon equivalent exceed the value qualified by more than 0.03.

For nonpipeline welding, a change in the pipe manufacturing process, an increase in the base material specified minimum yield stress of more than 50 N/mm^2 or an increase in the carbon content or carbon equivalent of 0.03% or greater, should constitute an essential variable.

3.4.1.2 *Joint Design*

A major change in joint design (for example, from V groove to U groove) or any change outside the specified joint design tolerances constitutes an essential variable. Changes within the specified tolerances of the joint design (angle of bevel, root face thickness, and root gap) are not essential variables.

3.4.1.3 *Pipe Diameter and Wall Thickness*

A change in outside diameter outside the range 0.75 D to 1.5 D and/or any change in thickness outside the range 0.75 t to 1.5 t constitute essential variables, unless otherwise specified by the company.

3.4.1.4 *Filler Metal*

The following changes in filler metal constitute essential variables:

1. A change from one filler metal group to another (see Table 1).

2. A change from one consumable manufacturer and/or trade name, or AWS classification to another.
3. A change in the diameter of electrode or filler metal.
4. A change in the minimum specified yield strength of the filler metal.

3.4.1.5 Shielding Gas and Flow Rate

A change from one shielding gas to another or from one mixture to another constitutes an essential variable. A flow rate of the shielding gas outside the range specified and qualified also constitutes an essential variable.

3.4.1.6 Shielding Flux

1. Any combination of flux and electrode in Group 4 may be used to qualify a procedure. The combination should be identified by its complete AWS classification number, such as F71-EL-12 or F63-EM12K. Any change in either wire or flux manufacturer and/or AWS classification number should constitute an essential variable.

   A change in the flux size grading should also constitute an essential variable.

3.4.1.7 Speed of Travel

A change in the range of speed of travel or electrode run-out length should constitute an essential variable.

3.4.1.8 Preheat and Interpass Temperature

A change in preheat or interpass temperature ranges should constitute an essential variable.

3.4.1.9 Welding Current or Heat Input

Any change outside the specified ranges of currents or heat inputs should constitute an essential variable.

3.4.1.10 Number and Sequence of Weld Beads

A change in the minimum number of the weld beads deposited or the sequence of deposition should constitute an essential variable.

3.4.1.11 Number of Welders

A change in the number of root pass or hot pass welders should constitute an essential variable.

3.4.1.12 Type and Removal of Line-Up Clamp

Removal of the line-up clamp at a stage earlier than the approval procedure, and change in type of line-up clamp should constitute an essential variable.

3.5 TESTING OF WELDED JOINTS—BUTT WELDS

3.5.1 Preparation

3.5.1.1 Nondestructive Testing of Test Welds

On completion of welding, all procedure qualification test pieces should be left cold for at least 48 hours and should then be subjected to NDT in accordance with Sections 3.9 and 3.11. This should be carried out prior to sectioning for mechanical testing. Postweld heat treatment, if required, should be performed after 48 hours has elapsed, but before NDT is performed.

The NDT should consist of:

- Visual examination, with the aid of optical instruments where necessary, to determine the dimensions of indications.
- Magnetic particle testing as specified in (Section 3.11.2).
- Radiographic testing in accordance with (Section 3.11.1.1), supplemented, if specified by the company, by ultrasonic testing

in accordance with (Section 3.11.3). This supplementary ultrasonic testing should always be carried out for test welds made in whole or in part by the GMAW, GTAW, or FCAW processes.

Acceptance criteria for the NDT should be as stated in Section 3.11. If a test weld is found to be unsatisfactory following NDT, it should be rejected and not be submitted for destructive/mechanical testing.

3.5.1.2 Destructive Testing of Test Welds

Following satisfactory acceptance by NDT, all procedure qualification test welds should be sectioned at the locations shown in Figure 3 and Fig. 3.1.

The minimum of test specimens and the tests to which they should be subjected are given in Table 2 and in Sections 3.5.2 and 3.5.3. The specimens should be prepared, as shown in Figures 4, 5, 6, or 7.

Unless specified otherwise, tests should be performed at ambient temperature. For pipe less than or equal to 1 5/16 in. (33.4 mm) in diameter, one full-section specimen may be substituted for the four reduced-section nick-break and root-bend specimens. The full-section specimen should be tested in accordance with (Section 3.5.2.2) and should meet the requirements of (Section 3.5.2.3).

3.5.2 Macroscopic Examination and Hardness Tests

3.5.2.1 Preparation

Specimens should be prepared for macroscopic examination by grinding to a 600-grit paper finish. The prepared surfaces should be etched using a suitable etchant (e.g., 3% Nital or ammonium persulfate) to reveal the grain structure.

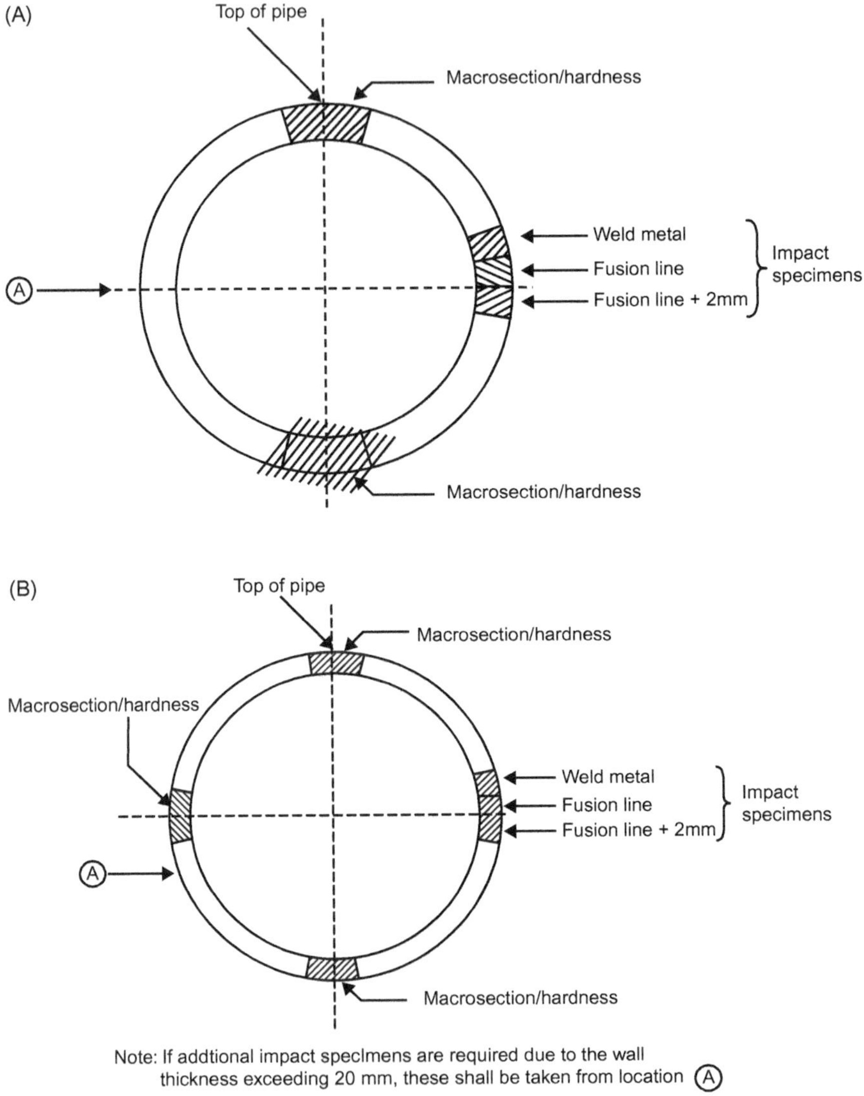

FIGURE 3.1 Location of macrosection and impact test specimens procedure qualification of butt welds. (A) Pipe diameters up to and including 4 in. (114.3 mm), (B) Pipe diameters greater than 4 in.(114.3 mm).

The sections of the weld taken for macroscopic examinations should be used for hardness testing.

3.5.2.2 Method

The hardness should be measured in accordance with ASTM E 92 using a Vickers instrument with a 10-kg maximum load. For pipe butt welds, hardness traverses should be carried out on lines 2 mm from the inner and outer pipe surfaces on the weld cross-sections, and also a line through the mid wall if the pipe thickness is greater than 16 mm.

Lines of indentations should give at least three values in each of the weld metal, the HAZ each side of the weld, and the base metal. One HAZ impression of each side of the weld should be within 0.5 mm of the weld junction.

The macro-specimens should be examined at 5× magnification.

3.5.2.3 Requirements

The macrospecimens should not show defects exceeding the acceptance standards given in Section 3.11. Each specimen should exhibit a smooth and regular profile and the reinforcement should blend smoothly with the parent metal. Slight intermittent undercuts should be permitted provided the depth does not exceed 0.4 mm. Excess penetration should not exceed 3 mm. Joint misalignment should not exceed 1.6 mm.

The maximum hardness levels attained in each of the three zones, i.e., parent metal, HAZ and weld metal zone, should not exceed 280 HV 10. For welds in components or pipe designated for sour service, the hardness should not exceed 248 HV 10.

3.5.3 Charpy V-Notch Impact Testing

3.5.3.1 Preparation

Impact testing should be carried out when the nominal pipe wall thickness exceeds 5 mm. Three values should be obtained from each of the weld center line, fusion line, and the fusion line +2 mm. The specimens should be taken from the mid-thickness with the notch in a radial orientation. When the pipe dimensions preclude, the preparation of a rectangular (5 mm × 10 mm) cross-section specimen should be prepared with the maximum feasible thickness.

In each case the impact energy should satisfy the requirements specified for a 5 × 10 mm specimen. When the wall thickness is more than 20 mm, a test series at the same locations should also be made at the root side of the weld.

Testing should be carried out at a temperature determined in accordance with Table 3.1.

The minimum design temperature, T, should be stated in the contract documents. If no such information is available, the impact testing should be carried out at 0°C.

The dimensions, preparation, and testing of the impact test specimens should be in accordance with ISO 148. Subsized specimen may be used when standard specimens cannot be prepared.

TABLE 3.1 Charpy V-Notch Test Temperature

Nominal Wall Thickness (t) (mm)	Test Temperature (°C)
$t \leq 16$	T
$25 \geq t > 16$	T-10
$t > 25$	T-20

Note: T, minimum design temperature.

Minimum required impact values are given in Table 3.2 for API Spec 5 L grades of line pipe. Requirements for equivalent materials should be determined by correlating the minimum specified yield strength with standards.

3.5.4 Retests

If the results of the WPQT are unsatisfactory due to defective preparation of the specimens or due to a local weld defect, the company may allow the procedure below to be followed.

3.5.4.1 Tensile Test and Bend Test Specimens

If a tensile or bend test specimen does not meet the requirements, two additional tensile tests or bend tests should be made, both of which should meet the prescribed requirements.

3.5.4.2 Impact Test Specimens

If one of the specimens gives an unsatisfactory result that is clearly caused by a local defect, a further test specimen may be taken and tested; only one such replacement should be permitted per set of three specimens.

3.5.5 New Procedure Requirement

If the test joint fails to meet the minimum requirements, a new WPS should be established and the WPQT should be repeated to the satisfaction of the company.

3.6 TESTING OF WELDED JOINTS—FILLET WELDS

3.6.1 Macroscopic Examination and Hardness Tests

Specimens for macroscopic examination should be extracted from fillet-weld-qualification welds at the same locations as shown for butt welds.

Hardness traverses should be carried out across the root of the fillet weld and also the cap region at a depth of 2 mm.

TABLE 3.2 Charpy V requirements (J)

Steel Grade (API Spec 5 L)	Specimen Size (mm)	Charpy-V Requirement (J)	
		Min. Average	Min. Single
B	10 × 10	27	22
	10 × 6.7	21	15
	10 × 5	18	13
X42	10 × 10	29	22
	10 × 6.7	23	17
	10 × 5	19	15
X46	10 × 10	32	24
	10 × 6.7	25	19
	10 × 5	21	16
X52	10 × 10	36	27
	10 × 6.7	28	21
	10 × 5	24	18
X56	10 × 10	39	29
	10 × 6.7	30	23
	10 × 5	26	20
X60	10 × 10	41	31
	10 × 6.7	32	24
	10 × 5	27	21
X65	10 × 10	45	34
	10 × 6.7	35	27
	10 × 5	30	23
X70	10 × 10	48	36
	10 × 6.7	37	28
	10 × 5	32	24

3.7 QUALIFICATION OF WELDERS

The welder should perform the qualification test welding in accordance with the approved WPS and the qualification of welders should be conducted in the presence of the company's representative.

A welder qualification should be valid for a period of 6 months. If it can be shown by way of the results of nondestructive examination carried out in accordance with the requirements of this specification that during this period, he has successfully produced welds in accordance with the approved welding procedure, the period will be automatically extended by another six months.

3.7.1 Single Qualification

A pipeline welder should qualify for welding by performing a test on contract material. He will be qualified to weld only in the same position as the test weld.

For nonpipeline applications, a welder who successfully passes a butt-weld qualification test on contract material in the fixed position with the axis inclined 45 (± 10) degrees to the horizontal plane (6 G position) should be qualified to make butt welds on contract material in all positions. This is applicable only to manual welding. Automatic or mechanical welding procedures should be qualified in the same position as they are applied. The use of segments of pipe nipples for welder qualification should not be permitted.

3.7.2 Records and Welder Identification System

A Welders Competence Certificate, which includes references to the corresponding WPS number, the essential variables, and the test results, should be issued for each welder or welding operator and for each test.

When production welding, the welder/welding operator should always be identifiable by a badge bearing his name, his photograph, and his identification number. Welders and welding operators not wearing their badges should be suspended from production welding. In the event that a welder ceases working on the contract his identifying mark and number should not be assigned to another welder.

3.8 PRODUCTION WELDING

Limitations imposed by the essential variables of the procedure qualifications should be adhered to in production welding. No welding should be carried out before the WPSs and WPQTs are completed, nor before the welders have been qualified, and approved by the Company. Only qualified welders as defined in Section 3.7 may be employed. Preparation and welding of pipeline components should be in accordance with the appropriate qualified WPS.

The surfaces to be welded should be smooth, uniform, and free from laminations, tears, scale, slag, grease, paint, and other deleterious material that might adversely affect the welding.

If work is to be carried out in the vicinity of equipment already installed, before any welding commences adequate protection should be provided to prevent damage from weld spatter, flame cutting droplets, etc. Care should be taken to avoid overloading or damaging any of the pipeline components at all stages of the work.

Current return cables of welding equipment should be connected directly to the pipe on which the welding is to be done.

If the pipe size exceeds 12 in., at least two welders should weld simultaneously around the pipe circumference.

Arcs should be struck only on fusion faces or on striking plates provided as an aid to arc starting. Stray arc strikes should be removed by grinding away all material which has been affected by the arc heat. Where this results in the minimum thickness being below tolerance, the section of pipe containing the arc strike should be removed. Weld repairs or build-up should not be made on the base pipe.

A circular cap to prevent entry of foreign material, of a design which will not damage pipe ends, should be used to cover the open ends of the pipe and should be placed on the line during interruptions in the work expected to last more than 2 hours. Caps should not be removed until recommencement of the work. All open ends of pipe strings should be capped off and sealed when welding is completed.

3.8.1 Alignment

The pipeline components should be firmly supported in both the vertical and horizontal plane, and no welding should be carried out until as much of the pipeline system as will be so stiffened has been properly aligned.

Misalignment should be reduced to a minimum by rotation of the pipes to obtain the best fit, or by other approved methods. When a pipe with one longitudinal seam is used, this seam should be within the top 120 degrees of the circumference and the longitudinal seams of adjacent pipes should be offset by a circumferential distance of at least half the pipe diameter.

The alignment of the abutting ends should be set so as to minimize the offset between surfaces. If the offset exceeds 1.6 mm, provided it is caused by dimensional variations within the specified tolerances, the pipe with the smaller diameter should be trimmed to a taper not steeper than one in four.

Alignment of two pipeline components of different nominal thickness and the same outside diameter should be carried out by tapering the inner surface of the thicker component with a taper not steeper than one in four.

Tack welds should be avoided wherever possible. Where required, bridge tacks should be used. Tack welding should be performed by qualified welders using the same qualified welding procedure as will be used for the main welds.

Hammering of pipe to obtain proper lineup should be kept to a minimum plastically deforming the pipe or bevel to obtain proper alignment is not permitted.

3.8.2 Use of Line-Up Clamp for Butt Welds

Line-up clamps, holding devices, etc., should be used for butt welds to avoid tack welding in the groove and to optimize alignment. Internal line-up clamps of a type acceptable to the company should be used for pipe sizes of 8 in. and larger. External line-up clamps may be used for pipe sizes 6 in. and smaller. For tie-in welds, external line-up clamps may be used for all pipe sizes.

Internal line-up clamps should remain in place at least until the root pass is completed around the full circumference.

External line-up clamps should not be removed until a minimum of 50% of the root pass, uniformly spaced around the circumference, has been completed.

Root bead segments used with external line-up clamps should be cleaned, ground down to a feather edge at both ends, and visually inspected prior to completion of the root pass. Any such segment which is not in accordance with the acceptance standards given in Section 3.7 should be removed before completion of the root pass.

The pipe joint should not be moved until after the second weld pass (hot pass) has been made.

3.8.3 Field Bevel

The equipment used for edge preparation and cleaning (e.g., cutting, grinding, gouging, brushing, etc.) should cause no detrimental metallurgical effects to the edges to be welded.

Pipe ends should be beveled by machining or grinding. Preparation of weld edges by gas cutting should, wherever practical, be done with a mechanically guided torch. Edges should be left free of slag and the cut surface should be ground to a smooth uniform surface by removing approximately 0.5 mm of metal.

After grinding, the beveled edges should be visually examined to ensure freedom from defects. Any beveled edge that has been damaged should be restored to within the tolerances required by the welding procedure to be applied. Restoration involving welding is not permitted. Should laminations or split ends be discovered in any pipe, the full pipe joint should be removed and subjected to full ultrasonic examination for the presence of laminations in accordance of the line pipe purchase specification before partial reuse is considered.

3.8.4 Cleaning Between Beads

Pneumatic grinders and deslagging tools should be designed such that the exhaust air does not impinge on a hot weld region.

3.8.5 Roll Welding

3.8.5.1 Alignment

Roll welding should be restricted to fully automatic welding and then only when the pipe can be adequately supported on rollers with drives coupled electrically to the automatic welding machine.

3.8.5.2 Identification of Welds

Prior to starting the root pass, the welder or welding operator should clearly mark the pipe adjacent to his weld with the identification mark assigned to him in his qualification certificate. Tack welding of components should not be so marked.

Marking should be done with weather-proof chalk, paint, crayon, or felt pen. Die stamps should not be used for marking the welds. The welder/operator who makes the root pass should write his code at the top of the pipe. If, however, two welders/operators weld the root pass, each welder/operator should mark ~op o~ on the pipe with his identification code on the side on which he has worked. Subsequent welders/operators should write their identification codes below the first code in the sequence in which they work. The identification marks should not be removed until after the welds have been both visually and nondestructively inspected and accepted.

3.8.6 Preheating and Interpass Temperature Control and Measurement

Preheating should be carried out using either electrical resistance or induction heaters or using gas burners specifically made and shaped for this type of operation. Torches intended for flame cutting or gouging should not be used. Induction heating coils can cause an arc blow during welding; therefore, power to these coils should be off when welding is in progress.

Where induction heating is proposed for the application of either preheat or postweld heat treatment, the equipment and the procedure to be used should be approved by the company. In particular, the Contractor should demonstrate to the satisfaction of the company that heating rates and required temperatures can be properly controlled and that adequate precautions have been taken against overheating.

The preheating temperature should cover an area of at least 75 mm width on either side of the weld and should be maintained over the full length of the weld until the weld is completed.

The weld area should be protected from draughts, and insulation should be provided on adjacent areas where this is necessary to maintain the required temperature of preheating during welding.

The temperature measurement for preheat and interpass temperature may be determined by thermocouples or thermosticks, or a combination of both.

Where thermosticks are used, these should be of the type that melt when the required temperature is reached.

3.8.6.1 Temperature Requirements

The minimum preheat should be calculated from BS 5135. The heat input value used in the calculation should be the minimum value indicated in the procedure specification. The carbon equivalent value should be the highest to be encountered when using the weld procedure, as determined from a survey of the relevant material certificates. Scale A of BS 5135 should be used for cellulosic SMAW electrodes and Scale C for basic low hydrogen SMAW electrodes. The scale to be used with other processes should be subject to the approval of the company. In all cases, a minimum preheat of 50°C should be used.

For the welding of "sweepolets" the preheat should be 50°C higher than that indicated by BS 5135, using the above parameters.

Preheating to a minimum temperature of 100°C is required when the thickness of any one of the sections to be welded exceeds 32 mm. The interpass temperature should not be permitted to drop below the minimum preheat temperature (if required). The interpass temperature should not be allowed to exceed 300°C.

3.8.6.2 Tackwelding

Preheating for tack welding should comply with the general requirements.

The material, over a zone 75 mm wide round the position of the intended bridge tack and through its full thickness, should be heated to the required temperature. For tack welding the minimum preheating temperature should be 50°C above any preheat temperature specified in the WPS for production welding, with a maximum of 300°C.

3.8.7 Postweld Heat Treatment (Stress Relieving)

The necessity for postweld heat treatment should be determined from the applicable equipment design code or standard. In the absence of specific requirements from such codes or standards, the requirements of ANSI/ASME B3 1.8 para 825 should be followed.

Stress relieving of welds in pipes of API Spec 5 L Grades x60 and above, should be carried out in the range of 560–600°C. Stress relieving of welds in all other grades of carbon steel should be carried out in the range of 580–620°C. The holding time should be 2.5 min/mL wall thickness, with a minimum time of 1 hour. The heating and cooling rate should be a maximum of 300°C per hour.

Pipe ends should be covered when stress relieving the welds to avoid draughts inside the pipe. Threads and gasket surfaces should be protected from oxidation during heat treatment, if there is no subsequent machining operation which will remove any damage.

A sufficient number of thermocouples should be used in order to give a reliable temperature measurement. On pipe less than 8 in. nominal diameter, one thermocouple is regarded as sufficient; those with a diameter of 8–20 in. should have at least two at 180 degrees between each other, and pipe with a diameter of above 20 in. should have at least three thermocouples at 120 degrees between each other. In no circumstance should the distance between two thermocouples, measured around the pipe circumference, be greater than 800 mm.

Postweld heat-treatment procedures should be submitted to the company for approval along with the WPS and WPQT record, prior to the commencement of work.

3.8.8 Separation of Girth Welds

The minimum allowable distance between girth welds should be the external diameter of the pipe, or 500 mm, whichever is larger.

3.8.9 Control of Welding Consumables During Production Welding

Ovens or storage cabinets with automatic heat controls and temperature readout equipment should be provided to maintain low-hydrogen welding electrodes and welding flux at the required temperature.

After removal from an oven, all low-hydrogen electrodes should be kept in a storage cabinet or quiver. No electrodes should be left exposed to the atmosphere. Issue of low-hydrogen electrodes from the storage cabinet should be controlled so that all electrodes are used within 4 hours of issue.

To maintain the low-hydrogen content, electrodes should be baked immediately before use at 300°C (±30°C) for 1 hour or in accordance with the manufacturer's recommendations. This requirement may be waived when the electrodes are removed from an hermetically sealed package immediately before use. Plastic wrapped cartons are not considered to be hermetically sealed.

After drying as described above, electrodes may be transferred to an intermediate storage cabinet maintained at approximately 150°C. Upon removal from such drying or storage the electrodes should be transferred in small numbers to heated ~ quivers with a minimum temperature of 70°C and used within 4 hours.

Electrodes not used within 4 hours, or for some reason exposed to adverse atmospheric conditions, should be baked again in accordance with the above conditions. Electrodes may be baked twice only. If these electrodes are not used within 4 hours after the second baking treatment, they should be discarded.

3.8.10 Weld Interruption

The deposition of each weld should generally be a continuous operation. However, in the case that welding must be discontinued, this should not take place before at least the root and hot passes are completed.

Before resumption of welding, the join should be reheated to a temperature within the interpass temperature. Interrupted welds should be inspected by radiographic examination in accordance with relevant standards.

3.8.11 Weld Finishing

Welds should be left as welded and should not be treated with a flame torch or by any mechanical means to change their appearance, other than the cleaning defined in the WPS. Welds should not be peened.

3.8.12 Surface Finishing

When welding is completed, all surfaces adjacent to the welds should be cleaned to allow proper radiographic or ultrasonic inspection, and to remove all detrimental burrs and other marks. Cleaning procedures should be indicated on the welding procedures or drawings. Any damage should be rectified prior to NDT.

3.9 INSPECTION AND TESTING OF PRODUCTION WELDS

3.9.1 Methods of Inspection

The welds should be evaluated on the basis of Section 3.10 requirements. Written NDT procedures as required in Section 3.12 should be prepared and submitted to the Company for approval. All operations should be performed in accordance with these procedures on welds in their final condition.

3.9.2 Radiographic Examination

For the purposes of this chapter, radiography means the use of X-ray examination. Gamma-ray examination may be used only with the approval of the company.

The radiographs should be free from imperfections due to processing or other defects which could interfere with interpretation.

Radiography should be supplemented by ultrasonic testing and magnetic particle testing where there is reason to suspect planar defects such as cracks or lack of fusion.

3.9.3 Ultrasonic Examination

In addition to radiography, a limited ultrasonic examination should be made in the case of the use of mechanized or semiautomatic GMAW techniques. The minimum extent of ultrasonic examination should be 25% of the first 100 production welds of the same type with 5% of similar welds selected at random thereafter.

Ultrasonic examination may be used instead of radiographic examination if the wall thicknesses is equal to or greater than 12 mm and with the approval of the company. The method should be selected such that the progress of the construction is not affected. For pipeline girth welds, only mechanized ultrasonic techniques which provide a permanent registration of the results of the examination (hard-copy) should be used.

All ultrasonic examinations should be carried out in accordance with the methods described in acceptance criteria should be in accordance with standards.

3.9.4 Sectioning of Welds

The company should have the right to request a test weld to be made at any stage during normal production welding to allow a metallurgical examination of the deposited weld. Similarly, when NDT cannot be carried out or gives inconclusive results, the company should have the right to have a pipe section, including the weld, cut out, and removed for testing. Subsequent mechanical testing should be carried out in accordance with standards as appropriate.

3.9.5 Certification of Nondestructive Testing Personnel

3.9.5.1 Procedures

Nondestructive testing personnel should be certified in accordance with ASNT Recommended Practice SNT-TC-IA for the test method used or an equivalent certification scheme approved by the company.

3.10 EXTENT OF INSPECTION AND TESTING

3.10.1 Prewelding Inspection

All materials to be welded should be subjected to visual inspection for surface defects, laminations, etc., for compliance with the requirements in the relevant line pipe or fitting specification.

All weld preparations and repaired preparations should be inspected visually.

Edge preparations for tie-in welds should also receive magnetic particle examination.

3.10.2 Inspection During Welding

During production welding, the welding parameters should be checked against the WPS.

3.10.3 Inspection After Welding

All welds should be visually inspected, and cracks, craters, pinholes, weld spatter, residual slag, or arc strikes are not acceptable.

Butt welds should be inspected radiographically, and, where appropriate, inspected ultrasonically. All fillet welds should be tested using wet magnetic particle inspection.

Where radiography is not practical or where the company requires a further examination of a weld to assist in the evaluation of suspected defects, ultrasonic examination and magnetic particle inspection should be used. The frequency of the radiographic inspection should be determined by the company; however. the following is recommended.

CATEGORY 1
For special cases in production welds:
One hundred percent

1. For tie-ins, similar welds and welds performed in bellholes.
2. For complete or partial welds (excluding cap repair).
3. For welds performed on pipeline sections intended for crossings (freeways, rail, road, and river crossings)
4. For welds between pipes of different grades and for joints of pipes having same outside diameter and different wall thickness.
5. For welds performed on insertion into line.
6. When there is a change in welding team.
7. When the company considers that performance conditions have changed, either because of the location or because of climatic conditions (for example, wind or sand storm).

CATEGORY 2
For normal production, as a minimum 30% of welds should be radiographed.

3.11 ACCEPTANCE STANDARDS FOR NONDESTRUCTIVE TESTING

3.11.1 Radiographic Testing

3.11.1.1 Inadequate Penetration Without High-Low

Inadequate penetration without high-low (IP) is defined as the incomplete filling of the weld root. This condition is shown schematically in Figure 13, and is unacceptable.

3.11.1.2 Inadequate Penetration Due to High-Low

Inadequate penetration due to high-low (IPD) is the condition that exists when one edge of the root is exposed (or unbonded) because adjacent pipe or fitting joints are misaligned. This condition is shown schematically in Figure 14, and is unacceptable.

3.11.1.3 Incomplete Fusion

Incomplete fusion (IF) is a discontinuity between the weld metal and the base metal that is open to the surface. This condition is shown schematically in Figure 16, and is unacceptable.

3.11.1.4 Incomplete Fusion Due to Cold Lap (IFD)

Areas of IFD which are separated by a distance not exceeding the length of the smaller indication should be treated as a single IFD.

For welds other than girth welds, the acceptance criteria for the individual and aggregate length of defects should be one half of the values indicated in **1** to **3**.

3.11.1.5 Burn Through and Excessive Penetration

Root penetration should not exceed 3 mm.

3.11.1.6 Cracks

Cracks of any type are unacceptable.

3.11.1.7 Accumulation of Imperfections

Any accumulation of (otherwise acceptable) imperfection should be unacceptable when any of the following conditions exists:

3.11.2 Magnetic Particle Testing

Relevant indications should be unacceptable when any of the following conditions exist:

1. Linear indications are evaluated as cracks, including crater cracks or star cracks.
2. Linear indications are evaluated as IF.

Rounded indications should be evaluated according to the criteria as applicable. The maximum dimension of a rounded indication should be considered as its size for evaluation purposes.

Note: When doubt exists about the type of discontinuity disclosed by an indication, other nondestructive testing methods should be used for verification.

3.11.3 Ultrasonic Testing

All indications that produce a response greater than 20% of the reference level should be investigated to determine the location, shape, extent, and type of reflectors arid should be evaluated according to the specified criteria.

3.11.4 Ndt Evaluation Report

An evaluation report should be made stating the identification number of each weld inspected and the name of the welder/operator together with the description of the inspection method and, if applicable, the reason for rejection of the weld. The "I.I.W. Collection of Reference Radiographs of Welds in Steel" should be used for identifying the defects found by radiography. For other inspection techniques, a description of the defect indication should be given.

3.11.5 Welder Performance Records

The contractor should maintain an accurate record of the performance and repair rate of each welder. The repair rate should be expressed as a percentage of repaired welds over the total number of welds produced.

The company should have the right to remove any welder from the job if his performance is considered to be of an unacceptable standard.

3.11.6 Assessment of Radiograph Indications in the Pipe Material Adjacent to Welds Being Inspected

Indications that the pipe material next to the weld has been damaged in any way (e.g., arc burns, mechanical damage, weld beads, defects in the longitudinal seam of longitudinally welded pipe, etc.) should be reason to reject the weld.

3.12 REPAIR AND REMOVAL OF DEFECTS

3.12.1 Cracks

A weld containing shallow crater cracks or star cracks, which are located at the stopping point of weld beads and are the result of weld metal contraction during solidification, may be repaired provided the length of the crater cracks does not exceed 4 mm. With the exception of these shallow crater cracks, any weld containing cracks, regardless of size or location should be removed.

3.12.2 Defects Other Than Cracks

Welds containing defects outside the limits given in Section 3.10 may be repaired only if approved by the company, and only one such repair may be made. A weld with unacceptable defects may be repaired once only. If the repair is then not acceptable, the complete weld, including the heat-affected zone, should be removed.

All repairs should meet the standards of acceptability given in Section 3.10 and should be executed by qualified welders according to approved repair procedures (see Section 3.3.1).

3.12.3 Repair Procedure

Before repairs are made, the defects necessitating the repair should be entirely removed to sound metal. All slag and scale should then also be removed.

The removal of weld metal or portions of the base metal may be done by machining, grinding, chipping, oxygen gouging, or air carbon-arc gouging. When thermal gouging methods are used the appropriate pre-heating (determined from the WPS) should be applied. Thermal gouging methods should not be used for removal of the weld root. The unacceptable portions of the weld should be removed without substantial removal of the base metal and in such a manner that the remaining weld metal or base metal is not nicked or undercut. Where thermal gouging is used, the edges should be dressed to remove the hardened surface.

The repair weld grooves should be free from scale and should have acceptable contours. They should be visually and magnetic particle inspected prior to rewelding to ensure the defect has been completely removed.

If planar defects in welds are to be repaired, every effort should be made to prevent propagation of the defect during its removal. During the final stages of removal, grinding and not gouging should be used. Magnetic particle inspection should be carried out to check for the complete removal of the defect.

3.12.3.1 Repair Procedure for Subsurface and Root Defects Other Than Cracks

A repair procedure should be formulated incorporating a prequalified repair-welding procedure in accordance with Section 3.3 of this chapter. In addition to welding details, the repair procedure should include:

1. Method of examination of the defect area.
2. Method of defect removal.
3. Requirements for interpass NDT, where applicable.

3.12.3.2 Weld Size Adjustments

For overlap or excessive convexity, excess weld metal should be removed by grinding or machining.

Additional weld metal to compensate for any deficiency in size should be deposited using procedures as qualified for the original weld. The surfaces should be thoroughly cleaned before depositing the additional weld metal.

3.13 PROCEDURES FOR NONDESTRUCTIVE TESTING

3.13.1 Radiographic Test Methods

As for practical matters, internal sources (crawlers) should be used. This is possible at least for pipelines of 8 in. diameter and larger. In cases where the sources cannot be placed inside the pipe the double-wall/single-image technique should be applied.

3.13.1.1 Film Radiography

j. Identification: the system(s) used to identify the weld, and the location of the radiograph in relation to the weld, should be given in the procedure.

Radiographs should be made using one of the film types defined in Table 3.3.

TABLE 3.3 Definition of Film Types

Speed	Contrast	Grain	Examples	
Slow	Very high	Very fine	Kodak Industrex M	
			Kodak	Industrex
			TMX	
			Kodak Industrex T	
			Agfa D2	
			Agfa D4	
			Fuji 50	
Medium	High	Fine	Kodak	Industrex
			AX	
			Kodak	Industrex
			AA	
			Agfa D5	
			Agfa D7	
			Fuji 100	

Prepacked films, e.g., roll pack, and rigid or flexible cassettes may be used.

3.13.1.2 Type of Penetrameters

Wire type penetrameters (image quality indicators, IQIs) in accordance with BS 3971 may be also used.

3.13.1.3 Selection of Penetrameters

3.13.1.3.1 Wire Type Penetrameters

Add the following:

In case of applying BS 3971, radiographic sensitivity should be two percent or better in the welded area. This should be determined by the use of an IQI in accordance with BS 3971.

For double-wall/double-image techniques, the diameter of the wires employed should be based on the nominal double-wall thickness of the pipe plus the measured thickness of the external weld reinforcement.

For double-wall/single-image techniques, the diameter of the wire should be based on the nominal single-wall thickness of the pipe plus the measured thickness of the external weld reinforcement.

3.13.1.4 Production Radiography

Only Level II or III radiographers should interpret the radiographic images of production welds.

Radiographs should be taken within 24 hours of weld completion. As soon as they have been taken, they should be developed and dried and interpreted by the contractor. Thereafter all radiographs should be submitted to the company for examination and approval.

With each batch of radiographs submitted to the company for examination and approval, the contractor should give a list of the radiographs in duplicate. This list should contain the contractors interpretation of each radiograph. The radiography list should be clearly marked to show which welds are repairs or rewelds. The company will state on this list the acceptability, or otherwise, of the weld.

The company should state whether the weld has passed, is to be repaired, is to be cut out, or requires additional NDT. The Contractor should repair or cut out and reweld the welds, or carry out additional NDT in accordance with the statements on the list returned to him.

3.13.1.5 Identification of Images

Images should be clearly identified by the use of lead numbers, letters or markers so that the correct weld and any discontinuities in it can be quickly and accurately located. The following techniques should be used for this purpose:

On pipe diameters over three inches, a tape measure with lead numbers every 100 mm should be placed adjacent to the weld. The zero point should be on the top of the pipe and the divisions should run clockwise in ascending order, viewed in the direction of pipeline laying progress.

On diameters fewer than 3 in., each shot should be designated by a lead letter placed on the pipe, i.e., A, B, C.

Where the weld has been dressed, markers in a form to be approved by the company should be placed alongside the weld, but just clear of the heat affected zone, to identify its position.

All radiographs should be clearly identified with reference to the welds they represent. All measurements used for defect location should be stated in metric units. The pipeline reference weld number and section number should be included in the radiograph identification.

Each weld should be marked using indelible material to provide reference points for the accurate relocation of the position of each radiograph.

3.13.1.6 Film Density

Film should be exposed so that the density through the weld metal is not less than 1.8 for X-rays or 2.0 for gamma rays, and should not be greater than 3.0 for single film viewing.

Radiographs with a density up to 3.5 may be acceptable if adequate viewer capacity is readily available.

Films should not be viewed when wet.

3.13.1.7 Image Processing

On-site administration, storing, and recording of radiographic films prior to and after exposure should be the contractor's responsibility.

Conditions of storage, temperature, and humidity should comply with the recommendations of the film manufacturer.

Before completion of the contract, all exposed film should be indexed, cataloged, boxed up and handed over to the company in accordance with the company's requirements.

3.13.1.8 Image Processing Area

The image processing room and all accessories therein should be equipped to handle the processing of all radiographs taken and should be kept clean and dust-free at all times.

Viewing illuminators should be used to produce sufficient light intensity so that all portions of the radiograph of the weld and base metal will transmit sufficient light to reveal the pertinent details of the radiograph. The light intensity should be compatible with the density of the radiograph specified herein.

3.13.1.9 Radiation Protection

The radiographer should be responsible for the protection of all persons in the vicinity of the radiographic equipment. The Contractor should familiarize himself with the safety requirements as set out in regulations, which contains recommendations for protection against hazards from sealed sources and equipment producing ionizing radiation. This document should form the basis of safety precautions to be applied. The contractor should satisfy the company that all aspects of relevant safety procedures are adequately covered and have been implemented.

The contractor should identify the areas where radiography is being performed by means of signs, symbols, etc., and should be responsible for the policing of such areas.

The contractor should be responsible for effectively coordinating the radiographic functions so as to minimize interference with other primary activities.

The maximum permissible accumulated dose for every person involved in radiography is as follows:

5 rem in any one year (50 msv)
3 rem over any calendar quarter (30 msv)
100 millirem per week (1 msv)
20 millirem per day (0.2 msv)

Note: Rem is symbol for roentgens equivalent mean whereas msv is symbol for millisievert which is SI unit.

For personnel not directly involved in radiography the maximum radiation received should not exceed 1.5 rem per year warning signs installed for public should be made to prevent people entering area with radiation expecting 0.75 millirem (0.0075 msv) per hour.

Radiation surveys are an integral part of the safe operation of X-ray machines or radioactive sources which must be conducted by contractor.

These surveys are conducted to determine the extent of radiation hazard in any given area and are vitally important when working in a populated area. The survey meter is a rate instrument which indicates the exposure received per unit time. The most commonly used instruments are the Geiger–Müller counter (G-M) and chamber meters. The effect of distance on the intensity of radiation is calculated by the inverse square law, represented by the following equation.

$$\frac{I_1}{I_2} = \frac{(D_2)^2}{(D_1)^2}$$

where:

I = intensity of radiation
D = distance from the source of radiation.

Film badges should be sent to relevant organization on a monthly basis to measure radiation exposure of each person involved in radiography. Record of the result should be kept in the relevant file.

3.13.2 Magnetic Particle Test Method

If approved by the company, liquid penetrant testing may be substituted for magnetic particle testing.

3.13.3 Ultrasonic Test Method

Calibration for surface breaking and near-surface defects should be carried out in accordance with API Spec 5L, using an N1O reference notch.

3.14 AUTOMATIC WELDING WITHOUT FILLER METAL ADDITIONS

3.14.1 Acceptable Processes

Automatic welding without the addition of filler metal should not be used.

ABBREVIATIONS

AWS	American Welding Society
FCAW	fluxcored arc welding
GMAW	gas metal arc welding
GTAW	gas tungsten arc welding
HAZ	heat-affected zone
HV 10	Vickers Hardness (10 kg load)
IF	incomplete fusion
IFD	incomplete fusion due to cold lap
IIW	International Institute of welding
IP	inadequate penetration without high-low
IPD	inadequate penetration due to high-low
IQI	image quality indicator
NDT	nondestructive testing
SMAW	shielded metal arc welding
WPQT	Welding Procedure Qualification Test (ing)
WPS	Welding Procedure Specification

Chapter 4

Transportation Pipelines Pressure Testing

4.1 INTRODUCTION

This chapter covers the minimum requirements of a hydrostatic pressure test to be carried out on off-plot piping.

Upon completion of the pipeline and before purging and commissioning operations, the line should be high-pressure tested in order to prove the strength of the pipeline, its tightness (absence of leaks) and the integrity of weldments and materials.

The test is also intended to confirm acceptability of pipeline for the service.

4.2 MATERIALS, EQUIPMENT, AND PERSONNEL FOR TESTING

Equipment for the hydrostatic test should be properly selected and in good working order. Equipment affecting the accuracy of the measurements used to validate the specified test pressure should be designed to measure the pressure to be encountered during the hydrostatic test. Equipment and personnel for conducting the hydrostatic test may include the following:

1. Compressor, centrifugal filling pump with required filters to fill the section of the line to be tested with the required filling rate;
2. Portable reciprocating test pump to provide test pressure to the section of the line to be tested. The pump should be suitable to provide maximum pressure indicated in the hydrostatic test pressure diagram;
3. Flow meters and measuring containers as required;
4. Portable water tank, if needed;
5. Two sets of circular chart pressure recorder for 24 hours pressure test recording Portable type, 300-mm (12-in.) chart with stainless steel pressure element, suitable range for the test pressure required, one week mechanical spring loaded chart winding clock with flexible capillary inking system. Minimum accuracy should be 1% with minimum sensitivity of 0.5%. The recorder should be complete with ½-in. process connection and adequate numbers of charts.

Oil and Gas Pipelines and Piping Systems. DOI: http://dx.doi.org/10.1016/B978-0-12-803777-5.00004-6

6. Two sets of circular chart temperature recorder for 24 hours temperature recording, portable type, 300-mm (12-in.) chart recording with temperature element complete with bellows/and capillary, fully compensated, range 0–60°C, one week mechanical spring loaded chart winding clock with flexible capillary attached to the temperature element. The capillary tubing should be 5 m long. The temperature recorder should provide a recording to 0.5°C. The recorder should conform to standards. Inking system should be similar to that described for item **5** in Section 4.2;
7. Direct reading 150-mm (6-in.) pressure gages. The gage should comply with requirements of standards with a span of 1.5 times of the maximum test pressure;
8. Dead weight tester with a valid calibration certificate and a suitable range for the test pressure required, minimum accuracy 0.1% of reading. The tester should conform to standards;
9. Several thermometers suitable for measurement of 0–60°C. The thermometer should conform to standards;
10. Temporary Connections and scraper traps, branch and service lines, loop lines, end caps and manifolds suitable to withstand the expected maximum test pressure;
11. Filter and all spare parts required;
12. Pig and spheres when requested should be equipped with a device enabling the fitting of an acoustic or depleted radioactive source;
13. Water, air, electricity, fuel, and lubricants as required;
14. Corrosion inhibitors, drying chemicals, together with the means for injection and measuring of these chemicals;
15. Means of transport and telecommunication between test and check sites;
16. The necessary equipment and qualified personnel and technicians to be utilized in conducting the pressure tests and dealing rapidly with an emergency repair.

The engineer's approval should be obtained prior to use of materials, equipment, products, and apparatus intended for the execution of the pressure test. Therefore, the executor is required to prepare a list of all items to be used in execution of the testing and submit it to the engineer for his approval at least 1 month before starting the test.

The engineer should have the right to reject any item which, in his opinion, does not conform to the required specification and the executor should replace any item rejected by the engineer.

4.3 TESTING PROCEDURE AND PROGRAM

Before commencing hydrostatic testing, the executor should prepare and submit for the engineer's approval a detailed test procedure together with a test pressure diagram.

The executor should provide a testing technician to supervise all the executor's testing activities, record all test data, and provide liaison with the engineer throughout the testing operation.

The pipeline should be tested hydrostatically in sections prior to tie-in of block valves.

In addition to manufacturer's test certificate, each assembled valve should be hydrostatically tested prior to installation.

4.3.1 Testing Plan and Procedure

Testing procedures should be based on requirements of this chapter. Other than establishing a profile of the line, the following factors should be taken into account when a detailed hydrostatic test procedure is prepared by the executor:

1. Design pressure anticipated throughout the life of the line;
2. The length and location of the sections to be tested. A detailed analysis of the profile to determine static and dynamic pressures while the pipeline is being tested should be performed so that the pipeline will not be over pressured at points that are at low elevations;
3. Location of pipe and other piping components in the test section by size, wall thickness, material grade, or pressure rating;
4. Specified maximum and minimum test pressure as well as maximum and minimum stress to be imposed in the piping;
5. Pressure rating and location of all pipeline valves (if its presence in the test section is practically inevitable) and air vents as well as connections to the test sections;
6. Each sectioning test schedule with drawings showing the distribution of all test equipment such as vent valves, pressure measuring instruments (recorder and gages), and the temperature measuring instruments (recorder and gages) along the line;
7. Source of water to be used for the test;
8. Any requirements for inhibition, purification, or treatment of water to be used for the test;
9. Procedure for cleaning, gaging, and filling the line;
10. Procedure for pressurization of the test sections including location of injection point(s);
11. Minimum duration of time for test sections;
12. Anticipated temperature of test water in over- and under-ground piping;
13. Procedure for water evacuation from the pipeline, method of its disposal, and location for disposal of test water;
14. Safety precautions to be taken and safety practices to be adopted;
15. The complete schedule of proposed equipment and materials and where they will be installed;

16. The list of personnel and their qualifications, responsible for carrying out the test program.

In the course of preparing test procedure, it should be borne in mind that after testing operations, each test section should be dewatered and dried, and the test water may be transferred from one test section to the other.

During preparation of a test procedure, the executor should consider that no test section should be allowed to stand partially full of water or filled with water-saturated air.

The executor should take into account the fact that he is responsible for any damage and loss caused from improper disposal of test water.

In the course of preparing a test procedure, it should be borne in mind that all major river crossings should be pretested at 95% of SMYS before installation. During such pretesting, the test pressure should be held for 4 hours.

Consideration should also be given to the fact that, when preparing the procedure, pipeline sections that have been assembled and tested separately, such as crossings, should be retested along with the remainder of the entire pipeline.

4.4 TEST PREPARATION

All sections to be tested should be isolated by blind flanges, weld caps, or blanking plates with a design pressure exceeding the maximum test pressure.

Testing should be carried out only when the engineer or his authorized representative is present to witness the test.

Provisions should be made for filling, bleeding, and complete drainage of the test water from each test section. Drain points should be at the lowest points and bleed-off points should be at the highest points in each test section, if practical.

Prior to commencement of the test, a thorough check should be made to ensure all fittings, caps, flanges, etc., are in place. All flanges and flanged fittings should be bolted and bolts should be properly torqued.

The executor should obtain sufficient and satisfactory water to hydrostatically test the pipeline. Bore water should not be used, except as approved by the engineer for cases in which surface water is not practically available.

The executor should pump, filter, and measure the fill water required for hydrostatic testing.

The executor should, at his own expense, carry out the water analysis at each water supply point and hand over the analysis results to the engineer. The executor should treat the water, if necessary, at each water supply point with chemicals as directed by the engineer.

The executor should supply all chemicals necessary for water treatment at his cost.

Before water is taken by the executor from any source for testing, the company will obtain the necessary permission or grants from the requisite authorities, public or private. The executor should submit request for the permission 1 month in advance of the test date.

Water should be filtered before entering the pipeline with a filter arrangement in which the filter can be cleaned without disconnecting the piping. The filter should be capable of removing 99% of all particles that are 140 μm or more in diameter.

Measuring equipment for pressure and temperature should be supplied complete with their calibration certificate from a laboratory acceptable to the engineer.

The executor should ensure that all piping components and accessories within the test section are correctly positioned, that all end caps on the test section including those on off-takes are adequately braced to withstand any movement, and that elbows within the test section are adequately padded or otherwise supported to prevent movement.

Before commencement of test on any section, the executor should give the engineer a written notice at least 1 week in advance of the test date. Any changes to the test date should be relayed to the engineer as soon as such changes are known.

Check valves used in liquid petroleum pipeline should be a full-opening, swing type to permit running pigs.

4.5 CLEANING AND GAGING OPERATIONS

All debris such as soil, welding rods, hand tools, etc., introduced into the line, accidentally or intentionally, should be removed by running pigs. Dents should be located and rectified by gaging operation, prior to commencement of hydrostatic test.

The cleaning and gaging operation should be carried out only when:

1. The trench containing the section to be tested has been back-filled, and a major clean-up of the right of way has been completed.
2. The pipeline has been securely fixed into the supports at exposed sections and the concrete blocks properly cured.
3. The headers have been supplied and installed by the executor on either end of the test section.
4. All equipment and materials mentioned in Section 4.2 are ready for the operations.

The executor should insert and run air-propelled scraper pigs to clean the pipeline section of all debris and foreign matters.

After the pipeline section has been cleaned to the satisfaction of the engineer, the executor should insert and run an air-propelled scraper with an attached gaging plate.

Scrapers which become lodged in the pipeline and cannot be forced through by the application of increased air pressure should be located and removed from the test section by cutting out the pipe. The defective portion of pipe causing the scraper to stick should be cut out and a new piece of pipe welded into the section.

The scraper with a gaging plate should then be rerun through the entire length of the test section. All expenses in this respect should be borne by the executor.

If the pig successfully travels through the line in an entire length of the test section, the condition of the gaging plate should be examined. The gaging plate should be in good condition without any sustained damage, it should have no sharp edges and concavities, and its condition should be acceptable to the engineer.

If, however, the gaging plate sustains damage, then it must be assumed that the line contains a fault. The executor should then locate the fault and determine whether a reduced pipe diameter, dent, or obstruction exists in the pipeline.

The executor should furnish all labor, equipment, and materials for making the necessary temporary connections for inserting, propelling, and removing the pig and for repairing all defects in the pipe as determined by the running cleaning and gaging pig.

The executor may elect to clean and gage the line with the same pig.

The executor may elect to run the scrapers by inserting water.

4.6 TESTING OPERATIONS

Filling the pipeline with water should be made in a controlled way at a reasonably slow rate and in a manner that ensures all air is excluded from the system. Air should be vented at the high points in the system.

Inclusion of air leads to inaccurate test results. Filling should be carried out using pigs to give adequate control and efficient removal of air. Water should contain suitable corrosion inhibitor, approved by the engineer, and biocides especially in cases when the project execution requires the pipeline to be left flooded with water for an extended period after the completion of testing.

4.6.1 Filling Line With Water

Filling should be made in a manner to ensure absence of air in the line. Cleaned filtered water should be used for line filling and conducting the test. The filling unit should be also equipped with a flow meter to measure the amount of water pumped into the pipeline.

The executor should install the necessary temporary launcher and receiver scraper traps. During the filling operation, a sufficient back pressure should be held at the discharge test header to prevent the fill pigs from running away from the fill water.

If the next section is to be tested by the water used in the previous section, the executor should also provide interconnection piping for transferring water from one section to the next section.

Cleaned, filtered fresh water with no more than 500 ppm suspended solids should be used for the test. If the temperature of the water is likely to fall to 0°C or below, glycol or any other antifreeze, approved by the engineer, should be used.

If river water is intended for use, the executor should carry out, at his own expense, a water analysis and submit the result to the engineer prior to the commencement of the test. Corrosion inhibitor should be added to the filling water in the suction of water filling pump at the rate recommended by the manufacturer of the inhibitor. Before use, the executor should obtain approval of the engineer for the use of a proposed inhibitor. Where site conditions make the use of seawater inevitable, the executor may use seawater to fill the line for the test provided that prior approval of the engineer is obtained.

Under such circumstances, the use of a special corrosion inhibitor containing additives to prevent marine growth (bactericide) is essential.

Here again the approval of the engineer for the inhibitor and additives to be used is required.

The executor should pump water into the test section ahead of the fill pigs to fill the line for a distance of approximately 100 m; then pump water behind the fill pigs at a rate agreed by the engineer.

The pumps used to fill the line should have minimum capacity in order to fill the line at a rate of about 1600 m/h.

However, the filling rate should be such as to run the pigs at a constant velocity to be decided by the engineer, depending on the profile of the terrain and the diameter of the line.

The test section should be filled continuously until reasonably clean water is obtained at the discharge test header. All temporary riser valves should be closed and blind flanges should be installed on all unused connections. Any entrapped air should be vented.

4.6.2 Installation of Instruments

4.6.2.1 Pressure Recorder

At least one such recorder is to be used on each test section located at the test section control point. The recorder is to be connected to the test section with a ½-in. pipe connection with block and bleed valves. The manifold is to be capable of isolating all instruments from the section of pipe to be tested.

4.6.2.2 Temperature Recorder

At least one such recorder is to be used at each test section located at the test section control point in accordance with Fig. 4.1.

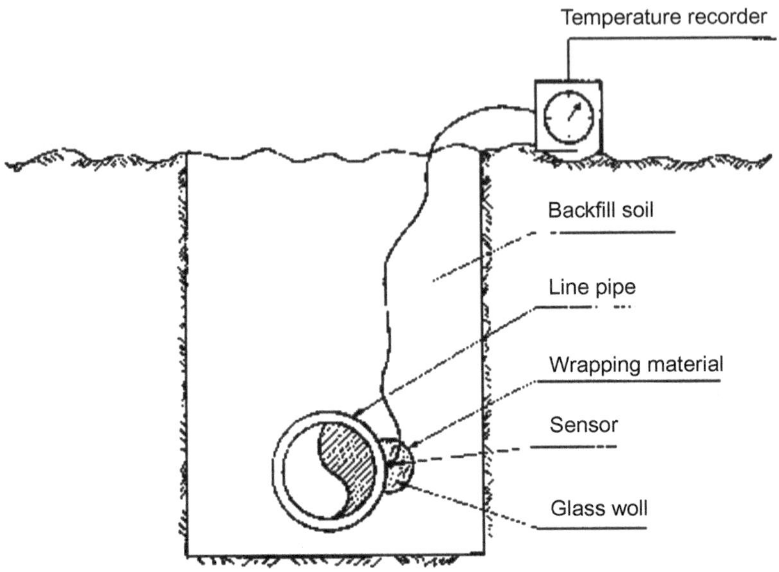

FIGURE 4.1 Connection of temperature recorder (to be installed within 30 m of either end of test section).

The engineer's representative should approve the test rig before commencing tests on the first section.

4.6.3 Calibration of Instruments

The dead weight tester used should have a valid calibration certificate.

Pressure recorder is to be calibrated before the start of the test, midpoint of the test, and at the end of the test before releasing pressure. Calibration to be "in situ" by the dead weight tester.

Pressure gage: To be calibrated as per pressure recorder.

Temperature recorder: To be calibrated before start of test by laboratory check and during test by comparison with mercury "in situ" thermometers.

Measuring tanks and/or meters should be calibrated.

The engineer may request calibration certificates or other additional tests before or during testing.

4.6.4 Pretest Requirements

4.6.4.1 Temperature Stabilization

The temperature of the line-fill water should be stable before testing commences, see also Section 4.6.4.2. For longer test sections, typically in hot climates, the line-fill water temperature may take several days to stabilize.

Prior to commencing the hydrostatic test, the water temperature should be within 1.0°C of ground temperature. This should be determined as the difference between average pipe temperature and average ground temperature over the test section length.

The calculation of the temperature stabilization period based on the expected line-fill water temperature and ambient temperature should be detailed in the test procedure.

Pressure and temperatures, including ambient, should be recorded every hour during the stabilization period.

The test section temperature and the ambient temperature (ground/air) should be plotted against time during the temperature stabilization period.

4.6.4.2 Pressurization

The test engineer should carry out a plot of pressure/added volume (P/V plot, see Fig. 4.2) using measurement of volume added either by pump strokes or by flow meter and instrument reading of pressure gage plus dead weight tester. A tank should be available to enable checks to be made of the volume rating of the pressure pump or flow meter.

The rate of pressurization should be constant and not exceed one bar/min. until a pressure of 35 bar or 50% of the test pressure, whichever is lesser, has been attained. During this period, volume and pressure readings should be recorded at 1 bar intervals.

The person operating the pressurizing equipment should immediately report to the test engineer any variation in the rate of pressure increase by the same volume of added water.

During pressurization, all potential leakage points should be checked.

When the pressure of 35 bar or 50% of the test pressure, whichever is lesser, has been reached, the air content should be determined as specified in Section 4.6.4.3.

When the air content is within the maximum allowable limit of 0.2% of the test section volume, pressurization should continue, at the rate given above, as follows:

- If flanges are installed, pressurization should continue up to 70 bar or to the MAOP, whichever is the lesser and be held while checking flanges for leaks. This is the highest pressure permitted to tighten up any leaking flanges.
- If flanges are not installed or the flanges have been checked, pressurization should continue up to 80% of the test pressure and be held at this level for not less than 2 hours.

During this water stabilization period, accessible flanges, if installed, should again be checked for small leaks. If any are found, the test section should be depressurized at not more than 2 bar/min to 70 bar or to MAOP, whichever is lesser, prior to any bolt tightening.

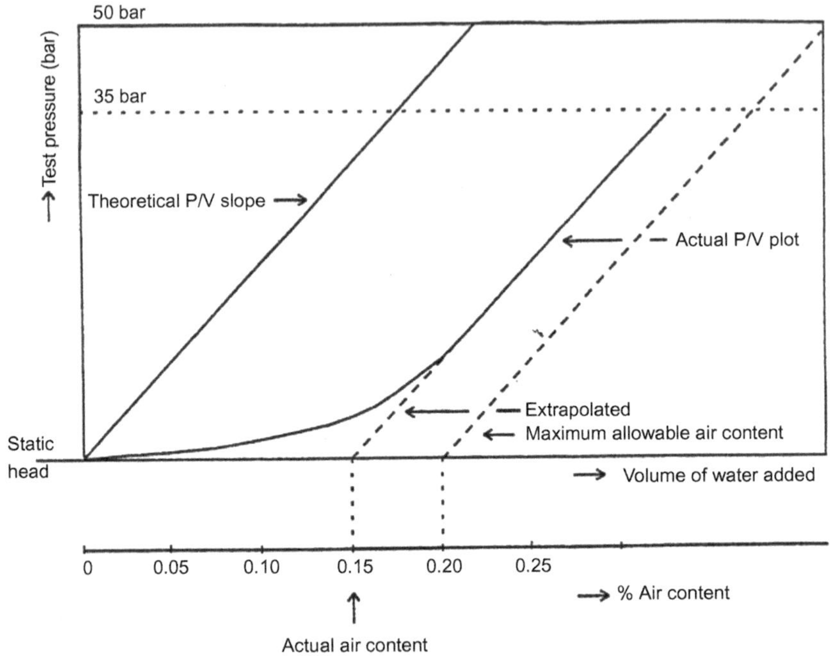

FIGURE 4.2 Determination of residual air volume.

The volume and pressure readings should be recorded at 10-minute intervals until the test pressure has been reached.

After stabilization, the pressure should be raised to 95% of the test pressure and held for 30 minutes, then the pressurizations should be continued to the specified test pressure at a rate not exceeding 0.5 bar/min.

The pressures and added volumes should be continuously plotted until the specified test pressure has been achieved. The plot should be constantly checked and, in the event of any deviation of 10% or more from the theoretical line corresponding to 100% water content, the test should be terminated. The pressure should be released and an investigation should be carried out to determine the cause of the deviation prior to depressurization.

4.6.4.3 Air Content Determination

The air content of the filled line should be determined during initial pressurization by using the pressure/added volume plot as specified in Fig. 4.2. The linear section of the curve should be extrapolated to the volume axis, which should correspond to static head pressure. The volume of air can be read from the intersection of the line with the volume axis and should be used to calculate the air content thus:

$$\text{Percentage air content} + (\text{Volume of air}/\text{Volume of line})100$$

If the air content exceeds 0.2% of the line volume, testing should be terminated and an investigation should be carried out to determine the cause. The test section should be emptied and refilled at the discretion of the company.

4.6.4.4 Hydrostatic Pressure Test

4.6.4.4.1 Duration of Hydrostatic Test

The minimum duration for the hydrostatic pressure test should be a 4-hour strength test followed by a 24-hour leak tightness test.

The assembly testing should comprise a 4-hour strength test followed by visual examination at the leak tightness test pressure.

4.6.4.4.2 Strength Test

The test pressure at any point of the test section should at least be equal to the test pressure required in the ANSI/ASME B31.4 or B31.8 as applicable, or to the pressure creating a hoop stress of 90% SMYS of the line pipe material, based on the minimum wall thickness, whichever is higher, or unless otherwise specified by the company.

During the hydrostatic pressure test the combined stress should not exceed 100% SMYS of line pipe material based on minimum wall thickness. The combined stress should be calculated in accordance with ANSI/ASME B31.4 or B31.8.

The margin between the hoop stress of 90% SMYS and the combined stress of 100% SMYS allows for elevation differences in the test section and/or longitudinal stresses, e.g., due to bending. However, the elevation differences in each test section should be limited to a value corresponding to 5% of SMYS of the line pipe material or to 50 m, or as specified in the scope of work.

The test engineer should confirm that the test pressure does not exceed the pressure to which the pipe has been subjected during the mill test and that it is not higher than the design pressure of the fittings specified for the pipeline.

The combined stress for the hydrostatic pressure test condition should be calculated in accordance with ANSI/ASME B31.4 and B31.8.

The calculation should include major residual stresses from construction and longitudinal stresses due to axial and bending loads, e.g., at unsupported pipeline spans. The combined stress during the hydrostatic pressure test should be limited to 100% of SMYS based on the minimum wall thickness. If the calculated combined stress is higher than 100% of SMYS, special measures should be taken to reduce the longitudinal stresses in the test section.

The pressure should be maintained during the strength test at TP ± 1 bar by bleeding or adding water as required. The volumes of water added or removed should be measured and recorded.

During the test, TP should be recorded continuously, and the dead weight tester readings and air temperatures should be recorded at least every 30 minutes. The pipe and soil temperature should be recorded at the beginning and end of the 4-hour test period.

The test section temperature and the ambient temperature against time plot created for the stabilization period should be maintained.

4.6.4.4.3 Leak Tightness Test

The leak tightness test should commence immediately after the strength test has been completed satisfactorily. No water should be added or removed during the tightness test. The test is intended to demonstrate that there is no leak in the pipeline.

If it can be ensured that the pressure variations due to temperature fluctuations are in the specified limits (Section 4.6.4.4.2), then a combined strength/leak tightness test at the strength test pressure, without water addition or removal, should be carried out. To allow for pressure variations caused by temperature fluctuations during the test duration, the test pressure should be set to a level of 80% of TP.

During the test, the pressure should be recorded continuously, and the dead weight tester and air temperature readings should be recorded every 30 minutes. The pipe and soil temperature should be recorded at maximum 3-hour intervals. The temperature recording interval should be reduced to a 1-hour duration for the first and last 3-hour periods of the 24 hours to assist with the pressure/temperature variation calculation. The test section temperature and the ambient temperature against time plot created for the stabilization period should be maintained.

If the pipeline consists of several sections that have already been strength-tested, the pressure for the final leak tightness test should be based on 80% of the lowest TP of the tested sections. The difference in elevation over the test section should be taken into account.

4.6.4.4.3.1 Acceptance Criteria To determine whether any pressure variation is a result of temperature changes or whether a leak is present, the pressure/temperature changes should be calculated from the pressure/temperature equation Formula 2 (2a for restrained test section or 2b for unrestrained test section) or unless otherwise approved by the company.

Formula 1. Volume/pressure equation

$$\frac{\Delta V}{\Delta P} = V\left[\frac{D}{Et}(1 - \upsilon^2) + \frac{1}{B}\right]$$

Formula 2. Pressure/temperature equation

2a) For restrained test sections

$$\frac{\Delta P}{\Delta T} = \frac{\gamma - 2(1 + \upsilon)\alpha}{\frac{D}{Et}(1 - \upsilon^2) + \frac{1}{B}}$$

2b) For unrestrained test sections

$$\frac{\Delta P}{\Delta T} = \frac{\gamma - 3(1 + \upsilon)\alpha}{\frac{D}{Et}(1 - \upsilon^2) + \frac{1}{B}}$$

where:

V = incremental volume (m^3)
P = incremental pressure (bar)
T = incremental temperature (°C)
V = pipeline fill volume (m^3)
D = pipeline outside diameter (m)
E = Young's elastic modulus of steel (for carbon steel, $E = 2.07 \times 10^6$ bar) (bar)
t = nominal pipe wall thickness (m)
U = Poisson ratio (for steel, $\upsilon = 0.3$)
B = bulk modulus of water (Fig. 4.3 for fresh water or Fig. 4.4 for seawater) (bar)
γ = volumetric expansion coefficient of water (Fig. 4.5 for fresh water or Fig. 4.6 for seawater) ($°C^{-1}$)
α = coefficient of linear expansion of steel (for carbon steel, $\alpha = 1.17 \times 10^{-5}\ °C^{-1}$) ($°C^{-1}$)

Notes:

1. B is very sensitive to temperature and also sensitive to antifreeze.
2. For fresh water g changes significantly at low temperatures because the density of water is greatest at 4°C.
3. Above-ground test sections, if not anchored, are normally unrestrained.
4. Figs. 4.3 and 4.5 may be used for demineralized water.

The pressure, the test section temperature, and the ambient temperature should be plotted against time during the period of the leak tightness test.

The tightness test should be deemed acceptable if any pressure change can be accounted for by a corresponding temperature change using Formula 2. Allowance should be made for any recording inaccuracy.

In the event of any doubt by the contractor or the company about the leak tightness of the line, e.g., when temperature and pressure trends are differentiated, should be extended until such time as the acceptability is demonstrated.

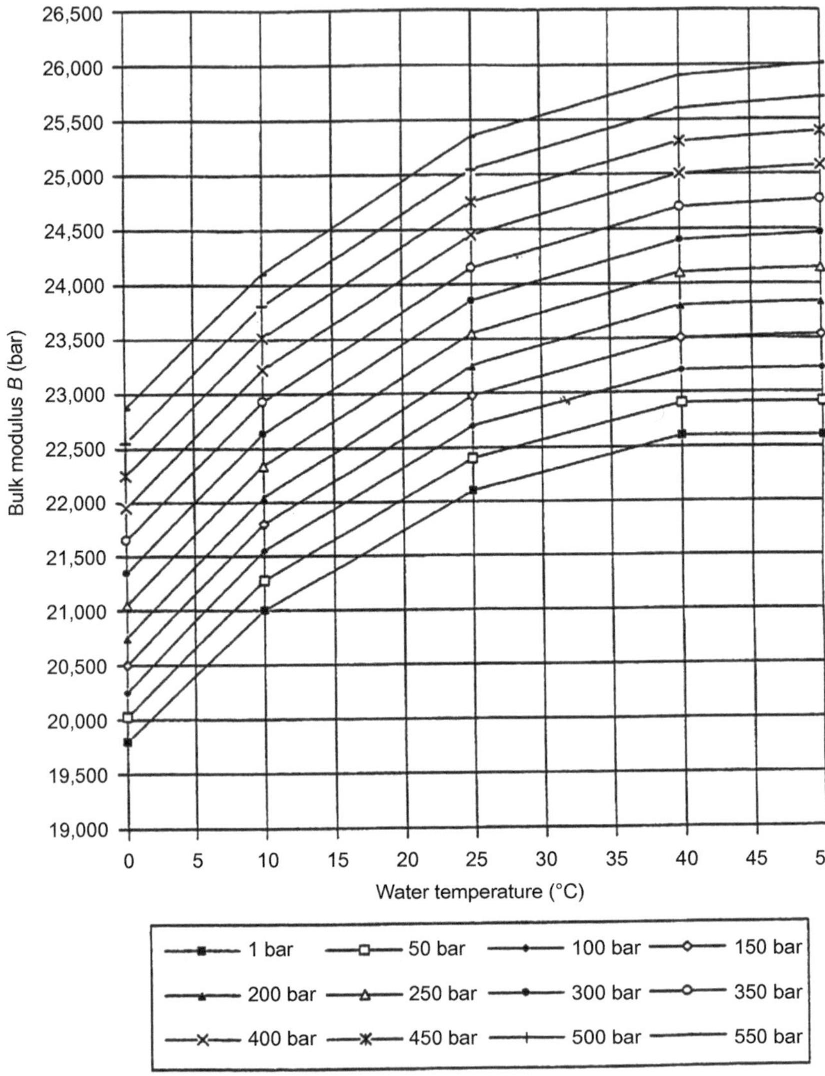

FIGURE 4.3 Bulk modulus of fresh water.

4.7 DEFECTS, REPAIR OR RENEWALS, AND RETESTING

Should a failure occur under this test or the difference in pressure is appreciable, the executor should make a thorough survey along the line, locate, uncover, and repair all pipe failures and backfill all pipe repairs.

A test after repair and temperature/pressure stabilization has been completed should be performed.

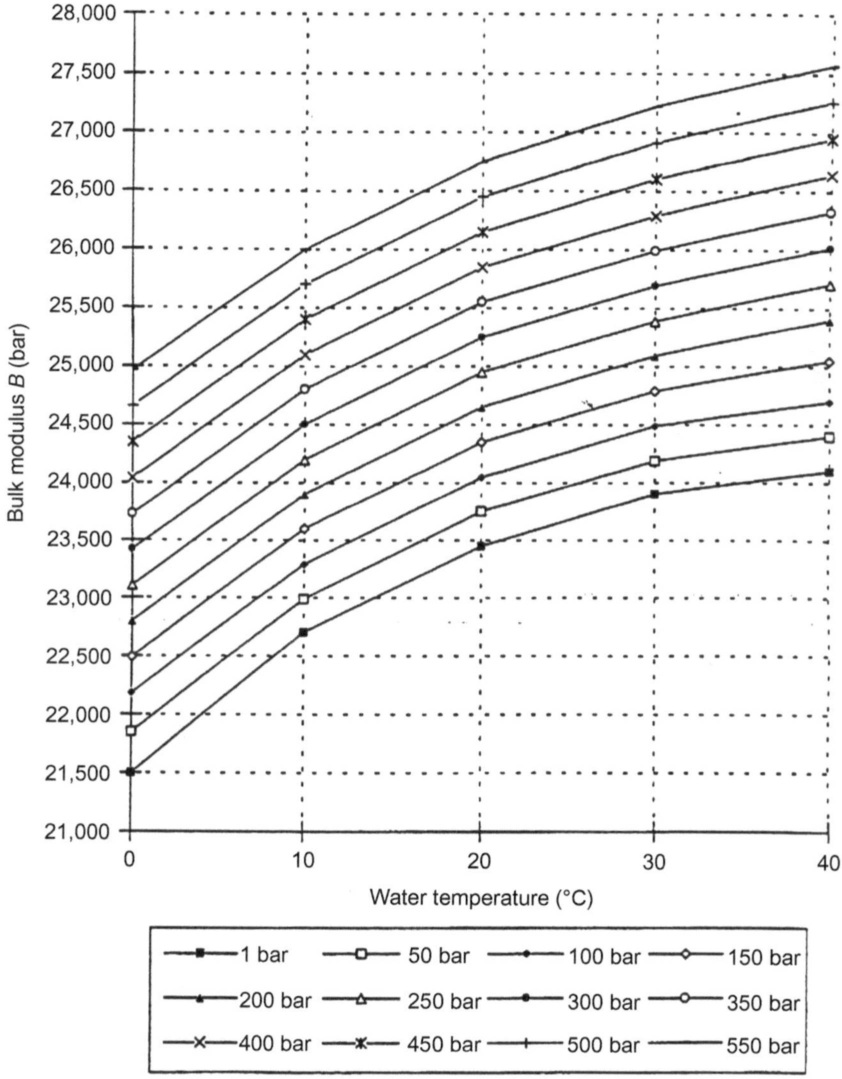

FIGURE 4.4 Bulk modulus of seawater.

If a leak occurs in the test section but does not immediately reduce the test pressure to the corresponding 90% of SMYS, the executor should immediately reduce the test pressure to that level by bleeding water from the test section. A pressure range corresponding to 70–90% SMYS should be maintained while the executor searches for the leak.

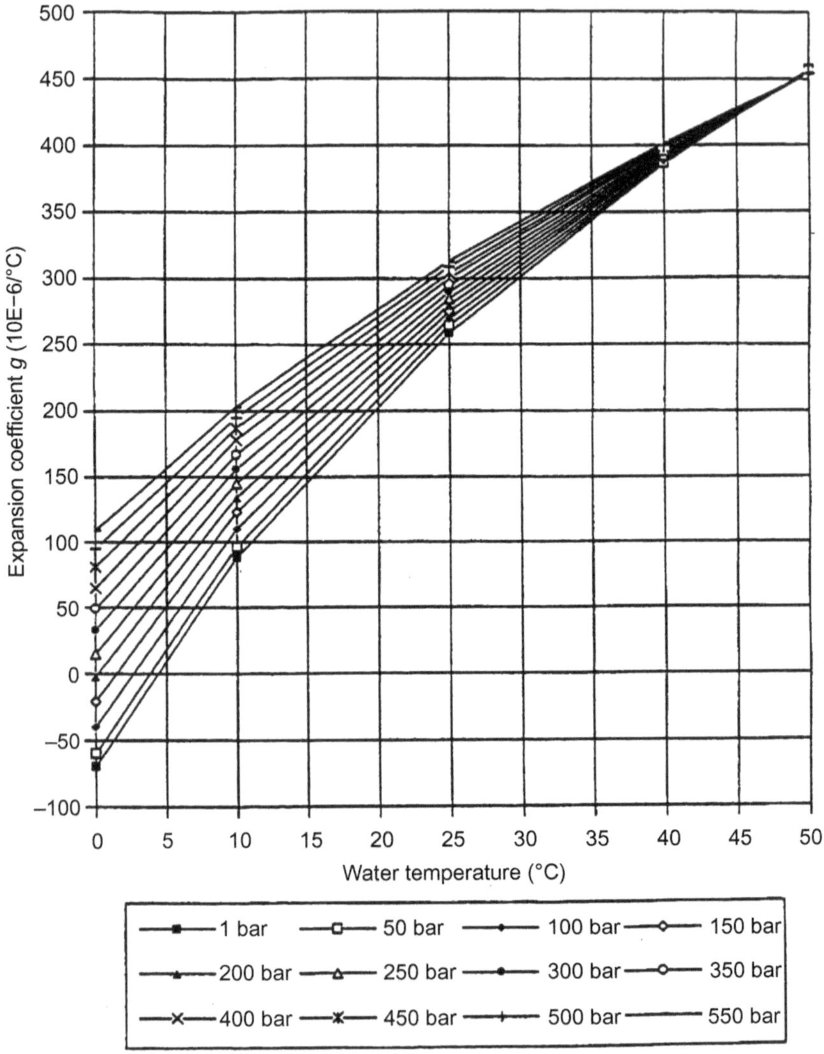

FIGURE 4.5 Volumetric expansion coefficient of fresh water.

The executor should employ whatever methods, labor, and equipment as required to locate pipe failures. In the event of small leaks these methods may include:

1. Patrolling the test section on foot;
2. Progressively sectioning and repressurizing the line;
3. Use of sonic equipment;
4. Use of detectable additives in the test water.

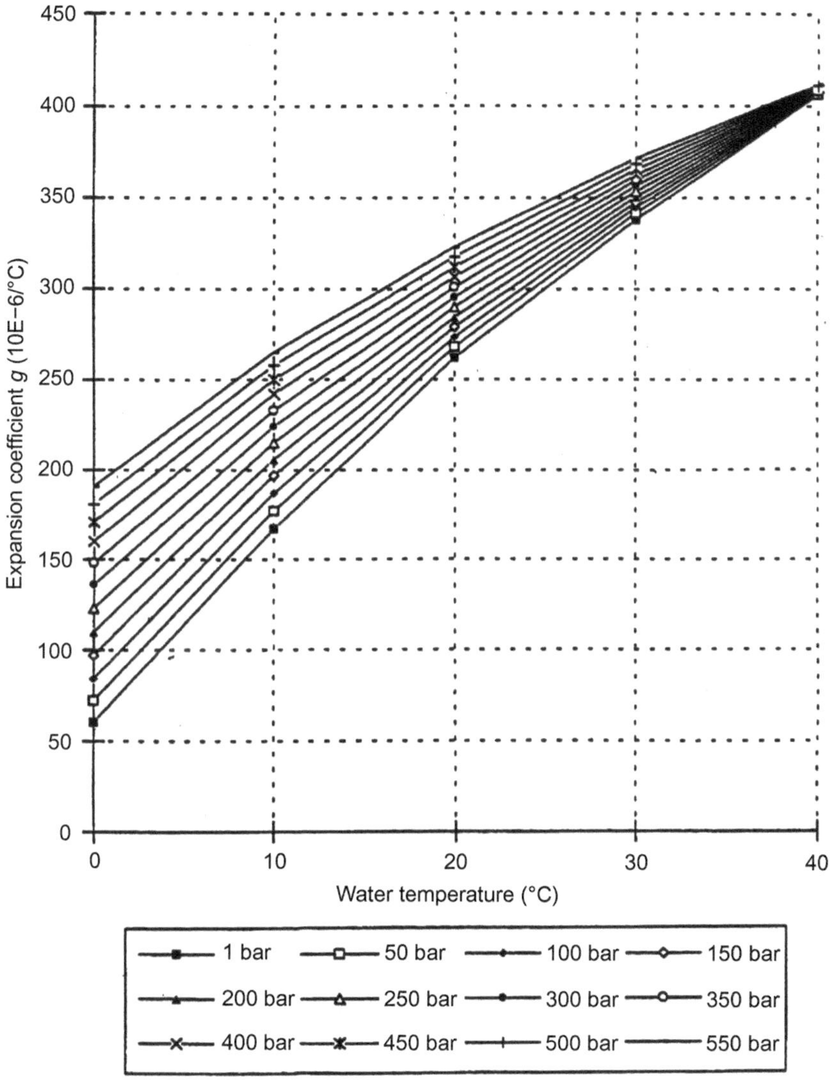

FIGURE 4.6 Volumetric expansion coefficient of seawater.

The executor should remove the defective pipe as directed by the engineer. The pipe removed should be marked for orientation with respect to its position in the trench and with the approximate kilometer post and survey station of the failure. The executor should not cut nor damage the edge of the pipe failure. The failure should be photographed. The executor should transport damaged pipe to the company's warehouse as directed by the engineer.

If any pipe defect causing failure is determined to be of mill origin, the executor should be reimbursed for the time the executor's personnel, equipment, and material used in locating, uncovering, and repairing the pipe failure; coating and backfilling the pipe repair; replenishing the fill water; venting and/or repigging the test section, if necessary; and raising the pressure to the pressure level at which the failure occurred.

If a defect is attributed to the executor's activities, he should bear all expenses involved and mentioned here above.

After repair or renewal, the section should be tested again using the same procedure until satisfactory results are obtained. A report documenting the failure and the suspected reasons for the failure should be completed. A typical form is shown in Form No. 1.

4.8 FINAL TESTING

This test is not mandatory and should be carried out at the engineer's discretion and option.

After the completion of tests on all sections of the mainline and following the completion of tie-ins between all tested sections, the engineer may decide to conduct a leak test on the entire installation. Upon receiving such instruction, the executor should carry out a leak test under requirements specified for conducting a leak test of individual test sections.

4.9 TESTING OF UNCHECKED JOINTS, FABRICATED ASSEMBLIES, AND VALVES

All scraper trap assemblies and block valve assemblies should be tested separately at the test pressure of line with a duration of not less than 4 hours, provided that they have not been tested with the line.

Valves should be in fully open position during the test. Testing against a closed valve is not permissible. When the valve manufacturer's rating is lower than the test pressure, a spool piece should replace the valve during the test.

4.10 DEWATERING AND DRYING

Upon the completion of testing, the test pressure should be bled off to achieve 0 bar gage (1 bar abs) in the test section. Dewatering may be accomplished using air compressors and swabbing pigs.

Air compressors will be required to remove the water once the testing is complete. The compressors will need to have sufficient capacity to remove the water at rate agreed by the engineer.

Swabbing pigs of highly flexible material, e.g., polyurethane foam, should be used for effective removal of test water.

The hydro test or fill water should be removed sufficiently to allow commissioning and subsequent effective operation of the pipeline. The degree to which this water removal must be taken depends upon the eventual service of the pipeline.

With the gas transmission pipelines the water removal and effective drying is more severe when it is compared to oil pipelines, due to freezing of water and formation of gas hydrates.

Therefore, a type of drying operation must be carefully selected by the engineer when taking into consideration the water dew point of the gas specification and the amount of dryness required.

A combination of two methods may be used to achieve a dry gas pipeline (drying with air and methanol).

Prior to dewatering operation, the executor should ensure that drain and vent connections are installed on the line at locations indicated on the drawings or designated by the engineer. He should make sure that some temporary connections or blinds, if required to facilitate operation, are removed and/or certain vent and drains are opened and no water is trapped at low points, bends, or valve bodies.

Proper disposal of test water should include removal or neutralization of inhibitor, hydrazine, ammonia, and/or magnetite from the test water before it is released to the environment so that contamination of river, agricultural, or livestock is obviated.

The location and rate of disposal is subject to the written approval of the engineer. However, the executor should be responsible for all the damages done to public property by wrongdoing with respect to dewatering.

4.10.1 Dewatering and Drying of Gas Line

Unless otherwise specified, compressed air should be used for dewatering and drying.

4.10.1.1 Dewatering of Gas Line

1. If dewatering is accomplished from scraper trap assembly to scraper trap station, it should take place only after hydrostatic testing and tie-ins have been accepted for all test sections, between the scraper traps.
2. In case dewatering should be done in a continuous operation, water should be displaced by a pig train consisting of two four-cup pigs. Pig cups will be furnished by the executor and the executor should install pig cups and make sure that the pigs remain in satisfactory condition.

TABLE 4.1 Volume of Slug of Methanol Required for Dewatering 1 km of Gas Line of Various Sizes (If Use of Gas Is Specified for Dewatering)

Pipe Dia.		Methanol Slug(L)
NPS	DN	
6	150	5.7
8	200	9.8
10	250	15.3
12	300	22
14	350	26.8
16	400	35.5
18	450	45.5
20	500	56.7
24	600	82.6
26	650	97.5
30	750	130.8
32	800	149.2
36	No equivalent	187
40	1000	231
42	1050	254.7
48	1200	321.8
56	1400	341

The volume of the methanol slug should be located between two pigs. The volume of the methanol slug to be furnished by the executor should be in accordance with Table 4.1.

3. If the executor elects to move test water from test section to test section, dewatering should take place immediately after the hydrostatic test has been accepted for the upstream test section. In such a case, the executor should furnish all equipment and material as well as compressed air.
4. The executor should properly dispose the test water at the receiving end as directed by the engineer.
5. The executor should furnish and install sufficient pipe to transport the water away from the pipeline as directed by the engineer.

TABLE 4.2 Volume of Slug of Methanol Required for Drying 1 km of Gas Line of Various Sizes

Pipe Dia.		Methanol Slug (L)
NPS	DN	
6	150	3.8
8	200	6.5
10	250	10.2
12	300	14.7
14	350	17.9
16	400	23.7
18	450	30.3
20	500	37.8
24	600	55.1
26	650	65
30	750	87.2
32	800	99.5
36	No equivalent	124.8
40	1000	154.2
42	1050	170
48	1200	214.6
56	1400	226.9

Methanol requirement is calculated on the basis of 0.2 L for each cubic meter of pipeline capacity.

4.10.1.2 Drying of Gas Line

1. For drying, the executor should launch a pig train consisting of two four-cups pigs and a slug of methanol. The volume of the methanol slug should be located between the two pigs. The volume of methanol slug to be furnished by the executor should be in accordance with Table 4.2.
2. After receiving the first drying pig train and blowing down the pipeline to atmospheric pressure, the executor should launch a second drying pig train consisting of two pigs and a slug of methanol. This practice should be repeated to achieve water dew point of −5°C.

The executor should continue to blow air or gas through the test section until tests performed by the engineer on the expelled air (or gas) indicate that the test section is satisfactorily dried. If the use of natural gas is specified, it will be supplied by the company.

4.10.2 Dewatering and Drying of Petroleum Liquid Line

Following the successful completion and acceptance of the hydrostatic test, a batching pig should be launched with air for dewatering purposes.

After a complete drainage of the water has been carried out, the executor should blow hot compressed air into the section that is subject to dewatering to get the line free of water and moisture.

4.11 RECORDING TEST DATA AND REPORTING TEST RESULTS

The executor should prepare test reports and pipeline failure reports for the complete line and submit to the engineer in three copies. The reports should be made in the form which will be indicated by the engineer in the detailed testing procedure manual.

All test data should be accurately recorded by the executor. Test documentation submitted by the executor for each test section should include the following:

1. A comprehensive test summary, on a form approved by the engineer that also indicates test procedure adopted.
2. Log of dead weight pressure readings, ambient temperature, and elapsed time.
3. Log of test water temperature observed at the same time observing pressure figures (item **2** in Section 4.11).
4. Recording charts for pressure and temperature for duration of the test.
5. Log of volumes of water added or removed from the line during the test (quantities and time).
6. Pressure–volume plot and calculations of temperature corrections.
7. Failures developed during and following the test together with reason for the failures.
8. Profile of the entire test section of pipeline showing the elevation, test sites, and maximum and minimum test pressure.
9. Determination and plot of air content.

All pressure and temperature charts should clearly indicate:

1. The date and hour the chart was placed on and taken from the recorder;
2. The location of the recorder;

3. The test section number or the code signifying the appurtenance;
4. The signatures of the executor's representative and the engineer's testing supervisor. The test chart should be furnished to the engineer within 24 hours of the completed test.

An accurate and complete documentation of test data becomes a permanent record that must be retained for as long as the facility tested remains in operation. In addition to the aforesaid information, a typical test report similar to that shown in Form No. 2 should be submitted by the executor.

4.12 SAFETY CONSIDERATION DURING TESTS

All testing of pipelines after construction should be done with due regard to the safety of personnel, the public, and property. Testing procedures prepared by the executor and approved by the engineer should contain full details of the arrangements to be made and precautions to be taken. Warning signs should be posted at all points of access to the right-of-ways and the areas where lines are exposed. Areas should be roped off where necessary. Only personnel required for the testing or directly involved with the testing should be allowed to enter the areas during the test duration.

Form No. 1: Pipeline Failure Report Form

PIPELINE FAILURE REPORT FORM

1	Company : ....
2	Section tested : ....
3	Time of failure : .... a.m./p.m., Date ....
4	Location of failure: ....
5	Pressure in Bar at point of failure: ....
6	Description of failure: Leak.................... Break............. Length of failure......................
	If leak, fill in blanks:...................................... liter lost per hour
8	Describe any peculiarities of defects on failed part, such as mill defects, corrosion or evidence of prior damage, etc.:
9	Possible cause of failure : ....
10	Pipe size, DN:.................. WT..... , grade............. Mfg by ....
11	Repairs: pipe installed, DN:...... WT......., grade....... Mfg by............ length of joint or pup
12	Date repaired :.................. by ....
13	Damages to property, persons injured etc.: ....
14	Other remarks:

Form No. 2: Typical Test Report Form
Field Pressure and Test Report

TYPICAL TEST REPORT FORM
FIELD PRESSURE AND TEST REPORT

Page.........of..............

Executor: ....

Pipeline Description: ....

Section Tested: from:.. to: ....

Test Section No.:.. Length ....

Type and size of pipe DN:...................... w.t.................. , grade.......................................

Manufacturer: ....

Pressure measuring/recording unit location:...................................... water source..................

Time and date test started: $\frac{a.m.}{p.m.}$ .... Test pressure (maximum

.... Bar g. Test pressure (minimum).. Bar g

Time and date test ended: $\frac{a.m.}{p.m.}$ .... Pressure volume plot Yes.......... No...........

Section Accepted :.................. section leaking............................section ruptured..........................

Temperature	Beginning	Maximum	Minimum	Completion
Ambient				
Skin				

The theoretical slope should be calculated from Formula 1 of Section 4.6.4.4.3.1 and plotted in the actual P/V plot by the test engineer before pressurizing commences. The bulk modulus for the line-fill water should be taken at the average test section temperature and at a pressure of 35 bar from Figs. 4.3 or 4.4.

Note:

1. To ensure accuracy, the P/V plot should be made only up to 50 bar.
2. If the P/V plot has not become linear at 50% of the MAOP, it should be continued up to a maximum value of 35 bar or the test pressure, whichever is the lesser.

4.13 HYDROSTATIC TESTING OF ABOVE-GROUND PIPELINES

Except when specified otherwise in the project specification procedure, the test pressure for hydrostatic testing of buried lines should apply to above-ground pipeline. The procedure includes cleaning, gaging, filling, strength test, leak test, and dewatering as indicated in previous sections of this chapter. Being an above-ground, acoustic or depleted radioactive source is not required to locate a pig.

Due attention should be paid to the effect of the temperature change on pressure during the test period. Use of a temperature recorder is not necessary for a pressure test of above-ground pipelines.

Pipes and connections (flanged or screwed) must not be painted before completion of the test.

For hydrostatic testing, all pipe supports should be in position and completed before testing is undertaken.

Large adjacent lines should not be tested simultaneously where the weight of the combined test water load may exceed the load taking of supports.

Care must be exercised to avoid overloading any parts of supporting during hydrostatic testing.

In certain cases (e.g., piping intended for gas service) temporary or additional supporting may be required to adequately support the pipeline against the weight of the testing medium. After lines have been drained, the temporary piping supports should be removed.

Chapter 5

Inside Pipe Chemical Cleaning

5.1 INTRODUCTION

This chapter covers the minimum requirements for Internal Chemical Cleaning of Piping Systems on the upstream side of process machineries and lube oil/seal oil systems for which removal of rust, mill scale, grease, and foreign matter is essential.

The executor should furnish all equipment and materials such as pumps, filters, chemicals, piping, hoses, etc., and provide technical personnel and labor to supervise and perform the job.

The executor should be responsible for the disconnection of any piping from equipment, if required, prior to cleaning.

The executor should be responsible for the removal of items such as valves with trim, which should be attacked by the cleaning agents.

All work should be coordinated with and executed under the direction of the engineer. The executor should consult the engineer prior to mobilization and seek his advice on adequacy of preparation activities before performing the work.

The executor should be responsible for the removal and proper and lawful disposal of all chemicals used for cleaning from the working area and its surrounding without any cost incurring to the company and to the satisfaction of the engineer. The executor should not dispose acids, spent acids, or other chemical cleaning agents into existing drainage systems.

5.2 CLEANING PROCEDURE

5.2.1 Normal Cleaning

Prior to the chemical cleaning operation, normal cleaning should be performed by flushing and/or mechanical means, as applicable.

5.2.2 Chemical Cleaning

All piping indicated on the drawing or otherwise specified should be chemically cleaned in accordance with the standard specification.

Oil and Gas Pipelines and Piping Systems. DOI: http://dx.doi.org/10.1016/B978-0-12-803777-5.00005-8

5.2.2.1 Preparation for Chemical Cleaning

1. Generally, installed piping should be cleaned in place after hydrostatic testing has been completed; the systems should be checked for leaks, prior to cleaning by chemicals.

   Cleaning may be performed in a bath and/or by soaking or, alternatively, portions of piping systems may be connected together in a predetermined location and cleaned using the circulation method unless field conditions dictate otherwise. However, after cleaning has been performed, no further welding work should be carried out.
2. Piping to be cleaned should be disconnected or blanked off from mechanical equipment. Chemicals should not be circulated through bearings, cylinders, oil filter cartridges, instruments, screens, and filters. Valves and equipment may be included in the circulation system provided that their materials are resistant to chemical solutions involved and prior approval of the engineer is obtained. Otherwise, the executor is responsible for any possible damage that may occur due to chemical reaction between the parts and the chemical; damaged parts should be replaced at his own expense.
3. The executor, in preparing his circulating system, should ensure that circulation is complete and that there are no dead ends or branches that will create traps. The executor should make sure that there are adequate existing high-point vents and low-point drains in the circulation system. If additional high- and low-point vents drains are required, they should be installed by the executor after obtaining the engineer's approval.
4. The portable pumping set provided by the executor should be capable of safe handling of the chemicals to be used. The pump design should be such to have the desired capacity and should be capable of circulating the chemical at a velocity required for cleaning the piping system.
5. Prior to pickling, all heavy deposits of oils, greases, soils, or foreign matter other than mill-scale, rust, or rust scale should be removed by precleaning as cited in Section 5.2.1.
6. Exposed flange faces and pipe threads should be protected from chemical solutions.
7. All solutions, during the recirculation sequence should be filtered. Full-flow filter or equipment should be equipped with screen capable of removing particles that are 4 μm or more in diameter.
8. The executor should prepare chemical cleaning procedure.

   This procedure should be approved by the engineer prior to implementation. The procedure must include, but not be limited to, the following items.
   a. The result of the tests on composition of precipitates.
   b. Precleaning.
   c. Exact compositions (percentages) of cleaning solutions and inhibitors.
   d. Inhibitors to be used to prevent corrosion and hydrogen brittleness.
   e. Temperatures of the solutions during cleaning.

f. Circulation times.
g. Indication when the solutions will be renewed: concerning iron content, acid concentration pollutions, etc.
h. Corrosion to be expected (degree of attack and uniformity).
i. Flow velocities.
j. Neutralization.
k. Passivation.
l. Flushing and drying.
m. Preservation after cleaning.

5.2.2.2 Consideration to Be Given in Chemical Cleaning

Only clean fresh water (e.g., tap water) should be used in preparing the acid concentration where pickling is performed in carbon steel or low-alloy steel piping. Distilled or demineralized water should be used in preparing the acid concentration where stainless steel piping is subject to chemical cleaning.

The pickling time should be kept to a minimum.

The acid solution must not be allowed to stand undisturbed in any part of the system at any time.

Acid, inhibitors, and neutralizers should not harm the various materials of the system to be cleaned and should be in accordance with accepted practice.

Before changing from one to another fluid, the piping must be allowed to drain sufficiently so that as little acid or water as possible is mixed with the subsequent fluid.

Process pumps should not be used to circulate the acid solution.

All heating has to be external to the systems being cleaned. Live steam should never be introduced into any part of the system.

Passivation should be performed immediately after neutralization process.

In addition to all instruments (e.g., flow meters, thermowells, pressure gages, etc.), equipment, valves, and other materials that may be attacked by the acid, corrosion coupons other than those that are required for pickling should be removed from the system intended for cleaning.

The number and installation point of corrosion coupons required for the pickling should be approved by the engineer.

Suitable acid-resistance gate valves should be installed at the vent and drain points by the executor.

Installation point of instruments should be tightly plugged after removal of the instruments.

Suitable spools should be installed in place of removed valves and equipment.

During pickling with the circulating system, the velocity of the flushing, rinsing, acidizing, neutralizing, and passivating should be kept within the range of 90–150 cm/s (3–5 ft/s).

The power and capacity of the pump(s), size of connections, and their setup should be so that the above-mentioned velocity range can be maintained in all parts of the system to be cleaned; there should be no existing dead ends.

The metal composition of piping to be cleaned should be provided to the executor prior to pickling for selection of suitable acid and pickling procedure.

The volume of the systems to be pickled should be calculated by the executor prior to commencement of the pickling.

The volume of the tank that should be used for the circulation process should be at least 10% more than that of the system that has the largest volume and is subject to pickling.

For taking samples during the pickling process, suitable points near the inlet and outlet of cleaning system should be selected. These points should be approved by the engineer, and sampling facilities should be installed at these points.

The executor must include in his procedure the methods by which he intends to check acid solution in order to maintain the recommended concentrations (i.e., hydrometer, titration, etc.).

5.2.2.3 Tests to Be Performed Prior to Pickling

Prior to pickling, a sample should be taken from a deposit of each system to be pickled and analyzed by the executor. The percentage of the following items in the deposit should be determined using test methods (ASTM D-887).

- Iron content (including Fe^{2+} and total Iron)
- Silica
- Carbonates
- Phosphates
- Hydrocarbons

Based on result of the tests, a decision should be made for the type and concentration of acid to be used and pickling procedure.

5.2.2.4 Cleaning of Carbon Steel Piping by Pickling

The procedures outlined below indicate the requirements needed for various processes of pickling. The cleaning procedure proposed by the executor should include, but not limited to, these requirements.

The circulation process should be adopted for chemical cleaning of carbon steel piping unless there are special conditions where this process appears to be impractical; in such instances, approval of the engineer should be obtained for using another process of pickling.

If equivalent chemicals are intended to be used, it should be clearly stated in the executor's proposed procedure.

5.2.2.4.1 Pickling by Circulation Process

The following steps should be taken in circulation process:

1. **Flushing**

   Prior to pickling, the system should be flushed with clean fresh water (potable water is accepted). The system should be checked for leaks. If any leak is observed, the executor should take remedial action to stop the leak, and flushing should be continued until the visual inspection of outlet water indicates it has same appearance of inlet water.

   During flushing, the water velocity should be kept within the range of 90–150 cm/s (3–5 ft/s); it is preferred to stick to about 150 cm/s.

   During flushing, the system should be packed with the water and no air should be present in high points.

2. **Degreasing**

   After completion of flushing, degreasing should commence to eliminate any presence of oil and/or paint that's stuck to the inside surface of pipe. A solution of soda ash, with a concentration of 1–3% together with 0.05% penetrating agent and 0.5–1% three sodium phosphate as an additive, should be used for degreasing. During the degreasing operation, the solution temperature should be maintained between 70°C and 80°C. The following solutions may also be used as an alternative provided that approval of the engineer is obtained.

   a. Ten percent by weight solution of technical grade sodium hydroxide with 0.05% penetrating agent at a temperature range of 65–80°C.
   b. Alkaline degreasers. The degreasing solution should be circulated in the system for 2 h and then should remain stagnant for 12 h; thereafter, the solution should be circulated for 1–2 h and then finally drained.

      During degreasing, tests should be made for the following at 1-h intervals:

      i. Soda content of the outlet fluid.
      ii. pH value of the outlet fluid.
      iii. Oil content of the fluid.

   Presence of sufficient active soda in the system outlet is indicative of a satisfactory completion of the degreasing operation.

3. **Rinsing**

   After completion of degreasing, flushing with clean fresh water in ambient temperature with the same procedure as described in item **1** of Section 5.2.2.4.1 should be conducted. Rinsing should be continued until a pH value of 7–8 is achieved as indicated on hydrion paper or equivalent.

4. **Acid wash**

   After rinsing, an acid wash should be started, especially where system valves are included in the operation; it is necessary to use an inhibited phosphoric, hydrochloric, sulfuric, or less aggressive organic acid. The type of acid should be chosen based on the result of a test on that deposit that is to be removed.

The executor should exercise care and take necessary precautions against the following:

a. The executor should ensure that there are no alloys other than carbon steel in the system that is subject to acid wash.

b. The executor is responsible to make sure that all spent acid solutions are effectively inhibited.

c. The duration of the acid wash operation must not be of such length to give greater than 5 μm (0.2 mil) general metal loss on the most vulnerable alloy-, weld-, and heat-affected zones or galvanic couples in the composite metallurgical system. No pitting will be accepted. Metal loss may be determined by thickness measurements on pipe body or installation of corrosion test coupons.

d. The difference between iron ions at the beginning and during acid wash should not exceed 5000 ppm by weight. If this difference exceeds 5000 ppm, it means that no proper acid and/or inhibitor is used and the pipe metal is being solved in the acid. In this case the acid wash should be immediately stopped and the acid be drained from the system followed by proper rinsing until the pH of 7–8 is achieved. Then the type and concentration of acid and inhibitor should be changed. This change should be approved by the engineer prior to resumption of acid wash.

The concentration of the acid should be chosen so that 80–95% of the precipitates on the pipe body can be dissolved in it within a period of 6 hours.

Based on the type of acid, suitable inhibitor should be used. If the tests mentioned in Section 5.2.2.3 indicate presence of silica in deposit, hydrofluoric acid, or one of its salts with suitable concentration (as indicated earlier) should be added to the solution.

The acid should be injected into the system and circulated at velocity of 90–150 cm/s (3–5 ft/s) for 6–8 hours. During the acid wash, the following tests should be conducted. Time intervals of the tests can be 1 hour, but it should be reduced to half an hour after 6 hours of circulation. The tests should determine followings:

- pH value of acid
- Percent of acid
- ppm of iron (Fe^{2+} and total iron)
- Percent of silica

At the end of the acid wash, all rust, mill scale, and foreign materials should have been removed from the system. If visual inspection reveals the presence of any of these substances or materials, the acid wash should be repeated with a time interval that should be approved by the engineer.

5. **Rinsing with fresh water or removal of acid by Nitrogen gas**

   After acid wash, the system should be rinsed using clean fresh water. Nitrogen gas may also be used for removal of acid. If fresh water is used for rinsing, it should be circulated until pH value of 7 is achieved.

   If nitrogen gas is used for acid removal, it should be injected from vent (highest point), and the spent acid be drained from the lowest point, until all fluids are drained.
6. **Rinsing with organic acid**

   After rinsing with fresh water or acid removal by nitrogen gas, the system should be rinsed with citric acid with concentration of 1% for removal of free iron ion which is not removed by water rinsing (or nitrogen gas). The duration of this step should be 1–3 hours at ambient temperature.
7. **Rinsing with fresh water or displacement with nitrogen gas**

   After step **6** the system should be rinsed once again with fresh water or displaced with nitrogen gas as indicated in step **5**.
8. **Neutralization and passivation**

   After step **7**, neutralization and passivation should be performed as one step. The solution should be 1–3%by weight soda ash with 0.5–1% three sodium phosphate, or 1% NaOH or 0.6% sodium nitrate (nitron). It should be noted that nitron is preferred to other additives.

   The solution should be circulated for 2–4 hours with a temperature of 50–60°C.
9. **Flushing and drying**

   After steps **1**–8 are completed, the system should be flushed with hot dry air until the system is completely dried. The dew point of spent air should be lower than the minimum ambient temperature.

5.2.2.4.2 Pickling by Soaking Process

In chemical cleaning by a soaking process, the same concentrations and temperatures specified in the circulation process should be used, and the same steps should be followed except that the fluid(s) should be injected from the lowest point of the system (drain), and air (or previous spent fluid) should be vent from the highest point (vent). Since there is no movement of fluid(s), the duration of acid wash with soaking process should be kept longer than that of the circulation process, but in any case the end of each step should be determined as indicated in the circulating process.

5.2.2.4.3 Pickling by Dipping Process

In the dipping process, a bath should be prepared with a size that allows the biggest portion of the piping system to be immersed into it. Then fluids with the same concentration and temperature as indicated in Section 5.2.2.4.1

should be used, and the same steps as those cited for circulation process should be followed for dipping the piping system into the fluid baths.

5.2.2.5 *Pickling Stainless Steel Piping*

Pickling of stainless steel piping should be avoided, but if the project specification calls for it, stainless steel piping may be chemically cleaned as per detail procedure taking the following exceptions into account.

Only distilled and/or demineralized water should be used in flushing, rinsing, preparation of acids, neutralizing, passivating, and degreasing fluids.

Only phosphoric acid, sulfuric acid, and/or organic acids should be used for acid wash. Requirements in respect of concentration, temperature, and duration of pickling should be adhered to.

Hydrochloric acid should not be used in any circumstances.

Chloride concentration during pickling process should never exceed 1 mg/kg (1 ppm by weight) before, during, and after pickling.

5.2.2.6 *Protection of Cleaned Piping*

After the completion of the acid wash and subsequent drying of the piping system, the hot air should be completely displaced with nitrogen gas. The system should be tightened to obviate leakage of nitrogen gas to open atmosphere, and a positive pressure should be maintained until the commissioning process gets started.

5.3 INSPECTION

The engineer reserves the right to inspect and/or supervise the cleaning at any step of the operation and to make any checks deemed necessary to ensure that the desired degree of cleanliness has been achieved.

During the chemical cleaning in addition to the tests the attack of the surfaces should be checked by using pretreated steel plates.

5.4 DOCUMENTATION REQUIREMENTS

Chemical cleaning procedures and test results should be submitted to the engineer on completion of chemical cleaning and prior to commissioning of the system. The documents should include but not necessarily be limited to the following:

Approved chemical cleaning procedure.

Flow diagram of the system(s) that have been chemically cleaned.

A certificate signed and approved by the engineer showing that the entire piping system that requires internal cleaning has been satisfactorily cleaned, based on the approved procedure.

The test results conducted before, during, and after chemical cleaning.

The preservation system including the type of purging fluid and the pressure of the system(s).

5.5 TESTING OF PICKLING BATHS

5.5.1 Iron Content

The method described below gives accurate results only if all the iron present is in the form of ferrous salts.

This is generally the case if the sample is taken from a pickling bath that is in operation (or in which pickling has taken place previously for only a short while). The sample must then be tested immediately after having been taken. In all other cases part of the sample (approximately 25 mL) must first be reduced (e.g., with zinc amalgam) until all the iron is present in the form of ferrous salts, after which the sample is tested according to the following method.

5.5.1.1 Chemicals

4N sulfuric acid.

85% wt phosphoric acid.

Manganese sulfate solution, obtained by dissolving 70 g $MnSO_4\ 4H_2O$ in 500 mL distilled water and adding 125 mL 85% phosphoric acid and 125 mL concentrated sulfuric acid, and then making the solution up to 1 L with distilled water.

0.1N potassium permanganate.

5.5.1.2 Procedure

Dilute a suitable quantity, V mL, bath fluid with distilled water to about 100 mL. If the bath fluid does not contain any hydrochloric acid, add 10 mL sulfuric acid and 0.5 ml phosphoric acid. If it does, add 25 mL manganese sulfate solution. Titrate with potassium permanganate until the pale pink color remains for 15 seconds (a mL).

Calculate the iron content of the bath fluid in g/L from $\frac{\alpha}{V} \times b \times 56$, in which b represents the titer of the potassium permanganate.

5.5.2 Acid Content

The titration method with methyl orange as indicator described below is simple to carry out, but has the drawback that the correct change of color is not clearly perceptible to everyone.

Therefore a method with phenolphthalein as indicator is described as an alternative, but this is less accurate.

The titration can also be carried out potentiometrically (not further described here). This requires a relatively expensive apparatus, but the method is more suitable, particularly for conducting larger series of tests.

5.5.2.1 Chemicals

1N caustic soda solution and 0.1N caustic soda solution.
Methyl orange.
Phenolphthalein.

5.5.2.2 Procedure

1. **Acid Content of Pickling Baths**
   a. **Methyl Orange Indicator**

Dilute a suitable quantity, V mL, bath fluid with distilled water to about 100 mL. Add a few drops of methyl orange and titrate with 1N caustic soda solution until the color changes from red to yellow (a mL).

Calculate the acid content of the pickling bath in g/L from:

$$\frac{\alpha}{V} \times b \times 36.5 \text{ g/L} \quad \text{if the acid is hydrochloric acid}$$

$$\frac{\alpha}{V} \times b \times 49 \text{ g/L} \quad \text{if the acid is sulfuric acid}$$

$$\frac{\alpha}{V} \times b \times 98 \text{ g/L} \quad \text{if the acid is phosphoric acid}$$

in which b represents the titer of the caustic soda solution.

   b. **Phenolphthalein Indicator**

Dilute a suitable quantity, V mL, bath fluid with distilled water to approximately 100 mL. Add a few drops of phenolphthalein and titrate with 0.1N caustic soda solution until the color changes to red (a mL). If the liquid contains a lot of iron, it is practically impossible to detect the change in color. If this is the case, the following steps must be taken.

First carry out a preliminary determination to ascertain how much caustic soda solution is roughly required for the color change.

For accurate titration add in one portion enough caustic soda solution to bring the liquid very close to the expected end point and then titrate further with small quantities of caustic soda solution. Shake the Erlenmeyer flask well after each addition of caustic soda solution and add a few drops of phenolphthalein.

Keep the flask tilted, allow the precipitate to settle and observe whether the clear layer that separates out is pink or red.

Calculate the acid content of the pickling bath in g/L from:

$$\left(\frac{\alpha}{V}\times b-2\times\frac{c}{56}\right)\times 36.5\ \text{g/L}\quad \text{if the acid is hydrochloric acid}$$

$$\left(\frac{\alpha}{V}\times b-2\times\frac{c}{56}\right)\times 49\ \text{g/L}\quad \text{if the acid is sulfuric acid}$$

$$\left(\frac{\alpha}{V}\times b-2\times\frac{c}{56}\right)\times 49\ \text{g/L}\quad \text{if the acid is phosphoric acid}$$

in which b represents the titer of the caustic soda solution and c the iron content of the bath fluid in g/L (test 1.1 of this attachment).

2. **Acid Content of Rinsing Baths**

Dilute a suitable quantity, V mL, bath fluid with distilled water to about 100 mL. Add a few drops of methyl orange and titrate with 0.1N caustic soda solution until the color changes from red to yellow (a mL).

Calculate the acid content of the rinsing bath in g/L from:

$$\frac{\alpha}{V}\times b\times 36.5\ \text{g/L if the acid is hydrochloric acid}$$

$$\frac{\alpha}{V}\times b\times 49\ \text{g/L if the acid is sulphuric acid}$$

$$\frac{\alpha}{V}\times b\times 98\ \text{g/L if the acid is phosphoric acid}$$

in which b represents the titer of the caustic soda solution.

5.5.3 Reduction in Weight and Attack of the Surface During the Pickling of Steel (Draft Specification of the "Vereniging Metaalbeits") (Metal Pickling Association)

The test is intended as a check on the proper functioning of the pickling bath, i.e., that the attack of the steel is minimal and that it occurs uniformly.

The rate of corrosion of the steel is ascertained by pickling pretreated steel plates for a certain length of time and determining the resultant reduction in weight; the plates are then examined visually in order to establish whether a uniform attack has taken place.

5.5.3.1 Apparatus

Plates, bright steel Qmc 37, with a hole for suspension, dimensions 50 mm × 50 mm × 5 mm.

Balance, weighing capacity 200 g, sensitivity 2 mg. Magnifying glass, magnification at least 5× and at most 8×. Two glass beakers, each with a capacity of 1000 ml.
One pair of crucible tongs, for inserting and withdrawing the test plates.
Sulfuric acid, chemically pure, diluted (50 g H_2SO_4/L).
Alcohol, denaturated (methylated spirits).

5.5.3.2 Procedure

1. **Pretreatment of the Test Plates**

   At least three test plates are required for the test; however, it is recommended to pretreat a few extra plates.

   1.1 Remove grease from the plates with an organic degreasing agent, e.g., tri- or perchloroethylene.

   1.2 Pickle the plates for 2 minutes in diluted sulfuric acid at a temperature of 60°C.

   If a fresh quantity of sulfuric acid is not used for each test, the fluid must be renewed as soon as the iron content exceeds 5 g/L.

   1.3 Rinse the plates in clean, cold, and fresh running water (e.g., tap water).

   1.4 Dip the plates in alcohol.

   1.5 Dry the plates in air.

   1.6 Weigh each plate to the nearest 2 mg.

   1.7 Inspect the surface of the plates for corrosion with the aid of the magnifying glass. Only noncorroded plates may be used for carrying out the test.

2. **Execution of the test**

   Use at least three pretreated plates, in accordance with **1.7**, that are shown to be uncorroded.

   2.1 By means of a nylon thread or steel wire with a suitable plastic covering, suspend the plates in the pickling bath, together with a charge of the material to be pickled.

   Suspend the plates at about half the depth of the pickling fluid in such a way that they are evenly distributed over the length of the charge. Ensure that the pickling fluid has free access everywhere.

   Pickle the plates for the same length of time as the charge, but no longer than 30 minutes.

   2.2 Then rinse the plates one by one in clean, cold, and fresh running water (e.g., tap water).

   2.3 Dip the plates in alcohol.

   2.4 Dry the plates in air.

   2.5 Weigh each plate to the nearest 2 mg.

2.6 Calculate from the reduction in weight and the area the loss of weight in mg/dm^2 for each plate as follows:

$$\text{Loss of weight } (\text{mg}/\text{dm}^2) = \frac{\text{Loss of weight (mg)}}{0.6}$$

2.7 Inspect the surface of the plates for irregular corrosion with the aid of the magnifying glass.

In practice it will not always be possible to conduct the test on the spot in the manner described above because of the lack of the necessary apparatus. If the engineer considers that in those cases, the test can be conducted at a laboratory elsewhere.

For this purpose 1-L samples must be taken at those places in the pickling bath where the plates would have been suspended if the test had been carried out normally: the temperature at those places should be measured at the same time. The samples must be clearly labeled with the data on the temperature and the pickling time of the charge. The samples should then be dispatched to the laboratory for testing.

5.5.3.3 *Procedure for the Laboratory Test*

Pour the samples of pickling fluid into wide 1500-mL glass beakers. Heat the contents of these beakers to the temperature indicated on the samples concerned and keep the fluid at this temperature for the entire test. In each beaker suspend a plate, pretreated and weighed as described above (see **1.1**–**1.7** inclusive) in such a way that it hangs roughly in the middle of the fluid. Cover the beaker well to prevent evaporation losses. Pickle each plate for the pickling time of the charge marked on the samples, but not longer than 30 minutes. Then treat the plates further, as described above under **2.2**–**2.7** inclusive.

Chapter 6

Pipe Supports

6.1 INTRODUCTION

This chapter covers the requirements for selection, design, fabrication, and erection of supports and supporting elements used for on-plot and off-plot piping in oil, gas, and petrochemical industries.

Both insulated and uninsulated piping are covered by this chapter.

Pipe supports should be designed, manufactured, fabricated, inspected, and/or selected for installation in accordance with ASME B31.3, B31.4, B31.8 and also MSS SP-58-2002, MSS SP-69-2002, MSS SP-89-1998, UL 203-2005, and other standards.

The executor should prepare an index of pipe supports with an identification number together with a bill of materials.

Welding of support elements to galvanized steel structures is not permitted.

Mechanical devices such as snubbers, sway braces, and sway struts should be:

- designed with 50% decreased allowable material stresses when subjected to vibration;
- designed to withstand the specified loads without buckling, provided with self-aligning spherical ball bushings at both ends of the assembly, permitting a minimum or 10 degrees rotation in any plane;
- furnished in such a way that an adjustable specified length of ± 40 mm is possible and fitted with a secure locking device.

Snubbers should have a clearly readable travel scale.

Snubbers should be able to operate in the frequency range from 3 to 33 Hz.

6.2 STRUCTURAL ATTACHMENTS

6.2.1 Integral Attachments

Integral lugs, plates, angle clips, pipe stanchions, pipe dummies, trunnions, etc., used as part of an assembly for the support or guiding of pipe, may be welded directly to the pipe provided the materials are compatible for welding and the design is adequate for the temperature and load.

Oil and Gas Pipelines and Piping Systems. DOI: http://dx.doi.org/10.1016/B978-0-12-803777-5.00006-X

Field-welded supports should be set correctly in place and adjusted to the final position before welding to the pipe.

Reinforcing pads (with a vent hole) for trunnions and pipe stanchions should be a full plate rather than a ring.

6.2.2 Structural Connections

The load from piping and pipe-supporting elements (including restraints and braces) should be suitably transmitted to a pressure vessel, building, platform, support structure, foundation, or to other piping capable of bearing the load without deleterious effects.

6.3 SUPPORTS FOR INSULATED PIPES AND ATTACHMENTS

Insulated lines running in pipe trenches should be supported high enough to assure the insulation will remain above the highest expected storm water levels.

Clamped cradles or pipe shoes may be used on the following insulated lines:

- piping lined with glass, rubber, plastics, etc.;
- piping requiring postweld heat treatment;
- expensive materials such as titanium, Hastelloy, Monel, etc.;
- piping with corrosion-resistant coating (e.g., galvanized piping). For all other insulated lines welded cradles or pipe shoes should be used.

6.3.1 Insulated Lines for Hot Service

Pipe shoes for insulated hot piping should include slots, as directed by the engineer, to secure the insulation weatherproofing jacket near the shoe.

6.3.2 Insulated Lines in Cold and Dual Temperature Services

Supporting system for the above-mentioned lines should be in accordance with standard.

6.3.3 Uninsulated Lines

Cradles or pipe shoes, if specified, may be of the clamped or the welded type.

6.4 ELEVATED PIPE SUPPORTS (H SUPPORTS)

Elevated pipe supports for different size of pipelines and different height should be in accordance with standards.

Pile-driven supports should have a minimum bearing capacity of 15 tons.

6.5 FABRICATION AND INSTALLATION OF PIPE SUPPORTS

This section is a supplement to the reference standard MSS SP-89, 1998. For ease of reference, the clause or section numbering of reference standard has been used for the supplement.

6.5.1 Forming

Cold forming of plate and flat bars may be performed on materials 12.7 mm (0.5 in.) thick or less to a minimum inside radius of 1× the stock thickness. Cold forming may be performed on material over 12.7 mm (0.5 in.) thick to a minimum inside radius of 2.5× the stock thickness. Material over 12.7 mm (0.5 in.) thick may be cold formed to an inside radius less than 2.5× but not less than 1× the stock thickness provided heat treatment is performed.

Round bars of 19 mm (0.75 in.) diameter and smaller may be cold formed to a minimum inside radius of ½× the bar diameter. Round bars greater than 19 mm (0.75 in.) diameter may be cold formed to a minimum inside radius of 2½× bar diameter. Forming is not permitted on threaded areas.

Heating material to 704°C (1300°F) or less to facilitate the operation should be considered cold forming.

Hot forming of plate and flat bars may be performed on materials of any thickness to an inside radius not less than 1× the stock thickness within the following surface temperature ranges (no holding time required):

Carbon steel	760°C Min. (1400°F)	1093°C Max. (2000°F)
Chrome-Moly alloy steel	843°C Min. (1550°F)	1093°C Max. (2000°F)
Austenitic Stainless Steel	760°C Min. (1400°F)	1148°C Max. (2100°F)

Material should not be heated in bundles or closed stacks in other than induction-type furnaces but should be separated to allow good circulation within the furnaces. Materials should not be heated above the maximum temperature shown. No hot forming operation should be performed below the minimum temperature shown. Carbon steel and chrome-moly alloy steel should be cooled in still air. Water quenching is not permitted. Cooling of stainless steel other than still air cooling may be accomplished as per ASTM A403/A403M-04.

Round bars of any diameter may be hot formed to a minimum inside radius of ½× the bar diameter within the temperature ranges. Forming is not permitted on threaded areas.

Heat treatment, when required of carbon steel and chrome-moly alloy steel, should be done within the temperature ranges shown below. The material is to be held at the designated temperature for 1 h/in of thickness, but not less than 1 hour, followed by slow cooling in furnace or still air.

Carbon steel	593°C Min (1100°F)	676°C Max. (1250°F)
Chrome-Moly Alloy steel	704°C Min. (1300°F)	760°C Max. (1400°F)

A carbide solution heat treatment of austenitic stainless steel, when required by design specification, should be performed as prescribed by ASTM A403/A403M-04.

Incremental bending by braking is not an acceptable method of forming.

Formed components may be furnished in "as-formed" condition without any further mechanical work.

6.5.2 Welding

Unless otherwise specified by the job specification, welders and welding procedures should be qualified in accordance with the ASME Boiler and Pressure Vessel Code, Section IX.

Welder's performance test results and Welding Procedure Qualification Records should be available, upon request, to authorized inspection personnel.

When tack welds are to become a part of the finished weld, they should be visually examined and ground or feathered, if necessary. Defective tack welds and tack welds made by unqualified welders should be removed.

Attachments welded directly to the pipe should be of an appropriate (compatible) chemical composition and be able to withstand the anticipated loads at the piping temperature. The method of attachment to the pipe should meet all the preheating, welding, and postweld heat–treating (PWHT) requirements of the pipe.

Preheating and PWHT requirements for pipe hangers should be as outlined in ASME B.31.3.

Unacceptable welds should be removed by flame or arc gouging, grinding, chipping, or machining. Welds requiring repair should be welded in accordance with the requirements of the original weld. Base metal irregularities requiring repair by welding should be repaired in accordance with the material specification or ASTM A6/A6M:2005, as applicable. Welders and welding procedures used in making repair welds should be qualified.

6.5.3 Surface Discontinuities

Only those surface discontinuities that are detrimental to the strength or function of a product should be cause for rejection.

Surface discontinuities of welds should be evaluated in accordance with the applicable code or job specification requirements.

6.6 METALLIC COATINGS

Metallic coatings for corrosion resistance may be applied by electroplating, pregalvanizing, hot-dip galvanizing, or mechanical plating.

Electroplating should be in accordance with ASTM B633-98e1 or ASTM B766-86 (2003) for the specific coating used. To avoid difficulty in assembling threaded parts that are plated, it is recommended that female machine threads be tapped oversize by an amount equal to 4× the maximum plating thickness. It is not permissible to rethread male parts after plating. It is standard practice for female threads to be uncoated.

Pregalvanized sheets should meet the requirements of ASTM A653/A653M-04a.

Hot-dip galvanizing should be done in accordance with ASTM A153/A153M-05 or ASTM A123/A123M-02. To avoid difficulty in assembling threaded parts, it is recommended that male parts be shaken, spun, or hand-brushed to remove spelter lumps from the threads. Female machine threads may be tapped oversize to accommodate the male thread.

Mechanical plating should be done in accordance with ASTM B695-04 and ASTM B696-00(2004)e1.

Repair of galvanized surfaces may be performed by any suitable cold galvanizing compound or by hot-spray metallizing.

Protective shields may be manufactured from a pregalvanized sheet meeting the requirements of ASTM A653/A653M-04a.

- **Chromic Acid Dropping Test**

  This test should be performed according to UL 203-2005.

6.7 NONMETALLIC COATING

Nonmetallic coatings should be selected by types for specific purposes. The application of coatings should be in accordance with the coating manufacturer's recommendations. In general, only such coatings with good adhesive quality, namely, that do not lift, peel, or chip when scratched and that will withstand reasonably rough handling, should be used.

Nonmetallic coatings that are intended for threaded products may be applied before assembly.

Nonmetallic coatings, jackets, and liners to prevent abrasion of glass or plastic pipe, etc., should be applied in accordance with the manufacturer's recommendations.

Nonmetallic coatings, jackets, and liners for electrolytic resistance should have dielectric strength suitable for the intended use.

6.8 TESTING OF PIPE HANGER AND SUPPORT COMPONENT

A hanger should be fabricated to fit the appropriate rod sizes and rod attachments specified in standards for the size of pipe shown and should have sufficient strength to support the test loading, resistance to vibration, and corrosion resistance (metallic coatings).

6.8.1 Test Classification

Testing of hanger and support components falls into five categories, given below:

1. Design proof test,
2. Qualification test,
3. Calibration test,
4. Pull test,
5. Vibration test.

6.8.2 Pull Test

Pull tests should be performed according to UL 203-2005.

6.8.3 Vibration Test

This test should be performed in accordance with UL 203-2005.

6.8.4 Quality Control

Quality control should be exercised over the procurement of raw material, fabrication procedures, and dimensions to assure the continued validity of design proof and qualification tests.

Except for small items that have restricted space for die-stamping (e.g., hanger rods, eye nuts, etc.), each standard support component should, in addition to paint marking, be die-stamped with its size and identification mark.

Hot-dip galvanized supports should be die-stamped with their respective identification mark and serial number before galvanizing.

After hot-dip galvanizing, these marks and numbers should be marked by paint.

6.9 HANGER INSTALLATION

6.9.1 Rod Hangers

Rod hangers, either rigid or spring type, are adjustable vertical assemblies consisting of structural attachment, hanger rod (with or without intermediate components), and pipe attachment.

Spring assemblies should be shipped to the job site with the springs compressed in the installed position as indicated on the pipe support detail drawing and/or requisition. They should have caution tags attached that warn that the spring locks must be removed before the line is put in service. Additionally, the spring supports should have the marks "C" or "H" on the casing load indication scale:

- Mark "C" indicates the cold position of the spring when the line is at ambient temperature but filled with its actual service fluid. For tank lines filled with the actual service fluid, it indicates the position when the tank is empty and has zero settlement.
- Mark "H" indicates the operation positions of the spring. For tank lines, it indicates the position when the tank is filled and has settled.

Spring supports should be installed with the spring locks in place. These spring locking plates or pins should not be removed before hydrostatic testing and insulation of the piping system is completed.

Normally, all construction aids such as spring locks, temporary supports, welding tracks, etc., should be removed prior to commissioning. However, when the spring force on the empty line will cause possible damage to connected vulnerable equipment, the spring locks should remain in position until the line is filled with the actual service fluid.

The relevant support and support drawing should bear the warning "Lock Against Empty Conditions" and the locks should be attached with the spring support during operation.

Hanger rods for lines subject to expansion/contraction of more than 75 mm should be set out of plumb, equal to half of the calculated travel of the pipe at the point of support, in the opposite direction to the travel as indicated on the support detail drawing, the piping arrangement drawing, and/or the isometric drawing.

Guides may be sliding, rolling, or others.

Saddles or pipe shoes should be attached to piping to prevent damage to insulation.

Chapter 7

Gaging, Cleaning, and Removing Liquids From Pipelines

7.1 INTRODUCTION

This chapter covers minimum technical requirements for the design, manufacture, quality control, testing, and finishing of launching and receiving traps that should be installed in oil, gas, and petrochemical industries under the conditions stated in this chapter.

7.2 BASIC DESIGN, CONSTRUCTION, AND RATING

The following description is intended to indicate the general and minimum requirement of pig launching and receiving traps and does not relieve the supplier of his full responsibility for design, fabrication, performance, and safety of the equipment.

Trap assemblies should be suitable for launching or receiving pigs for the purpose of gaging, cleaning, and removing liquids from a pipeline that may contain water and/or liquid hydrocarbons and impurities such as sand and scale. Electronic and inspection pigs, batching-corrosion-inhibition service, and special pigs may require different trap assemblies that should be checked with manufacturers before purchase or rental.

Pig traps with a nominal diameter of 20 in. and above should normally be provided with pig lifting facilities, such as a runway beam, unless it can be easily accessed by cranes. Provision of a trolley with a push rod and pulling line should be considered to assist loading or removal of pigs from the trap. The use of these facilities, including the possible use of internal trays, should be agreed upon with the company.

7.3 PIG TRAP SYSTEM COMPONENT

7.3.1 Barrel, Reducer, and Spool

The spool piece should have the same nominal size as the connecting pipeline, flanged or beveled end, as specified in the requisition.

Oil and Gas Pipelines and Piping Systems. DOI: http://dx.doi.org/10.1016/B978-0-12-803777-5.00007-1

The barrel should have a sufficient length to accommodate three pipeline cleaning pigs.

Barrels for use in a gas transmission system should be designed in accordance with ANSI B 31.8, and those for use in a liquid hydrocarbon transmission should be in accordance with ANSI B 31.4.

A thermal relief line should be provided at locations where shut-in pressure of trapped fluid could exceed the design pressure.

The reducer should be eccentric for launcher and be concentric for receiver.

7.3.2 Supports

Permanent supports should be used to support and restrain the pig traps, and these should be designed to carry the weight of the pig trap system filled with water (or other fluids if their density is greater than that of water) together with the weight of pigs. The support under the barrel should normally be of the sliding/clamp type to compensate for expansion of the unrestrained part of the pipeline. The frictional force should be reduced to a minimum by appropriate means. The height of the saddle should be such that the bottom of the barrel is located 800 mm above ground level. Supports should be positioned such that the pig trap valves can be removed for maintenance or replacement without removal of barrel.

Design of supports should comply with the requirements of BS-5500.

If specified by the purchaser, base plate, anchor bolts, etc., required for foundations should be supplied by the manufacturer and all the applied forces and moments on the saddles should be specified by the manufacturer.

Note:

Where there would be a possibility of corrosion occurring under clamps, welded clamps should be used with no direct welding on to the pipeline except for circumferential welds.

7.3.3 Connections

The trap should be completed with the connections as shown in Appendices C & D.

Weldolet and Threadolet are only allowed for connections equal or smaller than 2 in. Connections larger than 2 in. should be extruded outlets or sweepolets.

7.3.4 End Closure

The end closure should conform to the general requirements of ASME VIII division 1 section UG-35(b) (quick actuating closures). Attention is drawn to the requirements for a fail-safe design of the opening mechanism; specifically, the failure of any part of the opening mechanism should leave the closure closed rather than open.

The end closure should be of the quick-acting type, lever or hand wheel operated, and hinged vertically.

The quick-acting design should allow the opening and closing by one man in approximately 1 minute, without the use of additional devices.

The design of the end closure should be suitable for permanent location in open environment.

Closure 18" and larger should be hand wheel operated.

The activation of the seals should be such that the fluid within the pig trap at any pressure between 1 bar (abs) and the pig trap design pressure be sufficient for this purpose.

End closures with exposed screw expanders or captive ratchet braces should not be used because of the high maintenance requirements and the nonfail-safe aspects of some opening mechanism designs.

The end closure should have the following safety devices:

- A pressure locking device to prevent opening of the door when the pig trap is pressurized.
- A safety bleeder that when released will alert the operator to a possible hazard unless pressure in the pig trap is relieved completely. Opening of the door should not be possible unless the bleeder is released. Engaging the bleeder should only be possible when the closure is closed. The bleeder should be designed such that there is no risk of blockage.

7.3.5 Materials

Pipe for barrel or spool should be in accordance with the line pipe specification standards.

Flanges should be welding neck in accordance with standards.

Fittings should be in accordance with standards.

The selected nonmetallic material should be suitable for the long-term exposure to the transported fluid at the design pressure and temperature conditions. Elastomeric material should resist explosive decompression.

Studs and nuts should be aluminum coated or electroless nickel plated. The preferred materials for standards applications are ASTM A193/A193M grade B7 and ASTM A194/A194M grade 2H for nonsour service conditions, and ASTM A193/193M grade B7M and ASTM A194/A194M grade 2HM for sour service conditions.

All components in sour service should conform to the requirements of NACE MR 0175. In addition, seamless pipe should have a maximum sulfur content of 0.010 wt%. The base material and welds should have a maximum hardness of 248 HV10.

Spool pipe should be compatible with the line pipe with respect to weldability, wall thickness/material grade transitions, and dimensions. Dimensional considerations include actual internal diameter, ovality, and wall thickness, transition taper angles.

The carbon equivalent of the steel should be less than or equal to 0.43% calculated by the following formula:

$$CE - c + Mn/6 + (cr + Mo + V)/5 + (Ni + cu)/15$$

The carbon content should not exceed 0.23% for forgings. The base material and welds should have a maximum hardness of 325 HV10 for nonsour and 248 HV10 for sour conditions. Base metal hardness readings should be made in accordance with ASTM E92 on each heat at five random locations. The actual yield to tensile strength ratio should not exceed 0.90 for nonsour and 0.85 for sour services.

7.3.6 Fabrication

Welding processes, welding procedures, welder qualifications, weld repairs, welding electrodes, thermal stress relief, and heat treatment, etc., should conform to ASME VIII Division 1 and ASME Section IX. Only welders and welding operators who are qualified in accordance with Section IX of ASME should be employed in production.

Weld repairs, if required, should not be permitted after heat treatment without approval by the purchaser. Repaired welds should be heat treated according to the codes stated above.

Welding should be carried out using procedures and welders/welding operators qualified in accordance with ASME IX. Welding procedure qualification should include hardness testing of the weld, HAZ, and base metal. The hardness should be measured in accordance with ASTM E 92 for pipe butt welds; the hardness traverses should be carried out on Lines 2 mm from the pipe surfaces on a weld cross section. Each traverse should have at least three hardness measurements taken in each weld metal, the HAZ each side of the weld, and the base metal for each HAZ; one of the hardness measurements should be within 0.5 mm of the weld-fusion line. The hardness should not exceed 325 HV10 under nonsour conditions and 248 HV10 under sour conditions.

All main seam welds should be full penetration and where possible double-sided. All nozzle-to-body welds should be full penetrant.

The inside of the trap should be free from obstructions that could prevent the free rolling of spheres, or travel of pigs or carriers.

Wall thickness transitions should meet the welding configuration requirements as specified in the design code ANSI B31.4 and ANSI B31.8.

Notes:

1. tD, the maximum thickness for design pressure, should not be greater than 1.5*t*, where *t* is the nominal thickness of thinner plate.
2. Pipe with a wall thickness less than 4.8 mm should not be used.

The seam welds of a trap should not interfere or coincide with outlets welded to it.

End profiles of the pipes to be butt-welded should be in accordance with ANSI/ASME B16.25.

The requirement for heat treatment should be determined in accordance with ASME VIII, division 1, subsection C. Procedures to be applied should be in accordance with ASME VIII, division 1, part UW-40.

7.3.7 Inspection and Testing

Inspection and testing should be performed before any coating or paint is applied.

All components should be visually examined in accordance with ASME VIII division 1, part UG93.

Each weld on the trap and pig signaler should be examined by 100% radiography (RT). In addition, carbon steel welds should receive 100% magnetic particle examination (MT) and stainless steel welds should receive 100% liquid penetrant examination (PT). RT should be in accordance with ASME V articles 2 and 22, with acceptance criteria in accordance with ASME VIII, division 1, part UW-51. MT should be in accordance with ASME V, article 7 and 25, with acceptance criteria in accordance with ASME VIII, Appendix 6. PT should be in accordance with ASME V, articles 6 and 24, with acceptance criteria in accordance with ASME VIII, Appendix 8.

Surface examination of the trap should be done with wet magnetic particles unless agreed to otherwise by the Company.

For the end closure, all mating clamp- and flange-machined surfaces, door hinge, hinge attachments, and locking mechanisms should be subject to magnetic particle inspection. Any defects should constitute a basis for rejection.

The trap should be hydrostatically tested to 1.5× the design pressure. The test pressure should be held for a period of at least 4 hours. The acceptance criteria are no leakage or loss in pressure.

A functional test should be performed to demonstrate that closure door, pig signaler, loading, and unloading devices function satisfactorily.

Chemical analysis and mechanical and impact tests are required for the barrel, reducer, and neck pipe (when furnished) for each trap in accordance with the design codes. These items should be tested ultrasonically to the satisfaction of purchaser in accordance with Appendix 12 of ASME code Sec. VIII Div.1.

7.3.8 Surface Preparation

After the acceptance of the hydrostatic test, all external surfaces should be prepared and prime coated in accordance with standards. The suppliers' proposed method of surface preparation should be approved by the purchaser.

7.3.9 Pipeline Pig Passage Detector

The signaler should be an intrusive type and mounted on the trap/pipeline via a DN 50 branch connection.

For pipelines that cannot be depressurized for pig signaler maintenance, the signalers should be completed with a ball valve to isolate the pig signaler from the pipeline and a portable jacking tool for safe lifting of the transfer mechanism.

After the pig has passed the pig signaler, the internal mechanism should be reset automatically to the position required for indicating the passage of the next pig. The resetting of the signal mechanical indicator should be undertaken manually only. Resetting of the electrical switch should be automatic.

The mechanical signal flag or the electrical switch should not triggered by the flow or pressure of the pipeline fluid.

The trigger should not obstruct or damage a passing pig and the trigger should not be damaged by a passing pig.

The penetration of the trigger into the main pipe should be kept to a minimum to avoid unnecessary obstruction of fluid flow.

The design methodology as described in ASME VIII division 1 should be used for the design calculation of the pig signaler pressure housing. The design pressure of the pig signaler pressure housing should comply with the pressure/temperature rating classes for flanges as stated in ANSI/ASME B 16.5, based on material group 1.1 unless otherwise stated in the requisition, for a limited temperature range of $-20°C$ to $+120°C$.

The minimum wall thickness of the pig signaler pressure housing and any extension should be 4.8 mm.

The indicator should withstand all specified weather and climatic conditions and be of robust construction.

With the exception of a mild steel mounting base, all metal parts should consist of corrosion-resistant material.

The indicator should be bidirectional and function equally well in either direction.

Visual signals should be clearly visible from a distance of 50 m and should be manually resettable.

If carbon steel and stainless steel components are used in combination, the risk of galvanic corrosion at the contact areas should be minimized.

7.3.10 Electrical Signals

- Type: Microswitch, PDT (pig detector transmitter).
- Rating: 24 V-DC, 2 A.
- Load: Relay (inductive).
- Housing: Weather-proof/Dust-proof/Explosion-proof suitable for use in accordance with Institute of Petroleum Division 1 Group II, Gases and Vapors.
- Cable entry: ET with compression gland for PVC/LC/SWA/PVC cable.
- Mounting: Above ground nonextended.
- Pressure rating: The same as pig trap rating.
- Kind: Unidirectional.
- Elec./Signal: Auto reset.

7.3.11 Nameplates and Labeling

Each pig launching and receiving trap should be labeled with engraved stainless or noncorrosive alloy nameplates together with noncorrosive fixing materials, showing all data as called for in this chapter including, but not limited to, the following;

1. Purchaser's name and order number.
2. The year of manufacture.
3. Manufacturer's name or trademark.
4. Type of materials, size, serial number, and designation making it possible to obtain relevant information from the manufacturer.
5. Flange pressure rating.
6. Design and test pressure.
7. Dimensions and physical properties including weight.
8. Tag number.
9. Design temperature.

The nameplate should be legible and easily visible when fixed to nonremovable part of the frame.

The nameplate should be corrosion- and moisture-resistant and provided with indelible inscriptions.

The pressure housing of the pig signaler should be stamped with the pressure class rating as indicated in the Requisition sheet.

If sour service conditions are specified, the nameplate or marking should include "NACE MR 0175."

7.3.12 Tools and Testing Equipment

Special tools and equipment, if required for erection, commissioning, maintenance, and testing, should be shipped together with the assembly including sufficient washers, "O" rings, seals, lubricants, and others.

7.3.13 Provision for Handling and Erection Equipment

Each unit should be provided with hoisting facilities, bolts foundation clamps. Small materials required for erection on site should be packed inside the transport unit.

7.4 INSPECTION DURING MANUFACTURING

The purchaser or his nominee should have free access to the manufacturing plant engaged in the construction of the equipment to carry out the necessary inspections at any stage of construction.

The supplier should place at the disposal of the purchaser, free of charge, such instruments as are required at the inspection point to enable the purchaser to carry out his inspection of equipment efficiently in this respect. Such inspections in no way relieve the supplier of his responsibilities under the terms specified in this chapter and/or in other applicable relevant documents.

7.5 TESTS AND CERTIFICATES

The general requirement for test is described but not limited to the following:

The test procedure as proposed by the supplier should be agreed to and approved by the purchaser before tests are carried out.

The purchaser may require witnessed tests to be carried out in the presence of its nominated representative who should be informed at least 4 weeks in advance of the date of conducting the tests and confirmed 10 days before the test.

All the test equipment, labor, consumables, and other expenses should be provided by the supplier at no extra cost to the purchaser.

Test certificates should refer to the serial number of the equipment tested and must bear the purchaser's name and manufacturer's name and seal; the certificate should be approved by the purchaser before shipment instructions are given.

The certificates should specify the chemical and mechanical properties of material, destructive and nondestructive tests, heat treatment, design codes, and hydrostatic test.

7.6 FINISH

All unpainted surfaces, e.g., flange surfaces, should be properly protected against corrosion with an antirust compound that is easily removable by hydrocarbon solvents.

The pig launching and receiving trap, including pig loading and unloading devices as well as the pig signaler, should be cleaned and painted with two layers of antirust undercoat. A final layer of paint suitable for the specified environment should be applied on the purchaser's request.

The color of the final layer should be as specified in the requisition.

All unpainted surfaces (internal and external) should have a coat of moisture- and fungus-resistance varnish.

7.7 INFORMATION FOR MANUFACTURER/SUPPLIER

Further to the information included in other parts of this specification, relevant data sheets should be completed and furnished with the requisition as a part of this chapter specification.

The equipment may be rejected if measurements and inspection reveal any discrepancies between quoted figures resulting in requisition and those measured actually.

7.7.1 Packing and Shipment

The equipment should be suitably packed and protected against all damages or defects that may occur during handling, sea shipment to the port, and rough road haulage to the site, and extended tropical open-air storage. All items should be properly packed to comply with the requirements of BS-1133.

7.7.2 Guarantees and Warranties

The supplier should guarantee his equipment during commissioning and for one year of operation (starting from the date of completion of commissioning) against the following defects:

1. All operational defects.
2. All material defects.
3. All fabrication and design defects.
4. All defective parts should be replaced by the supplier in the shortest possible time, free of charge, inclusive of dismantling, reassembly at site, and all transportation costs.

The supplier should guarantee the provision of spare parts to the purchaser for a minimum period of 8 years from the date of dispatch. In the event the supplier cannot supply the required spares (whether of his own manufacturer or other's) within the period of time, the costs of complete replacement units will be borne by the supplier.

7.7.3 Spare Parts

All spare parts should comply with the same standards, specifications, and tests of the original equipment and should be fully interchangeable with the original parts without any modifications at the site. They should be correctly marked in accordance with the spare parts lists and interchangeability record and be prescribed to prevent deterioration during shipment and storage in humid tropical climate.

7.8 DOCUMENTATION LITERATURE TO BE SUBMITTED BY MANUFACTURER/SUPPLIER (INFORMATIVE)

At the quotation stage, the supplier should submit four sets of the following documents.

1. Report of experience background, major clients, and annual sale for the similar equipment.
2. Reference list showing the successful operation of similar equipment for at least 2 years, and the locations of equipment for at least 2 years, and the locations of equipment offered in major oil industries.
3. Typical type test certificate of similar equipment.
4. Declaration of confirmation with the set standards, and or clear indication of deviations from the standards and specification.
5. Spare parts and special tools requirements.
6. List of recommended commissioning spare parts with the price.
7. List of recommended spare parts for 3 years of operation.
8. List of special tools, testing devices, and instruments.
9. Shipping dimensions (length, width, and height) and weight, with shipping schedule.

At ordering stage, the supplier should submit five sets of the following documents.

1. Outline drawings of the floor plan, elevation and end view, giving complete sizes and dimensions, various connections to outside equipment, and recommended installation details for the purchaser's approval.
2. Design calculations and proposed test procedure for purchaser's approval.
3. Reproducibles (one set only) of above-mentioned drawings after approval duly certified by the supplier.
4. Prints of certified drawings as well as agreed to and approved test procedures.
5. The purchaser's comments or approval should be given within 21 days of the receipt of the relevant documents.

Before shipping the equipment, the supplier should submit 15 sets of the following documents, according to the following time table, to be received by the company.

1. Codes and standards compliance certificates, e8 weeks minimum.
2. Installation, operation, and maintenance manuals, 4 weeks minimum.
3. Final factory test certificates, including test data and calculated results, 3 weeks minimum.
4. Inspection certificate issued by the purchaser nominated inspector, 2 weeks minimum.
5. Final revision of illustrated and numbered part list and 3-year running spare parts list, 2 weeks minimum.

APPENDIX A: MAJOR PHYSICAL PROPERTIES

Flowing Contents	
Specific gravity	At 15°C
	At 4°C
Coefficient of expansion	(M/°C)
Viscosity (ES)	At 15°C
	At 4°C
Pour point (°C)	Winter
	Summer
Min. flash point (°C)	

Note:

Further to the above-mentioned information, the following properties should be included:

1. Design pressure.
2. Design temperature.
3. Chemical composition of flowing content.
4. Design factor.
5. Corrosion allowance (if required).

APPENDIX B: AMBIENT CONDITIONS

Maximum sun temperature (for calculating the maximum temperature rise of the equipment)................................
Minimum ambient temperature................................
Maximum recorded velocity of prevailing wind..............................

APPENDIX C: PIG LAUNCHING TRAP

Date:
1-Scraper Launcher
Mating Pipe Specification
2-Operating Pressure

ORIENTATION ACC.
TO VIEW A-A

SECTION A-A

ITEM NO.	A	B	C	D	E	F	G	H	I	X	Y	Z
***												

	No.	DLA	FLANGED/ THREADED	ORIENTATION	CLASS	DESCRIPTION
*	N1					MAIN LINE CONNECTION
	N2					KICKER CONNECTION
	N3					DRAIN
**	N7	*50	—			PIG SIGNALE
	N4					VENT
	N5	15	NPT			P.1.
	N6	25	NPT			TSV

* IF BEVELED END SPOOL PIECE IS REQUIRED SHOULD BE SPECIFIED.

** WILL BE INSTALLED ON MAIN LINE SCRAPER LAUNCHING, IF SPECIFIED BY THE PURCHASER.

*** DIMENSIONS "A" THRU "Z" EXCEPT "X" AND "Y" SHOULD BE SPECIFIED BY THE MANUFACTURER AND CONFIRMED BY THE PURCHASER.

APPENDIX D: PIG RECEIVING TRAP

Date:
1-Scraper Launcher
Mating Pipe Specification
2-Operating Pressure

ORIENTATION ACC.
TO VIEW A-A

SECTION A-A

ITEM NO.	A	B	C	D	E	F	G	H	I	X	Y	Z
***												

	No.	DLA	FLANGED/ THREADED	ORIENTATION	CLASS	DESCRIPTION
*	N1					MAIN LINE CONNECTION
	N2					KICKER CONNECTION
	N3					DRAIN
**	N4					VENT
	N5	15	NPT			P.I.
	N6	25	NPT			T.S.V.
	N7	25				PIG SIGNALER

* IF BEVELED END SPOOL PIECE IS REQUIRED SHOULD BE SPECIFIED.

** WILL BE INSTALLED ON MAIN LINE SCRAPER RECEIVING

*** DIMENSIONS "A" THRU "Z" EXCEPT "X" AND "Y" SHOULD BE SPECIFIED BY THE MANUFACTURER AND CONFIRMED BY THE PURCHASER.

Chapter 8

General Requirements for the Purchase of Pipes for Use in Oil and Gas Industries

8.1 INTRODUCTION

This chapter gives technical specifications and general requirements for the purchase of steel line pipe for use in oil, gas, and petrochemical industries under nonsour as well as sour conditions and is based on API Spec. 5L, 42nd edition, Jan. 2000, and should be read in conjunction with that document.

The purpose of this specification is to provide pipe suitable for use in conveying gas and oil in both the oil and natural gas industries.

This specification covers seamless and welded end and plain-end steel line pipe.

This specification supplements requirements established for product specification level two (PSL2). Those clauses that apply to PSL2 and are not amended by this chapter remain valid as written.

Grades covered by this specification are B, X-42, X-46, X-52, X-56, X-60, X-65, and X-70. Pipe should be ordered only to these grades; intermediate grades should not be used.

Note:

The grade designations used herein for grades B do not include reference to the specified minimum yield strength (SMYS). Other grade designations used herein comprise the letter X followed by the first two digits of SMYS.

Higher grade pipe should not be substituted for pipe ordered to a lower grade without prior approval of the purchaser, irrespective of strength level.

8.2 QUALITY ASSURANCE SYSTEM

The manufacturer should establish and maintain a quality assurance system in accordance with ISO 9001, or an approved equivalent. The purchaser's nominated inspector(s) or representative(s) should have the right to undertake such audits as he deems necessary to assess the effectiveness of the manufacturer's quality assurance system.

Oil and Gas Pipelines and Piping Systems. DOI: http://dx.doi.org/10.1016/B978-0-12-803777-5.00008-3

8.3 COMPLIANCE

Although sampling may be adopted to determine batch compliance, nevertheless the manufacturer is responsible to ensure and certify that all pipes meet the requirements of this chapter.

8.4 CONFLICTING REQUIREMENTS

In the case of conflict between documents relating to the order, the following priority of documents should apply.

1. Purchase order
2. This chapter

8.5 INFORMATION TO BE SUPPLIED BY THE PURCHASER

8.5.1 Additional Information to Be Supplied by the Purchaser to the Manufacturer at the Time of Order Are:

1. Minimum design temperature
2. Pipeline category (offshore or onshore) for length requirement
3. The requirements for testing in the simulated heat treated condition
4. The requirement for color code or marking to identify pipe mill and wall thickness
5. The requirement for bevel protectors
6. The requirements for preproduction welding procedure qualifications
7. Higher absorbed energy requirements for Charpy test
8. The requirements for DWTT
9. Purchaser inspection, if required, and to what extent
10. Suitability of line pipe for sour conditions

8.6 PROCESS OF MANUFACTURE AND MATERIAL

8.6.1 Seamless Process

Cold sizing and straightening are only permitted if the total strain in the seamless pipe does not exceed 3.0%.

8.6.2 Types of Pipe

8.6.2.1 Continuous Welded Pipe

Pipe manufactured by this process is unacceptable.

8.6.2.2 *Electric Welded Pipe*

Only HFW pipe with a minimum welding frequency of 150 kHz, made from hot-rolled coil is acceptable. A normalizing heat treatment of the weld and heat-affected zone should always be carried out, irrespective of grade and chemical composition. Full-body normalizing, normalizing and tempering, or quenching and tempering are also acceptable.

8.6.2.3 *Laser-Welded Pipe*

This type of pipe is unacceptable.

8.6.2.4 *Longitudinal Seam Submerged-Arc Welded Pipe*

The full length of the weld seam should be made by automatic submerged arc welding, using run-on and run-off tabs. The welding procedure should be approved by the purchaser. Welding should be checked at regular intervals to ensure that current, voltage, and travel speed remain within the ranges of the approved welding procedure.

8.6.2.5 *Gas Metal-Arc Welded Pipe*

This type of pipe is unacceptable.

8.6.2.6 *Combination of GMAW and SAW Pipe*

GMAW is only acceptable for making a continuous tack weld in SAW pipe, which is then considered as SAW pipe.

8.6.2.7 *Double-Seam SAW Pipe*

This type of pipe is unacceptable, unless specifically ordered by the purchaser.

8.6.2.8 *Double-Seam GMAW Pipe*

This type of pipe is unacceptable.

8.6.2.9 *Double-Seam Combination of GMAW and SAW Pipe*

This type of pipe is unacceptable, unless specifically ordered by the purchaser, in which case the conditions of Section 8.6.2.6 should hold.

8.6.2.10 *Helical Seam SAW Pipe (SPW Pipe)*

This type of pipe is acceptable, in which case all the requirements indicated in this chapter should be applied.

8.6.3 Types of Seam Weld

8.6.3.1 Tack Weld

Tack welds should be made in accordance with a qualified tack welding procedure using automatic SAW, GMAW, gas-shielded FCAW, or shielded metal-arc welding using low-hydrogen electrodes from which the diffusible hydrogen content of the resulting weldment should not exceed 10 mL/100 g of deposited metal.

8.6.4 Cold Expansion

Longitudinally SAW pipe should be mechanically cold-expanded between a minimum of 0.8% and a maximum of 1.5% of diameter. Suitable means it should be provided to protect the weld from contact with the internal expander during mechanical expansion. Nonexpanded SAW pipe should not be supplied unless explicitly stated on the purchase order together with any supplementary test requirements.

HFW&SPW pipe should not be cold-expanded.

8.6.5 Material

The steel should be made in a basic oxygen or electric arc furnace and should be fully killed and fine grained with a grain size of ASTM E-7 or finer, as defined in ASTM E-112.

For quenched and tempered pipe, this grain size requirement should not apply.

For sour service, the steel should be calcium treated and vacuum degassed.

8.6.6 Heat Treatment

The heat-treating process should be performed in accordance with a documented procedure.

SMLS pipe should be furnished in the hot-formed, normalized, normalized and tempered, or quenched and tempered condition.

For hot-formed pipe, the finishing temperature should be greater than 780°C. Pipe finished at a lower temperature than 780°C should be subjected to a further normalizing heat treatment, with a minimum holding time of 30 minutes.

SAW pipe should be furnished in the as-rolled, normalized, thermomechanically controlled process, or quenched and tempered condition.

HFW pipe should be furnished from hot-rolled coil and the entire weld plus HAZ should be normalized. Alternatively, the pipe may be full body normalized, normalized and tempered, or quenched and tempered.

Details of heat treatment should be agreed with the purchaser prior to the start of production.

8.6.7 Preparation of Edges for Welding

The edges of the plates or strip to be welded should be profiled by machining and at least 10 mm or 1.5× the wall thickness, whichever is greater, and should be removed from each side of the plate or strip either by machining or shearing. The abutting edges of the plate or strip should be aligned for welding, and adequate provisions should be made to ensure that the alignment is maintained during the progress of the welding operation and that any root gap is controlled within limits approved in the procedure test. All surfaces to be welded should be thoroughly cleaned of scale, oil, and other foreign matter before welding is started. The weld should be of uniform width and profile and should merge smoothly into the surface of the strip without appreciable deviation from the line of the joint. The forming procedure must ensure that there is a minimum of peaking and this should be demonstrated in the procedure test to be within the acceptable limits.

8.6.8 Manufacturing Procedure Qualification

The manufacturing procedure should be recorded and qualified in accordance with Standards. The procedure qualification tests should be witnessed by the purchaser. The purchaser may, at his discretion, accept the results of previously authenticated tests. The purchaser should reserve the right to require requalification in the case of a change in the procedure specification.

8.7 MATERIAL REQUIREMENTS

8.7.1 Chemical Composition

For each enquiry/order, the manufacturer should propose a chemical composition for the pipe to be supplied. The composition should be contained in the manufacturing procedure specification and, as determined by product analysis, should comply with the maximum allowable limits specified in Tables 8.1 and 8.2. The limitations on heat and product analysis should be those agreed upon following the acceptance of the manufacturing procedure specification.

The manufacturer should propose a nominal product analysis in the manufacturing specification. The range of acceptable variations in the product analysis is given in standards.

This should be applied to the chemical composition proposed by the manufacturer in the manufacturing specification.

The maximum variation on a agreed-upon composition is allowed, provided that the final maximum alloy content given in Tables 8.1 and 8.2 is not exceeded.

For sour services, Table 8.2 should be applied.

TABLE 8.1 Chemical Requirements for Product Analysis by Percentage of Weight (Nonsour Services)

Element	GR. B	X-42	X-46	X-52	X-56	X-60	X-65	X-70	Maximum Variation On Agreed Composition (See Section 8.7.1)	Notes
C	0.17	0.18	0.18	0.17	0.17	0.15	0.15	0.15	0.03	–
Mn	1.15	1.25	1.25	1.25	1.45	1.45	1.50	1.50	0.30	–
Si	0.40	0.40	0.40	0.40	0.40	0.35	0.35	0.35	0.25	–
P	0.020	0.020	0.020	0.020	0.020	0.020	0.020	0.020	–	–
S	0.015	0.015	0.015	0.015	0.015	0.015	0.015	0.015	–	1
V	0.08	0.08	0.08	0.08	0.08	0.08	0.08	0.08	0.02	2
Nb	0.05	0.05	0.05	0.05	0.05	0.05	0.05	0.05	0.02	2
Ti	0.04	0.04	0.04	0.04	0.04	0.04	0.04	0.04	0.02	2
Cr	0.20	0.20	0.20	0.20	0.20	0.20	0.20	0.20	0.05	3
Mo	0.10	0.10	0.10	0.10	0.10	0.10	0.10	0.10	0.05	3
Ni	0.30	0.30	0.30	0.30	0.30	0.30	0.30	0.30	0.10	3
Cu	0.25	0.25	0.25	0.25	0.25	0.25	0.25	0.25	0.10	3

Al	0.05	0.05	0.05	0.05	0.05	0.05	0.05	0.05	–	4
N	0.012	0.012	0.012	0.012	0.012	0.012	0.012	0.012	–	4
B	0.0005	0.0005	0.0005	0.0005	0.0005	0.0005	0.0005	0.0005	–	–
Ca	0.006	0.006	0.006	0.006	0.006	0.006	0.006	0.006	–	–
CE	0.39	0.39	0.39	0.39	0.39	0.39	0.39	0.39	–	5
Pcm	0.21	0.21	0.21	0.21	0.21	0.21	0.21	0.21	–	6

Notes:
1. For HFW pipes, sulfur content should not exceed 0.005%.
2. V + Nb + Ti should not exceed 0.12%.
3. Cr + Mo + Ni + Cu should not exceed 0.6%.
4. The total Al:N ratio should not be less than 2:1.
5. $CE = C + \frac{Mn}{6} + \frac{(Cr + Mo + V)}{5} + \frac{(Ni + Cu)}{15}$ (in case of carbon content greater than 0.12%)
6. $Pcm = C + \frac{Si}{30} + \frac{(Mn + Cu + Cr)}{20} + \frac{Ni}{60} + \frac{Mo}{15} + \frac{V}{10} + 5B$ (in case of carbon content less than or equal to 0.12%)

TABLE 8.2 Chemical Requirements for Product Analysis by Percentage of Weight (Sour Services)

Maximum Permitted Alloy Content (wt%)										
Element	GR. B	X-42	X-46	X-52	X-56	X-60	X-65	X-70	Maximum Variation On Agreed Composition (See Section 8.7.1)	Notes
C	0.17	0.18	0.18	0.17	0.17	0.15	0.15	0.15	0.03	–
Mn	1.15	1.25	1.25	1.25	1.30	1.30	1.30	1.30	0.30	1
Si	0.40	0.40	0.40	0.40	0.40	0.35	0.35	0.35	0.25	–
P	0.015	0.015	0.015	0.015	0.015	0.015	0.015	0.015	–	–
S	0.003	0.003	0.003	0.003	0.003	0.003	0.003	0.003	–	–
V	0.08	0.08	0.08	0.08	0.08	0.08	0.08	0.08	0.02	2
Nb	0.05	0.05	0.05	0.05	0.05	0.05	0.05	0.05	0.02	2
Ti	0.04	0.04	0.04	0.04	0.04	0.04	0.04	0.04	0.02	2
Cr	0.20	0.20	0.20	0.20	0.20	0.20	0.20	0.20	0.05	3
Mo	0.10	0.10	0.10	0.10	0.10	0.10	0.10	0.10	0.05	3
Ni	0.35	0.35	0.35	0.35	0.35	0.35	0.35	0.35	0.10	3

Cu	0.35	0.35	0.35	0.35	0.35	0.35	0.35	0.35	0.10	3
Al	0.05	0.05	0.05	0.05	0.05	0.05	0.05	0.05	–	4
N	0.012	0.012	0.012	0.012	0.012	0.012	0.012	0.012	–	4
B	0.0005	0.0005	0.0005	0.0005	0.0005	0.0005	0.0005	0.0005	–	–
Ca	0.006	0.006	0.006	0.006	0.006	0.006	0.006	0.006	–	5
CE	0.39	0.39	0.39	0.39	0.39	0.39	0.39	0.39	–	6
Pcm	0.21	0.21	0.21	0.21	0.21	0.21	0.21	0.21	–	7

Notes:
1. For SAW pipes in X-70 grade, the maximum MN content may be increased to 1.4 wt%.
2. V + Nb + Ti should not exceed 0.15%.
3. Cr + Mo + Ni + Cu should not exceed 0.60%.
4. The total Al:N ratio should not be less than 2:1.
5. Calcium should be 2 × sulfur content for sulfur in the range 0.0015–0.003%.
6. $CE = C + \frac{Mn}{6} + \frac{(Cr + Mo + V)}{5} + \frac{(Ni + Cu)}{15}$ (in case of carbon content greater than 012%)
7. $Pcm = C + \frac{Si}{30} + \frac{(Mn + Cu + Cr)}{20} + \frac{Ni}{60} + \frac{Mo}{15} + \frac{V}{10} + 5B$ (in case of carbon content less than or equal to 0.12%)

8.7.1.1 Elements Analyzed

For Grade B pipe only C, Mn, Si, S and P levels should be determined. For higher grades, analysis for all elements in Table 8.1 or Table 8.2 should be performed.

8.7.2 Mechanical Properties

8.7.2.1 Tensile Properties

Grades B, X-42, X-46, X-52, X-56, X-60, X-65 and X-70 should conform to the tensile requirements specified in Table 3B. For all pipes, the ratio of body yield strength to body ultimate tensile strength should not exceed 0.90.

The yield strength should be the tensile stress required to produce a total elongation of 0.5% of the gage length as determined by an extensometer. When elongation is recorded or reported, the record or report should show the nominal width of the test specimen when strip specimens are used and the diameter and gage length when round bar specimens are used, or should state when full-section specimens are used.

The required minimum tensile elongation should be determined according to the formula given in footnote a of Table 3B but should be not less than 20%.

8.7.2.2 Flattening Test Acceptance Criteria

Acceptance criteria for flattening tests of electric welded pipes should be as follows:

No cracks or breaks should occur in either weld or parent metal during flattening of the test specimen to 50% of its original OD. The specimen should be further flattened to one-third of original OD without cracks or breaks other than in the weld. The presence of lamination or burnt metal should not become apparent during the entire test.

8.7.2.3 Fracture Toughness Tests

For all pipes, Charpy V-notch tests should be performed on each test ring taken for tensile testing and tested in accordance with Standards, SR 5. Evidence should be required to show that the notch ductility of the pipe is adequate for service conditions.

If stress relieving is required for field welds, Charpy testing is also required in the simulated stress-relieved condition. The purchaser/company should inform the manufacturer at the time of enquiry/order of the need for testing in the stress-relieved condition.

For gas transmission lines with a pipe diameter of DN 400 (NPS 16) or greater and grade X-52 and higher, DWTTs should be carried out on one pipe from each heat of steel in accordance with the requirements of standards.

8.7.2.4 Metallographic Examination

Specimens for metallographic examination should be extracted from HFW and SAW pipe such that the weld, complete heat-treated zone, and parent material on both sides of the weld are visible over the full-wall thickness. A minimum of one specimen under nonsour service and two specimens under sour service should be microscopically examined from one pipe in each heat, or after each break in production, whichever is the more frequent. The examination should determine the adequacy of microstructure and heat treatment for HFW pipe and proper fusion throughout the full thickness of the joint for SAW pipe.

8.7.2.5 Hardness Testing

Hardness testing should be performed in accordance with Standards. The hardness of weld, HAZ, and base material should not exceed 280 HV10 for pipes under nonsour conditions and 248 HV10 for pipes under sour conditions.

8.7.2.6 Preparation of Samples

Samples removed for the determination of tensile, toughness, or microstructural properties should be prepared by machining. Where thermal cutting has been used to remove pipe coupons from which test specimens are prepared, the full extent of the heat affected region should be removed during machining of the specimen (see also Section 8.9.2).

For pipes under sour condition, HIC and SSC tests should be performed.

8.8 DIMENSIONS, WEIGHTS, LENGTHS, DEFECTS, AND END FINISHES

8.8.1 Diameter

The outside diameter of the pipe body, as measured by taping the circumference, should not deviate from the value given in standards by more than the tolerances given below:

Size Designation		Minus Tolerance	Plus Tolerance
<DN 60	(NPS 2 3/8)	0.8 mm	0.4
≥ DN 60 (NPS 2 3/8) and ≤ DN 400 (NPS 16)		0.75% OD	0.75% OD
>DN 400	(NPS 16)	3 mm	3 mm

For a length of 100 mm from each pipe end, the average internal diameter should not deviate from the nominal internal diameter by more than the tolerances given below:

Size Designation	Type of Pipe	Minus Tolerance (mm)	Plus Tolerance (mm)
≤DN 250 (NPS 10)	Welded	0.5	1.5
	SMLS	1.0	2.0
>DN 250 (NPS 10) and ≤DN 500 (NPS 20)	Welded	1.0	1.5
	SMLS	1.5	2.0
>DN 500 (NPS 20)	Welded	1.5	1.5
	SMLS	2.0	2.0

Notes:

1. The nominal internal diameter is defined as the outside diameter D (given in Metric Tables in Standards) minus twice the nominal wall thickness.
2. The internal diameter should be measured using an internal gage or a measuring tape inside the pipe. The method and equipment should be approved by the purchaser/company. For pipe of DN 200 (NPS 8) and smaller, the internal diameter may be calculated by measuring the outside diameter with a circumference tape and subtracting twice the actual wall thickness from this value.
3. The end of each pipe should be tested for out-of-roundness using an internal ring. For 32″ and higher the diameter of ring should be 0.6% of nominal O.D. less than nominal inside diameter and for less than 32″ the diameter of ring should be 5 mm less than nominal inside diameter of pipe. The gage should pass freely into each end of the pipe through 100 mm distance when held normal to the pipe axis.
4. On welded-expanded pipe, the internal diameter of one end of the pipe should not differ by more than 2 mm from that of the other end.

Any pipe found to be out of tolerance should be the cause for individual diameter measurement of all pipe back to the last, and up to the next two sequential pipes measured and found to be within tolerance.

8.8.2 Wall Thickness

Each length of pipe should be measured for conformance with the specified wall thickness requirements. For all sizes and grades of welded pipe, the wall thickness at any place in the pipe measured during inspection should not deviate from the nominal wall thickness by more than the tolerances specified in the following Table (except that the weld area should not be limited by the plus tolerance):

Wall Thickness, t (mm)	Minus Tolerance	Plus Tolerance
$t \leq 7$	0.35 mm	10%
$7 < t \leq 10$	5%	10%
$t > 10$	0.5 mm	10%

For all sizes and grades of seamless pipe, the wall thickness at any point should not deviate from the nominal thickness by more than +15% or −10%.

8.8.3 Length

Unless otherwise indicated in the purchase order, pipe should be supplied in the following lengths:

The average length of pipes in one order should be not less than 11.6 m with a minimum of 95% of pipes between 11 and 12.2 m in length.

No pipe should be less than 10 m in length.

No pipe should be greater than 12.8 m in length.

Note:

For heavy-wall seamless pipe, where supply of the pipe lengths stated above may not be possible, the purchaser and the manufacturer should agree on an alternative pipe length.

- **Straightness**

  Deviation from a straight line should not exceed 0.15% of the length.
- **Offset of plate edges**

  The radial offset of plate edges (misalignment) for pipe with a nominal wall thickness of 10 mm or less should be no more than 1.0 mm for SAW pipe and 0.5 mm for HFW pipe. For pipe with a nominal wall thickness greater than 10 mm, the maximum allowable radial offset should be 1.6 mm for SAW pipe and 5% of the nominal wall thickness for HFW pipe.
- **Out-of-line (misalignment) weld bead for pipe with filler metal welds**

  Any misalignment of the weld beads of SAW pipe should not exceed the values given below:

Specified Wall Thickness (mm)	Maximum Misalignment of the Weld Beads (mm)
$T \leq 20$	3
$T > 20$	4

8.8.4 Height of Outside and Inside Weld Beads—SAW

The height of outside and inside weld beads should not exceed 3 mm.

At pipe ends and other areas that will be radiographed, the reinforcement of both inside and outside bead should allow radiographic sensitivity requirements of Section 8.9.7 to be met.

8.8.5 Height of Flash of HFW Pipe

The inside flash of HFW pipe should not extend above the prolongation of the original inside surface by more than (0.3 mm + 5% of nominal wall thickness).

8.8.6 Hard Spots

Any hard spot with a hardness greater than 280 HV10 for pipes under nonsour conditions, and 248 HV10 for pipes under sour conditions should be rejected.

8.8.7 Laminations

Any laminations or inclusion extending into the face or bevel of the pipe should be considered as a defect.

8.8.8 Other Defects

The ends of each SAW pipe and two positions along the length should be checked for out-of-roundness at the position of the longitudinal weld. Templates with a minimum chord length of 75% of the pipe's internal diameter should be used for measurement of local irregularity in profiles. For pipes of DN 400 and above, a template of 300 mm minimum chord length should be used. The template profile should have a radius equal to the nominal radius of the pipe outer or inner circumference for measurement of the outer or inner surface, respectively. The nominal inner radius should be taken as the nominal outer radius minus twice the nominal wall thickness.

The template gaging surface should have an appropriate cut-out to accommodate the weld bead of the pipe. The cut-out should be at the center of the gaging surface and should have a width of less than the weld bead width plus 5 mm. Any local irregularity should be measured by a calibrated taper gage inserted in any gap between the template and the pipe surface. The local irregularity should not exceed 1.5 mm.

A cold-formed dent with a sharp bottom gouge and all sharp gouges (without dents) deeper than 1.0 mm should be considered defects requiring rectification or rejection.

8.8.9 Pipe Ends

8.8.9.1 Plain Ends

The entire end bevel should be machined with special care exercised to keep the root face as close to 1.59 mm as possible; i.e., with minimum use of ± 0.79 mm tolerance. If, however, the root face is less than 0.8 mm or more than 2.38 mm, it should not be brought into tolerance by filing or grinding, which means the entire bevel should be remachined.

For wall thickness greater than 22 mm (7/8″), the ends should be beveled (Fig. 8.1).

8.9 INSPECTION AND TESTING

8.9.1 Test Equipment

Measuring equipment for inspection and testing should be selected such that it has a resolution and accuracy at least 5 × finer than the tolerance of the parameter being measured. Similarly, standards against which a piece of

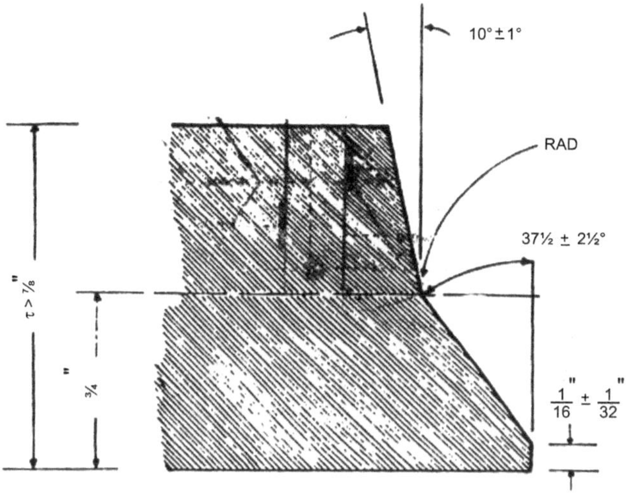

FIGURE 8.1 End preparation for pipe and fittings over 22 mm (7/8″) thickness.

equipment is calibrated should be at least 5 × as accurate as the equipment being calibrated.

Only measuring equipment that can be demonstrated to have been previously calibrated satisfactorily and still be within its documented calibration period or interval should be used for inspection and testing.

8.9.2 Testing of Chemical Composition

8.9.2.1 Sampling Methods

Samples may be taken using any of the methods indicated in API Spec. 5L provided that they are taken from finished pipe. In this instance, finished pipe should mean the pipe that has been formed and welded (if applicable) but before trimming to final pipe lengths (see also Section 8.7.2.6).

8.9.3 Testing of Mechanical Properties

8.9.3.1 Tensile Test Specimens

Tensile properties should be determined from specimens removed from pipe that has been subjected to all mechanical and heat-treatment operations. Where stress relieving of pipe is required to be performed, e.g., after field welding, additional tensile testing of parent metal and weldments should be performed on stress-relieved specimens. The company should specify on the purchase order if this requirement applies.

The testing procedure should be in accordance with ASTM A 370 (see API Spec. 5L, Clause 9.8.2.1).

8.9.3.2 Tensile Testing Frequency

Tensile tests should be performed on samples taken from two pipes per heat. For heats less than 100 tons, tests on only one pipe should be required.

8.9.3.3 Transverse Tensile Tests

The yield strength, ultimate tensile strength, and elongation values should be determined on flattened rectangular specimens.

8.9.3.4 Weld Tensile Tests

Weld tensile specimens should be taken from the same part of the pipe used for preparing parent metal tensile specimens. The weld reinforcements should be removed before tensile testing.

8.9.4 Fracture Toughness Tests

8.9.4.1 Charpy Test Specimens

Impact testing should be carried out using 10 × 10 or 10 × 7.5 or 10 × 5 mm cross-section specimens with or without tapered ends. The largest possible specimen should be used. Where the pipe dimensions are insufficient to extract a 10 × 5 mm. specimen, impact testing is not required.

For pipes of DN 150(NPS 10) or less, impact test specimens should be taken parallel to the axis of the pipe (i.e., longitudinal specimens should be taken).

For pipes greater than DN 250 (NPS 10), impact test specimens should be taken transverse to the axis of the pipe, except where the wall thickness prevents extraction of a 10 × 5 mm specimen, in which case longitudinal specimens should be taken.

For weld center line and HAZ impact tests, only transverse specimens should be used.

8.9.4.2 Charpy Testing Frequency

For all pipes, Charpy V-notch tests should be performed on each test ring taken for tensile testing and tested.

8.9.4.3 Drop-Weight Tear Test Specimen and Test Frequency

For gas transmission lines with a pipe diameter of DN 400 (NPS 16) or greater, DWTTS should be carried out on one pipe from each heat of steel.

8.9.5 Hydrostatic Tests

8.9.5.1 Hydrostatic Test Requirements

Each length of pipe should withstand, without leakage, an inspection hydrostatic test to at least the pressure specified in Section 8.9.5.2. Hydrostatic testing should be performed after cold expansion (if applicable). The test pressure for all sizes and types of pipe should be held for no fewer than 10 seconds.

8.9.5.2 Test Pressure

The test pressure for all types and sizes of pipe should be such that the hoop stress, calculated on the basis of the minimum specified wall thickness and including stresses from end loading, is at least 95% of the SMYS.

If applied, the end load compensation factor as determined by the formula given in Appendix K of API Spec. 5L should be used.

8.9.5.3 Visual Inspection

The full body and welds (if applicable) of every pipe should be examined, internally and externally, for surface defects. For internal examination of pipe DN 600 (NPS 24) and larger, the inspector should pass through the bore of the pipe. Adequate illumination should be provided to enable proper inspection.

8.9.6 Nondestructive Inspection

8.9.6.1 Methods of Inspection

All personnel performing NDT activities should be qualified in the technique applied, in accordance with ISO 9712 or equivalent.

A level II inspector is required for shift supervision, manual weld inspection, and calibration of all systems (both manual and automated).

A level I inspector is acceptable for all other NDT methods. A level II inspector is acceptable for supervision of all other NDT methods.

All NDT should be performed in accordance with written procedures.

These procedures should have the prior approval of the purchaser/company.

NDT for acceptance of the pipe (final inspection) should take place after all heat treating and expansion operations and, for welded pipe, after hydrostatic testing of the pipe. It may, however, take place before cropping, beveling, and end sizing.

Submerged arc welds should be inspected over their entire length, for both longitudinal and transverse defects, using ultrasonic examination. In addition, each end of the weld seam should be examined radiographically for a distance of at least 230 mm.

HFW pipe welds should be examined for longitudinal defects over their entire length by ultrasonic methods.

8.9.6.2 Ultrasonic Lamination Testing of Each Seamless Pipe Body

This should be performed using a helical pattern with at least 25% scanning coverage of the pipe surface.

Ultrasonic thickness testing of the pipe body and ends of seamless pipe should be performed by scanning along a helical or straight pattern in such a way that at least 10% of the pipe surface is covered.

The body and ends of all seamless pipes should be 100% ultrasonically tested for inside and outside surface defects as well as transverse, longitudinal, and inclined-embedded defects.

EMT may be applied for nominal wall thicknesses less than 6 mm.

8.9.6.3 Ultrasonic Lamination Testing of HFW Pipe

Each plate or strip rolled should be ultrasonically tested for laminations using an oscillating scanning pattern. The scanning coverage using this technique should be a minimum of 12.5%.

Alternatively, the scanning should be executed along straight, evenly distributed parallel lines with a scanning coverage of at least 25%. Coil for HFW pipe. HFW pipe may be tested after welding of the longitudinal seam by rotary ultrasonic testing of the pipe body. The coverage in this case should be 100%.

In addition, the longitudinal edges of a plate or coiled strip should be 100% ultrasonically tested, over a width of at least 25 mm from the trimmed plate/coil edge. This may be performed either before or after pipe forming. For HFW pipe subjected to 100% rotary ultrasonic testing of the pipe body, strip edge testing is not required.

8.9.6.4 Pipe Ends

After beveling, the complete circumference of the pipe end should be tested ultrasonically from the inside for laminations covering a width that includes the entire bevel. Alternatively, the pipe may be tested from the outside prior to beveling, in which case a band of at least 25 mm wide, to include the eventual beveled area, should be tested.

If UT has not been performed from the outside before beveling, and if UT from the inside is not feasible due to dimensional limitations, then MT or PT should be applied to the bevel face. Defects visible to the naked eye such as laps, cracks, or laminations should not be permitted.

8.9.7 Radiological Inspection

8.9.7.1 Radiological Inspection Equipment

The radiographic examination should be executed with X-ray equipment using fine-grain type film and lead-intensifying screens.

For acceptance of the radiographic films, the technique used should result in a sensitivity better than 2% of the thickness of the weld metal and in a relative film density of 2.0 to 3.5 in the weld metal.

The manufacturer should record on a review form accompanying the radiograph or within the mill computer system, the interpretation of each radiograph and disposition of the pipe inspected.

8.9.7.2 NDT Reference Standards

The penetrameter used should be of the wire type in accordance with ISO 1027. The selection of penetrameter wire diameters should be based on a sensitivity of 2% of weld metal thickness.

8.9.7.3 Imperfections Observed During Radiographical Inspection

In addition to the acceptance limits, the following limits should be deemed unacceptable, if exceeded:

1. Any total area of porosity projected radially through the weld equal to 6.5 mm^2 (or equivalent to three 1.5 mm in diameter) in any 645 mm^2 of projected weld area.
2. Any single slag inclusion with a length of 3 mm or a total cumulative length of slag inclusion equal to 6.3 mm in any 150-mm length of the weld. A slag inclusion of 1.6 mm in width.

The stringiest of all specified limits should prevail.

8.9.8 Ultrasonic and Electromagnetic Inspection

8.9.8.1 Ultrasonic Equipment

The automatic ultrasonic equipment should incorporate:

1. A device which monitors the effectiveness of the coupling.

   In case where a zero degree compression wave probe is used to monitor coupling, or where a through transmission technique is used for seamless or HFW pipe, loss of coupling exists when the sensitivity (echo height) decreases by more than 10 dB relative to the static calibration.

   In case where a through transmission technique through the weld seam is used for SAW line pipe, loss of coupling exists when the signal drops below the electronic noise level plus 10 dB at the position of the through transmission signal.

A clear acoustic warning system and an automatic paint spray system (or equivalent) should be activated when loss of coupling occurs.

2. An automatic paint-spraying device or equivalent system, which is activated when the received ultrasonic echo exceeds the preset acceptance limit. This alarm should operate without any interference of the ultrasonic operator and should be applied within a 25-mm advancement past the detected defect. The reset time of the alarm system, after detection of a defect, to be again available for detection, should be shorter than the time needed for 25 mm advancement in the scanning direction.
3. An automatic weld tracking system for correct positioning of the crystal s/probes with respect to the weld center of all welded pipes.

Entrance angles of shear wave probes should be as follows:

Seamless pipe	45 (40–48) degrees
HFW pipe	45 (40–48) degrees
SAW pipe—longitudinal defect detection	50–70 degrees
SAW pipe—transverse defect detection	45 (40–48) degrees (on weld bead) 50–70 degrees (X or K transmission)

Lamination testing may be performed in pulse echo or transmission mode. If pulse echo mode is employed for detecting lamination in plates or strips, the probe should be applied on the side of the plate or strip, which will eventually be on the inside of the finished pipe. Wall thickness measurement should be done only in pulse echo mode. The probe(s) used for wall thickness/lamination checks should satisfy the following requirements:

• Twin crystal probes	The focal length should be 50% of the wall thickness
• Single crystal probes	The near surface resolution should be better than 25% of the wall in pulse echo mode: thickness, measured at the primary reference sensitivity level

The transducer arrangement should be such that sound intensity in both the longitudinal and circumferential directions does not decrease by more than 3 dB at any point in the pipe wall, referred to the maximum sound intensity adjusted in the static calibration.

The equipment should be checked with an applicable reference standard (test piece) as described in Fig. 8.2A and B in Section 8.9.7.2 at every 4 hours and at the beginning and end of a batch in order to demonstrate the effectiveness of the inspection procedures and the correct functioning of the equipment.

In case discrepancies of more than 3 dB occur, then all pipes, plates, or strips inspected since the previous check should be reinspected. Proper functioning of the UT equipment and the linearity of the electronic instrumentation should be checked once every 6 months or when a change is made to the equipment.

From each pipe under test, an automatic "on-line" record should be made without operator intervention. For every pipe, a summary record should be made showing pipe identification number, time and examination results, including reexaminations.

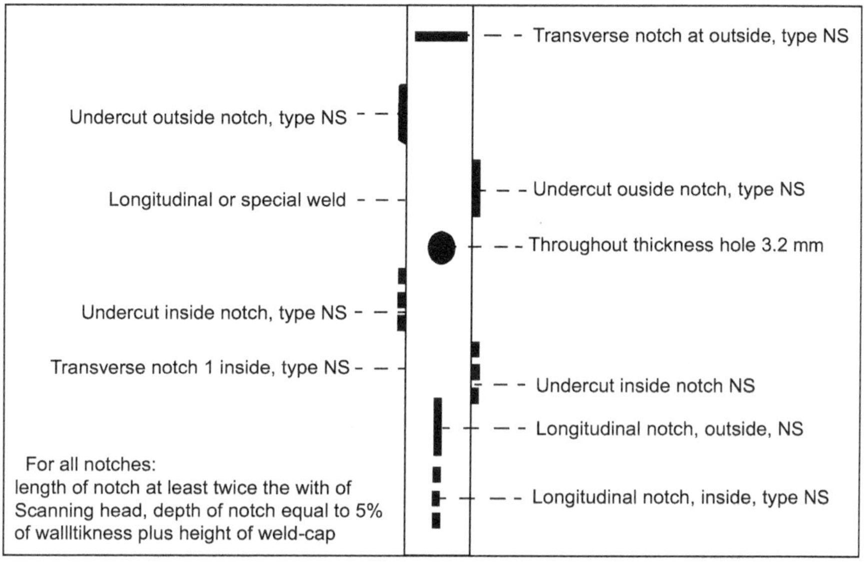

FIGURE 8.2 Reference standard pipe/plate for inspection of welds type and size of notches should be as given in Section 8.9.8.3. (A) Test plate or section for saw pipe**. (B) Test pipe or section for HFW pipe**. **The location of each reference standard may be at the manufacturer's option, provided that no interference will occur.

If parts of the finished pipe ends are not covered by an automatic UT system (untested area), manual ultrasonics should be carried out using approved procedures for manual ultrasonic examination based on the requirements given above.

The complete circumference of seamless pipe ends or rotary-tested HFW pipe ends should be tested manually over the length of the untested area plus 25 mm overlap of the automatically tested area.

8.9.8.2 Electromagnetic equipment

If permitted by the company, EMT methods such as Eddy current testing or magnetic flux leakage testing may be applied for surface defect detection in seamless pipe.

EMT should be performed in accordance with ASTM E570. Testing should be performed by automatic equipment over the entire surface of the pipe.

If parts of the finished pipe ends are not covered by an automatic EMT system (untested area), then manual ultrasonic examination based on the requirements given above should be performed. The complete circumference of the pipe ends should be tested by manual UT, over the length of the untested area plus 25 mm overlap of the automatically tested area.

8.9.8.3 NDT Reference Standards

The reference (calibration) standard should have the same specified diameter and thickness as the product being inspected and should be of sufficient length to permit calibration of ultrasonic inspection equipment at the speed to be used in production. The reference standard should also be of the same material type and have the same surface finish and heat treatment as the product to be inspected. It should be free from discontinuities or other conditions giving indications that may interfere with detection of the reference reflectors.

The reference standard should contain machined notches (N5 or N10) or radially drilled holes (3.2 mm), and/or flat-bottomed holes. The type and location of the notches and drilled holes in the reference standard for welded pipe, should be in accordance with standards.

The manufacturer may use a type of reference reflector not specified above, provided he can demonstrate to the company that the examination is at least as sensitive as prescribed in Standard. In such cases the manufacturer should obtain the approval of the company.

The primary reference sensitivity level should be adjusted on the following reference reflectors:

Type of Examination	Type of Pipe		
	SMLS	SAW	HFW
Lamination detection	FBH 6.3 mm	FBH 6.3 mm	FBH 6.3 mm
Surface defect detection	Notch N5		
Defect detection (body and pipe ends)	Notch N5		
Defect detection (weld)		RDH 3.2 mm	Notch N10
Defect detection (plate and axial defect)		Notch N5	

For all reference reflectors, except for RDH 3.2 mm, the acceptance-limit signal should be equal to the primary sensitivity level, i.e., equal to the height of the signal produced by the reference reflector. For the RDH 3.2-mm reference reflector, the acceptance limit signal should be 10 dB below the primary reference sensitivity level.

All sensitivity adjustments should be carried out dynamically.

Flat-bottomed holes for lamination detection should be drilled to the mid-wall position.

8.9.8.4 Acceptance Limits

For all examination types, indications exceeding the acceptance limit signals are unacceptable.

For lamination detection in plate/coil, seamless pipe body and pipe ends, the acceptance limits should be based on the lamination size and frequency and be in accordance with classification of SEL-072 (German Standards DIN), as described below:

Location	SEL-072 Lamination Acceptance Levels
Plate/coil body	Table 1 Class 3
Plate/coil edges	Table 2 Class 1
Seamless pipe body	Table 1 Class 3

8.9.9 Disposition of Defects

In all cases where grinding repairs are made as a result of imperfections being disclosed by NDT, the part of the pipe containing such repairs should be subjected to additional NDT using the same technique, and MT, after the grinding operation.

If more than 10% of the pipe in the final inspection on any one day fails to meet any or all of the foregoing requirements, production should cease pending establishment and rectification of the cause of defects.

8.10 COATING AND PROTECTION

8.10.1 Coatings

The pipe should be supplied bare and unoiled, protective coating or varnishing of the pipe identity markings is, however, permitted.

The company reserves the right to establish a color code or marking system to identify both the pipe mill and wall thickness.

8.10.2 Bevel Protectors

If specified in the purchase order, suitable bevel protectors should be used at both ends of the pipe to protect the bevels from damage under normal handling and transportation condition.

8.11 MANUFACTURING PROCEDURE AND WELDING PROCEDURE

8.11.1 Manufacturing Procedure Specification

The manufacturer should produce a manufacturing procedure specification that should be submitted for the company's approval at least two weeks prior to the start of production. The manufacturing procedure specification should include the following as a minimum:

- Steel Supply
  - Steelmaker

- Steelmaking and casting techniques. For pipes under sour conditions, the following details should be also specified:
  - Details of steelmaking process, including deoxidation and desulfurization practice, inclusion shape control method, and the use of vacuum degassing.
  - Details of casting process, i.e., ingot or continuous casting, including casting speed, tundish superheat, segregation control measures, etc.
  - Details of plate and strip manufacture, including slab-reheating temperatures, start and finish rolling temperatures and reduction rations.
  - Heat-treatment details
- Details of chemical composition including:
  - target chemistry;
  - ranges for deliberately added elements;

- Seamless Pipe
  - Pipe forming procedure
  - Pipe heat treatment procedure
  - Hydrostatic test procedure
  - NDT procedure
- HFW Pipe
  - Strip manufacturing method including details of rolling, skelp splitting, and any specialized cooling and heat treatment.
  - Strip NDT procedures
  - Pipe manufacturing method including details of methods used for preparing the edge of the strip for welding and for control of misalignment of edges and pipe shape.
  - Welding procedure including details of the following:
    - Methods to be used for heating plate edges and for the control and monitoring of power input in relation to the temperature of the pipe surface and to the speed of the pipe
    - Frequency (in KHz) of the welding power supply
    - Details of any protective atmosphere used for welding
    - Methods used to accomplish and control two upset-forge welding of the heated pipe edges
    - Method used for trimming of the weld bead
    - Weld seam heat-treatment procedure
    - Hydrostatic test procedure
    - NDT procedure
- SAW Pipe
  - Plate manufacturing method including details of specialized cooling and heat treatment
  - Plate NDT procedures
  - Pipe forming procedure

- Seam welding procedure including details of the following:
  - Method of alignment, clamping, and tack welding (if any) of the joints to be welded together with details of run-on and run-off tabs to be used and the method of their attachment to the pipe.
  - For pipe made by the cage-forming process, details of the methods used to maintain the alignment of the inside and outside welds
  - Welding process
  - Brand name, classification, size and grade of filler metal and flux
  - Speed of welding
  - Number of electrodes and polarity for each electrode
  - Welding current for each wire
  - Welding voltage for each wire
  - Dimensions of welding preparation
  - Number of weld passes and their disposition
  - Details of tracking system for both inside and outside welding and also method for checking the setup of the system
  - Limits on internal and external weld reinforcement
  - Repair welding procedure
- The method and degree of expansion to be applied
- Pipe heat-treatment procedure (when appropriate)
- Hydrostatic test procedure
- NDT procedure

8.11.2 Welding Procedure Qualification

The seam welding procedure for welded pipe may be qualified by the first day of production tests. If qualification prior to the start of production is required, the company should notify the manufacturer at the time of enquiry/order.

For qualification of the welding procedure, the following tests should be executed on a full-length test weld made in accordance with the manufacturing procedure specification.

8.11.2.1 UT

The weld seam should be examined in accordance with Section 8.9.8. This should be performed at least 48 hours after completion of the test weld.

8.11.2.2 RT

The complete welded seam of SAW pipe should be examined in accordance with Section 8.9.7.

8.11.2.3 PT or MT

The weld seam should be subjected to PT or MT in order to check for surface defects in the weld material in accordance with Section 8.12.2.3.

8.11.2.4 All Weld Tensile Tests (SAW Only)

One specimen of the weld should be subjected to an all-weld tensile test. Test results should meet the minimum specified requirements of the plate with regard to yield, tensile strength, and elongation.

8.11.2.5 Macrographic, Micrographic, and Hardness Testing

A specimen should be removed from the weld seam and subjected to macrographic, micrographic, and hardness testing.

8.12 FIRST-DAY PRODUCTION TESTS

Three of the completely finished pipes of the first day's production should be selected at random for testing to verify that the submitted manufacturing procedure results in fully acceptable pipe. If more than one heat is used in the first-day production pipes, at least two heats should be represented by the tests pipes. At the company discretion, the company may make the selection. For orders of fewer than 50 tons, first-day production tests are not required.

If the pipes have been made from coiled skelp, the pipes made from each end of the coil should be tested in addition to the above pipes.

The pipes tested as above should be considered to be test pipe(s) per heat or per shift as required by standards. The above, first-day production test should be repeated after any change in the manufacturing procedure or interruption to the program.

The manufacturer should submit to the company a report giving the results of all tests indicated below together with macrographs of the weld cross section and micrographs confirming the microstructure of the plate and seamless pipe.

8.12.1 Visual Examination

All pipes should be examined visually for dimensional tolerances and for surface defects.

8.12.2 Nondestructive Testing

8.12.2.1 UT

The weld seams of all pipes should be examined by means of an automatic ultrasonic scanning device in accordance with Section 8.9.8.1 and should meet the requirements of Section 8.9.8.4.

8.12.2.2 RT

The weld seams of all SAW pipes should be radiographically examined throughout their full length in accordance with Section 8.9.7.

8.12.2.3 PT or MT

The weld seams of all welded pipes greater than or equal to DN 600 (NPS 24) should be subjected to PT or MT, throughout their full length both inside and outside, to check for longitudinal and transverse surface defects in the weld material.

For pipe less than DN 600 (NPS 24), the full length of the weld seam outside surface, plus the equivalent length of one pipe diameter each end of the internal surface, should be examined. Seamless pipe should also be subjected to PT or MT over the entire outside pipe body.

PT should be in accordance with ASTM E165.

Cracks are unacceptable and their causes should be investigated.

8.12.3 Physical Testing

The physical properties of all pipes should be tested as specified below. Test results should meet the requirements for the specified grade and type of pipe.

8.12.3.1 Weld Seam

The weld seam of all selected welded pipes should be physically tested. For SAW pipe, in addition, an all-weld metal tensile test should be made including the determination of tensile strength, yield strength and elongation. For determination of the elongation value, the "Oliver" formula, as specified in ISO 2566-1, may be used. Results of the all-weld metal tensile tests should meet the minimum specified requirements of the plate from which the pipe is made. For SAW pipe, in addition, weld impact tests should be carried out in accordance with standards.

8.12.3.2 Pipe Material

Tensile tests should be carried out on the two pipes made from each end of a coiled skelp, or on two pipes made from different heats, as required by Sections 6.2, 9.3, 9.8, 9.9 and 9.10, except that for pipes greater than DN 200 (NPS 8) tensile tests should be performed in both the transverse and longitudinal directions.

8.12.3.3 Charpy Impact Test

Tests should be carried out on all selected pipes in accordance with standards. In addition, full transition temperature curves should be produced, showing impact energy (in J) and percentage shear (fibrous) of the fracture surface, plotted against temperature, over a temperature range sufficient to reproduce fracture acceptance from 10% to 100% fibrous shear.

8.12.3.4 Drop-Weight Tear Test

For pipe to be used in gas transmission lines, drop-weight tear tests should be carried out in accordance with standards.

8.12.3.5 HIC and SSC Tests

For pipe under sour services, HIC and SSC tests should be performed as specified in Appendices "N" and "O."

8.12.4 Macrographic, Micrographic, and Hardness Examination

8.12.4.1 SAW and HFW Pipe

For SAW pipe, three specimens should be extracted from one pipe at three locations along the weld and should be cross sectioned, polished, and etched for macroexamination, which should provide evidence that proper fusion has been achieved throughout the full thickness of the joint, the extent of interpenetration, and the alignment of internal and external weld passes.

For HFW pipe, a total of three specimens should be taken from the selected pipes for microexamination to provide proof that heat treatment of the weld zone has been adequate.

For SAW and HFW pipe, a series of Vickers hardness (HV 10) tests should be made on one of the etched specimens selected by the company. These series of readings should extend from unaffected base metal on one side across the weld to unaffected base metal on the other side. Three traverses should be made, one 2 mm from the outer edge, the second across the center, and the third 2 mm from the inner edge. The spacing between the hardness impressions should be 0.75 mm. The location of the hardness impressions for SAW pipe is shown in Fig. 8.3. The hardness impressions nearest the fusion line should be within 0.5 mm of the fusion line.

8.12.4.2 Seamless Pipe

Three specimens from one pipe should be extracted from locations 120 degrees apart from a position chosen by the company, polished and etched for examination and checked for microstructure. A hardness survey should be made on one of the above specimen selected by the purchaser. Three

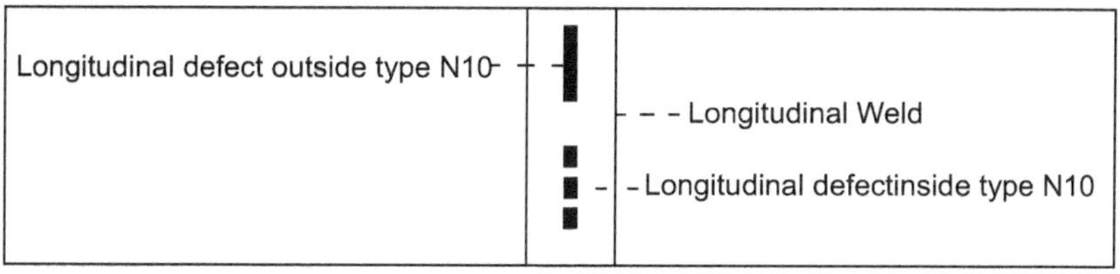

FIGURE 8.3 Longitudinal weld hardness survey.

traverses should be made, one 2 mm from the outer edge, the second across the center, and the third 2 mm from the inner edge. A minimum of 12 readings should be taken at 5-mm intervals.

8.12.4.3 Acceptance Criteria

No hardness measurement should exceed 280 HV10. For pipes under non-sour conditions and 248 HV10 for pipes under sour conditions.

8.13 REPAIR OF DEFECTS BY WELDING (NORMATIVE)

8.13.1 Seamless Pipe and Parent Metal of Welded Pipe

Welding repair on seamless pipe and on parent metal of welded pipe is not acceptable.

8.13.2 Weld Seam of Welded Pipe

Repair of the weld seam of SAW pipe is not acceptable within 200 mm of the bevel ends.

The nature of any weld defect indicated by nondestructive inspection should be ascertained before any repair is performed. Where necessary, complementary ultrasonic and radiographic inspections should be carried out to characterize the defect. Repair welding to rectify pipe welds containing cracks is not permitted.

Repairs to the weld seam should be limited to three per pipe. The length of repair weld should not exceed 5% of the total weld length on each pipe.

Weld repairs should not be carried out after cold expansion or hydrostatic testing of a pipe.

Repair welding should be executed using qualified procedures and in accordance with the requirements of standards.

The repaired area should be nondestructively tested by RT, manual UT and MT.

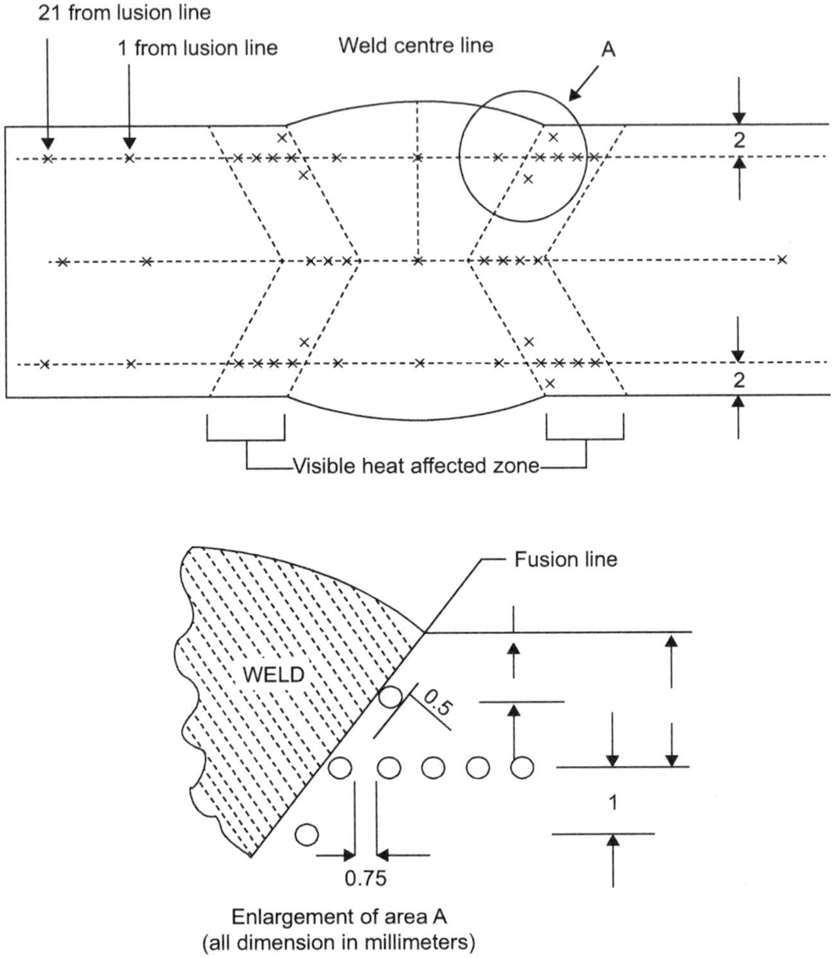

FIGURE 8.4 Location of Charpy V-notch specimens in saw pipe welds.

8.14 REPAIR WELDING PROCEDURE (NORMATIVE)

8.14.1 Charpy V-Notch Impact Test

Charpy V-notch impact testing should be performed on the repair welding procedure qualification test weld. Specimens should be taken from the locations shown in Figs. 8.4 and 8.5. The test temperature and acceptance criteria should be the same as those given for the pipe in standards.

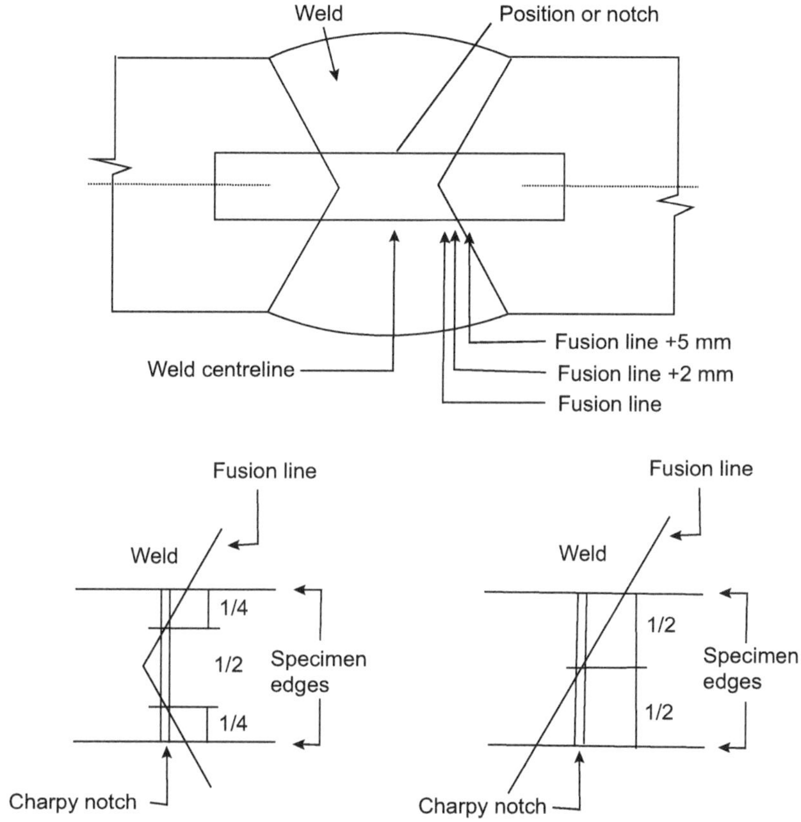

FIGURE 8.5 Detail of fusion line Charpy V-notch location. (A) Symmetric weld and (B) asymmetric weld.

8.15 SUPPLEMENTARY REQUIREMENTS (NORMATIVE)

8.15.1 Fracture Toughness Testing (Charpy V-Notch) for Pipe of Size 4½ or Larger

The fracture toughness of the pipes should be determined by Charpy V-notch impact testing in accordance with ASTM A-370. The impact test temperature should be lower than or equal to that specified in the Table below:

Nominal Wall Thickness Wt (mm)	Test Temperature (°C)	Maximum Test Temperature (°C)
Wt £ 16.0	*T*[a]	0
16.0 < Wt ≤ 25	*T*-10	0

25 < Wt ≤ 32	*T*-20	0
Wt > 32	*T*-30	0

[a]*T is the minimum design temperature, which should be specified in the purchase order. If no minimum design temperature is indicated, it should be taken as 0°C.*

Impact testing should be carried out using 10 × 10 or 10 × 7.5 or 10 × 5-mm cross-section specimens with or without tapered ends. The largest possible specimen should be used.

For pipes of DN 250 (NPS 10) or less, impact test specimens should be taken parallel to the axis of the pipe (i.e., longitudinal specimens should be taken).

For pipes greater than DN 250 (NPS 10), impact test specimens should be taken transverse to the axis of the pipe, except where the wall thickness prevents extraction of a 10 × 5-mm specimen, in which case longitudinal specimens should be taken.

For weld center line and HAZ impact tests, only transverse specimens should be used.

One set of three specimens should be taken from the mid thickness locations in the pipe wall at the following positions (see also Figs. 8.4 and 8.5).

Seamless pipe	– Pipe body
SAW pipe	– Pipe body at 90 degrees to the weld
	– Weld centerline
	– Fusion line
	– Fusion line + 2 mm
	– Fusion line + 5 mm
HFW pipe	– Pipe body at 90 degrees to the weld
	– Weld centerline

The minimum absorbed energy requirements for full size (10 × 10 mm) specimens taken transverse to the pipe axis are given in the Table below:

Grade	Minimum Average Value (J)	Minimum Individual Value (J)
B	27	22
X-42	30	24
X-46	32	25
X-52	36	30
X-56	39	31
X-60	42	35
X-65	45	38
X-70	50	40

For other specimen sizes and orientations, the above values should be multiplied by the following corresponding factors:

Size (mm)	Orientation	Factor
10 × 10	Longitudinal	1.5
10 × 7.5	Transverse	0.75
10 × 7.5	Longitudinal	1.125
10 × 5	Transverse	0.5
10 × 5	Longitudinal	0.75

The shear area at the fracture surface of the test specimens should be recorded. Each sample should exhibit no less than 50% fibrous shear.

The Charpy test requirements specified are based on crack-initiation principles. For gas transmission and two phase lines, higher-absorbed energy requirements may be specified to avoid the risk of running fractures. In this case the purchaser should state the required values in the purchase order.

8.15.2 Drop-Weight Tear Testing

Drop-weight tear testing on welded pipe sizes DN 400 (NPS 16) and larger, grade X-52 and higher.

Drop-weight tear tests are required on pipes of DN 400 (NPS 16) and larger.

Two transverse DWTT specimens should be taken from one length of pipe from each heat supplied in the order.

Tests should be performed at the minimum design temperature. If no minimum design temperature is specified, it should be taken as 0°C.

Full transition curves should be established for one heat out of ten, with a minimum of one.

All specimens should exhibit a minimum of 75% shear on the fracture surface.

8.15.3 Test Certificates and Traceability for Line Pipe

The manufacturer's certificate should state that the pipe complies with standards under sour/nonsour conditions.

8.16 PURCHASER INSPECTION (NORMATIVE)

The company should specify if, and to what extent, he will monitor the manufacturer's production, quality control, and inspection.

Sufficient fluorescent lighting both overhead and at pipe ends should be provided at the inspection area. Facilities should be provided for rolling each pipe joint for inspection. The manufacturer should make ultrasonic or other suitable equipment available for use by the company to check the remaining wall thickness where any defects have been ground out of the pipe.

If the company has to reject pipe repeatedly for any recurring cause, this should be reason to refuse further pipes for final examination until the cause has been rectified.

8.17 HYDROGEN-INDUCED CRACKING SENSITIVITY TESTS (APPLICABLE TO SOUR SERVICES ONLY)

8.17.1 Hydrogen-Induced Cracking Tests

Hydrogen-induced cracking (HIC) tests should be performed and reported in accordance with NACE TM 0284-96. Test Solution "A" should be used. Values of crack-length ratio (CLR), crack-thickness ratio (CTR), and crack-sensitivity ratio (CSR) should be reported and photomicrographs of specimens showing any blister, or alternatively dimensioned sketches, should be provided with the report.

Temperature	25 ± 3°C
H_2S concentration	(2300–3500 ppm) Saturated condition
pH Value-initial	2.9–3.3
pH Value-final	3.5–4.0
Test period	96 hours

8.17.2 Samples for HIC Tests

Samples for HIC testing should be taken in accordance with NACE TM 0284-96.

One pipe from each of the first three heats of pipe produced should be tested. One pipe out of every subsequent ten heats should be tested.

8.17.3 Acceptance Criteria

The following acceptance criteria should be met:

CLR	15%	Maximum
CSR	1.5%	Maximum
CTR	5%	Maximum

The maximum individual crack length on any section should not exceed 5 mm.

Blistering area should not be more than 1% of the exposed area of two wide faces of each coupon.

If any specimen fails to meet the above acceptance criteria, the heat of steel represented by the test should be rejected.

8.18 SULFIDE STRESS CRACKING TESTS (APPLICABLE TO SOUR SERVICES ONLY)

8.18.1 Sulfide Stress Cracking Test

Sulfide stress cracking tests should be performed in accordance with NACE TM 0177-96 Test Solution A. A four-point bend test piece in accordance with ASTM G-39 should be used, and a test duration should be 720 hours.

The test piece should be stressed to 0.72 of the SMYS.

8.18.2 Samples for Sulfide Stress Cracking Test

One sample should be taken from each test pipe provided for the manufacturing procedure qualification; three test pieces should be taken from each sample. One sample out of every 10 heats should be tested during production. Each test piece should be 115 mm long × 15 mm wide × 5 mm thick and should, for welded pipe, contain a section of the weld at its center. Samples may be flattened prior to machining test pieces from the inside surface of the pipe.

8.18.3 Acceptance Criteria

The sample should show no visible cracking that exceeds 0.1 mm in the through-thickness direction with magnification of 100×.

Summary of Testing and Inspection Requirements (Applicable to Nonsour Services)

Types of Test/ Inspection	First-Day Production Tests		During Production	
	Frequency	Remarks	Frequency	Remarks
Visual Inspection				
• Dimensions	All pipes		All pipes	
• Out of roundness at weld position	All SAW pipes		All SAW pipes	
• Pipe end squareness	All pipes		2 pipes per shift	
• Straightness	All pipes		Random	
• Surface defects	All pipes	External (plus internal ^{3} DN600)	All pipes	External (plus internal ^{3} DN600)
Ultrasonic Examination				
• Pipe ends	All pipes	25 mm of pipe ends	All pipes	25 mm of pipe ends
• Welded pipe				
• Plate/skelp	All plates/ skelp	25 mm of trimmed plate material	All plates/ skelp	25 mm of trimmed plate material
• Weld seam	All pipes	SAW pipe ends should be radiographed	All pipes	SAW pipe ends should be Radiographed
• Seamless pipe	All pipes	25% of surface	All pipes	25% of surface
Radiography				
• Weld seam	All selected SAW pipes	100% weld	All SAW welds	End 230 mm
• Weld repair areas on seam weld	All weld repairs		All weld repairs	

MT				
• Seam weld	All selected pipes			
• Seamless pipe body	All selected pipes			
• Bevel faces	All pipes	Only if ultrasonic testing is impossible	All pipes	Only if ultrasonic testing is impossible
Hydro test	All pipes		All pipes	
Physical Tests				
• Tensile test	Two selected pipes	Both longitudinal and transverse specimens for pipe greater than DN 200 (NPS 8)	Two pipes per heat, one pipe for heat less than 100 tons	
• Weld tensile test	All selected pipes	Welded pipe only	As above	
• All weld tensile test	One pipe	SAW only		
• Charpy V-notch				
• At temperature in Part II, Appendix F, SR.5.1	All selected pipes		As required for tensile test	
• Transition curve	One pipe			
• Drop-weight tear test				
• Transition curve	One pipe		One pipe per 10 heats, at least one pipe	
• At minimum design temperature	All heats		One pipe per heat (two specimens)	
• Flattening test			As per Fig. 5 of API-5L	HFW only
• Weld manipulation test			One pipe per 50 pipes	SAW only
• Weld ductility test			One test per lot (see 6.2.5 API Spec. 5L)	HFW only

• Macro, micro plus hardness	One pipe (3 specimens) Three pipes (3 specimens)	SAW & SMILS HFW	One pipe per heat or after each stop in production	HFW &-SAW Only
Hydro Test	All pipes	All pipes		
Chemical Composition				
• Ladle analysis	Once per heat		Once per heat	
• Check analysis	Twice per heat		Twice per heat	

Summary of Testing and Inspection Requirements (Applicable to Sour Services)

Types of Test/ Inspection	First-Day Production Tests		During Production	
	Frequency	Remarks	Frequency	Remarks
Visual Inspection				
• Dimensions	All pipes		All pipes	
• Out of roundness at weld position	All SAW pipes		All SAW pipes	
• Pipe end squareness	All pipes		2 pipes per shift	
• Straightness	All pipes		Random	
• Surface defects	All pipes	External (plus internal ≥ DN600)	All pipes	External (plus internal ≥ DN600)
Ultrasonic Examination				
• Pipe ends	All pipe	25 mm of pipe ends	All pipes	25 mm of pipe ends
• Welded pipe				
• Plate/skelp	All plates/ skelp	25 mm of trimmed plate material	All plates/ skelp	25 mm of trimmed plate material
• Weld seam	All pipes	SAW pipe ends should be radiographed	All pipes	SAW pipe ends should be radiographed
• Seamless pipe	All pipes	25% of surface	All pipes	25% of surface
Radiography				
• Weld seam	All selected SAW pipes	100% weld	All SAW welds	End 230 mm
• Weld repair areas on seam weld	All weld repairs		All weld repairs	

MT				
• Seam weld	All selected pipes			
• Seamless pipe body	All selected pipes			
• Bevel faces	All pipes	Only if ultrasonic testing is impossible	All pipes	Only if ultrasonic testing is impossible
Physical Tests				
• Tensile test	Two selected pipes	Both longitudinal and transverse specimens for pipe greater than DN 200 (NPS 8)	Two pipes per heat, one pipe for heat less than l00 tons	
• Weld tensile test	All selected pipes	Welded pipe only	As above	
	One pipe	SAW only		
• All weld tensile test				
• Charpy V-notch				
• At temperature in Appendix F, SR.5.1	All selected pipes		As required for tensile test	
• Transition curve	One pipe			
• Drop-weight tear test				
• Transition curve	One pipe		One pipe per 10 heats, at least one pipe	
• At minimum design temperature	All heats		One pipe per heat (two specimens)	
• Weld manipulation test			One pipe per 50 pipes	SAW only
• Macro, micro plus hardness	One pipe (3 specimens)	SAW & SMILS	One pipe per heat or after each stop in production	SAW

Hydro Test	All pipes	All pipes
HIC	One pipe from first three heats	One pipe per 10 heats
SSC	One pipe from first three heats	One pipe per 10 heats
Chemical Composition		
• Ladle analysis	Once per heat	Once per heat
• Check analysis	Twice per heat	Twice per heat

Chapter 9

Piping Material Selection

9.1 INTRODUCTION

This chapter contains piping classes primarily developed for petroleum refineries and petrochemical plants installed onshore. It is also intended for use in onshore exploration and production facilities as well as booster stations as far as applicable. Facilities covered by this chapter are all within the property limits as defined in ASME B31.3.

9.2 IDENTIFICATION OF PIPING CLASSES

Each piping class is identified from two alphabetical characters that precede a two-digit figure, e.g., AN04. The first alphabetical character indicates pressure rating of flange, i.e., ANSI or PN classes as follows:

Character	A	for ANSI rating	PN 20	(150)
Character	C	for ANSI rating	PN 50	(300)
Character	F	for ANSI rating	PN 100	(600)
Character	G	for ANSI rating	PN 150	(900)
Character	H	for ANSI rating	PN 250	(1500)
Character	J	for ANSI rating	PN 420	(2500)

The second alphabetical character indicates material group selected as follows:

Character	N	for carbon steel
Character	P	for low and intermediate alloy steel
Character	S	for stainless steel
Character	T	for aluminum and aluminum base alloy
Character	V	for copper and copper alloys
Character	W	for nickel and nickel base alloys
Character	X	for nonmetallic material
Character	Z	for carbon steel with lining

The two-digit figure indicates differing service conditions (e.g., a process fluid being handled and service temperature limits). The figure has not been selected on the basis of specific purposes and as such is not meaningful. However, piping classes which have identical figures with the same material group are for the same service condition.

Oil and Gas Pipelines and Piping Systems. DOI: http://dx.doi.org/10.1016/B978-0-12-803777-5.00009-5

Example for identification of piping class is given below: Piping Class AN04 is compiled from:

A	for	ANSI rating	PN 20 (150)
N	for	Material	Group carbon steel
04	for	Service condition related to this class	

Similarly, Piping Classes CN07 and FN07 are for ANSI rating 300 and 600, respectively. Also, both piping classes indicate same material group and same service.

9.3 SELECTION OF PIPING CLASSES

To select a piping class, the "Service Index" should be screened to see whether the intended service is listed. If so, the appropriate ANSI rating class should then be identified by matching the required design pressure and temperature with the design limits given in the piping classes.

For services not listed in the "Service Index," the "Piping Class Index" can be screened to see whether a piping class is available in which the materials are considered suitable for the intended service. The piping class selected may be used provided that company approval is obtained.

9.4 GENERAL BASES OBSERVED IN PREPARATION OF PIPING CLASS TABLES

9.4.1 Design Codes

- Piping classes have been designed in accordance with ANSI/ASME B31.3.
- The design limits specified in the piping classes have been derived from the pressure/temperature ratings for flanges given in ANSI/ASME B16.5 unless otherwise stated in the piping class notes.
- Where specified by ANSI/ASME B31.3, bolting calculations have been performed to verify the ability to seat the selected gasket and to maintain a sealed joint under the given P/T range, ASME Section VIII Division I Appendix 2 has been followed for this.

Nominal wall thicknesses and outside diameters of pipe, as specified in the piping classes, are in accordance with ANSI/ASME B36.10M and ANSI/ASME B36.19M.

9.4.2 Design Values

Allowable stresses for the materials specifications contained in the piping class have been established in line with ANSI/ASME B31.3, Paragraph 302.3.

For API 5L Grade B, a distinction has been made between seamless pipe and welded pipe with a weld joint factor $E = 0.95$ (in accordance with ANSI/ASME B31.3, Table 302.4). For all other materials, the selection of pipe wall thickness is based on allowable internal pressure calculations for seamless pipe only. Consequently, where welded pipe is used, a weld joint factor of 1 should be guaranteed.

In accordance with ANSI/ASME B31.3, Paragraph 302.2.2, not more than 87.5% of the nominal wall thickness has been used in calculations for butt welding fittings.

9.4.3 Branch Connections

Reinforcement requirements for 90-degree pipe-to-pipe branch connections have been checked against the design limits of the piping class. The check calculations were performed in accordance with ANSI/ASME B 31.3, Paragraphs 304.3.2 and 304.3.3. The additional reinforcement of the welds is not taken into account.

The branch fitting outlet and the butt welding fittings as listed in Page 4 of each piping class could replace the welded pipe-to-pipe connections.

9.4.4 Sour Service

The indications on Page 1 of a piping class identify the sour conditions for which the piping class is suitable. This is summarized below:

	Indication on Page 1 of the Piping Class		Condition for Which the Piping Class Is Suitable
	Sour	PWHT	Wet H_2S, high severity
Piping classes	Sour		Wet H_2S, moderate and low severity

9.5 DESIGN CONSIDERATIONS

This chapter should be used in conjunction with the following considerations:

All pipes should conform to requirements of related ASTM and API standards.

9.6 SELECTION OF BASIC MATERIAL

In consultation with the process designer or process engineer, the materials engineer should establish the preferred materials selection based on the process requirements such as medium, pressure, temperature, flow and the environment of the process facility. Design life and cost considerations should also be taken into account in this respect.

Carbon steel piping systems for below-mentioned services should be designed, and engineered specifically with considering their specific requirements:

1. Caustic soda as per concentration and temperature ranges specified in standards.
2. Sulfuric acid (H_2SO_4) concentration 65 wt% and above, temperature maximum 50°C. Sulfuric acid (sulfuric acid 65–75 wt%) plus hydrocarbons, temperature maximum 65°C.
3. Dry chlorine in either the liquid or gaseous phase at temperature between −50°C and −70°C. Dry chlorine contains less than 150 mg/kg of water.
4. Dry liquid and gaseous hydrogen fluoride at ambient temperature and following conditions and mixtures:
   a. Hydrocarbons, 33% HF and traces of water up to 70°C and 6 bar.
   b. Hydrocarbons, 4% HF and traces of water up to 160°C and 3 bar.
   c. Hydrocarbons, and traces of HF, up to 200°C and 3 bar.

If approved, a lap-joint flange may be substituted by a ring joint, welding neck, or raised-face-welding neck flanges.

Slip-on flanges may be considered in hydrogen services, subject to engineering approval. In such cases, the enclosed space between the OD of the pipe and the bore of the flange should be vented by drilling a 3-mm diameter hole in the flange hub.

Corrosion allowance different from that shown in the line class should be indicated on the related P&ID drawing.

When any carbon steel piping are to be designed and installed with carbon steel steam jacketing, (liquid sulfur and hot bitumen services), related piping classes for main piping and jacketing should govern for each system except for main piping corrosion allowance.

The main piping corrosion allowance should be increased by an amount equal to jacketing corrosion allowance.

9.7 SERVICE INDEX (SEE NOTE 1 OF THIS TABLE)

Service		Recommended		Remarks
Medium	Properties	Temp. Limits (°C)	Piping Class(es)	
Acetic acid	All concentration	0 to 100	S02	
Acetone		0 to 100	N01	
Acetonitrile		0 to 130	N01	
Air	Instrument-tool	0 to 200	N01	
Air	Instrument-tool	0 to 100	Z01	
Allyl chloride	Wet	0 to 20	X01	
Ammonia	Aqueous	0 to 165	N02	
Ammonia	Gas, dry, and wet	0 to 100	N02	
Aviation alkylate		0 to 120	N02	
Benzene		0 to 85	N01	
Butane	Gas	0 to 340	N01	
Butane	Liquid	0 to 40	N02	
Butane		−50 to 50	N07	
Butanol		0 to 200	N01	
Butylene	Gas	0 to 140	N01	
Butylene		−50 to 100	N07	
Calcium carbonate	Aqueous	0 to 50	N02	
Calcium chloride	Aqueous	0 to 50	N02	
Calcium hydroxide	Aqueous	0 to 50	N02	
Carbon dioxide	Dry	0 to 350	N01	
Carbon dioxide	Wet	0 to 100	S02	
Chlorine	Gas, dry	−35 to 70	N16	
Chlorine	Gas, wet	0 to 50	Z02	
Chlorine	Gas, wet	0 to 150	Z04	

(*Continued*)

(Continued)

Service		Recommended		Remarks
Medium	**Properties**	**Temp. Limits (°C)**	**Piping Class(es)**	
Chlorine	Liquid, dry	−35 to 70	N16	
Condensate	Steam nonaerated	0 to 250	N10	
Coolant	60/40% H_2O/methanol	−50 to 50	N07	
Diethanol amine	Dry	0 to 150	N02	
Diethanol amine	Water	0 to 150	S02	
Diethylene glycol	All concentrations—dry	0 to 150	N01	To prevent product contamination
Diethylene glycol		0 to 150	S02	
Dimethylketone	Water	0 to 100	N02	
Diphenyl propane		0 to 140	S02	Final product DPP plant
Ethane		−90 to 150	S04	
Ethanol		0 to 100	N01	
Ethylene		−140 to 150	S04	
Ethylene oxide	CO_2, water	0 to 150	S05	
Ethylene oxide		0 to 200	S05	
Foam, firefighting	Concentrate	0 to 50	S02	DN 15−80
Formic acids	All concentrations	0 to 30	S02	
Fuel gas	Wet H_2S		N02	Severity category VERY LOW
Fuel oil	Sulfur compounds	0 to 330	N01	
Gas oil		0 to 200	N01	
Gasoline		0 to 100	N02	
Heat transfer fluid	Downtherm	0 to 390	N03	
Hydrocarbons	Noncorrosive	0 to 330	N01	
Hydrocarbons	LPG	0 to 200	N02	

(Continued)

(Continued)

Service		Recommended		Remarks
Medium	**Properties**	**Temp. Limits (°C)**	**Piping Class(es)**	
Hydrocarbons	LPG with mod. sev. H_2S	0 to 200	N04	Moderate sour
Hydrocarbons	LPG with or without wet H_2S	−50 to 150	N07	Low temp, nonsour or moderate sour
Hydrocarbons	With mod. sev. wet H_2S	0 to 200	N04	Moderate sour
Hydrocarbons	With high severity HIC	0 to 200	N06	High sour
Hydrocarbons	With S and/or NA	0 to 240	N01	
Hydrocarbons	With S and/or NA	240 to 450	P04	Corrosion rate must be checked
Hydrochloric acid	All concentrations	0 to 30	X01	
Hydrochloric acid	All concentrations	0 to 150	Z04	
Hydrogen	With or without wet H_2S	0 to 230	N14	Moderate sour
Hydrogen	No H_2S	230 to 450	P06	Check API 1941 for pres./temp. limit
Hydrogen	With H_2S	230 to 538	S07	
Hydrogen chloride	Gas, dry	0 to 100	N01	
Hydrogen chloride	Gas wet	0 to 50	X01	
Hydrogen fluoride	With or without hydrocarbon		N12	
Hydrogen peroxide	Up to 98%		P02	
Hydrogen sulfide	Gas, dry	0 to 200	N04	Consult corrosion eng.
Isobutyl alcohol	Finished product	0 to 100	N01	
Isopropyl alcohol		0 to 100	N01	
Luboil and seal oil		0 to 200	S02	To prevent product contamination
Methane		−200 to 150	S04	
Methane		0 to 100	N02	
Methanol		0 to 100	N02	
Methanol		50 to 150	N07	
Methyl ethyl ketone		0 to 100	N01	

(Continued)

(Continued)

Service		Recommended		Remarks
Medium	Properties	Temp. Limits (°C)	Piping Class(es)	
Methyl hexanol			N02	Consult corrosion engineer
Methyl mercaptan		0 to 30	N02	
Naphtha		0 to 340	N01	
Naphthenic acid		0 to 95	N02	
Nitrogen	Liquid	200 to 150	S04	
Nitrogen		0 to 340	N01	
Nitrogen		0 to 200	N02	
Phenol		0 to 200	N01	To prevent product contamination
Phenol		0 to 150	S02	
Phosphoric acid	All concentration			
Polyols	Acidic or alkaline	0 to 200	S02	
Polypropylene		0 to 200	N01	
Potassium hydroxide				Consult corrosion engineer
Propylene	Liquid	−100 to 150	S04	
Seal oil			S02	
Secondary butyl alcohol		0 to 200	N01	
Sodium hydroxide (caustic soda)				Consult corrosion engineer
Steam	Saturated/superheated	0 to 400	N10	
Steam	Superheated	400 to 450	P02	
Steam		400 to 538	P05	
Sulfur	Dry	0 to 150	N01	
Sulfur trioxide	Dry gas	0 to 340	N01	
Sulfuric acid	>65 wt%	0 to 60	N05	
Water, boiler feed	Demineralized aerated	0 to 150	S02	

(*Continued*)

(Continued)

Service		Recommended		Remarks
Medium	Properties	Temp. Limits (°C)	Piping Class(es)	
Water, boiler feed	Treated noncorrosive	0 to 250	N10	
Water, boiler feed		0 to 75	Z02	For existing systems only
Water, cooling	Brackish-/seawater	0 to 50	Z03	
Water, cooling	Treated	0 to 60	N09	
Water, demineralized			S02	
Water, demineralized	Process use	0 to 80	Z02	For existing systems only
Water, firefighting	Brackish-/seawater	0 to 45	Z03	
Water, firefighting	Brackish-/seawater	0 to 50	Z05	For above ground systems only
Water, firefighting	Fresh	0 to 50	N09	Check oxygen and corrosion rate/life
Water, firefighting	Piping normally dry	0 to 50	Z01	For above ground systems only
Water, potable	Fresh, treated	0 to 20	Z01	
Water process	Demineralized	0 to 80	Z02	
Xylene		0 to 340	N01	

NA, Naphtha; HIC, hydrogen-induced cracking.

Notes:

1. Since concern of service index is material group and service conditions, the piping class indicated does not bear a prefix for pressure rating.
2. At higher temperatures, corrosion due to carbonyl formation may increase significantly; consult the materials/corrosion engineer.
3. Design temperature is limited in accordance with API publication 941 (Nelson curve).
4. For steam jacketing (piping class) should be referred to N10.

9.8 TEMPERATURE AND CAUSTIC SODA CONCENTRATION RANGES FOR METALLIC PIPE, FITTINGS, AND VALVES

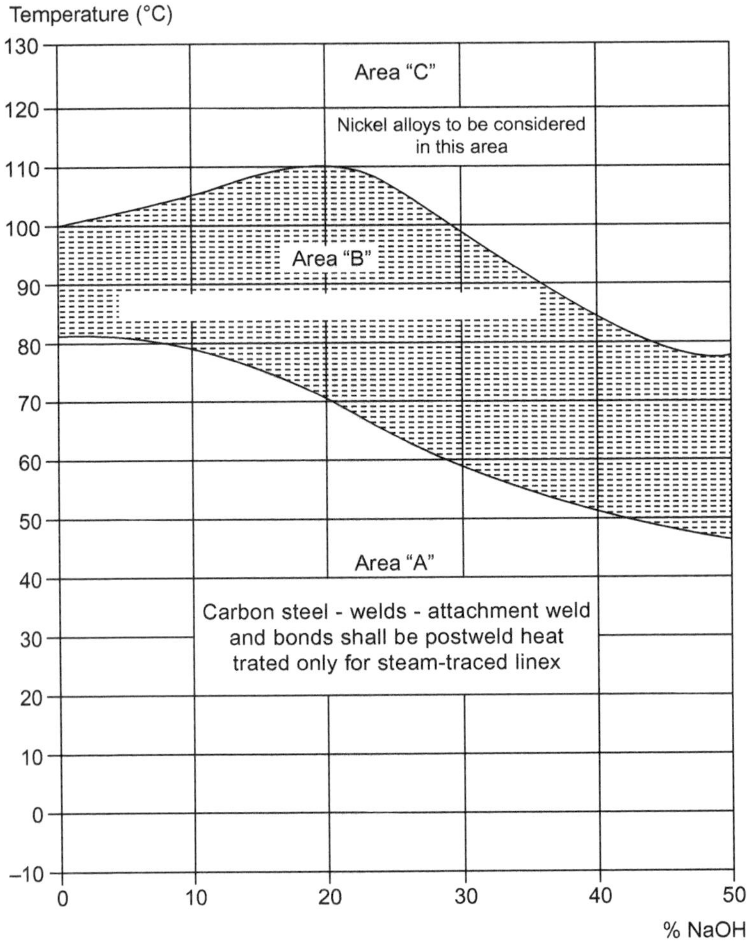

9.9 GENERAL REQUIREMENTS FOR CARBON STEEL PIPING IN CAUSTIC SODA SERVICE

Requirements to prevent caustic soda embrittlement in carbon steel piping are given below:

9.9.1 Design

For area "A," nonheated lines, no requirements.

For area "A," steam-traced lines, stress-relief for welds, attachment welds, and cold-formed bends.

For area "B," all lines, stress-relief for welds, attachment welds, and cold-formed bends.
Design piping systems for furnace postweld heat treatment (PWHT) preferably, i.e., with flanges, so as to enable placing in a furnace.
Particularly intricate parts such as valve and pump manifolds should be designed for furnace PWHT.
If field PWHT is unavoidable, ensure that areas and parts to be heat treated are readily accessible.
Restrict the application of cold-formed parts especially those that are cold-formed during fabrication.
Exclude hot spots by direct wall-to-wall contact in the case of steam or electrical traced lines by application of spacers (ceramic, glass fiber, or filled phenolic resin).
Include the design of steam or electrical tracers. Fixation points for tracers to be at a distance of 6.5 m maximum with special attention at bends and fittings.

9.9.1.1 Drawings

All drawings for the fabrication of carbon steel piping intended for caustic soda service should be clearly marked "CAUSTIC SODA SERVICE."

9.9.2 Fabrication

9.9.2.1 Welding

Inert gas or CO_2-shielded arc welding should be used for the root pass. Welds should be made without excessive penetration (max. 2 mm) and should be without grooves and/or craters. Pipework should be inspected after welding.

9.9.2.2 Installation

Hanger supports should be clamped around the pipes and bolted. Strips of CAF or glass fiber material should be applied between pipe and support. All indications given for the design of traced lines should be followed closely.

Ensure that all attachment welds are made before PWHT is applied.

9.9.3 Examination

Visual examination of all welded piping parts should be done during fabrication by an experienced inspector.

A minimum of 10% of all welds should be checked by radiography before PWHT. All defects should be repaired to fulfill minimum requirements.

9.9.4 Heat Treatment

Welds, attachment welds, and cold-formed piping parts should be given a stress-relief, PWHT.

This heat treatment should be carried out in a furnace or, if required, by electric-induction heating.

Heat treatment should be in the range of 580–620°C with a holding time of 3 min/mm thickness and a minimum holding time of 1 hour. Cooling should be controlled at a rate of 100°C/h down to 350°C.

The complete PWHT cycle should be recorded.

9.9.5 Testing

All piping parts should be hydraulically tested after heat treatment.

The hydrostatic test pressure to be used should be 1.5× the maximum allowable pressure at ambient temperature, as mentioned in the relevant piping classes.

Ensure that draining and drying after testing is carried out properly.

9.9.6 Identification

All piping fabricated in accordance with standards should be clearly identified by a suitable marking, either by painting or fixing an adhesive tape around the parts.

Pipe class number and line designation should be painted on each part.

9.10 GENERAL REQUIREMENTS FOR CARBON STEEL PIPING IN SULFURIC ACID SERVICE

Carbon steel piping systems in sulfuric acid should be designed in accordance with the rules given below.

9.10.1 Design

9.10.1.1 Flow Rate

Piping should be sized for a nominal velocity of 0.75 m/s in straight ends.

9.10.1.2 Flow Changes

Pipework design should be studied carefully to avoid sudden changes in the direction flow, turbulences, and extreme changes in velocity.

9.10.1.3 Drainage Falls

Horizontal runs of pipework should be avoided. Generous falls for self-draining purposes should be provided for, wherever possible. The fall should be minimum 1 cm/m.

9.10.1.4 Pipe Bends and Elbows

Pipe routings should be studied with the aim to reduce the number of bends and to restrict the number of elbows to the bare minimum.

Pipe bends should have a radius $R = 5D$, where D is the nominal pipe diameter. Standard elbows, which should be used for sweep-in connections, should be long-radius type $R = 1\frac{1}{2}D$. Short-radius elbows ($R = D$) should not be used.

9.10.1.5 Pipe Reducing

Reducers should be avoided as much as possible. Where a reduction is necessary, the reducer should be concentric, the reducing part should be smooth and the reduced diameter bore should correspond to the connecting part or be tapered to suit that bore.

If reducing is upstream, an eccentric reducer may be considered to ensure the required fall for drainage.

9.10.1.6 Junctions

Pipework should be designed to avoid 90-degree tee junctions, instead 45-degree laterals; Y-type or sweep-in junctions should be used.

9.10.1.7 Spool Pieces

Whenever turbulences or considerably higher velocities cannot be avoided, the use of spool pieces should be considered. These can be made out of solid, fully resistant alloys or carbon steel, lined with fully resistant material. Length of spool pieces to be $L = 20D$ where D is the nominal pipe diameter.

9.10.1.8 Postweld Heat Treatment

Pipe sections of intricate shapes, which are not accessible for finishing and/or where turbulences and velocity changes are most likely, e.g., valve and pump manifolds, should be designed as flanged sections, such to enable PWHT in a furnace.

9.10.1.9 Butt Welding

Pipework should be designed to restrict the number of butt welds that are not accessible for finishing and inspection of the inner surface. Misalignment of individual and adjoining pipe bores should not exceed 0.3 mm.

9.10.1.10 Flanges

Flanges should be installed to enable access to welds for finishing and to fabricate flanged pipe sections, junctions, reducers, and other special pipe parts. Welding neck flanges should be used. In exceptional cases the use of slip-on flanges may be considered.

9.10.1.11 Gaskets

Flat ring gaskets with ID dimensions exactly equal to the bore of the pipes should be used. The OD dimensions should be in accordance with ANSI/ASME B16.21. Gasket thickness to be 1.5 mm. Gasket material to be specified for sulfuric acid service. Attention should be given that not all acid-resistant type gaskets are suitable.

9.10.1.12 Drawings

All drawings for the fabrication of carbon steel piping intended for acid service should be clearly marked "SULFURIC ACID SERVICE."

9.10.2 Welding

Inert gas or CO_2-shielded arc welding should be used for the root pass. Where the weld metal penetrates to the bore of the pipe and/or fitting, great care should be taken to ensure full penetration without excess penetration. The internal bore at the location of the welds should be dressed flush with the inner pipe wall.

Permanent backing rings should not be used.

9.10.3 Examination

Visual examination of all welded piping parts should be done during fabrication by experienced inspectors.

A minimum of 10% of all welds should be checked by radiography. Radiographing should be done after the welds have been dressed. All defects should be repaired to fulfill the minimum requirements.

9.10.4 Heat Treatment

A normalizing heat treatment should be applied to sections of intricate shape, e.g., valve and pump manifolds, and sections where high heat input on welding or extreme stresses on forming have been introduced.

Heat treatment conditions:

normalizing temperature 900–930°C;
holding time 3 min/mm thickness with a minimum of 1 hour;
cooling in still air.

Attention should be paid to adequate support of the piping sections during normalizing to prevent excessive deformation and/or warping.

9.10.5 Testing

All piping parts should be hydrostatically tested after PWHT.

The hydrostatic test pressure to be used should be 1.5× the maximum allowable pressure at ambient temperature, as mentioned in the relevant piping classes.

Ensure that draining and drying after testing is carried out properly.

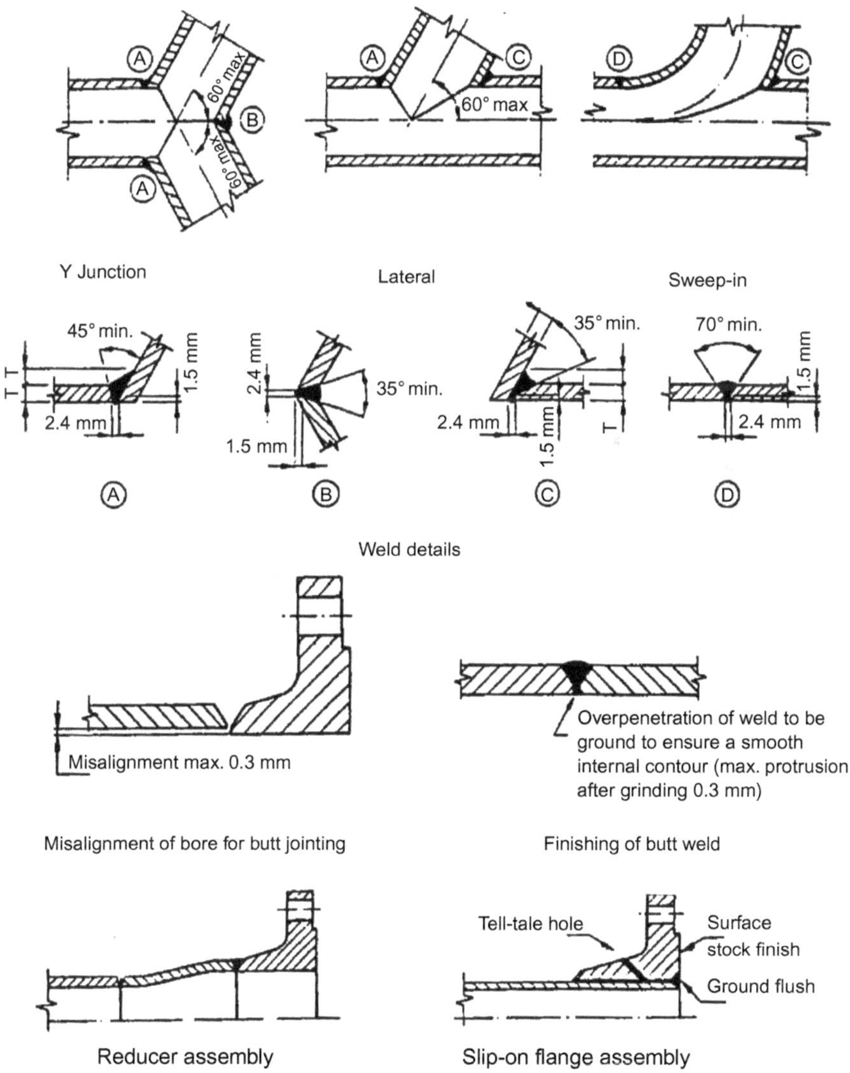

Construction and weld details for carbon steel piping in sulfuric acid service

9.11 GENERAL REQUIREMENTS FOR CARBON STEEL PIPING IN DRY CHLORINE SERVICE

This section provides requirements for the design and testing of piping systems for "dry" chlorine, in either the liquid or gaseous phase, at temperatures between −50°C and +70°C. "Dry" chlorine contains less than 150 mg/kg water.

9.11.1 Design

Materials for process piping in chlorine service are specified in piping Class CN16. Liquid chlorine should be considered to be a "(toxic) lethal" substance.

Only schedule 80 seamless pipe of minimum DN 20 (NPS ¾) should be used to ensure rigidity and protection against mechanical damage that may result in leaks.

Piping DN 25 (NPS 1) is adequate for all normal flows.

Piping arrangements should be as simple as possible, with a minimum of welded or flanged connections. For piping of DN 100 (NPS 4) and smaller, elbows should preferably be made by bending. Threaded joints should not be used. All welds should be butt welds. No socket welding fittings, bosses, and weldolets should be used where a fillet weld will be applied.

Liquid chlorine has a high coefficient of thermal expansion. If liquid chlorine is trapped between two valves and expands, enough pressure is created to burst the pipe. Therefore each line or line section should have an expansion chamber, a pressure relief valve or rupture disc discharging to a receiver.

The expansion chamber capacity should have at least 20% of the section volume and be based on a temperature rise of 27°C above the ambient temperature.

9.11.2 Manufacturing

9.11.2.1 Assembly of Piping Components

For the assembly of all piping components, pipe ends, fittings, and welding neck flanges to be butt-welded, a uniform root opening, as specified below, is required:

Nominal Pipe Size	Root Opening
Smaller than DN 50 (2 in.)	1.5 mm
DN 50–250 (2–10 in.)	1.5–2.5 mm
DN 300 (12 in.) and larger	2.5–3.5 mm

9.11.2.2 Alignment

Alignment should be in accordance with ANSI B31.3 but with the exception that the internal trimming should be 1:4, instead of 30°C.

9.11.2.3 Bending

1. Pipes may be bent in the hot or cold condition.
2. A normalizing heat treatment should be applied if the flattening deformation is more than 5%.
3. No heat treatment is required for hot-formed bends upon which the final forming operation is completed at a temperature above 620°C and below 950°C provided they are cooled in still air.

When hot bending is carried out outside the temperature range given above, a separate normalizing heat treatment at 900–950°C is required.

9.11.2.4 Welding

- Field welds should be kept to a minimum and should be carried out under fair weather conditions only.
- Permanent backing rings should not be used.
- PWHT is not required.
- Cracked tack welds should be removed using the same procedure.
- Temporary tack welds should not touch the root gap or the root face.
- For temporary tack welding, the use of "bridge pieces" is recommended to avoid damage to the root face of the gap area.

9.11.2.4.1 Welding Processes

For pipe sizes DN 50 (2 in.) and smaller, the "gas Tungsten arc welding process" should be applied. Larger sizes should be "gas Tungsten arc welding," but shielded metal arc welding may be used.

Procedures and welders should be qualified and approved by the principal before actual production welding starts.

9.11.2.4.2 Welding Consumables

An alternative possibility is the procurement of consumables directly from the manufacturer complete with certificates. These certificates should give information per batch on chemical composition and mechanical properties of the weld deposit.

Final approval of welding consumables for a particular job will follow through meeting the test requirements of the welding procedure qualification.

9.11.2.4.3 Inspection of Welds

All welds should be 100% radiographed.

The method of radiography to be employed for inspection should be in accordance with ASME Section V.

Acceptance criteria of welds should be in accordance with ANSI B31.3, Table 327.4.1A "Limitation on Imperfections in Welds."

9.11.3 Hydrostatic Testing

The hydrostatic testing should be carried out according to standards before the system is cleaned and dried. For piping Class CN16, the hydrostatic test pressure should be at least 45 bar (1.5× the pressure of the service limits).

9.11.4 Final Cleaning and Drying After Welding and Testing

Chlorine may react with oil. Therefore, in addition to requirements cited in standards, cleaning should be accomplished by pulling through each pipe section a cloth saturated with trichloroethylene or other suitable chlorinated solvent. Hydrocarbons or alcohol should never be used because remnants of these solvents react with chlorine.

Cleaning and drying are accomplished by passing steam through the line from the high end until the entire line is hot to the touch (approximately 60°C). Condensate and any foreign particles (such as oil or metal) should be drained out after the steam supply line has been disconnected and all pockets and low spots have been drained.

While the line is still warm, dry air should be blown through the line until the dew point of the discharged air is the same as that of the air blown into the system, e.g., −40°C or below. When drying is finished, the line should be kept closed in order to prevent reentry of atmospheric moisture.

9.11.4.1 Gas Testing

The dried piping system should be pressurized to 10 bar ga with dry air or nitrogen and tested for leaks by the application of soapy water to the outside of joints. Afterward, chlorine gas may be introduced and the system retested for leaks, as described below.

9.11.4.2 Detection of Chlorine Leaks

The location of a leak in a chlorine-containing system can be detected by the reaction of ammonia vapor with the escaping chlorine. The reaction gives a dense white cloud.

The most convenient way is to direct the ammonia vapor at the suspect leak by employing a plastic squeeze bottle containing aqueous ammonia. Do not squirt liquid aqueous ammonia on pipe and fittings.

9.11.4.3 Repairs

Repairing a leak may require welding. Before any welding is started, all piping should be thoroughly purged and checked for the thoroughness of the purge (inside and around the pipe) with an explosion meter. Carbon steel ignites in chlorine of 250°C, thus welding without purging could start a fire. Purge with dry air and continue a small flow of air during the welding operation.

9.12 GENERAL REQUIREMENTS FOR PIPING IN HYDROGEN FLUORIDE SERVICE

Mixtures of hydrogen fluoride (HF), hydrocarbons (and some water) as they occur in the HF alkylation process for the production of *iso*-octane and detergent alkylates. The following mixtures can be contained in carbon steel:

- Hydrocarbons 33% HF and traces of water up to 70°C and 6 bar
- Hydrocarbons 4% HF and traces of water up to 160°C and 3 bar
- Hydrocarbons and traces of HF, up to 200°C and 3 bar.

9.12.1 Design

Piping Classes:

Class CN12 ASTM A333Gr. 6, 3-mm corrosion allowance, temperature limits 0–200°C, specially meant for operation below 45°C

HF is considered a lethal substance, and all piping that operates below 45°C or may reach such low temperatures during any stage in operation should be made of notch-ductile materials. Classes CN12 fulfill that requirement.

The use of bellows is prohibited. Screwed connections should not be used, all connections should be welded or flanged. Flange connections should be reduced to the least possible number to avoid leakages.

Most bolting materials (Cr-Mo steel, stainless steel, Monel) may be susceptible to stress corrosion cracking if exposed to HF. Therefore flanged connections should not be permitted to leak and the edge of all flanges should be painted with one coat of Socony HF Detecting Paint No. 20-Y-15 for the detection of HF leaks. Hence, flanges should not be insulated.

For safe and easy handling during operation and (downhand) maintenance all valves and instruments should be located at an elevation of maximum 1 m above the floor.

To avoid that possible deposition of iron fluoride may hamper the operation, the valve types should be selected as follows:

- Globe valves with Monel trim and soft seats (PTFE) for valves in a normally closed position.
- Ball valves with Monel Ball and seats for valves in a normally open position.

All HF service piping should be installed above grade and should be self-draining to the necessary low-point bleeders.

To minimize the number of low-point bleeders, piping should drain into equipment, if possible (vessels, heat exchangers, pumps, control valves).

Process line size should be minimum DN 25. Drains, vents, bleeders, etc., may be made minimum DN 20. All control vales in HF service should be installed with block valves and bypass globe valve and should have a flush connection on either side of the control valves. Gage glasses should be provided with a protective cover of KEL-F or PTFE.

Equipment in HF Service is normally postweld heat treated; this is not required for piping.

9.12.2 Identification

All piping fabricated to the standards should be identified by a clear and suitable marking, either by painting or by fixing an adhesive band around the pipe.

Pipe class number and line designation number should be painted on each pipe piece.

9.12.3 Operation and Maintenance

The wearing of protective clothes, gloves, goggles, etc., should be prescribed and it should be ensured that all safety instructions are strictly observed.

Before breaking flanges of piping that have been in HF service, neutralization by means of ammonia or sodium bicarbonate is required to prevent HF contact with the skin. Even if no severe corrosion is experienced, fouling and heavy iron fluoride deposits are often present.

Neutralization of such thick fouling layers is rather difficult, and upon subsequent mechanical removal, acidic conditions may again be encountered underneath due to insufficient neutralization. If the acidity is such that it is unsafe to continue work, a second neutralization is recommended.

9.13 GENERAL REQUIREMENTS FOR RUBBER LININGS FOR PROCESS EQUIPMENT AND PIPING

9.13.1 Scope

This section covers both the general requirements for the purchase and related testing, inspection transportation and storage of vulcanized and non-vulcanized rubber-lined process equipment and piping, both shop fabricated and in situ.

9.13.2 Materials

The following rubber types are used for lining purposes (classification in accordance with ASTM D1418):

- Natural rubber (NR)
- Synthetic polyisoprene rubber (IR)
- Styrene-butadiene rubber (SBR)
- Chloroprene rubber (CR)

- Butyl rubber (IIR)
- Nitrile rubber (NBR)
- Ethylene propylene rubber (EPM and EPDM)
- Urethane rubber
- Chlorosulfonated polyethylene (CSM)[a]
- Fluoro rubber of the polymethylene type (FKM)[b]

[a]*Commercially available under the registered trade mark "Hypalon."*
[b]*Commercially available under the registered trade mark "Viton."*

Ebonites are rubbers with a hardness value of at least 60 degrees Type D Shore can be produced from NR, IR, SBR, NBR, or blends thereof.

9.13.2.1 *Material Selection*

The final selection of the type and thickness of the rubber lining, and the method of application, should be made in conjunction with the materials specialist and the lining contractor.

The following details should be included on the requisition of the equipment concerned:

• products to be handled • temperature • degree of vacuum or pressure • cycle of operations • abrasion and erosion aspects • immersion conditions	} Minimum, maximum, normal

9.13.2.2 *Quality of Rubber*

The grade of rubber should be specified on the requisition sheet. The lining contractor should state that the lining will satisfy the chemical and physical conditions specified with respect to the agreed service lifetime.

The manufacturer should supply the specification for the approved rubber compound and samples of the vulcanized rubber sheet for test and reference purposes. The specification of the rubber compound should not be changed without prior written approval from the principal.

9.13.3 Design and Fabrication

The fabrication of the equipment should be in accordance with BS 6374: Part 5 or DIN 28051 and DIN 28055.

The important points are:

- the surface must be accessible for manual working;
- the weld seams must be continuous, smooth and free from pores and, if necessary, machined or ground (Section 9.13.3.2);
- (steel) reinforcements should, if possible, be situated on the outside.

All branches should be flanged and the lining should be taken over the flange face to prevent ingress of the process liquid behind the lining.

For typical flanged connections in equipment, see Fig. 9.1.

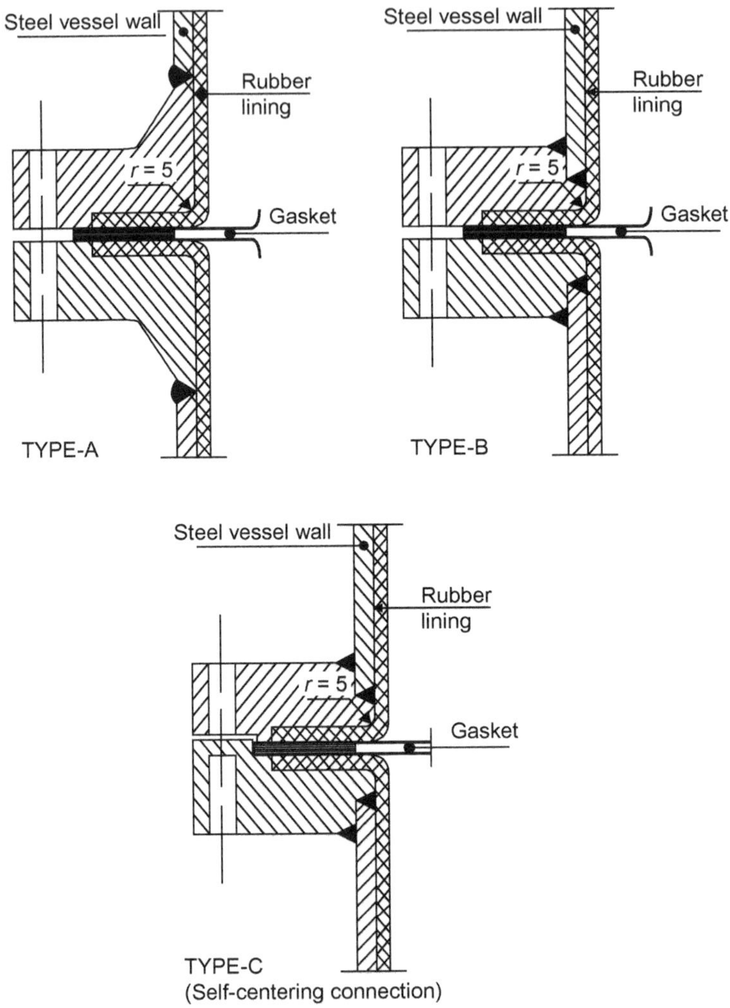

FIGURE 9.1 Typical details of rubber-lined flanges.

For typical flanged connections in piping, see Figs. 9.2 and 9.3.

For standard lengths and dimensions of piping and piping elements, see Figs. 9.4 and 9.5.

Sharp changes of contour in the surface to be lined should be finished to a suitable radius, such that the internal radius of the lining not less than 3 mm. Air vent holes to prevent air trapped in welded joints may sometimes be necessary (see Fig. 9.6).

9.13.3.1 Surface Finish of Substrate

The surface to be lined should be smooth, free from pitting, cavities, porosity, or other surface irregularities in accordance with DIN 28053.

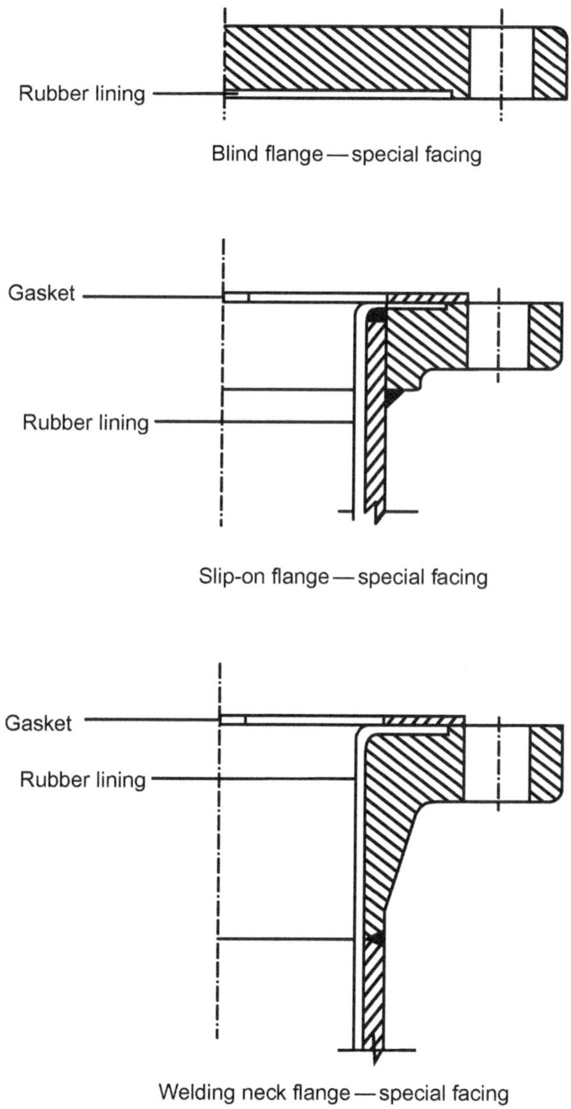

FIGURE 9.2 Flanged connections for hard-rubber-lined piping.

The surface should also be free from oil, grease, and other foreign matter. Metallic surfaces should be blast-cleaned to a surface finish corresponding to SA 2.5 in accordance with ISO 8501-1. After this operation the surface roughness should have a peak-to-valley height of 40–100 μm, with an average of 50 μm.

Immediately after the blast cleaning of the metallic substrate the grit, dust, etc., should be removed and a layer of adhesive primer with a dry-film thickness of approximately 30 μm should be applied.

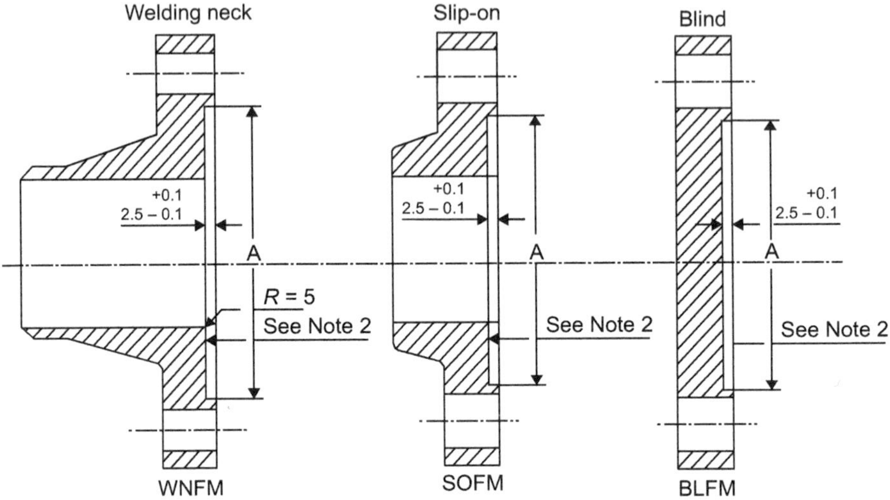

NOM size (DN)	A[1)] mm
25	58
40	78
50	98
80	132
100	166
150	218
200	274
250	330
300	396

Notes:

1) Tolerance (mm): +0.5 / −0.5

2) Facing finish : Ra 3.2 μm max. as per ANSI/ASME B46.1

3) Dimensions : Millimeters

FIGURE 9.3 Modified flanges for flanges, flat face with recess for rubber-lined piping system.

9.13.3.2 Welds in Metal Substrate

All metal-to-metal joints should be made by welding. Welds should be homogeneous and free of pores. Welds should be ground smooth and flush with the parent metal on the side to be covered. Wherever possible, they should be made from the side to be lined. Where this is not possible, the root should be chipped out and a sealing run should be applied. Internal corner and tee joints should be welded with full penetration.

Welds should be ground smooth and concave to the required radius. Welds should be examined according to applicable design codes. Fig. 9.6 shows acceptable welding details.

9.14.3.3 Joints in Rubber Lining

Overlap joints, as shown in figure 1 of Fig. 9.7, should be used when joining separate sheets of unvulcanized rubber. The total contacting surface between

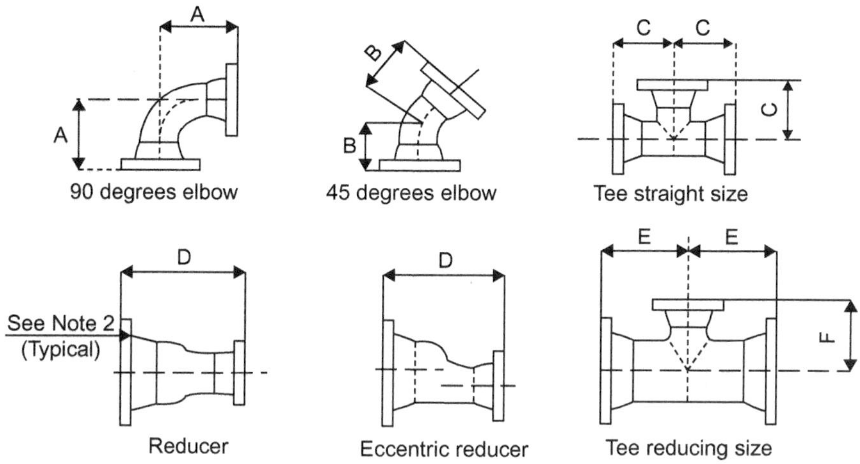

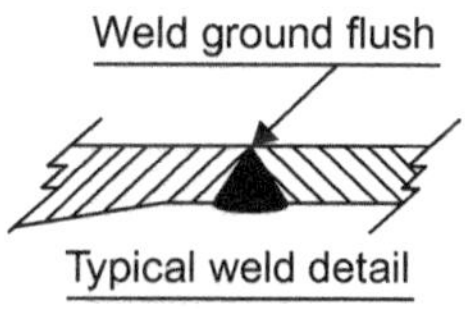

Nom size (DN)	A [1)] mm	B [1)] mm	C [1)] mm
25	93	77	93
40	118	90	118
50	139	99	127
80	183	120	155
100	227	139	160
150	317	183	231
200	405	228	279
250	482	260	317
300	570	303	367

Nominal size (DN)	D [1)] mm	E [1)] mm	F [1)] mm
40 × 25	180	118	112
50 × 25	194	127	106
50 × 40	200	127	121
80 × 40	219	155	134
80 × 50	221	155	139
100 × 40	238	180	147
100 × 50	240	180	151
100 × 80	246	180	167
150 × 80	297	231	193
150 × 100	303	231	205
200 × 100	328	279	231
200 × 150	341	279	256
250 × 100	354	317	259
250 × 150	367	317	282
250 × 200	380	317	304
300 × 150	404	367	307
300 × 200	417	367	330
300 × 250	417	367	342

Notes:

1) Tolerance (mm) : +4.0 / −4.0

2) Flange, welding neck : Flat face with recess see Figure 9.3

3) Flange & fitting mat : In accordance with Piping Class 1804.

FIGURE 9.4 Overall dimensions of flanged fittings for rubber lining.

the sheets should be a minimum of 4× the sheet thickness but should not exceed 32 mm at any point. Where applicable, overlaps should dictate the direction of the liquid flow.

When the total lining thickness is built up from more than one layer, only the joints in the top layer should be of the overlap bevel type, with the under

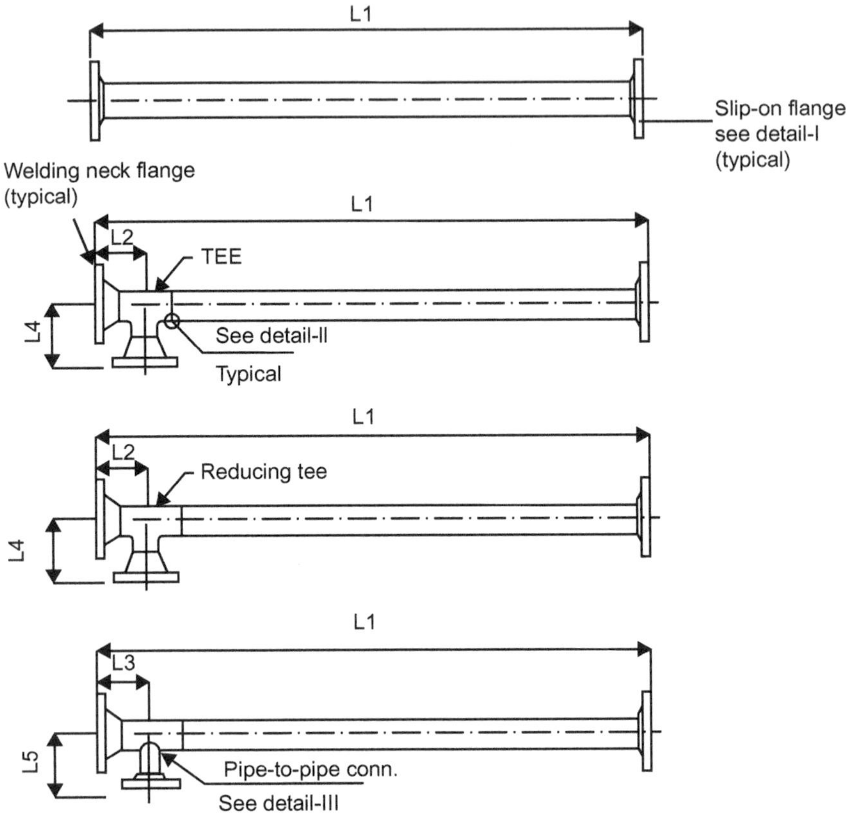

Run pipe Nom. size (DN)	1), 2) L1–mm	2) L2–mm	2) L3–mm	Outlet—Tee Nominal size (DN) 2) L4–mm									Outlet—pipe-to-pipe connection Nominal size (DN) 2) L4–mm				
				25	40	50	80	100	150	200	250	300	25	40	50	80	100
25	1000	93	—	93	—	—	—	—	—	—	—	—	—	—	—	—	—
40	3000	118	—	112	118	—	—	—	—	—	—	—	—	—	—	—	—
50	4000	127	—	106	121	127	—	—	—	—	—	—	—	—	—	—	—
80	5000	155	120	—	134	139	155	—	—	—	—	—	150	—	—	—	—
100	6000	180	120	—	147	151	167	180	—	—	—	—	165	—	—	—	—
150	6000	231	140	—	—	—	193	205	231	—	—	—	200	200	200	—	—
200	6000	279	150	—	—	—	—	231	256	279	—	—	225	225	225	225	—
250	6000	317	150	—	—	—	—	259	282	304	317	—	250	250	250	250	—
300	6000	367	175	—	—	—	—	—	307	330	342	367	275	275	275	275	275

FIGURE 9.5 Overall dimensions of flanged piping for rubber lining.

layers being flush-jointed, as shown in figure 2 of Fig. 9.7. The relatively weak flush joint (figure 3 of Fig. 9.7) is applied when the lining is used as a base for chemical-resistant brick lining. Joints in the different layers should be staggered.

Joints between rubber pipe linings and the rubber on the flange facing should not protrude so as to restrict the bore of the pipe or to prevent efficient sealing between the flange faces of adjacent lengths.

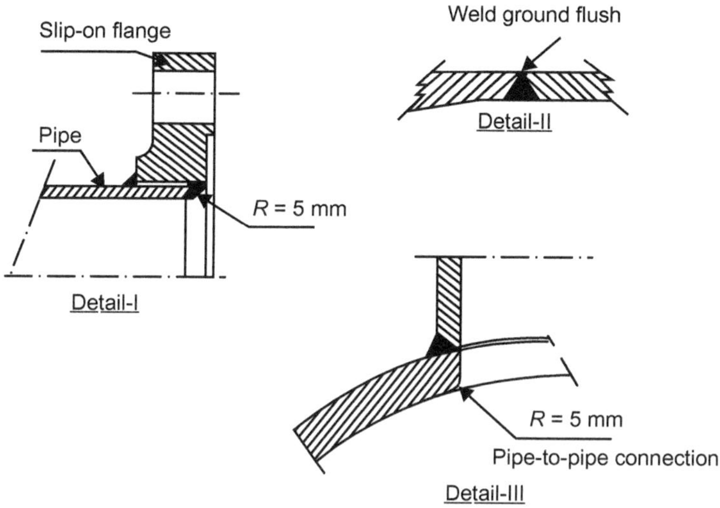

Notes:

1) Length	:	Based on max. allowable length for ruber-lined pipe.
2) Tolerace (mm)	:	+4.0
		−4.0
3) Flanges	:	Flat face with recess see Figure 9.3
4) Material	:	In accordance with piping class 1804.

FIGURE 9.5 Continued

9.13.3.4 Gaskets

The selection and application of the gasket material should be in accordance with the service conditions. For hard rubber linings generally a soft rubber gasket is used and for soft rubber linings a CAF gasket is used.

To prevent the gasket and lining bonding together, the rubber flange facing should be lightly rubbed with colloidal graphite.

9.13.3.5 Painting

Unless otherwise stated, all parts which are not rubber-lined should be degreased and blast-cleaned and painted with one coat of a suitable epoxy resin-based primer. This should be carried out after vulcanization of the rubber.

9.13.3.6 Lining Application

The manufacturer's procedures for the application of the lining should be adhered to. However, if shop vulcanization is used, the adhesive primer

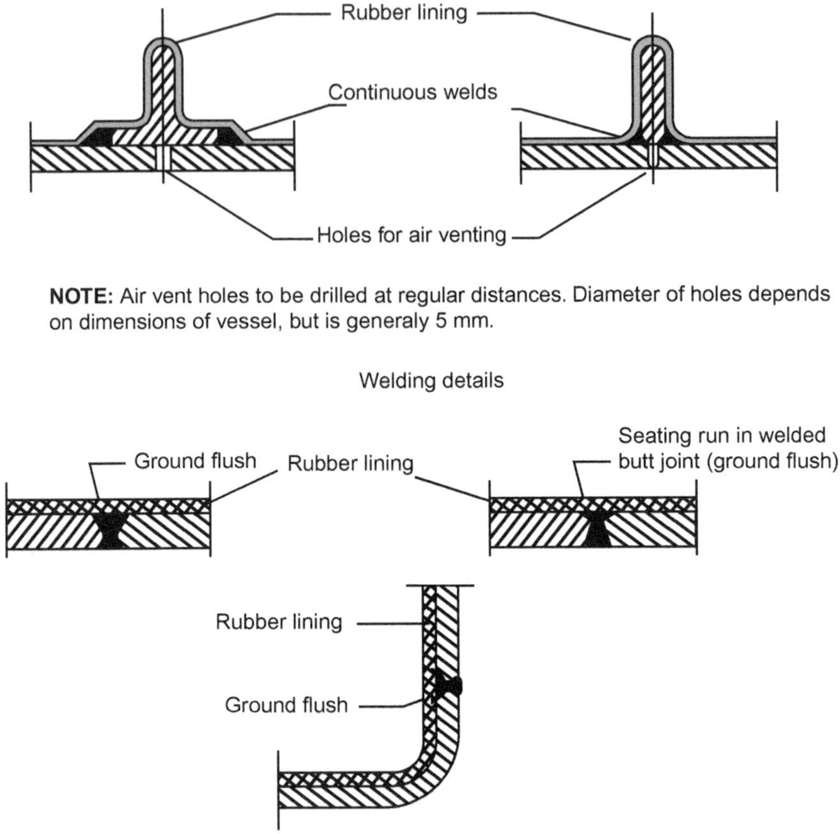

FIGURE 9.6 Air vent holes and welding details.

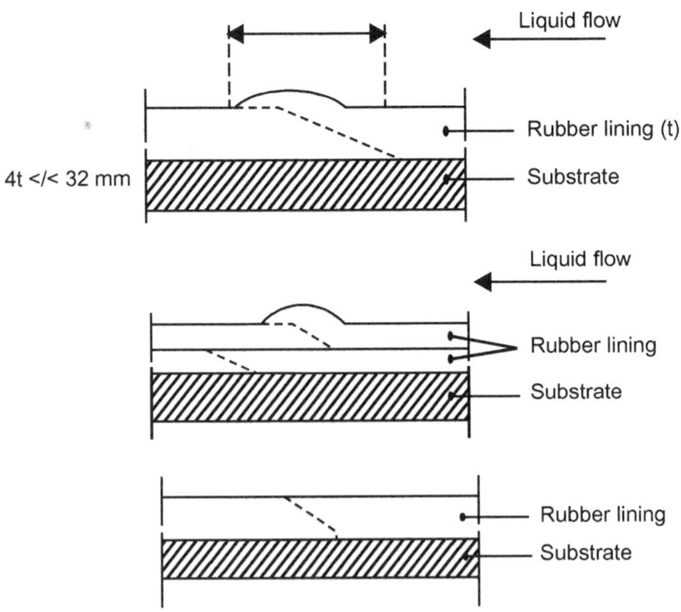

FIGURE 9.7 Overlap bevel joint and joints in rubber lining.

should be applied immediately after preparation of the substrate. The precut unvulcanized rubber sheets should be applied without inclusion of air and with specified joints within 96 hours.

If in situ vulcanization is carried out, the surface to be lined should have a temperature during the application of the unvulcanized rubber sheets not lower than 10°C and the relative humidity should not be higher than 75%, i.e., water condensation on the surface should be prevented during application.

The manufacturer should be responsible for the type of the adhesive system used to bond the rubber lining to the substrate. He should produce evidence that the adhesive system is suitable for the service conditions and will produce the bond required between rubber and substrate when the rubber is applied under the conditions of vulcanization.

9.14 QUALIFICATION TESTING

The principal will indicate at the time of enquiry or order whether qualification testing is required before delivery in order to establish the capabilities of the manufacturer or, e.g., because of time elapsed or new developments.

At the request of the manufacturer, and after approval by the principal, the tests required may be performed on products from current running stock. The number and size of the samples and the method of sampling should be established by agreement between the manufacturer and the principal.

Tests may be performed by the manufacturer or by an independent testing organization. In both cases a certificate stating the test results should be submitted. DIN 50049 3.1 B is acceptable for this purpose.

When the equipment is subject to a test pressure greater than 0.5 barg, the lining should be carried out after hydrotesting.

9.14.1 Rubber Lining

9.14.1.1 Rubber Quality

It should be verified that the type of rubber is correct (Section 9.13.2.2). Test method ASTM D3677 may be used for identification.

9.14.1.2 Performance of Rubber

The manufacturer should certify that the quality of the rubber will satisfy the chemical and physical conditions to which it will be exposed for the agreed operating life.

9.14.1.3 Physical Properties

The physical properties of the vulcanized rubber should comply with the values given by the manufacturer. The test methods as described in

BS 903:Part A2 are acceptable. These tests are carried out on separately supplied test samples. All hardness readings should conform to the specified value within ±5 degrees. A minimum of three readings should be taken for each square meter of lining. For large surfaces the maximum number of readings should be agreed upon by the manufacturer and the principal. In general, it is common to express hardness in Durometer A or Durometer D readings in accordance with ASTM D2240.

9.14.2 Rubber-Lined Parts

9.14.2.1 Surface Defects

Linings should be free from blisters larger than 10 mm in diameter, cracks, or other surface imperfections, porosity, voids, or inclusions.

9.14.2.2 Adhesion

For the determination of the adhesion to the metal substrate, samples should be prepared from the same rubber compound used for the lining and the same metal.

The pretreatment of the metal sample should be identical to the pretreatment of the surface of the equipment or piping. After the same vulcanization procedure, the adhesion should be determined according to method A (for ebonites) and to method B (for soft rubbers) as described in ASTM D429. The adhesion value calculated from the load at failure and the original bonded area should be as agreed upon in the specification. However, unless otherwise agreed, they should be at least $10\,N/mm^2$ (method A) and $4\,N/mm^2$ (method B) respectively.

9.14.2.3 High-Voltage Spark

The continuity of the lining on metallic substrates should be checked with a high-voltage spark test. Sparks should not be produced when the lining is tested with a direct-current apparatus, using a voltage which is determined by the following formulas:

$$6(1 + \text{thickness in mm})\text{kV which should not exceed } 30\text{ kV}$$

This voltage should be adjusted for high-carbon, black-filled (soft) rubbers to approximately 3 kV/mm thickness (exact voltage to be determined on a test sample). It is not possible to inspect antistatic linings with this test. In this case, after consultation with the principal, the "wet sponge test," a low-voltage holiday detector should be used.

Both the methods describing the continuity testing in BS 6374: Part 5 for high frequency with an AC source and the high-voltage test in DIN 55670 are acceptable to the principal.

9.14.2.4 *Thickness*

The thickness of the lining applied on substrates should be determined with a thickness meter and should conform to the agreed thickness with a minimum of 90% of that thickness. A minimum of three measurements per square meter should be made. For large surfaces the maximum number of readings should be agreed between the manufacturer and the principal.

9.15 ACCEPTANCE TESTS AND CERTIFICATION

9.15.1 Tests

The inspector representing the principal should check at random whether the rubber-lined process equipment and piping meets the requirements mentioned in Section 9.14. In addition, the following acceptance tests should be carried out.

If the requirements are not achieved, even after re-vulcanization, the rubber-lined item should be rejected.

9.15.1.1 *Visual Inspection*

The rubber lining should be free from cracks or any other imperfections. Blisters smaller than 10 mm diameter are acceptable, unless otherwise specified. Minor wrinkles and surface markings which will not have a significant effect on the performance of the lining are also acceptable.

The total amount of repairs should not be more than 100 cm^2/m^2 of lined surface. Lining repairs are not allowed in piping, on flange facings, or on nozzle necks of equipment.

Repair of damaged rubber linings should only be carried out by the contractor after consultation with and the agreement of the principal.

9.15.1.2 *Adhesion*

The adhesion between the rubber lining and the substrate should be homogeneous and without any defect. This may be investigated by lightly tapping the rubber lining with an appropriate wooden hammer. At areas where the adhesion is broken, a hollow sound will occur.

9.15.1.3 *Thickness*

The thickness of the rubber lining applied on carbon steel substrates should be determined and should conform to the thickness as mentioned on the requisition form with a minimum of 90% of that thickness. A minimum of three measurements per square meter should be made.

9.15.1.4 *Hardness*

The hardness should conform to the value specified on the requisition within a tolerance of ± 5 degrees. A minimum of three readings per square meter should be taken.

9.15.1.5 *Continuity of the Lining*

The continuity of the lining should be checked according to Section 9.14.2.3; however, it should be done using a reduced voltage, as determined by the following formula:

$$4(1 + \text{thickness in mm})\text{kV}.$$

9.15.1.6 *Hydraulic Testing*

If applicable, equipment and piping should be tested hydraulically at a pressure equal to the test pressure mentioned in the appropriate design code and at the maximum allowable service temperature for the particular lining. These conditions should be maintained for a period of 1 hour. At the end of the test, the lining should be visually inspected. Blisters, cracks, or other surface irregularities are not permitted. Thereafter the lining should pass the high-voltage spark test.

9.15.1.7 *Vacuum Testing*

If applicable, the equipment and piping should be tested at a vacuum of 130 mbar absolute at ambient temperature for a period of 1 hour. After this test no visible defects should be permitted in the lining.

9.15.1.8 *Flange Alignment*

The flatness of the lining applied on the gasket contact surfaces should be determined with a stretcher and should be within a tolerance of ± 0.3 mm.

9.15.1.9 *Certification*

The manufacturer should keep complete quality control and test reports. He should submit a certified record of inspection and testing, together with a statement of compliance with these requirements. These should include the certificates of the steel parts.

9.16 TRANSPORT AND STORAGE

Lined piping should be packed in a manner that will ensure that no damage to the lining, including its edges, can occur. Rubber-lined equipment and piping should not be transported or assembled if ambient temperature is below, or is likely to drop below, 0°C. The objects should be handled with care: hoisting should be carried out using nonmetallic slings. In particular branches, openings and flange facings are vulnerable and should be protected adequately, e.g., by wood.

All rubber-lined items should be clearly and permanently marked on the outside "RUBBER-LINED, HANDLE WITH CARE."

Rubber-lined equipment and piping should be stored indoors or under cover. Allowance should be made for free air circulation. Soft supporting material should be used, e.g., wood or rubber. The objects should not be exposed to direct heat or UV radiation. If this cannot be avoided owing to prevailing conditions, the items should be kept filled with water until taken into use. Freezing of this water should be prevented.

9.17 GENERAL REQUIREMENTS FOR CEMENT LINING OF NEW PIPELINES

9.17.1 Scope

This technical specification gives the minimum requirements for the design, application, inspection, installation, and testing of pipes from nominal size DN 100 (4 in.) up to and including nominal size DN 900 (36 in.) to be lined with cement mortar in the shop and the jointing of these lined pipes in situ, including the hand-applied lining of the field butt joints.

The in situ lining of piping systems, other than the abovementioned joints, is outside the scope of this specification.

Internal cement lining of pipelines is usually required for carbon steel cooling water or firefighting water piping systems containing/carrying seawater or brackish water at ambient temperatures to prevent internal corrosion. These systems can be installed above ground as well as underground. If an above ground lined piping system is drained for a longer period, special precautions are required to prevent cracking of the lining due to extreme high or low temperatures, and temperature or humidity variations. The special precautions to be taken are outside the scope of this specification.

9.17.2 Contract Strategy

9.17.2.1 Contractor Involvement

The contractor should prepare a detailed specification for the manufacturer. In addition, detailed drawings should be prepared if the details are not covered by the drawings. The aforementioned documents should be approved by the principal.

9.17.2.2 Selection of Cement Lining Contractor

The manufacturer selected should be able to design and install complete systems, including the procurement of all required piping materials together with technical services related to the products. In addition, the manufacturer should be responsible for the proper application of the cement lining to the field-joints. The manufacturer should have provided systems to at least one group operating company, or prequalification tests should be performed in accordance with Section 9.18. In both cases the final selection is subject to the principal's approval.

Damage to the lining may occur during the application of the shop painting, handling, transport, and storage of the pipes. Therefore the manufacturer of the cement lining should also apply the shop painting and/or wrapping to the outside of the pipe.

9.17.3 Design of Cement-Lined Piping Systems

Pressure design of pipes and fittings should be in accordance with ASME/ANSI B31.3.

9.17.3.1 Service Condition

In general the fluid to be conveyed through cement-lined pipes is seawater or brackish water at ambient temperatures. However, the contractor should obtain from the principal a chemical analysis of the fluid to be conveyed.

The temperature of the fluid to be transported should not exceed 45°C.

The composition of the lining should be determined taking into account the nature, operating temperature, and velocity of the fluid to be handled. The composition of the lining should be agreed with the principal.

The lining applied should have a design life of minimum 25 years.

9.17.3.2 Sizing Criteria

The diameter of the cement-lined piping systems should be sized so that the maximum water velocity inside the lined piping should not exceed 3 m/s.

The minimum outside diameter for piping and fittings to be cement lined is DN 100 (4 in.). The maximum diameter should be DN 900 (36 in.). The straight length of the pipes should be between 6 and 13 m.

Where carrying capacity is of importance, calculations should be made to determine the maximum friction and loss of waterhead.

Unless an increased thickness has been specified, the recommended lining thickness after curing should be as follows:

Nominal Pipe Size (in.)	Minimum Lining Thickness (mm)	Tolerance (mm)
4	5.0	+3
6	5.0	+3
8	6.0	+3
10	6.0	+3
12	8.0	+3
14	8.0	+3
16 up to 36	10.0	+3

9.17.3.3 Limitations on Pipe

9.17.3.3.1 Tolerances

Tolerances on dimensions of pipe ends should be in accordance with API 5L (including the ends of fabricated fittings).

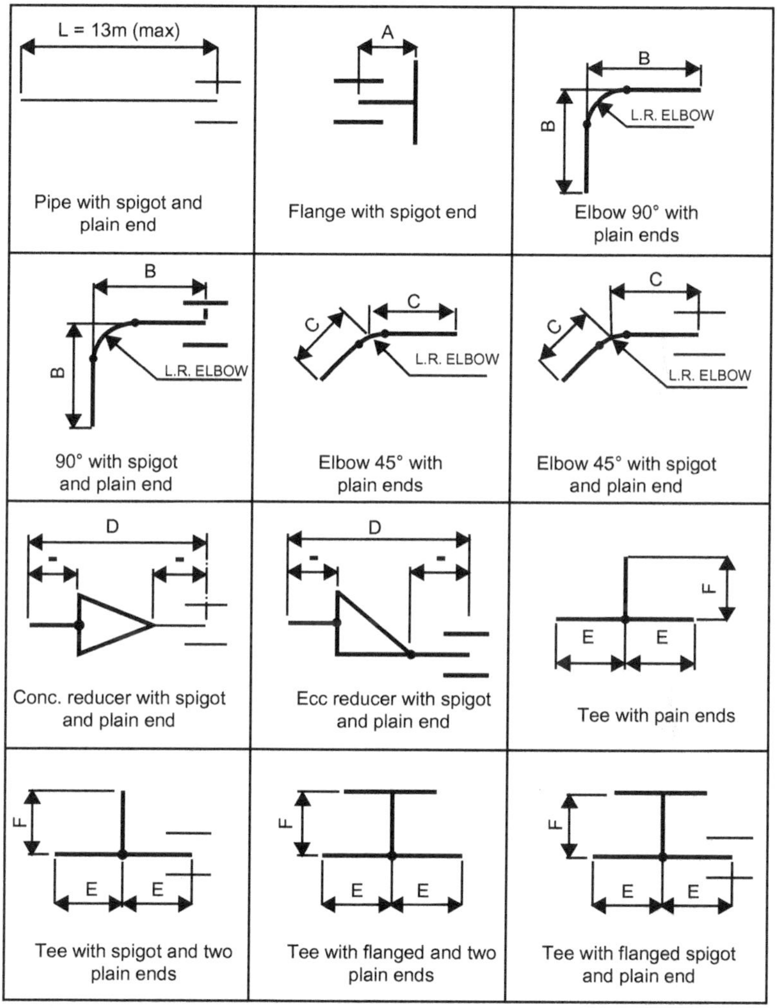

Notes:

1. Butt-welding elbows and reducers to ASME/ANSI B 16.9.
2. Flanges (slip-on) raised face or flat face to ANSI B 16.5.
3. For dimensions of sleeves and to lerances of pipe ends see Figure 9.11.
4. For flanged ends see Figure 9.12.
5. For typical detail of set-on branch see Figure 9.13.

FIGURE 9.8 Dimensions of cement-lined fittings nom. size DN 100 through DN 600.

9.17.3.4 Fittings, Flanges, Gaskets, and Valves

9.17.3.4.1 Fittings

For the selection and overall dimensions of fabricated fittings, see Figs. 9.8 and 9.9. For typical details of a set-on branch (for cement-lined fittings), see

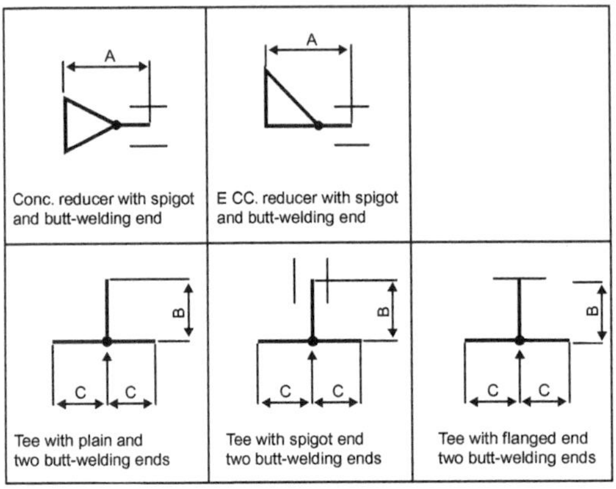

Nom. size (DN) mm	A mm	C mm	B mm
650 × 300	—	495	536
650 × 350	—	495	559
650 × 400	—	495	559
650 × 450	760	495	584
650 × 500	760	495	602
650 × 600	760	495	635
700 × 300	—	521	562
700 × 350	—	521	584
700 × 400	—	521	584
700 × 450	760	521	610
700 × 500	760	521	628
700 × 600	760	521	660
750 × 300	—	559	587
750 × 350	—	559	610
750 × 400	—	559	610
750 × 450	—	559	635
750 × 500	760	559	653
750 × 600	760	559	685
800 × 350	—	597	635
800 × 400	—	597	635
800 × 450	—	597	661
800 × 500	—	597	678
800 × 600	760	597	711
850 × 400	—	635	660
850 × 450	—	635	686
850 × 500	—	635	704
850 × 600	760	635	736
900 × 400	—	673	686
900 × 450	—	673	712
900 × 500	—	673	729
900 × 600	760	673	762

Notes:

1. Butt-welding reducers to ASME/ANSI B 16.9, or may be made from plate.
2. Flanges (slip-on)raised face or flat face to ANSI B 16.5.
3. For dimensions of sleeves and tolerances of pipe ends see Figure 9.11.
4. For flanged ends see Figure 9.12.
5. For typical detail of set-on branch see Figure 9.10.

FIGURE 9.9 Dimensions of cement-lined fittings nom. size DN 650 through DN 900.

Fig. 9.10. For dimensions of sleeves, see Fig. 9.11. Factory-made butt-welding fittings should be in accordance with ASME/ANSI B16.9.

All sleeve-joint connections of the piping should be prefabricated prior to installing the lining. All pipes, T-pieces, and other fittings should be prepared with beveled ends. When sleeve couplings should be applied, for piping with a diameter of DN 600 (24 in.) or less, the pipes should be prepared with one end plain and the other end with a sleeve coupling, externally welded to the pipe. T-pieces, valves, and other fittings should be prepared with sleeve couplings welded to both ends. In this case, one of the straight pipes connected to the aforementioned should be provided without sleeves.

Joints for piping with diameters DN 650 (26 in.) up to and including DN 900 (36 in.) should be butt-welded.

For shop-welding, welding procedures and welders should be qualified in accordance with ASME Section IX. The procedures should be submitted to the principal for approval.

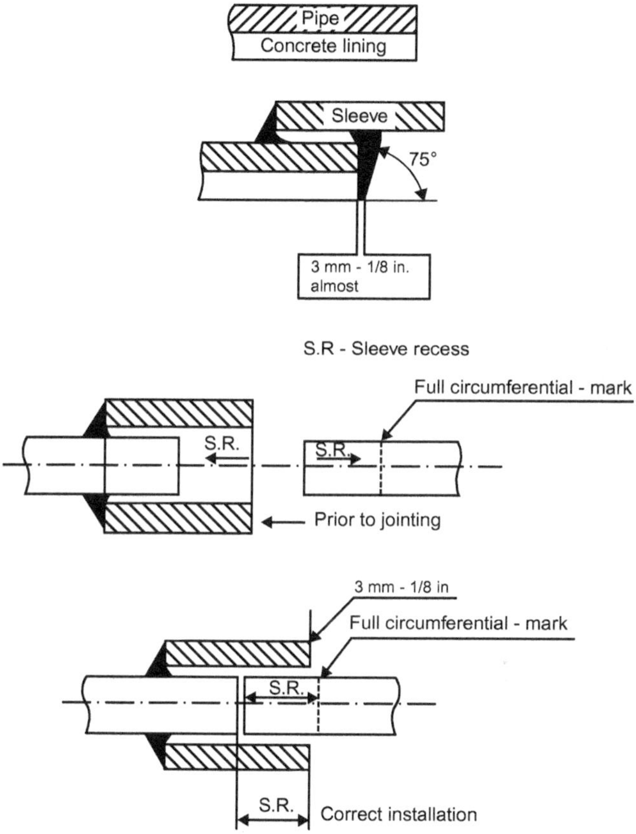

FIGURE 9.10 Typical detail of set-on branch for cement-lined fittings.

9.17.3.4.2 Flanges

All flanges should be class 150, raised face, except those connected to flat face flanges of glass-fiber-reinforced epoxy or equipment with cast iron flanges, etc. In these cases, a suitable matching flange should be used.

Slip-on flanges should be installed in pipe sizes from DN 100 through DN 600. For pipe sizes DN 650 and larger, welding neck flanges should be used. Flanges from DN 100 through DN 600 should be in accordance with ASME/ANSI B16.5, and flanges from DN 650 and larger should be in accordance with MSS SP-44.

Flange facing should be smooth finish between R_a 3.2 and 6.3 μm.

For flanged ends (slip-on) for cement-lined pipe and fittings, see Fig. 9.12. For shop-welding, refer Section 9.17.3.4.1.

9.17.3.4.3 Gaskets

For design pressures up to and including 10 bar ga, the gaskets should be 3-mm thick, reinforced chloroprene rubber with a Shore A hardness of 70.

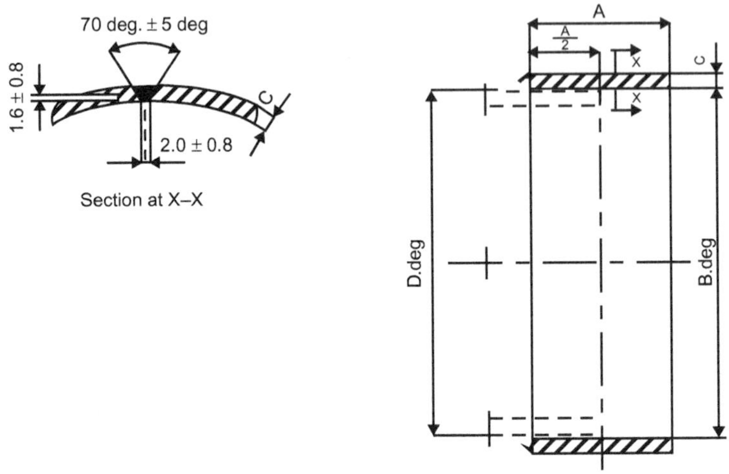

Nom. size (DN) mm	A mm	B Min. mm	B Max. mm	C Min. mm	D Min. mm	D Max. mm
100	100	120	122	6	113.9	115.9
150	150	174	176	8	167.9	169.9
200	150	225	227	8	218.7	220.7
250	150	279	281	8	272.7	274.7
300	150	329	331	8	323.1	326.3
350	150	361	363	8	354.8	358.0
400	150	412	414	8	405.6	408.8
450	150	462	464	10	456.2	459.4
500	150	513	515	10	507.2	510.4
600	150	616	618	10	609.2	612.4

Notes:

- Pipe ends must have the tolerance shown in the table for a distance of 100 mm.
- Sleeves can also be manufactured from sized pipe.

FIGURE 9.11 Sleeves for cement-lined pipe and fittings.

For design pressures above 10 bar ga, CAF gaskets 3 mm thick should be used.

The inside diameter of the gasket should be equal to the inside diameter of the cement lining.

Note that on some pipe-to-pipe valve (e.g., butterfly) connections the cement lining I.D. is tapered to equal the pipe I.D. at the gasket position to prevent possible interference of the valve disc with the cement lining (see Fig. 9.13).

9.17.3.5 Selection of Materials

9.17.3.5.1 Pipes

Seamless : DN 100 through DN 400-API 5L GR.B-(see Notes 1–4 of this table)
Welded : DN 450 through DN 900-API 5L GR.B-(see Notes 1, 2, 4, 5 of this table) (Submerged-arc weld)

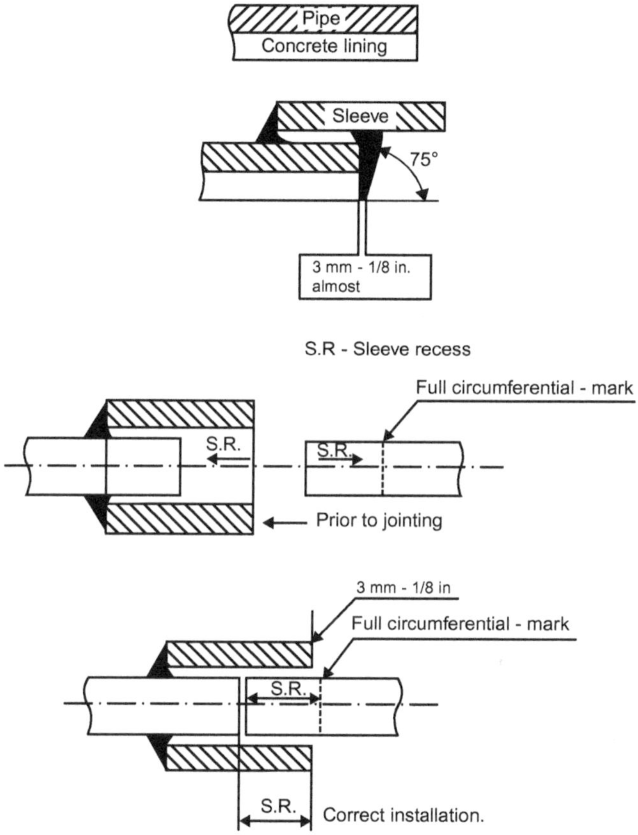

FIGURE 9.12 Flanged ends for cement-lined pipe and fittings.

Notes:

1. Carbon content 0.23% max.
2. Rimming steel not permissible
3. Nonexpanded pipe
4. Jointers not acceptable
5. Cold expansion is acceptable up to a maximum of 1.70%

9.17.3.5.2 Fittings

Fittings: ASTM A234-WPB should have the following restriction: carbon content of 0.23% max., and the manganese content may be increased to 1.3% max.

Base material:

DN 100 through DN 400: seamless pipe—ASTM A106 GR. B
DN 450 through DN 900: plate—ASTM A515 GR. 65 (or seamless pipe)

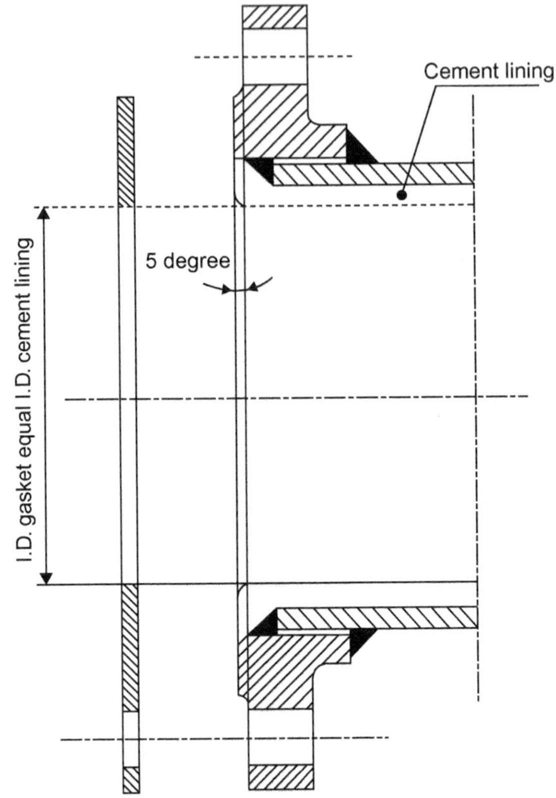

FIGURE 9.13 Typical details of flanged pipe-to-pipe connection.

9.17.3.5.3 Flanges

Flanges: ASTM A105 with the following requirements:
Normalized
Marking to A105-S9
Carbon content 0.25% max.
Manganese content may be increased to 1.20% max.

9.17.3.6 Compatibility of Materials

The possibility of galvanic corrosion should be taken into account when different metals are coupled together. The coupling should be broken by flange insulation made from nonconducting material.

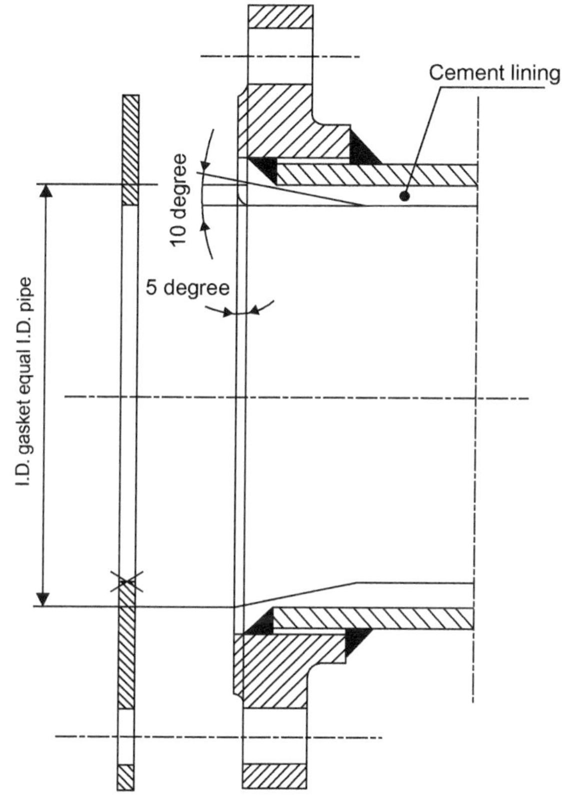

FIGURE 9.13 Continued.

9.17.4 Application of Cement Lining

9.17.4.1 Surface Preparation

The inside of the pipe should be cleaned of all grease, mill scale, loose rust, or other foreign materials, by blast cleaning to Sa2 in accordance with ISO 8501-1 or by power-tool cleaning to St3.

9.17.4.2 Composition of the Cement Mortar

9.17.4.2.1 Sand

Sand should be natural sand, manufactured sand, or a combination thereof and should conform to ASTM C33 or BS 882/1201 within the following limits:

Sieve	Percent Passing
No. 8 (2.36 mm)	100
No.16 (1.18 mm)	50–95
No. 30 (0.6 mm)	25–65
No. 50 (0.3 mm)	10–35
No. 100 (0.15 mm)	2–15

Deleterious substances should not exceed the following limits:

Substances	% By Mass	Standard
Clay lumps and friable particles	3.0	ASTM C142
Material finer than No. 200 sieve	5.0	ASTM C117
No. 1 and No. 2 combined	6.0	
Acid soluble chloride (Cl)	0.06	Analytic
Acid soluble sulfate (SO_3)	0.4	Analytic
Magnesium sulfate soundness	15	ASTM C88

The maximum size of sand-grain should not exceed one third of the thickness of the lining. The fineness modulus of the sand should be not less than 2.3 or more than 3.1.

The sieve analysis should be performed in accordance with ASTM C136.

9.17.4.2.2 Cement

For general application, the cement should be ordinary Portland cement in accordance with BS 12 or ASTM C150 or equivalent.

When high sulfate resistance is required, Portland cement type V in accordance with ASTM C150 or sulfate-resisting Portland cement in accordance BS 4027 with a maximum tricalcium-aluminate content of 3% should be used.

When moderate sulfate resistance or moderate heat of hydration is required, other types of cement can be used, for example

Portland cement type II in accordance with ASTM C150 or tricalcium-aluminate-free Portland cement.

Fly-ash and raw or calcined natural pozzolan according to ASTM C618 can be used as a mineral admixture in Portland cement.

Blast furnace slag should conform to the applicable parts of ASTM C595.

The type of cement including all admixtures or chemical additives should be approved by the principal.

The source of cement should not be changed without prior written approval of the principal. If bagged cement is used, all bags should be marked with the name of the manufacturer, type of cement, and volume. Similar information should be provided on the bills of lading accompanying each shipment of bulk cement.

A manufacturer's test certificate, showing results of the laboratory chemical tests and physical tests, should be submitted by the supplier to the principal not later than the day of delivery of the cement.

9.17.4.2.3 Water

Water should be potable and not contain chlorides (Cl) in excess of 500 mg/kg nor sulfates (SO_3) in excess of 500 mg/kg.

The water should not contain dissolved solids in excess of 2000 mg/kg or sugars, phosphates, and harmful impurities (e.g., oil).

The pH of the water should be between 5.0 and 8.0.

Testing should be carried out in accordance with BS 2690 and BS 3148 or equivalent.

9.17.4.2.4 Cement Mortar Mix

Unless otherwise agreed the cement/sand ratio and the water/cement ratio should be as follows:

- The cement/sand ratio should be one part by weight of cement and one and a half part by weight of dry sand (1:1.5) for linings not exceeding 6 mm.
- For linings with a thickness >6 mm, the cement/sand ratio should be 1:1 (parts by weight).
- The water/cement ratio should be between 0.3 and 0.4.

9.17.4.3 Installation of Shop Cement Lining

Straight sections of pipes, diameters from DN 100 (4 in.) up to and including DN 900 (36 in.), length 6–13 m, should be lined with the spinning method. The interior surface should be smooth, straight and true, and the sand/cement particles should be equally distributed throughout the lining thickness, after completion of the lining process.

9.17.4.3.1 Spinning Method

The lining should be applied by a spinning machine specifically designed and built for the purpose of applying cement mortar linings to the interior of steel pipe by means of centrifugal forces and rotation of the pipe. To prevent distortion or vibration during spinning, each section of pipe should, if required, be braced with external or internal supports. The entire quantity of mortar required for the lining of one section of pipe should be placed without interruption. The pipe should be rotated slowly until the mortar has been equally distributed along the inside periphery of the pipe.

Thereafter the rotation speed should be increased to produce a dense lining with a smooth surface and a minimum of shrinkage. Provisions should be made for removal of surplus of water.

9.17.4.3.2 Bends and Fittings

Bends and fittings that cannot be machine-lined in accordance with Section 9.17.4.3.1 may receive a hand-applied mortar lining. Hand-applied mortar should have a uniform surface. Cement mortar for hand work should be of the same consistency and material as the mortar for the machine method. Surfaces to be lined should be cleaned in accordance with Section 9.17.4.1 and damped with water immediately prior to placing the hand-applied mortar. Steel finishing trowels should be used for the hand application of cement mortar.

9.17.4.4 Curing

After application of the lining, the pipe and/or fittings should be sealed with plastic caps and left to cure in situ, or they may be transferred carefully to a curing yard.

The curing area should be sheltered, so that lined pipes and fittings are protected from harmful climatic conditions (e.g., exposure to direct sun, frost, etc.). Within 24 hours after application of the cement lining, the bores should be inspected and water added to aid curing, if required. After the inspection, the ends of the pipes and fittings should be re-capped with the plastic caps. These covers should not be removed within 14 days after cement lining in order to protect the lining from drying out.

Pipes or fittings should not be removed from the curing yard until the curing procedure is completed and the mortar has reached its specified strength.

Water to be used for curing should be in accordance with Section 9.17.4.2.3.

9.17.4.4.1 Normal Curing

The lining should be protected from drying out, as specified above for the entire period of the hardening process of the mortar in order to minimize shrinkage cracks. The minimum period of hardening should be 28 days.

9.17.4.4.2 Water Curing

The lining should be kept totally submerged for the total period of the hardening process of the mortar in order to minimize shrinkage cracks. The minimum period of hardening should be four days.

9.17.4.4.3 Steam Curing

Steam curing should only be applied if required and approved by the principal. Recording thermometers should be installed.

9.17.4.4.4 Membrane Curing

Membrane curing by application of any moisture-retaining liquid is not permitted.

9.17.4.5 Lining Repair

Dummy, spalled, and excessively cracked areas, etc., in fully accessible pipes should be removed and repaired by hand to the required thickness of the lining.

Cracks with a width less than 0.8 mm can be left, provided they will not impair the stability of the lining, as the self-healing effect will set them tight as soon as the pipes are in operation. Cracks with a width greater than 0.8 mm can be washed-in by means of soft brush with a liquid sand/cement mixture consisting of one part cement and one part fine sand (0.1 mm). The mixture should be liquid similar to heavy paint.

Larger local damages, other than cracks, should be repaired by removing all loose particles, old mortar, grease, and dirt by brushing with a stiff/wire brush. All traces of oil and grease should be removed with a suitable degreasing agent. The sides of the existing cement lining should be primed with a multipurpose adhesive, based on synthetic resin.

The damaged area should be filled with a ready-mixed mortar. The repaired lining should be finished by means of a trowel or a spatula and brushed flush with the original cement; it should be kept moist for at least 3 days. Minor damage can be repaired by means of a multipurpose adhesive.

9.17.5 Coating of Cement-Lined Pipes

The faces of all flanges to be used in cement-lined piping systems should be coated with Shell Ensis Fluid SDC, after which they should be provided with protective covers.

The shop painting of the outside surface of the pipe should be carried out after installing and curing of the cement lining, while the wrapping of the outside surface of the pipe (if required) should be done before the installation and curing of the lining.

9.17.6 Handling of Cement-Lined Pipe

Lined pipes and fittings should be handled carefully to avoid internal damage to the cement lining.

The end caps should be kept in place during transport and storage in order to prevent dust, dirt, foreign matter, etc., from entering the pipe.

For loading and unloading of very heavy pipes, it is recommended to use slings with cushion pads or a suitable fork arrangement placed at the center of the joints. Hooks or other devices that insert into the ends of the pipe should not be used. A flat-bed trailer provides the best support for the lined pipes during transport. During loading or unloading lined pipes should not be dropped onto or off the transporting vehicle.

To prevent bending of the pipes, which can cause damage to the lining, supports should be used during storage and shipping.

The distance between the supports should not exceed 3 m.

Pipes should be stored in supported tiers. The height of the tiers depends on the diameter of the pipes and should not be more than 10 pipes for diameters <DN 150 (6 in.), 6 pipes for diameters between DN 200 and DN 400 (8–16 in.), 4 pipes for diameters between DN 500 and DN 900 (20–36 in.).

9.17.7 Field Jointing

The pipes and fittings should be assembled in situ by means of field welds or flanged connections. The field welds should be butt joints for pipe diameters of 26 in. and above and sleeve joints for diameters below 26 in. However, butt joints should be internally lined in situ.

The contractor and principal should agree upon the type of multipurpose adhesive and the ready-mixed mortar to be used for the assembling of the pipes.

Alternative methods for field jointing of cement-lined pipes and fittings (for instance, with impregnated gaskets, etc.) are subject to approval of the principal.

9.17.7.1 Butt-Jointing

The beveled ends of the cement-lined pipes should be thoroughly cleaned, see Section 9.17.4.1, and all loose particles of the cement lining should be removed over 20 mm at both ends of the pipe, i.e., 20 mm at either side of the joint (see also figure 4 of Fig. 9.14). After the welding (see Section 9.17.7.4) has been completed, a hand-applied mortar should be used to finish the cement lining at the inside surface of the steel pipe at the location of field weld (see figure 5 of Fig. 9.14).

Unless another method has been agreed upon, the sides of the existing cement lining should be sealed with a priming coat consisting of one part of a multipurpose adhesive, based on synthetic resin (e.g., Conline "CEBOND", X-PANDO COMPOUND No. 2, SINMAST 121 or approved, equivalent bonding agent), and one part of potable water. After this priming coat has become tacky (20–30 minutes), the ready-mixed mortar should be applied.

The application of the adhesive and ready-mixed mortar should be in accordance with the manufacturer's specifications. The hand-applied mortar should be finished by means of a trowel or spatula and should be brushed flush after which a curing compound should be applied. Pressure testing of a pipe section with site-applied cement lining should be delayed until 28 days after application or until such time that the minimum compression strength has been reached (this will require an earlier compressive strength test in addition to the 28-day test, Section 9.19.3).

9.17.7.2 Sleeve-Jointing

The free access-length of the female part should be determined prior to the jointing and should be marked on the male part over the full circumference, see figures 2A and 2B of Fig. 9.14. The cut-end of the pipe inside the sleeve coupling should be thoroughly cleaned, dry and free of dust and should be

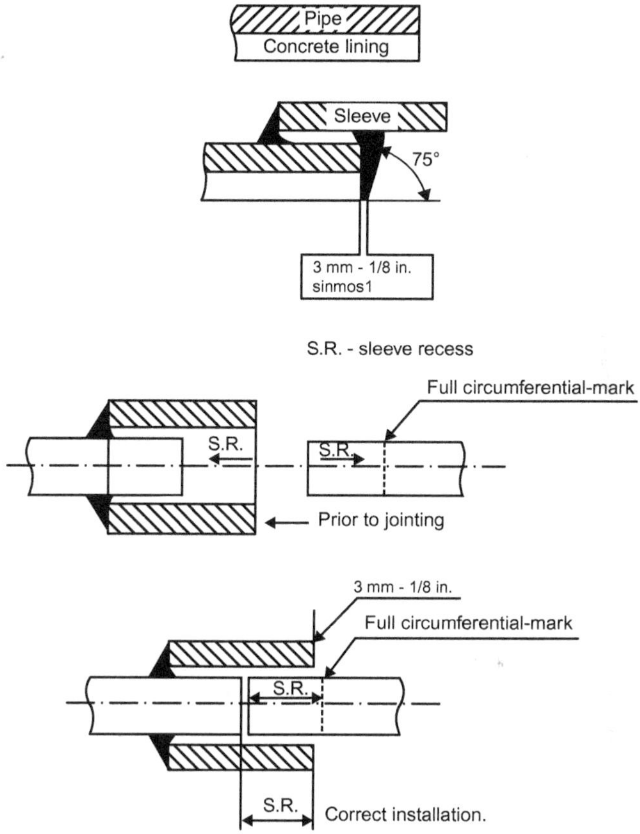

FIGURE 9.14 Pipe jointing—sleeve joints. Pipe jointing—butt joints.

provided with a concrete glue. This type of glue should be based on a two-component solvent-free epoxy resin (e.g., "SINMAST UW" or approved equivalent) suitable for application on moist mortar surfaces. Before installation of the male pipe end part, the concrete glue should be tampered with a minimum thickness at the cement lining edge of 3 mm and with an angle of 75 degrees (see also figure 2 of Fig. 9.14).

Immediately after the application of the concrete glue, the male part should be carefully pulled into the female part, without distorting the alignment (i.e., center-line) of the pipe and should be tack-welded in accordance with Section 9.17.7.4.

9.17.7.3 Cutting to Size In Situ

When cutting in situ is unavoidable, this should be carried out at the required position by means of a cutting disk of 3 mm.

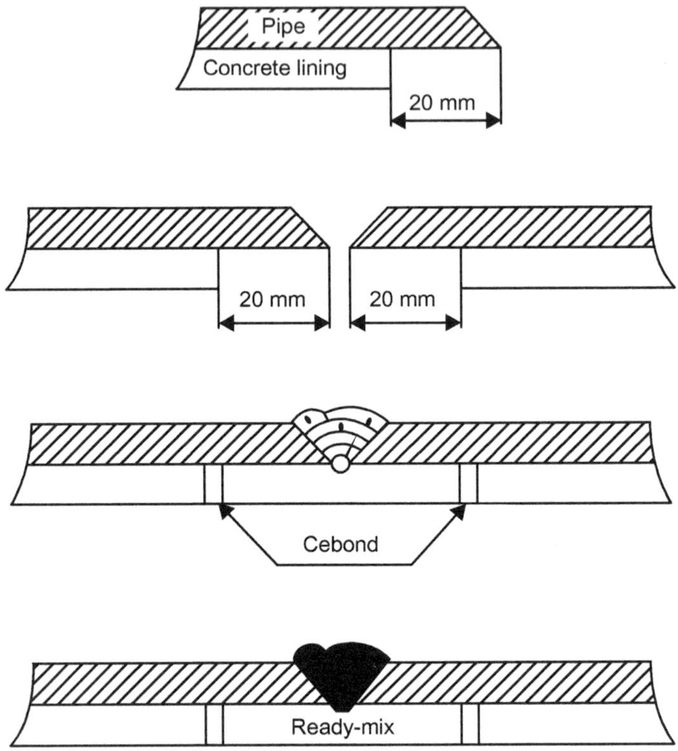

FIGURE 9.14 Continued.

Flame cutting is not allowed. The end of the steel pipe should be cut perpendicular to the pipe and beveled (if required) depending on the type of the field-joint. For a sleeve-joint, the cut pipe end should be used as the female part. For a butt-weld joint the cement lining should be cut over a length of at least 20 m/m at the end of the pipe. In both cases the cement lining should be cut perpendicular to the pipe. Typical details of the pipe ends are shown in Fig. 9.14.

9.17.7.4 Field Welding

The pipes should be tack-welded in three equidistant positions.

The pipe should be joined by the shielded metal arc welding process. The arc should not come in direct contact with the cement lining or seal material.

Starts and stops should be staggered so as not to start or stop more than once in the same place. Welding slag should be cleaned from all welds passed.

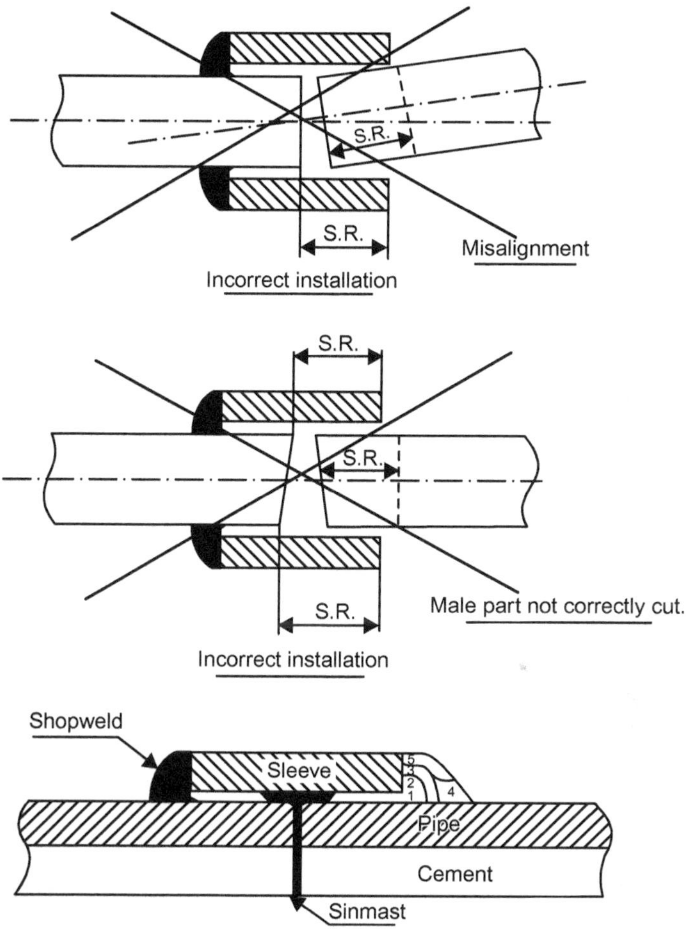

FIGURE 9.14 Continued.

Welding procedures and welders should be qualified in accordance with ASME Section IX; the procedures should be submitted to the principal for approval.

Welding materials used should be in accordance with the current list of approved welding consumables published by Lloyds Register of Shipping, Controls, or other internationally acknowledged bodies.

9.18 QUALITY CONTROL

Before the start of the actual cement lining production the manufacturer should make arrangements to execute line-up tests in order to demonstrate the suitability of the equipment for an uninterrupted production process.

All required materials should be supplied by the manufacturer. The testing should reflect the actual application conditions.

The entire process of applying cement mortar lining, at the manufacturer's works and at the construction site, should be subject to continuous inspection by a QC inspector appointed by the contractor, but such inspections should not relieve the manufacturer of his responsibility to furnish material and perform the work in accordance with this specification.

9.18.1 Procedure Qualification Test

Prior to the installation of the shop cement lining, or in case of any variation in the process or composition of the mortar or change in any components, the manufacturer should perform procedure tests to demonstrate that he is able to produce a lining system in accordance with the design requirements. The constituents, mortar, and finished pipe should be tested, the samples should be taken from one of the first finished test pipes or fittings, and testing should be carried out as indicated below:

Individual constituents of the mix

- cement/admixture
- sand
- water

Cement mortar test specimen

- density
- compressive strength
- flexural tensile strength
- water absorption

Finished product

- visual inspection.

The acceptance criteria for the tests should be in accordance with Section 9.19. Successful tests qualifies the procedure for the installation of the actual lining.

A record should be made of the complete test procedure, including:

- details of test piece,
- batch identification of cement mortar,
- test data and results,
- acceptance by principal's inspector.

9.18.2 Quality Control During Shop Application

During preparation of the cement mortar and subsequent application, a regular production sampling program should be established and maintained.

A logbook should be kept showing the portion of the completed lining that is represented by the sample and all information regarding the sample preparation and the operating parameters at the time of sample collection such as ambient temperatures, water content, cement/sand ratio, mixing times. Preparation of the samples should be witnessed by the QC inspector.

Immediately after the final spin an inspection of the cement lining should be carried out by looking through the pipe from each end, using a strong light. Defects in lining including but not restricted to sand pockets, voids, sags, oversanded areas, blisters, excessively cracked and dummy areas, and unsatisfactory thin spots should be removed before the initial set of the mortar.

Defective areas encompassing the full diameter of the pipe should be repaired by machine. Small defects in pipes >DN 600 (24 in.) should be repaired by hand to the full required thickness of the cement lining. In pipes less than DN 600 (24 in.), defective lining should be removed before the initial set of the mortar. Defective linings rejected after initial set should be replaced or repaired by the most practical method to be determined by the manufacturer in accordance with a procedure approved by the principal.

Most cracking occurs when the lining is allowed to dry out during curing, transportation and storage. The inspector should ensure that the lining is still moist after inspection and that air-tight end caps are placed and maintained on the pipe.

9.18.3 Quality Control During Field Jointing

Before assembling, the pipes and fittings should be inspected for possible cracks and damage. If required any cracks and damage should be repaired prior to the assembling of the piping system, see Section 9.17.4.5.

The cement should be the same as applied for the shop cement lining. Both sand and cement and quality should meet the requirements as described in Sections 9.17.4.2.1 and 9.17.4.2.2.

The lengths of pipe should be butted together and checked for alignment and good contact of the cement lining and pipe ends.

A ready-mixed mortar should be used for the butt joints and to repair linings of butt joints.

It is recommended to inspect the piping system 8 days after completion of the cement lining.

9.18.4 Production Tests

Production testing should be performed during manufacturing of the pipe lining and tests and inspection should be carried out in accordance with the table below. All test results should be reported and submitted to the Principal.

Test or Inspection	Frequency
Cement/admixture	Once per batch delivered
Sand/additives	Once per week
Water/cement ratio	Twice a day
Mixing ratio/times	Twice a day
Density	Twice a day
Compressive strength	Twice a day
Flexural strength	Twice a day
Ambient temperature	Once a day
Visual inspection	Continuously
Lining thickness	20% of pipes and fittings
Pipe and fitting ends	Each pipe and fitting
Surface condition	Each pipe and fitting
Lining structure	Once per week

The acceptance criteria for the tests should be in accordance with Section 9.19.

9.19 TESTS AND INSPECTION CRITERIA

All bare pipes should be inspected before cleaning and lining. The surface on which the cement lining is to be installed should be free from all grease, mill scale, loose rust, or other foreign materials prior to the installation of the cement lining.

Test samples of the cement lining mortar should be prepared by the manufacturer. Each sample should be clearly marked with the contractor's code numbers for that day/shift/crew and for the sequence of production. The manufacturer should be responsible for regular transport of samples to an independent qualified laboratory, subject to principal's approval.

If samples or completed lining do not meet the specified criteria, the installed cement lining represented by the failing sample(s) should be removed and replaced.

Completed lining not meeting the specified criteria under Sections 9.19.7/9.19.8/9.19.9/9.19.10 should be rejected.

9.19.1 Water/Cement Ratio

The water/cement ratio determined in accordance with DIN 1048: "Drying to constant weight," should be between 0.30 and 0.40.

9.19.2 Mixing Ratio

The dry-mix of lining materials should not contain less than 40% nor more than 50% of cement by weight.

9.19.3 Compressive Strength

The compressive strength after a curing period of 28 days, tested in accordance with ASTM C349, should not be less than 55 N/mm^2.

9.19.4 Flexural Strength

The flexural strength after a curing period of 28 days, tested in accordance with ASTM C348, should not be less than 6.5 N/mm^2.

9.19.5 Density

The density, measured in a saturated, surface-dry conditions, should not be less than 2160 kg/m^3.

9.19.6 Water Absorption

The water absorption of the sample, tested in accordance with ASTM C642, should not exceed 10%.

9.19.7 Lining Thickness

The thickness of the lining should be measured on the vertical and horizontal diameters of the cut faces at both pipe ends, that is, at clock positions 3, 6, 9, and 12, by direct measurement or by means of suitable electric instrument, e.g., a cover meter, calibrated before use. The values of lining thickness should be given with an accuracy of 0.10 mm. For tolerances of lining thicknesses, see Section 9.17.3.2 of this specification.

9.19.8 Pipe and Fitting Ends

Pipe and fitting ends of pipes and fittings with a diameter of 26 in. and above assembled in situ by means of field welding should have the cement lining removed over a length of 20 mm at either side of the pipe.

The ends of the lined pipes are considered to be defective if the lining end is:

1. not located as specified,
2. not perpendicular to the longitudinal axis of the pipe,
3. not square,
4. chipped or cracked,
5. separated from the steel pipe surface,
6. not to the specified thickness.

In addition to the above, the welding bevel should be free of cement.

9.19.9 Surface Condition

The surface condition of the finished cement lining should be smooth and even, not be flattened at individual spots, not have loose sand, not have dummy, spalled, or excessively cracked areas or show waves or grooves. Single waves or grooves are acceptable provided the minimum specified lining thickness is maintained. However, the maximum peak to trough height should not exceed 1.0 mm. Aggregate grains may only protrude at the surface sporadically. Hairline cracks and sporadically occurring surface cracks up to 0.8 mm are allowed.

Voids, being a place in the pipe where the cement lining is not continuous, are not acceptable. Voids occur during the spinning process when the cement does not distribute evenly.

Sags, appearing as large smooth lumps in the lining at the top of the pipe, are not acceptable.

9.19.10 Lining Structure

The polished section of a cement-lined pipe sample should not have visible pores and the individual grains of the sand should be surrounded on all sides with the cementing agent.

9.20 FABRICATION REPORT

After finishing the work, as defined in the purchase order, the contractor should provide a fabrication report with the following contents:

- project references such as: location, project number, piping system,
- bill of materials including lining,
- reference drawings and specifications,
- registration of date and time of application of all phases,
- registration of the produced samples and tests results,
- final inspection results,
- welding procedures used.

This report should also include certificates for the following materials:

- piping, fittings, flanges, gaskets and valves,
- welding consumables,
- sand.

Cement mill test certificates should be provided for each shipment of:

- cement,
- admixtures and additives,
- adhesives, concrete glue, and ready-mixed mortar.

Nom. size (DN) (mm)	A (mm)	B (mm)	C (mm)	D (mm)	E (mm)	F (mm)
100	135	254	164	–	165	165
150 × 100	–	–	–	440	200	200
150	145	379	245	–	200	200
200 × 100	–	–	–	452	230	230
200 × 150	–	–	–	452	230	230
200	150	455	277	–	230	230
250 × 100	–	–	–	478	230	280
250 × 150	–	–	–	478	230	280
250 × 200	–	–	–	478	280	280
250	165	531	309	–	280	280
300 × 100	–	–	–	–	230	305
300 × 150	–	–	–	503	230	305
300 × 200	–	–	–	503	280	305
300 × 250	–	–	–	503	305	305
300	180	607	340	–	305	305
350 × 100	–	–	–	–	230	355
350 × 150	–	–	–	630	230	355
350 × 200	–	–	–	630	280	355
350 × 250	–	–	–	630	305	355
350 × 300	–	–	–	630	355	355
350	180	683	372	¾	355	355
400 × 100	–	–	–	–	230	380
400 × 150	–	–	–	–	230	380
400 × 200	–	–	–	656	280	380
400 × 250	–	–	–	656	305	380
400 × 300	–	–	–	656	355	380
400 × 350	–	–	–	656	380	380
400	190	760	404	–	380	380
450 × 100	–	–	–	–	230	420
450 × 150	–	–	–	–	230	420
450 × 200	–	–	–	–	280	420
450 × 250	–	–	–	681	305	420
450 × 300	–	–	–	681	355	420
450 × 350	–	–	–	681	380	420
450 × 400	–	–	–	681	420	420
450	205	836	436	–	420	420
500 × 100	–	–	–	–	230	455
500 × 150	–	–	–	–	230	455
500 × 200	–	–	–	–	280	455
500 × 250	–	–	–	–	305	455
500 × 300	–	–	–	808	355	455
500 × 350	–	–	–	808	380	455
500 × 400	–	–	–	808	420	455
500 × 450	–	–	–	808	455	455
500	215	912	468	–	455	455
600 × 100	–	–	–	–	230	560
600 × 150	–	–	–	–	230	560
600 × 200	–	–	–	–	280	560

(Continued)

(Continued)

Nom. size (DN) (mm)	A (mm)	B (mm)	C (mm)	D (mm)	E (mm)	F (mm)
600 × 250	–	–	–	–	305	560
600 × 300	–	–	–	–	355	560
600 × 350	–	–	–	–	380	560
600 × 400	–	–	–	808	420	560
600 × 450	–	–	–	808	455	560
600 × 500	–	–	–	808	560	560
600	230	1064	531	–	560	560

9.21 REQUIREMENTS FOR GLASS-FIBER-REINFORCED EPOXY PIPES AND FITTINGS

This specification covers the general requirements for the purchase, inspection and transportation of pipes, fittings, and flanges made from glass-fiber-reinforced epoxy (GRE), which belongs to glass-fiber-reinforced thermosetting plastics (GRP).

Described are pipes, fittings, and flanges made by the filament-winding, centrifugal casting, or pressed-sheet molding process.

Section 9.21.4 describes the qualification testing program to which a manufacturer is subjected before his first delivery of GRP pipes and fittings.

Section 9.21.5 describes the minimum number of production tests to be carried out for each subsequent delivery of GRP pipes and fittings by a manufacturer whose products have been qualified successfully.

Section 9.21.6 describes the qualification testing procedure.

A number of requirements in this specification are comparable to those of API Specification 15 LR.

This specification does not cover high-pressure piping, as defined by API Spec. 15 LR (approx. 70 bar, (1000 psi)), casing and tubing, and reinforced plastic mortar piping. The company should be contacted for those applications.

The requirements of this specification should be adhered to, except where national and/or local regulations exist in which specific requirements are more stringent.

The contractor should determine by careful scrutiny which of these requirements are the more stringent and which combination of requirements will be acceptable as regards safety, economics, and legal aspects.

In all cases the contractor should inform the company of any deviation from the requirements of this specification considered to be necessary to comply with national and/or local regulations. The company may then negotiate with the authorities concerned with the object of obtaining agreement to follow this specification as closely as possible.

9.21.1 Base Materials

All base materials should be new and unused, and should be free from all contaminations and imperfections. The base materials, e.g., resins, glass-fiber reinforcing materials, pigments, and other materials, when combined as a composite structure, should produce pipe, fittings, and flanges that meet the requirements of this specification. All base materials should be specified in writing by the pipe manufacturer and certified by the raw materials supplier(s) per delivery.

9.21.1.1 Epoxy Resins

Unless otherwise agreed, the pipes, fittings, and flanges should be made from a bisphenol A epichlorohydrin epoxy resin, e.g., "EPIKOTE" 828 and an aromatic or cyclo-aliphatic amine-type curing agent. The manufacturer should describe the type of resin and curing system chosen.

9.21.1.2 Glass-Fiber Reinforcement, Fillers, and Pigments

Glass-fiber reinforcement for the reinforced wall should be made of E-glass (i.e., low-alkali glass) meeting an internationally accepted standards such as BS 3691, BS 3396 and should have a finish (coupling agent) that is compatible with the epoxy resin.

Fillers are not acceptable. Thixotropic additives added to the resin/curing agent mixture for viscosity control should not exceed 2% by weight.

Pigments are only acceptable as long as they do not affect the performance of the components as defined in Section 9.21.4 or if agreed with the company in order to fulfill special application requirements.

9.21.1.3 Lining Materials

Unless otherwise agreed Section J.3.1, flange surfaces and all pipes and fittings which are to be exposed to the fluid, should have a smooth uniform resin-rich lining consisting of:

- a surfacing mat (tissue) or a veil, which may be either a C-glass (i.e., chemical-resistant glass) or a synthetic fiber, e.g., linear polyester fibers of polyacrylonitrile fibers;
- the same resin which is used for the fabrication of the pipe, fitting, or flange.

9.21.1.4 Adhesives

Adhesive for adhesive-bonded connections should be of an epoxy type, formulated to be resistant to the product to be conveyed, and for the service temperatures and pressures. It should be of the type and quality regularly supplied by the pipe manufacturer for the duty intended, and as used for the qualification test Section 9.21.4.3.1 and should have a proven record of good service.

The adhesives should be provided in a kit containing at least epoxy resin and curing agent (separately in the recommended proportions) and mixing stick, joint cleaner, sandpaper and brush, together with instructions.

The adhesive kit should have been date stamped at the time of the packaging and should indicate the required storage conditions and date of expiration of shelf life.

The adhesive kit shelf life at 40°C should not be less than 6 months from the date of shipment or 12 months from the date of production.

9.21.1.5 Rubber Sealing Rings

The sealing rings for the spigot and socket connections should be made of a rubber type resistant to the product to be conveyed and for the service temperatures and pressures. The manufacturer should state the type of rubber, providing evidence for its suitability in the proposed application.

9.21.1.6 Fixation Rod

The fixation rod for thrust-resistant spigot and socket connections with rubber sealing rings should be made of a flexible thermoplastic material, resistant to the particular service conditions, such as temperature, surrounding environment, ultraviolet exposure, etc. The manufacturer should state the type of thermoplastic material, providing evidence for its suitability in the proposed application.

9.21.2 Design and Dimensions

The manufacturer should submit a piping stress analysis based on data for his specific brand and in accordance with ANSI/ASME B31.3.

The components should also be designed and manufactured to ANSI/ASME B31.3.

If the influence of chemicals is to be taken into account, the manufacturer should state the maximum allowable operating conditions for continuous chemical service.

The company should specify those applications where the presence of a lining is not mandatory, as in cases where less severe chemical resistance is required or for electrically conductive piping.

The manufacturer should provide proper installation instructions and, if requested by the company, adequate supervision at all stages of installation.

9.21.2.1 Dimensions

9.21.2.1.1 Pipes

Diameter The pipe standard for filament-wound pipe may be based on either the inside diameter (Type A) or the outside diameter (Type B). Standard diameters for both types of pipes are given in Table 9.1.

TABLE 9.1 Standard Diameters of GRP Pipes (See Note 1 of This Table)

Nominal Pipe Size DN		Type A (Based on Inside Diameter)		Type B (Based on Outside Diameter)	
in mm	in inch	Type A1 D_i (mm)	Type A2 D_i (mm)	Type B1 (see Note 2 of this table) D_o (mm)	Type B2 D_o (mm)
25	1	25	27	33.7	30
40	1.5	40	40	48.3	45
50	2	50	53.1	60.3	55
80	3	80	81.8	88.9	86
100	4	100	105.2	114.3	106
150	6	150	159	168.3	157
200	8	200	208.8	219.1	208
250	10	250	262.9	273	259
300	12	300	313.7	323.9	310
350	14	350	344.4		
400	16	400	393.7	429	412
450	18	450	433.8		
500	20	500	482.1	532	514
600	24	600	578.6	635	616
700	28	700		738	718
750	30	750	723.1		
800	32	800		842	820
900	36	900	867.9	945	922
1000	40	1000		1048	1024
1200		1200		1255	1228
1400		1400		1462	1432
1600		1600		1668	1636
1800		1800		1875	1840
2000		2000			2044
2400		2400			2452
2800		2800			2860

(Continued)

TABLE 9.1 (Continued)

Nominal Pipe Size DN		Type A (Based on Inside Diameter)		Type B (Based on Outside Diameter)	
in mm	in inch	Type A1 D_i (mm)	Type A2 D_i (mm)	Type B1 (see Note 2 of this table) D_o (mm)	Type B2 D_o (mm)
3200		3200			3268
3600		3600			3676
4000		4000			4084

Notes:
1. These dimensions are not yet accepted and approved by the ISO Council. Some manufacturers fabricate pipes with outside diameters that are completely differ from those mentioned in the table.
2. These dimensions are equal to the dimensions given in API Spec. 15 LR.

Centrifugally cast pipe is based on the outside diameter (Type B).

The inside diameter (D_i) of type A1 pipes is equal to the nominal diameter.

The A2 and B1 series are based on a commercial need for pipes with the outside diameters equal to those of pipes made from other materials, e.g., cast iron and steel so as to enable joints to be made to existing pipes without special jointing adapters.

The dimensions of the B1 series are equal to those of Spec. 15 LR.

The B2 series have their outside diameter (D_o) related to the nominal diameter (DN) by the equation $D_o = 1.02\ \text{DN} + 4$ mm.

Wall Thickness The reinforced wall thickness of the pipe should be sufficient to withstand the temperatures, pressures, and service conditions of the particular application. It should be at least 1.8 mm.

Liner/Top coat If a lining is specified, its thickness for filament-wound pipe should be at least 0.5 mm. For a centrifugally cast pipe, the thickness of this lining should be at least 1 mm. Pressed-sheet molding compound (SMC) fittings have no liner but a press skin which should be at least 0.2 mm thick.

All piping should have a smooth resin-rich top coat.

Ovality The difference between the largest and smallest diameter (ovality) in each cross section should be not more than 0.007 D_i (Type A) or 0.007 D_o (Type B).

Ends The pipe should be supplied with plain ends, with shaved ends, with spigot ends, or with one spigot end and one (integral) socket end or with flanged ends as stated by the company.

If pipe is to be furnished with threaded ends, threading should be to API Std. 5B, unless otherwise agreed.

9.21.2.1.2 Fittings

Fittings should be supplied with plain ends, spigot ends, integral socket ends, threaded adaptors, or flanges as stated by the company.

The reinforced wall thickness of the fittings should be sufficient to withstand the temperatures, pressures, and service conditions of the particular application. It should be at least 2.2 mm.

The difference between the largest and smallest measured inside diameter (ovality) should not be more than 0.007 D_i (Type A) or 0.007 D_o (Type B).

9.21.2.1.3 Flanges

The outside diameter and drilling template of flanges should be in accordance with ANSI B 16.5 class 150. The flange face should be flat-type.

9.21.2.2 Prefabricated Piping Systems

Prefabricated piping systems may have adhesive-bonded socket/spigot connections or hand-laminated butt and strap joints or integral spigots and/or sockets for connections with rubber sealing rings.

The surfaces without a liner at adhesive-bonded connections exposed to the product should be covered by the adhesive.

The butt and strap joints should be laminated over a length of at least the pipe diameter on the outside, and if the diameter allows, also on the inside.

All machined or cut surfaces, except the spigot ends, should receive a coat of resin type, formulated to be resistant to the product to be conveyed and for use at the service temperatures and pressures.

9.21.3 Fabrication

Unless otherwise agreed by the company, the following fabrication/construction methods should be adhered to.

9.21.3.1 Filament-Wound Pipe

Filament-wound pipe should be manufactured by winding a resin-impregnated continuous fibrous glass strand roving or woven glass roving tape on to the outside of a mandrel in a predetermined pattern under controlled tension.

9.21.3.2 Centrifugal Cast Pipe

Centrifugal cast pipe should be manufactured by applying resin and reinforcement to the inside of a mold that is rotated and heated, subsequently polymerizing the resin system.

9.21.3.3 Fittings and Flanges

Fittings should be of a filament-wound construction. Flanges should be of a filament-wound construction or a pressed-SMC construction. In the latter case the length of the reinforcing fibers should be at least 12 mm. The application of fittings and flanges of another design should be specifically agreed upon between company and manufacturer.

9.21.4 Technical Requirements

Pipes and fittings purchased to this specification should meet the requirements as stated in Sections 9.21.1, 9.21.2, and 9.21.3 and should further be in accordance with the technical requirements specified in this section.

The raw materials should be checked against the sales specification as given by the manufacturer of these materials.

The manufacturer should check for each production batch per shift of 8 hours the mixing ratio of resin and curing agent. He should also record permanently the mixing ratios.

9.21.4.1 Finished Products

The following qualification requirements apply to the finished products. Manufacturers complying to this Spec. 15 LR should contact the company upon the acceptability of tests carried out.

All tests to be carried out at room temperature, unless otherwise indicated.

9.21.4.2 Appearance

Unless otherwise agreed, the inside of pipe and fittings should have a smooth and uniform lining and be in accordance with ASTM D2563 level I.

The other parts of pipes and fittings should be classified according to level II of ASTM D2563, with the following exceptions:

- air bubble : maximum 2 mm; 3 bubbles/1000 mm^2
- pimple : level III
- pit : level III, but depth less than 10% of the wall thickness

Pipes and fittings should be uniform in composition and structure, density and other physical properties.

All ends of pipes and fittings should be cut at right angles to the axis and any sharp edges should be removed.

9.21.4.2.1 Curing

The degree of curing of GRE pipe and fittings should be determined by boiling samples in acetone (dimethyl ethyl ketone) for 3 hours. After boiling and drying to constant weight the samples should not show more than 2% loss of weight.

The degree of curing may also be assessed by determination of the transition temperature by differential scanning calorimetry or differential thermal analysis in accordance with ASTM D3418. The glass transition temperature should be at least 110°C.

9.21.4.2.2 Glass Content

Filament-Wound Pipe The glass/resin ratio should be tested in accordance with EN 60 or ASTM D2584. The glass content of the filament-wound pipe should be at least 65% by weight, whereas for filament-wound fittings the glass content should be at least 55% by weight.

Centrifugally Cast Pipes For the structural wall of centrifugally cast pipes and molded fittings, the figures should be at least 45% by weight and 30% by weight respectively.

The maximum glass content should in all cases be 77% by weight.

9.21.4.2.3 Consistency of the Pipe Material

Three samples should be taken from three places situated 120 degrees apart in the same cross section. The glass content of each sample should be determined in accordance with EN 60 or ASTM D2584. The difference in the glass content between two samples should be not more than 5% by weight.

9.21.4.2.4 Water Absorption

Pipes and fittings should not show evidence of delamination of other impairment when tested in accordance with ISO 62 or ASTM D570.

9.21.4.2.5 Hydrostatic Design Stress, Pressure Ratings, and Hydrostatic Pressure Test

The long-term hydrostatic strength of pipe, fittings, and joints should be determined in accordance with Procedure A or B of ASTM D2992. The manufacturer should select the procedure and the size for these tests. Adhesive joints, if any, should be included utilizing both the factory and field adhesives and their respective joining and curing procedures, if different. The samples tested should carry the full end load due to pressure.

Testing should be conducted at at least 20°C.

The hydrostatic design stress determined in accordance with Procedure A of ASTM D2992 should utilize a service (design) factor of 1.0.

The hydrostatic design stress determined in accordance with Procedure B of ASTM D2992 should utilize a service (design) factor of 0.5.

Pressure ratings for pipe should be calculated using the hydrostatic design stress for the specific pipe material and the ISO formula for hoop stress in section 3.2.1 of ASTM D2992. The minimum reinforced wall thickness should be identified.

Flanges should be pressure rated and hydrostatic pressure tested in accordance with ASTM D4024.

The pipe, pipe spools, or pipe joints should be subjected to a bi-axial loaded hydrostatic pressure test as described in ASTM D1599.

The GRE pipe, pipe spool or pipe joints should not display a weeping effect below a hoop stress value of 150 MN/m^2.

The test pressure should subsequently be increased to 3× the design pressure and be maintained for at least 5 minutes.

During this pressure test the test specimens should not show any sign of breakage.

Note:

This test will cause irreversible deformations in the material, so that the test specimens should be discarded.

9.21.4.2.6 Stiffness of Pipe

The minimum specific tangential initial stiffness (STIS) should be not less than 2500 N/m^2 for 10 bar piping,

Alternatively the pipe stiffness at 5% deflection in accordance with ASTM D2412 can be determined. In that case the pipe should be free from cracks or delaminations at a minimum stiffness of:

20 $lb/in.^2$ for 10 bar piping.

9.21.4.2.7 Beam Deflection

The modulus of elasticity of the GRE pipe should have a minimum of 7000 MN/m^2 at 90°C when tested in accordance with ASTM D2925.

9.21.4.2.8 Impact Resistance

A steel ball 50 mm in diameter and weighing approximately 550 g should be dropped perpendicularly on to the surface of the test pipe with a free fall (which may be guided) of 300 mm.

The ball should be caught or deflected after the hit so that the rebound does not strike the pipe again. The pipe should be filled with water containing, if possible, a soluble fluorescent dye. The test should be made at room temperature and the pipe should be supported on its bottom axis on a solid flat support. Four drops should be made on randomly selected areas that are separated by a minimum length of one pipe diameter from each other. The test should be repeated on the same pipe but with the pipe pressurized at the pressure class of the pipe. Four drops should be made on different areas from those previously used. The pipe should then be pressurized to 2× the pressure class rating at 25°C for 5 minutes.

Fittings should be tested in the same manner except that the drops should be reduced from four to one in each test.

The pipe or fitting should not show any porosity or visual delamination when examined, e.g., with an ultraviolet or normal lamp.

9.21.4.2.9 Linear Thermal Expansion

The manufacturer should state in his qualification testing report the coefficient of thermal expansion of a pipe length as determined in accordance with ASTM D696.

9.21.4.3 Piping Systems

Upon request of the company the pipe manufacturer should provide the following certified documentation for mutually agreed piping diameter(s).

9.21.4.3.1 Adhesive-Bonded Piping Systems

The relevant requirements for the specific component should also apply for the adhesive-bonded piping system.

It is not allowed to apply an additional overlap laminate to the joint to obtain the necessary strength.

9.21.4.3.2 Spigot and Socket With Rubber Sealing Rings Joint Piping Systems

A spigot and socket with rubber sealing rings joint piping assembly should meet the following requirements:

- No leakages should occur during hydrostatic pressure testing at 1.5× the design pressure during 10 minutes, whilst the test sections are deflected angularly in such a way that the center-line of one section of the assembled specimen makes an angle[1] of 1.5 degrees with the center-line of the other section.
- The joint assembly should withstand a combination of a bending force and an internal hydrostatic pressure of 1.5× the design pressure.

1. Indicated angles not to be used for actual design calculations.

The bending force should be applied in the middle of the joint and should be calculated depending on the length of the specimens and the support distance. The applied bending force should be such that the sum of the occurring axial stresses due to the internal hydrostatic pressure of 1.5× design pressure plus the occurring axial stress due to the bending force should be 2× the nominal axial stress.

After having applied the calculated bending force, 10 cycles of 0–1.5× design pressure should be performed.

The pressure cycle time should be 10 minutes (5 minutes without pressure, 5 minutes at design pressure).

The joint assembly should withstand a combination of a shear force (in N) of 20× the inside diameter (in mm) and an internal hydrostatic test pressure of 2× the design pressure. The test pressure should be cycled from zero to the test pressure, 10×, while holding the shear force. The time for one pressure cycle should be 10 minutes (5 minutes without pressure, 5 minutes at design pressure).

The test sections should be deflected while the pipe units are in horizontal position by applying a load vertically at the spigot end of the joint. The shear force should be uniformly applied over an arc of not more than 180 degrees along a longitudinal distance of one pipe diameter or 300 mm, whichever is the smaller, from the sealing of the assembled joint, at the unsupported spigot end of the pipe. The specimens in the test should be supported on blocks, placed immediately behind the bell. Instead of the applied load by external force, the use of own weight of the filled specimen can be chosen.

The joint assembly should withstand an internal vacuum of 0.74 bar absolute during 10 minutes when

- deflected angularly in such a way that the center-line of one section of the assembled specimen makes an angle[1] of 1.5 degrees with the center-line of the other section, and when
- deflected in a horizontal position by a shear force vertically applied at the spigot end of the joint over an arc of not more than 180 degrees along a longitudinal distance of one pipe diameter or 300 mm, whichever is the smaller, from the sealing of the assembled joint, at the unsupported spigot end of the pipe.

Note:
Pressure stabilizing for 30 minutes is allowed.

9.21.4.3.3 Flanged Piping Systems

Flanges should withstand, without any visible sign of damage, a bolt torque of at least 1.5× that recommended by the manufacturer at the design pressure.

For this test a flanged section should be bolted against a flat face steel flange. The bolts should be tightened in 7 N m increments according to the recommended practice.

Two flanged sections should be bolted together using the gasket and bolt torque for standard field installation as recommended by the manufacturer. This assembly should meet the following requirements:

- No leakages should occur during hydrostatic pressure testing at 1.5× the design pressure during 168 hours.

Retorquing to the manufacturers specified level after initial pressurization is permitted.

- No rupture of the flanged connection should occur during hydrostatic pressure testing at 2× the design pressure for 10 minutes. Leaking past the gasket interface is permissible during this test. Bolt torque may be increased, if necessary, during the test in order to minimize gasket leaking and to achieve the pressure necessary to cause flange failure.

9.21.4.3.4 Threaded Piping Systems

The qualification testing of threaded piping systems should be in accordance with the requirements given by the company.

9.21.5 Inspection and Testing

This section describes the minimum number of acceptance tests required for each delivery of GRP pipes and fittings purchased to this specification from a manufacturer whose products have been qualified successfully (Section 9.21.4).

Additional tests may be established by mutual agreement between the manufacturer/contractor and the company prior to any contract award.

If the material fails to pass any of these tests, this may constitute sufficient cause for rejection.

9.21.5.1 Acceptance Tests

9.21.5.1.1 Visual Inspection

All pipes and fittings should be visually inspected.

9.21.5.1.2 Dimensions

The dimensions of all pipes and fittings (Section 9.21.2.1) should be checked in accordance with ASTM D3567.

9.21.5.1.3 Curing

The degree of curing of each lot (see Note 1 in Section 9.21.5.1) of pipe and fittings (Section 9.21.4.2.1) should be checked at random by means of a

Barcol impressor (ASTM D2583 or EN 59) and should have a minimum value of 40.

9.21.5.1.4 Glass Content (see Note 2 in Section 9.21.5.1)

The glass content of each lot (see Note 1 in Section 9.21.5.1) of pipe and fittings should be checked (see Section 9.21.4.2.2).

9.21.5.1.5 Hydrostatic Pressure Test

All pipes furnished under this specification should be subjected to a hydrostatic pressure test at room temperature. The test pressure should be equal to 1.5× the pressure class rating and be maintained for at least 5 minutes.

During the pressure test the pipes and/or fittings should not show any sign of leakage.

Unless otherwise agreed all fittings, pipe spools, and prefabricated piping should be hydrostatic pressure tested at 1.5× the pressure class rating of the pipe.

The company should be contacted for those cases where testing of pipe spools would result in damage of the pipe ends caused by the end caps.

9.21.5.1.6 Impact Resistance (See Note 2 in Section 9.21.5.1)

The impact resistance of each lot (see Note 1 in Section 9.21.5.1) of pipe and fittings should be checked (Section 9.21.4.2.8).

Notes:

1. Unless otherwise agreed, a lot of pipe should consist of 900 m or a fraction thereof and a lot of fittings of one fitting, both of one size, wall thickness, and grade.
2. These tests are destructive tests; if appropriate, a deviation of test frequency should be established by agreement between the manufacturer and the company.

9.21.6 Qualification Testing

The full program of qualification testing is required before a manufacturer will be allowed to deliver for the first time. The company may require to repeat, completely, or in part, the qualification testing of a certain make, e.g., because of time elapsed or new developments.

Changes in the design and/or method of manufacture of pipes and/or fittings will in any case require new or additional qualification tests.

The qualification test should be carried out on products with representative diameters. The type of product, its pressure and temperature rating and number, etc., should be mutually agreed with the company.

Representative diameters and products:

Representative Products	**Pipe, Elbow 90 Degrees, Equal Lateral and Reducer**			
Representative Diameter (mm)	**50**	**150**	**300**	**600**
Diameter range (mm)	25	80	200	350
	40	100	250	400
	50	150	300	450
				500
				600

Qualification testing as described in Section 9.21.4 should be carried out by the manufacturer and witnessed and certified by an independent authority recognized by SIPM. Alternatively, testing and certification may be carried out by an independent testing organization. This should be confirmed by submitting a certificate stating the test results.

The company should be contacted for those cases where the material will be accepted and released, pending some time-consuming qualification tests (e.g., the beam deflection test). Such tests may be accepted on their satisfactory completion and the material will then receive the final clearance.

The manufacturer should state in his qualification testing report, the coefficient of thermal expansion of a pipe length as determined in accordance with ASTM D696.

9.21.7 Documentation

The manufacturer will be evaluated for ability to perform adequate and sufficient quality control (including inspections and tests performed at sufficient intervals before and during production) to ensure that proper and correct base materials are being used, that the finished material meets physical and chemical specifications, and that the finished product meets all dimensional and performance requirements.

BS 5750 will be used as a guideline in this respect.

In order to assure traceability of materials and products, the manufacturer should keep a record of all quality control tests performed and should maintain this record for a minimum period of 5 years from the date of manufacture.

9.21.7.1 Manufacturers Drawings

9.21.7.1.1 Pipes

The following pipe dimensions and tolerances, when applicable, should be stated by the manufacturer and should be in accordance with the certified manufacturer's drawings:

- pipe inside/outside diameter,
- minimum total wall thickness,

- overall pipe length,
- effective pipe length,
- outside/inside diameter of the end,
- length of joint,
- conical form of spigot/socket,
- spigot or socket chamber for rubber sealing rings and for fixation rod,
- shear length, i.e., distance between chamber for rubber sealing rings and fixation rod.

9.21.7.1.2 Fittings

The following fitting dimensions and tolerances, where applicable, should be stated by the manufacturer and should be in accordance with the certified manufacturers drawings:

- fitting inside/outside diameter,
- minimum total wall thickness,
- overall fitting length,
- effective fitting length,
- conical form of spigot/socket taper,
- length of joint,
- spigot or socket chamber for rubber sealing ring and for fixation rod,
- shear length, i.e., distance between chamber for rubber sealing ring and fixation rod.

9.21.7.1.3 Flanges

The following flange dimensions, where applicable, should be stated by the manufacturer and should be in accordance with the certified manufacturer's drawings:

- thickness of the flange,
- rating,
- bolt hole circle and diameter.

9.21.7.1.4 Prefabricated Piping Systems

The following flange dimensions, where applicable, should be stated by the manufacturer and should be in accordance with the certified manufacturers drawings:

- items as indicated in Sections 9.21.7.1.1, 9.21.7.1.2, and 9.21.7.1.3,
- face-to-face, center-line-to-face, and center-line-to-center-line,
- lateral off-set of flanges,
- flange face alignment.

9.21.7.2 Certification

The manufacturer should keep complete quality control and test reports. He should submit a certified record of inspection and testing together with a statement of compliance with the requirements. These should also include the certificates of the steel parts, if any.

If appropriate, he should issue a list, showing each deviation from the purchase order.

9.21.8 Marking and Packaging

9.21.8.1 Marking

All pipes and fittings should be permanently marked with the manufacturer's name or trade name, the pressure class, the nominal diameter, and the vendor's identification code.

The marking should remain legible under normal handling and installation practices.

Markings for identification purposes should be made in such a manner as not to impair the integrity of the pipe/fittings material.

9.21.8.2 Packaging

The pipes and fittings should be packed in a manner that will ensure arrival at destination in a satisfactory condition and which will be acceptable to the company. Pipe ends should be protected with suitable protective covers. The covers should be securely attached.

Fastening is necessary where container transportation is used, to ensure immobilization of pipe joints.

The bottom of crates should be provided with skids to facilitate handling by forklift truck.

9.21.9 Pipe Components—Nominal Size

The purpose of this section is to establish an equivalent identity for the piping components-nominal sizes in an Imperial System and an SI System (Table 9.2).

9.22 PIPE FLANGES PRESSURE TEMPERATURE RATING

The purpose of this section is to establish an equivalent identity for the pipe flange nominal pressure temperature ratings in an Imperial system and an SI System (Table 9.3).

TABLE 9.2 Nominal Size

DN (mm)	NPS (in.)
15	1/2
20	3/4
25	1
32	1¼
40	1½
50	2
65	2½
80	3
90	3½
100	4
125	5
150	6
200	8
250	10
300	12
350	14
400	16
450	18
500	20
600	24

DN, diameter nominal; NPS, nominal pipe size.

TABLE 9.3 ANSI Rating (Class)

PN (bar)	ANSI Rating (Class) (Pounds per Square Inch, Gage)
20	150
50	300
100	600
150	900
250	1500
420	2500

PN, pressure nominal.

9.23 EXTENDED SERVICE LIMITS FOR PIPING CLASSES AT ELEVATED TEMPERATURE

- In piping classes, the use of bolting material to ASTM A193-B7 and ASTM A194-Gr.2H is limited to a maximum temperature of 450°C. However, for a number of applications the temperature limit can be as high as 538°C because of service experience in the given situations. For those higher temperatures, the flanged connections and bolting should be uninsulated.
- For piping classes with a temperature allowance to 538°C, the extended service limits should be as shown below:

Piping class CP04—extended service limits:

Temperature (°C)	475	500	525	538
Pressure (bar ga)	31	27	20	12

Piping class FP04—extended service limits:

Temperature (°C)	475	500	525	538
Pressure (bar ga)	63	55	40	25

9.24 VENT, DRAIN, AND PRESSURE INSTRUMENTS CONNECTIONS ASSEMBLIES

This section contains sketches of assembly configurations. The figure number used in the assembly, appears on Page 5 of the piping class and consists of a two-digit code: the first digit identifies the type of valves, and the second digit identifies the type of connection.

Symbol Legend

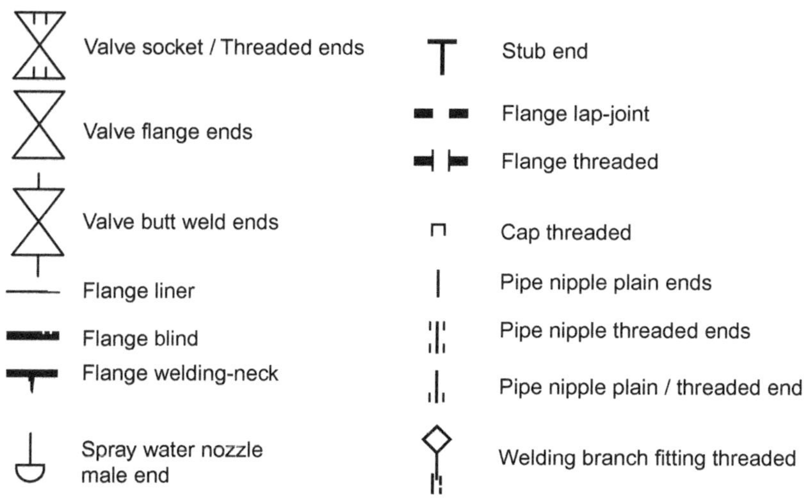

Note:

The above symbols, exclusively are used for this section.

Valve Type Identification Legend

0: Gate valve (except threaded ends)
1: Globe valve (except threaded ends)
2: Reduced bore ball valve
3: Full bore ball valve
4: Plug valve
5: Ball valve (in combination with O-ring groove in counter flange)
6: Threaded globe/gate valve
7: Diaphragm valve

Vent Connection Sketches

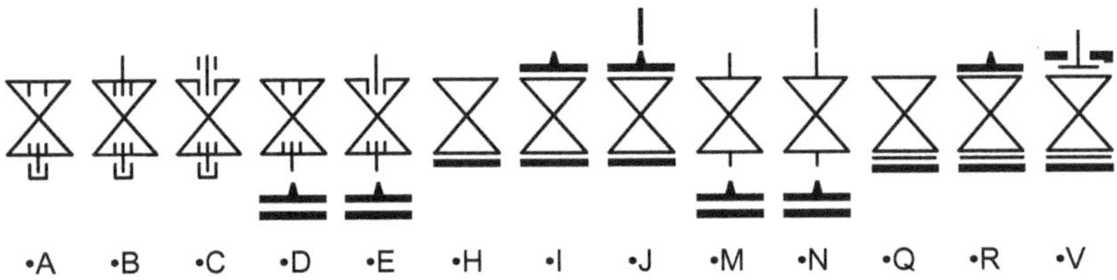

Pressure Instrument Connection

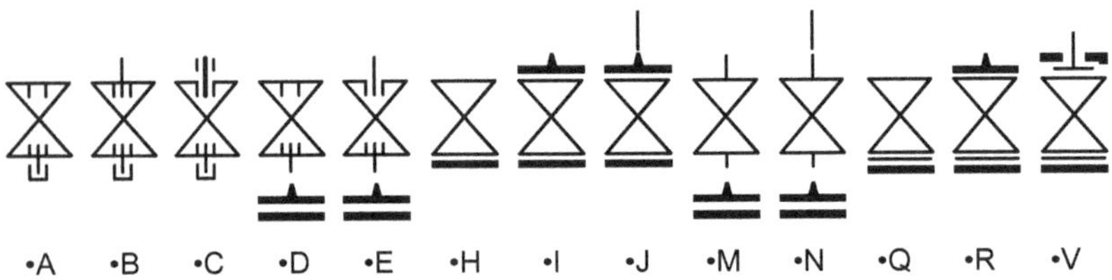

Chapter 10

Plant Piping Systems

10.1 INTRODUCTION

All materials included in the finished piping systems should be undamaged and fully in accordance with the piping material indicated on the isometric/piping plan drawing. Substitutions including heavier or thicker materials are not permitted without written approval of the engineer.

All weld numbers and the welder's identification number should be painted close to the weld to enable traceability of each weld and each welder.

To allow easy and quick reference during handling and storage, the executor should maintain the color coding on piping. All shop or field fabrication, assembly, and the installation process and utility piping system in the oil, gas, and petrochemical plants should be performed according to relevant sections of ASME/ANSI B 31.1 and B 31.3, as applicable.

10.1.1 Documentations

All documents cited hereunder should be submitted to the engineer for his review and/or approval.

10.1.1.1 Documentations to Be Prepared Before the Commencement of Pipework

The documents should include but not be limited to the following:

10.1.1.1.1 Quality Plan

The quality plan should include details and the sequence of all examinations that will be performed for control of the executor's work. The names of the individuals responsible for the implementation of all quality assurance and quality control functions should also be included.

10.1.1.1.2 Recording System

The executor should establish and maintain documented procedures for identification, collection, indexing, access filing, storage, maintenance, and disposition of the quality record.

Oil and Gas Pipelines and Piping Systems. DOI: http://dx.doi.org/10.1016/B978-0-12-803777-5.00010-1

10.1.1.1.3 Procedures

The procedures should include but not be limited to the following:

1. material take over, handling, and storage;
2. material and consumable control;
3. welding;
4. Non destructive test (N.D.T.);
5. mechanical working;
6. heat treatment;
7. mechanical cleaning;
8. chemical cleaning;
9. hydrostatic testing;
10. precommissioning and commissioning.

10.1.1.2 Documentations to Be Prepared During Execution of Pipework

The executor should maintain the following records:

1. material and consumable controls;
2. marked-up isometric drawings;
3. visual and dimensional inspection reports;
4. N.D.T. reports;
5. postweld heat treatment reports;
6. remedial action reports;
7. hydrotest reports;
8. any agreed deviation from job specification;

10.1.1.3 Documents to be Prepared After Completion of Pipework

1. as-built drawings,
2. certificate of compliance with job specification.

On completion of pipework, all documents mentioned in this chapter should be submitted to the engineer in numbers specified in the contract.

All pipework should be identified by indelible marking, free from sulfur, chloride, and other halogens. When spools will be subject to postweld heat treatment, a suitable titanium oxide, pigmented, heat-resisting paint should be used.

All applied markings should have a life of at least 1 year under site condition.

The marking applied should identify the material and the fabricator and include an item number enabling the spool to be traced to the relevant isometric drawing.

10.2 FABRICATION

All materials included in the finished piping systems should be undamaged and fully in accordance with the piping material indicated on the isometric/piping plan drawing. Substitutions, including heavier or thicker materials, are not permitted without written approval of the engineer.

All weld numbers and welder's identification number should be painted close to the weld to enable traceability of each weld and each welder.

To allow easy and quick reference during handling and storage, the executor should maintain the color coding on piping.

The executor should provide identification marks on leftover pipe length whenever marked-up pipe lengths have been fabricated/erected.

All protective coverings of piping for shipment and shipping containers should be of sturdy construction to withstand normal shipping abuse.

Piping should be stored in a relatively clean, dry or well-drained area on elevated dunnage and protected against contact with salts or salty water.

On all lines DN 80 (NPS 3) and over, pipe clamps should be used to maintain alignment when welding pipes together, both in the executor's pipe fabrication shop and on site of over-ground piping.

All piping should be fabricated in strict accordance with isometric spool drawings. If spool drawings are not furnished, piping should be fabricated to the dimensions shown on the piping arrangement drawings.

All "FW" located by dimension should be held to dimensions noted. Additional field welds, other than those indicated on the spool drawings, which may be required to suit handling, may be added by the executor.

The executor should be responsible for working to the dimensions shown on the drawings. However, executor should bear in mind that there may be variations between the dimensions shown in the drawings and those actually occurring at site due to minor variations in the location of equipment, inserts, etc. The executor should take care of these variations.

Isometrics, if supplied, may have the field welds marked on them. However, it is the responsibility of the executor to provide adequate number of "FW."

Wherever errors/omissions occur in the drawings and bills of material, it should be the executor's responsibility to notify the engineer prior to fabrication or erection.

10.3 DIMENSIONAL TOLERANCES

The tolerances listed in the following paragraphs are permissible maximums. These tolerances pertain to all piping including alloy.

General dimensions such as face to face, face or end to end, face or end to center, and center to center should be ± 3 mm (see Fig. 10.1, Item A). Tolerances should not be cumulative.

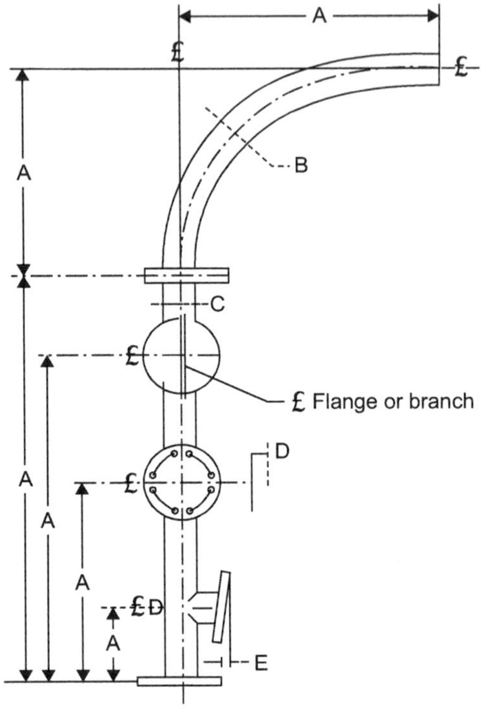

Item	Normal service conditions	Operation temp. > 460°C PN ≥ 150 (rating 900)
A	±3 mm max. from indicated dimension for face to face, center to face, location of attachments	
B	Max. 8% of dia. (for Int. press) Max. 3% of dia. (for Ext. press)	Max. 2% of dia.
	Flattening measured as difference between the max. and min. diameter at any cross section of bends.	
C	±1.5 mm max. lateral translation of branches or connections	±0.75 mm max. lateral translation of branches or connections
D	±1.5 mm max. rotation of flanges from the indicated position measured as shown	
E	4 mm/m	2 mm/m

FIGURE 10.1 Dimensional tolerances for fabricated pipework.

Flange bolt holes should straddle the vertical, horizontal, or North–South centerline unless otherwise noted. Rotation of flange bolt holes should not exceed 1.5 mm measured across the flange face parallel to a centerline and between the holes nearest to it (see Fig. 10.1, Item D).

Inclination of flange face from true, in any direction, should be 4 mm/m (see Fig. 10.1, Item E).

Displacement of branch connection from indicated location should be ±1.5 mm (see Fig. 10.1, Item C).

The difference between maximum and minimum diameter at any cross section of bends performed by the executor should not be more than 8% of the diameter (for internal pressure) and more than 3% of the diameter (for external pressure) (see Fig. 10.1, Item B) in this respect.

10.4 PIPE JOINTS

Longitudinal seams in adjoining lengths of welded pipe should be staggered over a distance of at least 5× the wall thickness of pipe measured over the circumference of the pipe or by approximately 30 degrees offset so that they do not form a continuous line at a butt welding joint.

Longitudinal welds should be located at the top 90 degrees of the pipe spool and should also clear branch connections and other welded attachments.

The toes of adjacent circumferential butt welds should be no closer than 4× the nominal thickness of the pipe, in the case of DN 300 (NPS 12) and below, with a minimum acceptable separation of 50 mm. For a nominal diameter greater than DN 300 (NPS 12), the minimum acceptable separation should be 100 mm.

10.5 WELDING

End preparation, alignment, and fit-up of pipe pieces to be welded, preheating, welded, postheating, and inspected after postweld heat treatment, should conform to standards.

Austenitic stainless steel weld deposits should have a ferrite content of 3–10%. One deposited weld metal sample should be taken for every 30 linear meter of welding and should be checked for carbon, chromium, nickel, silicon, molybdenum, manganese, and columbium content. These analyses should be used to determine the ferrite content by the Schaeffler Diagram. When approved by the engineer, the ferrite scope may be used as an alternative method to verify ferrite content. The ferritescope should be used prior to weld heat treatment.

Branch and nonpressure, part-attachment welds should not cross longitudinal seams or circumferential butt welds and should be subject to the toe-to-toe separation distance specified for circumferential butt welds.

Where such intersections are unavoidable the main weld should be subject to nondestructive examination prior to making the attachment weld. The extent of examination should be at least twice the diameter of the branch pipe measured from the center line of the branch.

Joints involving the intersection of more than two welds should be avoided.

Joints to be seal-welded should be made up clean and without the use of tape or any compound. Welding should be performed in accordance with the qualified procedure by a qualified welder. All exposed threads should be covered by the seal weld.

10.6 SCREWED PIPING (THREADED JOINTS)

If threading of piping is performed, the threads should be standard taper pipe threads, concentric with the outside diameter of the pipe in accordance with ANSI B1.20.1.

Threads should be clean cut, without any burrs or stripping, and the ends should be reamed. Threading of pipes should be done preferably after bending, forging, or heat treating operations. If this is not possible, threads should be gage checked and chased after welding, heat treatment, etc.

During assembly of threaded joints, all threads of pipes and fittings should be thoroughly cleaned of cuttings, dirt, oil, or any other foreign matter.

A thread compound or lubricant should be used for all assemblies except where seal-welded, in particular to prevent galling with stainless steel bolting. It should be suitable for the service conditions and not react unfavorably with the service fluid, the bolts, gaskets, or piping material.

10.7 FLANGED JOINTS

All flange facings should be true and perpendicular to the axis of pipe to which they are attached.

Slip-on flanges, when specified, and reducing flanges should be welded both inside and outside. If the inside weld extends beyond the face of the flange, it should be finished flush. Flange faces should be free from weld spatter, mars, and scratches.

Orifice flange taps should be located in the exact orientation shown on the spool drawing and the inside surface of orifice flanges should be made smooth and clear of any weld spatter that has penetrated through. The sections of pipe to which the orifice flanges are attached should be smooth and free from blisters and scale.

10.8 BENDING AND FORMING

Galvanized carbon steel piping, if required, should be cold-bended so as not to damage galvanized surfaces.

Bending of pipes should only be done where indicated on the isometric drawings.

If the pipe contains a longitudinal weld, this weld should be located in the neutral zone.

A detailed bending procedure should be supplied by the executor for the engineer's approval. Bending operations should not start until procedure has been approved by the engineer.

The bending procedure should include the following as a minimum:

1. bending process;
2. heating method, the heating rate, the minimum and maximum bending temperatures;
3. traveling speed of the pipe;
4. method of cooling;
5. inspection procedures, dimensional, visual, and nondestructive testing;
6. heat treatments, if applicable.

10.8.1 Cold Bending

Cold pipe bending, normally up to DN 80 (NPS 3), should be done in pipe-bending machines or presses using formers. No wrinkling, excessive thinning, or flattening is allowed. Excessive scratches, gages, or die marks should be grinded for thickness measurement; if the remaining thickness is less than minimum requirement, it should be rejected.

Cold bending of ferritic materials should be done at a temperature below the transformation range. The maximum allowable temperature for cold bending is shown in Table 10.1.

When heating is applied during cold bending of stainless steel pipe, the heat should be applied uniformly and should be carefully controlled. Local heating by handheld torches, or water cooling, is not permitted.

10.8.2 Hot Bending

Hot bending should be done at a temperature above the transformation range and in any case within a temperature range consistent with the material and

TABLE 10.1 Max. Temperature for Cold Bending

Material	Max. Temp. for Cold Bending (°C)
Carbon ½% Mo Alloy Steel	640
1% Cr ½% Mo Alloy Steel	640
1¼% Cr ½% Mo Alloy Steel	640
2¼% Cr 1% Mo Alloy Steel	680
4–6% Cr ½% Mo Alloy Steel	690
6–8% Cr ½% Mo Alloy Steel	690
8–10% Cr ½% Mo Alloy Steel	690
Austenitic stainless steel	425
Duplex stainless steel	300

the intended service. This temperature is normally between 900°C and 1100°C.

For hot bending, preferably the induction bending process should be used, which is defined as a hot-forming process involving bending machines in which the pipes are passed through an induction coil where successively a narrow band of pipe material is rapidly heated at the required bending temperature using the high frequency induction method:

1. It should be assured that the bending temperature is uniform for the complete circumference of the bend. (Induction ring with square profile is to be used.)
2. The adjacent pipe material should be kept cool by water and/or air jets, thus providing the necessary support for the heated area without the need of mandrels.
3. The proposed hot-bending process should be prequalified by a test bend, which should be mechanical and nondestructiveinspected. These prequalifications should be included in the bending procedure to be submitted to the engineer for approval.

Previous production records that include mechanical and nondestructive testing can also be used as a prequalification.

When a hot-bending technique is used, the bend should be subsequently heat-treated, i.e., annealed or normalized and tempered, according to standards.

10.8.2.1 Cutting and Trimming of Standard Fittings

Fittings like elbows, coupling, etc., should be cut/trimmed wherever required to meet fabrication and erection requirements, as per drawings or instructions of the engineer.

10.8.2.2 Jacketed Piping

The preassembly of jacketed elements to the maximum extent possible should be accomplished at the shop by the executor.

Position of jumpovers and nozzles on the jacket pipes, fittings, etc., should be marked according to pipe disposition and those should be prefabricated to avoid damaging of inner pipe and obstruction of jacket space.

10.8.2.3 Shop-Fabrication/Prefabrication

Piping spools, after fabrication, should be stacked with proper identification marks, so as to facilitate their withdrawal at any time during erection. During this period, all flange faces (gaskets contact surfaces) and threads should be adequately protected by coating with a removable rust preventive.

Care should also be taken to avoid any physical damage to flange faces and threads.

10.8.2.4 Piece Marking

Each fabricated spool piece should have a mark number painted adjacent to 50-mm wide color bands running completely around the pipe, except that for austenitic chrome nickel, nickel, or high nickel alloy, the painted mark number should be replaced by a metal tag securely attached to the pipe with metal straps.

10.9 FIELD INSTALLATION

During installation of stress-relieved lines, care must be taken to avoid heating, peening, or the development of stress concentration from any cause.

Before erection, all prefabricated spool pieces, pipes, fittings, etc., should be cleaned internally and externally.

Piping to be field pickled, sandblasted, etc., as noted on the spool drawings should be handled per standards. The executor should provide ample protection on all such cleaned piping to insure that it will be free and clear from all rust and corrosion products during the interim period between installation and start-up.

10.9.1 Piping Routing

No deviations from the piping route indicated in drawings should be permitted without the approval of the engineer.

Pipe to pipe, pipe to structures/equipment distances/clearances as shown in the drawings should be strictly followed, as these clearances may be required for the free expansion of piping/equipment. No deviations from these clearances should be permissible without the approval of the engineer.

In case of fouling of a line with other piping, structure, equipment, etc., the matter should be brought to the notice of the engineer and corrective action should be taken as per his instructions.

When the term "Field Route" is used on small screwed piping, the executor should route the piping in a neat and orderly manner consistent with good piping practice.

Slopes specified for various lines in the drawings should be maintained by the executor. Corrective action should be taken by the executor in consultation with the engineer wherever executor is not able to maintain the specified slope.

10.9.2 Cold Spring

Wherever cold spring is specified in a drawing, the executor should maintain the necessary gap.

Before performing the final tie-in, the executor should obtain a written confirmation from the engineer, indicating that the gap between the pipes is in accordance with drawing dimensions, which have been adjusted to compensate for cold spring.

If cold spring is not called for, lines should not be sprung and forced into place. If prefabricated piping does not fit, it must be corrected by straightening or rewelded.

Stress relieving of the weld (if necessary) should be performed before removing the gadgets for cold pulling.

10.9.3 Delivery, Handling, and Installation of Expansion Joints

All expansion joints should be installed in accordance with following specifications and installation drawings, if any, supplied to the executor.

Upon receipt, the executor should check for any damage occurred during transit.

The executor should bring to the notice of the engineer any damage done to the bellows/corrugations, hinges, tie rods, flanges/weld ends, etc. Each expansion joint should be blown free of dust/foreign matter with compressed air or should be cleaned with a piece of cloth. For handling and installation of expansion joints, great care should be taken while slinging. An expansion joint should never be slinged with bellows corrugations/external shrouds, tie rods, angles, etc. An expansion joint should preferably be slinged on the end pipes/flanges or on the middle pipe.

All expansion joint should be delivered to the executor at "installation length" as will be indicated on the drawings. The "installation length" is maintained by means of shipping rods, angles welded to the flanges or weld ends, or by wooden or metallic stops.

The pipe ends in which the expansion joint is to be installed should be perfectly aligned or should have specified lateral deflection as noted on the relevant drawings.

The pipe ends/flanges should be spaced at distance that will be specified in the drawings.

The expansion joint should be placed between the mating pipe ends/flanges and should be tack welded/bolted for checking correct alignment of the mating pipes.

After the expansion joint is installed, the executor should ensure that matching pipes and expansion joints are in correct alignment and that the pipes are well supported and guided.

The expansion joint should not have any lateral deflection. The executor should maintain parallelism rings or belows convolutions.

10.9.4 Precaution to Be Taken During Welding

For carrying out welding, earthing lead should not be attached with the expansion joint.

The expansion bellows should be protected from arc weld shots and welding spatter.

When an internal sleeve is provided, the bellows should be installed in the vertical position with the sleeve pointing downwards and the convolutions should be self-draining.

10.9.4.1 Insulation

Insulation should not be applied directly to the bellows convolutions.

10.9.4.2 Lubricants

The use of molybdenum disulfide lubricants should be avoided on external tie bars, etc., if the bellows operate at a high temperature.

10.9.4.3 Hydrostatic Test

Hydrostatic testing of the system having expansion joint should be performed with the shipping lugs in position.

These lugs should be removed after the testing and certification process is over.

10.10 INSTALLATION OF FLANGES

Extra care should be taken for flange connections to pumps, turbines, compressors, cold boxes, air coolers, etc. The flange connections to this equipment should be checked for misalignment, excessive gap, etc., after the final alignment of the equipment is over. The joint should be made up after obtaining approval of the engineer.

Temporary protective covers should be retained on all flange connections of pumps, turbines, compressors and other similar equipment, until the piping is finally connected, so as to avoid any foreign material from entering the equipment.

The executor should apply molycoat grease mixed with graphite powder (unless otherwise specified in piping classes) on all bolts and nuts during storage, after erection and wherever flange connections are broken and made up for any purpose whatsoever.

On lines and equipment where the operating pressure of hydrogen (H_2) mixtures will be 20.7 bars (300 psig) and over, bolts should be tightened using torque spanners. Any necessary retorquing should be carried out after the line is put in service.

10.11 INSTALLATION OF VALVES

Valves should be installed with spindle/actuators orientation/position as shown in the layout drawings. In case of any difficulty in doing this or if the spindle orientation/position is not shown in the drawings, the engineer should be consulted and work should be done as per his instructions. However the location of the valve hand wheel and/or stem should not obstruct walkways or platforms. In determining the valve stem position the following points should be considered.

No horizontally positioned stems in low-temperature service is allowed.

Butterfly valves should not be installed with the spindle in the vertical position for services where collection of dirt in the lower shaft bearing could occur.

To avoid accidental blocking owing to a loosened wedge, gate valves installed around safety/relief valves and in flare lines should be positioned with the stem pointing horizontally.

Care should be exercised to ensure that globe valves, check valves, and other unidirectional valves are installed with the flow direction arrow on the valve body pointing in the right direction. If the directional arrow is not marked on such valves, this should be done in the presence of the engineer before installation.

Fabrication of stem extensions, locking arrangements, and interlocking arrangements of valves should be carried out as per drawings/instructions of the engineer.

When installing socket welding or seal welding of ball valves, care should be taken to avoid damage to the valve seats.

10.12 INSTALLATION OF INSTRUMENT AND RELATED PIPING

Installation of in-line instruments (i.e., thermowells-restriction orifices, safety valves, control valves, rotameters, orifice flange assembly, venturimeters, flowmeters, etc.), should form part of the piping-erection work.

Care should be exercised and adequate precautions taken to avoid damage as well as entry of foreign matter into instruments during transportation, installation, and testing.

10.12.1 Instrument Air Piping

Piping from air header to different field instruments should be installed with the following considerations.

Where threaded connections are not seal-welded they should be sealed by the use of thread compound or P.T.F.E. (Teflon) tape. P.T.F.E. tape should not be used where temperatures exceed 230°C (450°F). Wherever thread compound is used on screwed fittings, it should be applied to the male thread only.

All pipe ends should be cut square, reamed of all burrs, and cleared of all foreign material.

Cutting oil should be used in cutting all threads on galvanized pipe.

10.13 VENTS AND DRAINS

High-point vents and low-point drains should be provided as per the instructions of the engineer, even if these are not shown in the working drawings. The details of vents should be as per piping material specifications.

10.14 PUMP, COMPRESSOR, AND STEAM TURBINE PIPING

1. Piping terminations at pumps or compressors should be installed so that mating flanges are parallel, concentric, and in contact prior to bolting the piping in place.
2. Auxiliary piping should be neatly routed along the baseplate and should not extend across the operating floor. This piping should not obstruct operation handling and inspection covers, bearing caps, upper halves of casing, etc.
3. Luboil lines should be separated from hot process and hot utility lines in order to avoid a fire hazard, e.g., autoignition at 260–320°C.
4. The temporary strainers should be installed as close to the machinery as possible for initial start-up and commissioning.

10.15 PIPING THROUGH WALLS AND CONCRETE FLOORS

Sleeves or holes through walls, floors of buildings, and table tops should have a size permitting the passage of a flange of the relevant pipe size to facilitate the installation of prefabricated piping and to permit insulating work.

Holes through walls and floors should be sealed after piping installation.

10.16 BURIED PIPING

Buried piping should be kept at a distance from electric power-, lighting-, and instrument-signal cables as instructed by the engineer.

For buried piping systems the executor should excavate and maintain the trench in which the piping system is to be laid.

The trench should be sufficiently wide for the pipe to be laid without damaging the pipe protective coating. The minimum trench width should not be less than pipe diameter plus 400 mm.

The trench should be excavated to a minimum depth of 100 mm below the bottom of the pipe, and the full width of trench should be graded and padded with sand or other suitable material approved by the engineer.

The sides of the trench should be free of rock, loose stones, blasting debris, or other spoil likely to fall or be dislodged, blown or swept under, around, or on top of the pipe.

Where Rock Shield or similar overwrap is used in accordance with standards of "coating". The 100-mm depth of padding may contain loose gravel and rock fragments provided that, in the opinion of the engineer, no damage to the coating would result from the inclusion of such gravel and rock fragments.

10.16.1 Protective Coating

Buried pipelines should have protective coating applied in accordance with standards "coating." The grade of protective coating will be as specified in the above-mentioned standard.

10.16.2 Laying

All brush, skids, pipe, pipe protectors, rocks, large clods, sticks, protecting rocks, and other hard objects should be removed from the bottom of the trench into which the coated and wrapped pipeline is to be lowered, so that the protective coating should not be punctured or abraded.

Pipe should normally be lowered into the trench immediately after the coating and wrapping has been passed by the engineer. Wide, nonabrasive slings or belts should be used at all times in handling the pipeline.

All coated and wrapped pipe that has been supported in any manner on padded skids or lowering devices should be subjected to close inspection by the executor to see that the coating is undamaged before the pipe reaches the bottom of the trench. Walking on coated pipe is absolutely forbidden. Backfilling should be carried out immediately after the pipeline is lowered into the trench, but the executor should first obtain the approval of the engineer. If any backfilling is carried out without approval of the engineer, he will have the right to require the executor to remove the backfill for examination of the coating and wrapping.

The initial backfill around the pipe and to a level of at least 100 mm (4 in.) above the top of the pipe should be sand- or earth-free from loose rock, large gravel sticks, branches, or other rubbish that may damage the pipe or its coating.

Where Rock Shield or similar overwrap is used, the initial backfill may contain loose gravel and rock fragments provided that, in the opinion of the engineer, no damage to the coating would result from the inclusion of such gravel and rock fragments.

Backfilling of trenches through roads should be carried out immediately after the pipe has been laid and the backfill should be compacted and finished level with the road surface. Such sections should be maintained by the executor until the work is completed and the road surface should be finally restored, as far as possible, to the same condition as before work started.

10.17 WINTERIZING AND STEAM TRACING

Steam tracing should be installed in accordance with standard. Piping connections to steam and condensate headers will be shown on the piping-arrangement-plan drawings and isometric drawings.

Bends should be used wherever practical and fittings kept to a minimum. Unions should be used when an item is traced and its removal is required for frequent maintenance.

No provision should be made for expansion movement of 13 mm or smaller tracers, since the sag or offset will take care of this amount of expansion.

For tubing or piping tracers larger than 13 mm, anchoring should generally be made at the midway point, and the piping arrangement at the ends of the tracers should be sufficiently flexible to allow for expansion of tracers.

Where it is impossible to allow for end movements, or in cases where for special reasons the unanchored length of pipe tracer exceeds 40 m, expansion loops should be provided. Minimum radius of expansion loop should be 6× the outside diameter of the tracers at bends of loop.

Insulation should be slotted at expansion loops and at anchored tracer ends where the tracers leave the pipe.

Anchors or guide clips should be installed on tracers near valves, flanges, expansion loops, and turns to avoid damage to insulation due to tracer expansion.

Tracers should be held in place with steel bands or 1.5 mm (#16 ga) soft galvanized wire loops spaced 1 m apart. On tracers 19 mm and larger spacing may be increased to 1½ m.

Lines, instruments, and instruments piping in low bubble point, acid caustic, amine or phenolic water service, or plastic or plastic lined pipe that require tracing should receive special attention to prevent general or localized overheating. Insulated spacer blocks may be installed to keep the tracer 13 mm away from the pipe.

Steam traps should be installed at low points only. All piping should as far as practicable be self draining, and steam flow through tracers should always follow a generally descending route, where practical.

Each tracer should have its own steam supply valve, and steam trap.

10.17.1 Internal Cleaning

Internal cleaning procedures should be submitted by the executor and approved by the engineer for the following:

After completion of construction works and before pressure testing, the inside of the piping system should be cleaned either mechanically or by flushing.

If so specified in project specification, chemical cleaning should be performed after completion of hydrostatic testing in accordance with standards.

10.18 INSPECTION AND TESTING

Prior to initial operation and as the piping erection progresses, the piping installation should be inspected to the extent necessary to assure compliance with the engineering design, drawings, materials, fabrication, assembly, and the test requirements of the codes and specifications. In addition to the inspection to be performed by the executor, all work is subject to inspection by the engineer or his appointed representative. Such inspection should not relieve the executor of his responsibilities as specified above.

10.18.1 Material Check

All materials should be checked by the executor or his nominee to assure compliance with the project documents.

Prefabricated pipe should be dimensionally checked by the executor or his nominee against isometric drawings.

10.18.2 Pressure Test

After completion of all erection works, all piping systems should be pressure-tested in accordance with standards.

Chapter 11

Welding of Plant Piping Systems

11.1 INTRODUCTION

This chapter covers the minimum requirements for welding work to be carried out for installation of on-plot piping in oil, gas, and petrochemical industries.

The standard relates to the requirements pertaining to welding techniques to be used, the qualification of a welder/welding operator, and welding procedures together with testing and recording involved.

It also deals with inspection, testing, limit of acceptability, and heat treatment of production welds, if required.

Facilities to which this chapter applies are indicated in ANSI/ASME B31.3 and B31.8.

11.2 QUALIFICATION OF WELDING PROCEDURE AND WELDER PERFORMANCE AND TEST RECORDS

No production welding should be carried out before Welding Procedure Specification (WPS), Procedure Qualification Record (PQR), and Welding Performance Qualification (WPQ) are qualified in accordance with requirements of ASME, Section IX.

The WPS proposed by the executor should be submitted to the engineer for his review and approval at least 2 weeks before the date of the qualification test.

Tests required for qualification of WPS should be organized by the executor, taking the following into consideration:

All materials and equipment required for preparation of test pieces and welding, including measuring instruments, etc., should be supplied by the executor unless otherwise specified.

After the completion of welding, the executor should prepare test specimens for delivery to an approved test center.

After the completion of the mechanical test, three copies of WPS(S) and related PQR(S) approved by the inspector should be submitted to the engineer.

Oil and Gas Pipelines and Piping Systems. DOI: http://dx.doi.org/10.1016/B978-0-12-803777-5.00011-3

> Weld performance, preparation of test specimens, and conducting of the mechanical test should be in the presence of the inspector.

All expenses involved in preparation of test welds, test specimens, and conducting mechanical tests, should be incurred by the executor unless otherwise specified.

The executor should submit to the engineer, at least 1 week before date of the test, a list containing the names of welders to be qualified.

All activities for the welder(s) test should be organized by the executor who is responsible for the costs involved.

A welder whose sample welds fail to meet the acceptance requirements of this chapter may be retested, at the engineer's discretion, after he has had further training. The retest should consist of one sample weld in each position in which the welder failed in his previous test.

Requalification of a welder may be required whenever there is reason to question his ability to make welds that meet the requirements of this chapter or when he has not performed welding of a similar qualification for a period of 2 months.

Test pieces to qualify the welder(s) should be selected from pipe diameter, wall thickness ranges, and positions of welding that will be involved in production welding.

The required welder qualification tests are divided into a nondestructive and a destructive test series within the following restrictions:

> For all materials with P-numbers indicated in ASME Code, Section IX, i.e., ferrous-, aluminum-, copper-, nickel-, and titanium-based material, the welder qualification test may be performed by either a destructive or nondestructive method.
>
> In respect of materials that are not covered in the above paragraph, qualification of welders should be established by a destructive test only.

The welder qualification test may be terminated at any stage of the testing whenever it becomes apparent to the inspector supervising the test that the welder is not following the welding procedures or does not have the skill required to produce satisfactory results.

High-quality, root-pass welds should be made using a butt joint without backing and a root-shielding gas. Backing rings, when used, should conform to requirements of ANSI/ASME B31.1 Clause.

11.3 PRODUCTION WELDING

1. Before any production welding is started, a detailed WPS and PQR should be qualified and/or established "related to the intended specific project."

   All essential variables of the WPS should be listed.

Under no circumstances may a welder perform any welding on piping systems in any position other than those for which he has been successfully qualified. Any such weld will be completed, removed, and replaced at the executor's expense by a qualified welder.

2. The executor should protect all electrodes from any deterioration or damage. Electrodes that show signs of deterioration or damage should be rejected and replaced at the executor's expenses. Welding machines should be operated within the amperage and voltage ranges recommended by the manufacturer for each size and type of electrode. Any welding equipment which does not meet these requirements should be repaired or replaced upon the engineer's instruction.
3. Joint preparation should be made according to ANSI B31.3 and ANSI B31.8 as applicable.

11.3.1 End Preparation for Welding

Material to be welded should be cut to the required size and shaped for welding pipe such as thermal, cold, and plasma cutting. For high-alloy steels, only plasma cutting is acceptable. The cut edges should be dressed back by machining or grinding to meet the following requirements:

Pipe ends should be free of any foreign materials. The surface to be welded should be smooth, uniform, free of laminations, tears, scale, slag, grease, paint, and other deleterious material that might adversely affect the welding.

End preparation is acceptable only if the surface is reasonably smooth and true, and slag from oxygen or arc cutting is cleaned from thermally cut surfaces. Discoloration remaining on a thermally cut surface is not considered detrimental oxidation.

Note:

Before welding, all foreign matter should be removed from the beveled ends. If any of the ends of the pipe joints are damaged to the extent that satisfactory welding contact cannot be obtained, the damaged pipe ends should be cut and beveled to the satisfaction of the engineer with a beveling machine.

The cost of all beveling should be borne by the executor should lamination, split ends, or other defects in the pipe be discovered and if the pipe joints containing such defects should be repaired or rejected as directed by the engineer.

Weld edges of stainless steels, low-alloy steels and nickel-based pipes prepared in the manner mentioned above should be inspected before welding by either magnetic particle or a die-penetrant method.

11.3.2 Alignment

In socket weld assembly, the pipe should be inserted into the socket to the maximum depth and then withdrawn approximately 2 mm away from contact between the end of the pipe and the shoulder of the socket. In sleeve pipe joints without an internal shoulder, there should be a distance of approximately 2 mm between the butting ends of the pipe.

The fit between the socket and the pipe should conform to applicable standards for socket weld fittings and in no case should the inside diameter of the socket or sleeve exceed the outside diameter of the pipe by more than 2 mm.

The space between abutting pipe ends, when aligned for welding, should be such as to insure complete penetration without burn-through. For pipe having the same dimensions the spacing should be approximately 1.5 mm. The alignment of the abutting pipe ends should be such as to minimize the offset between pipe surfaces. For pipe of the same nominal wall thickness the offset should not exceed 1.5 mm.

Flanges should be attached to piping so that the bolt holes straddle the established centerlines (horizontal, vertical, or layout centerlines) but should meet the orientation of equipment.

11.3.3 Production Welding Operation

Documents relating to qualified welding procedures and result of tests on the performance of welders/operators should be available before production welding gets started.

No welding should be performed if there are undesired weather conditions including rain, snow, high wind, blowing sand, etc.; windshields may be used when practical.

Preheating should be performed if specified in the relevant WPS.

11.3.3.1 Seal Welds

Where seal welding of threaded joints is performed, threads should be entirely covered by the seal weld. Seal welding should be done by qualified welders.

11.4 INSPECTION OF PRODUCTION WELD

11.4.1 Visual Inspection

11.4.1.1 Visual Inspection Before Performing Production Weld

1. Pipe end
   Inspect that the form and dimensions of the pipe end are in accordance with the WPS using appropriate measuring devices.

2. Cleanness
   Inspect immediately prior to welding to ensure that the fusion faces and adjacent material have the level of cleanness required. Wire brushing, dry grinding, or other mechanical means or solvent as appropriate, may be used for cleaning.
3. Fit-Up
   Check that fit-up (gap and alignment) of the parts to be welded, including any backing material, is in accordance with WPS, using appropriate measuring devices.
4. Welding consumables
   Check the classification of the welding consumables against what is cited in the welding procedure.
5. Preheating
   When preheating is required, check that the conditions specified in the welding procedure are observed so as to give a satisfactory temperature distribution around and through the joint to be welded without interfering with the access for welding.

11.4.1.2 Visual Inspection During Progress of Production Weld

1. Back gouging
   When back gouging is required by a qualified welding procedure, check that the back of the first pass is gouged out by suitable means to clean sound metal before welding is started on the gouged outside. The shape and surface of the resulting groove should be such as to permit complete fusion of the run to be deposited.
2. Interpass
   Check that each pass of weld material is cleaned before it is covered by a further pass; particular attention should be paid to the junctions between the weld metal and the fusion faces. Visual examination should be made for any visible defects such as cracks, cavities, or other deposited faults so that remedial action can be taken before further weld metal is deposited.

   In the case of multipass welds, check that the conditions specified in the welding procedure for interpass temperatures are observed.

11.4.1.3 Visual Inspection Following Completion of Production/Repair Weld

1. Cleaning and dressing
   All slag should be removed by manual or mechanical means, otherwise flaws may be obscured. When dressing of the weld face is required, it should be ensured that overheating of the joint due to the grinding action, grinding marks, and uneven finish are avoided.

In the case of fillet welds and butt welds that are to be dressed flush, ensure that the joint merges smoothly with the parent metal without under-flushing.

2. Penetration and root examination
   In the case of butt welds made from one side only, entire joints should be inspected to ensure that inadequate penetration, any root concavity, burn-through, or shrinkage grooves are within the acceptable limits using appropriate measuring devices and optical or other aids, if necessary, from the access point of view.
   In the case of butt welds made from two sides, partial penetration butt welds and fillet welds, penetration cannot be checked visually after welding. When necessary, alternative inspection methods should be used.
3. Contour
   The contour of the weld face and the height of the excess weld metal should be checked using appropriate measuring devices. Surface of the weld should be regular and the pattern and pitch of weave marks should present an even and satisfactory visual appearance.
4. Weld width
   The weld width should be consistent over the whole of the joint and should be in accordance with the dimensional requirements given on the working drawing. In the case of butt welds, the weld preparation should be at least completely filled.
5. Undercut
   Any undercut should be measured with appropriate measuring devices and should be checked against the acceptance criteria.
6. Overlap
   Toes of the weld where the weld width is excessive should be carefully inspected for weld fusion.
7. Weld flaws
   Weldment and heat-affected zones should be inspected using optical aids (if necessary).

11.4.2 Inspection by Nondestructive Testing

All nondestructive testing should be performed in accordance with the requirements and methods specified in ASME, Section V.

Nondestructive testing and examination of welds should be carried out according to detailed written procedures prepared by the executor and approved by the inspector.

Personnel responsible for various aspects of nondestructive testing, including testing operation, interpretation, evaluation, and reporting, should have qualifications and experience acceptable to the engineer.

Selected welds for non destructive testing (N.D.T.) should include a representative sample of each welder's work. Selection of welds for N.D.T. should be made by the inspector.

The welds should be accepted in the undressed condition unless dressing is necessary, in the opinion of the inspector, to effect satisfactory nondestructive testing.

The executor should submit to the inspector a certificate of calibration for all NDT equipment to be used for weld inspection.

11.4.2.1 Liquid Penetrant Examination

11.4.2.1.1 Initial Procedure

Liquid penetrant examination should be performed in accordance with Article 6 of ASME, Section V, Edition 1989.

11.4.2.1.2 Procedure Revision

A revised procedure should be prepared by the executor and approved by the inspector for the following cases:

1. Whenever a change or substitution is made in the type or family group of penetrant materials (including developers, emulsifiers, etc.) or in the processing techniques;
2. Whenever a change or substitution is made in the type of precleaning materials or processes;
3. For any change in part processing that can close surface openings of discontinuities or leave interfering deposits, such as grinding, grit blasting, and power-brush cleaning or acid treatments.

11.4.2.1.3 Techniques

Either a color contrast penetrant technique or a fluorescent penetrant technique should be used. For each technique, one of the following three types of penetrant systems should be used:

1. Water washable.
2. Postemulsifying.
3. Solvent removable.

11.4.2.2 Magnetic Particle Examination

11.4.2.2.1 Procedure

Examination procedures should be prepared in accordance with ASME, Section V. Article 7, Edition 1989.

11.4.2.2.2 Method of Examination

Examination should be done by the continuous method; that is, the magnetizing current remains on while the examination medium is being applied and while excesses of the examination medium are being removed.

11.4.2.2.3 Material of Particle

The ferromagnetic particles used as an examination medium may be either wet or dry and may be either fluorescent or nonfluorescent. If dry particles are used, prior approval of the inspector should be obtained.

11.4.2.2.4 Techniques

One or more of the following five magnetization techniques may be used, provided that prior approval of the inspector is obtained.

1. prod technique;
2. longitudinal magnetization technique;
3. circular magnetization technique;
4. yoke technique;
5. multidirectional magnetization technique.

11.4.2.2.5 Calibration of Equipment

Calibration should be made according to ASME Section V.

11.4.2.2.6 System and Sensitivity Evaluation

The overall performance and sensitivity of examination system (i.e., combination of the magnetic particle material/magnetic particle equipment as well as the sequence of operation and the level of magnetizing field) should be monitored with the test block at regular intervals to assure that the system performance is properly maintained: a reference block or fabricated test piece with known discontinuities should be prepared for the above demonstration.

11.4.2.2.7 Demagnetization

When residual magnetism in the part could harmfully interfere with the subsequent processing or usage, the part should be demagnetized after completion of the examination.

The final pass of attachment welds to P-4, P-5, and P-6 materials should be magnetic particles and examined after final postweld heat treatment (PWHT).

11.4.2.3 Ultrasonic Examination

The ultrasonic testing of the weld should be carried out by manual scanning using an A-scan and should be performed with written procedures. The procedure should include but not be limited to the following information:

1. Type of ultrasonic flaw detector.
2. Weld type and welding procedure.
3. Joint design.
4. Surface condition.
5. Type of standard block.
6. Reference block and its relative reflectors.
7. Type of probes.
8. Method of sensitivity setting for parent metal testing.
9. Method of sensitivity setting for weld metal.
10. Scanning techniques.
11. Reporting requirements.
12. Acceptance standard.
13. Operators qualification.

11.4.2.3.1 Equipment

Frequency The equipment should be capable of working at a test frequency within the range 1–5 MHz.

Time Base Linearity The linearity of the time base should be within 2% over the entire range.

Amplifier Linearity The amplifier should be linear to an accuracy of ±1 dB at any point within the range 20–80% of full screen height.

The amplitude control of flaw detector should be made according to Appendix I, Article 5, ASME, Section V.

11.4.2.3.2 Operators

Operators should be certified by the requirements of ASNT recommended practice SNT-TC-1A Level II or III.

If required, the operator should demonstrate his ability to perform the test, using the actual equipment and the technique to be employed.

11.4.2.3.3 Scanning

Parent Metal Parent metal at both sides of welded joints should be scanned, to the extent necessary for a weld examination, using a straight beam technique for:

1. locating any flaws, such as laminations and tears.
2. determining actual material thickness.

Weld Metal The weld examination should consist of scan from both sides:

1. of the weld root;
2. of the side fusion faces;
3. of the weld body;
4. in addition to scanning for defects lying transverse of the weld.

11.4.2.3.4 Sensitivity Setting

Sensitivity setting for straight and angle beam probes should be made according to Part 1, BS 3923 (1978).

11.4.2.4 Radiographic Examination

Production welds should be inspected by radiographic examinations according to approved procedure, and carried out after PWHTs, when required.

11.4.2.4.1 Procedure

Radiographic procedure should include, but not be limited to, the following:

1. material and thickness range;
2. type of X-ray tube or isotope;
3. strength of isotope used or X-ray voltage;
4. radiography technique;
5. film type;
6. intensifying screens used;
7. type of image quality indicator, place and numbers;
8. sensitivity;
9. density;
10. processing time and temperature;
11. reporting requirements

11.4.2.4.2 Techniques

A single-wall exposure technique should be used for radiography whenever practical. When it is not practical to use a single-wall technique, a double-wall/single-image technique should be conducted, and for pipes with DN less than 90 (NPS 3½) a double-wall/double-image technique should be used.

11.4.2.4.3 Film Type

Radiography should be made using industrial radiography film type Class II ASTM, equivalent to D7 normally and D4, if required in the opinion of the inspector, in special cases. Fluorescent and fluorometallic screens are not acceptable.

11.4.2.4.4 Image Quality Indicators

Wire type penetrameter should be used to measure radiographic sensitivity. In special cases, other types of penetrameters may be used. Maximum sensitivity should not exceed 2, where:

$$\text{Sensitivity} = \frac{\text{Size of thinnest wire}}{\text{Thickness of specimen}} \times 100$$

Number of penetrameters and placement should be made according to SE-142 ASME, Section V.

11.4.2.4.5 Film Density

Film should be exposed so that the H&D density through the weld metal should not be less than 1.8 and not greater than 3.0 for transparent-based film.

11.4.2.4.6 Selection of Radiation Sources

X-ray Minimum voltage of X-ray tubes should not be less than 200 kV.

Gamma Ray Minimum source strength should be 10 Ci for IR 192 and 5 Ci for cobalt 60.

Darkroom Processing should be carried out in a darkroom with the following facilities as a minimum requirement:

1. Automatic or manual processing devices with a temperature-indicating controller.
2. Loading bench.
3. Red light (subdued light).
4. Adequate ventilation.
5. Drying bench.
6. Clean and washable floor.

Processing Processing should follow a standard technique with five separate stages: development, stop bath, fixing, washing, and drying. Development time and temperature should be controlled according to standard (4 min at 20°C) or in accordance with the film manufacturer's recommendations.

Excess density due to additional developing time or higher temperature should not be allowed.

The inspector or the engineer is allowed to examine processing conditions to evaluate the quality and density of radiographs.

Personnel The personnel employed in carrying out radiography conforming to this chapter should be certified in accordance with the recommendations of ANST-recommended practice SNT-TC-1A. Radiographic personnel should obtain a certificate of competence for working with radiation sources from Atomic Energy Organizations.

Quality of Radiographs All radiographs should be free from mechanical, chemical and other blemishes, and should overlap sufficiently to ensure no portion of the joint remains unexamined.

Identification

1. Consideration should be given to following an image identification procedure. A system should be used to produce a permanent identification on the radiograph of a weld. Consecutive letter/number series and the date of the radiograph should be plainly and permanently included on the radiograph.
2. Markers, usually in the form of lead arrows or other symbols, should be placed alongside but clear of the outer edges of the weld to identify its position.
3. In general, permanent marking of the workpiece should be used to provide reference points for the accurate relocation of the position of each radiograph.

Interpretation Only level II or III NDT personnel in accordance with SNT-TC-1A for radiography should interpret images. The radiographer should examine each radiograph and should determine the acceptability of each weld based on Chapter VI ANSI/ASME B31.3-1990. The radiographer should describe to the inspector those weld defects that he considers cause for rejection of the weld. The inspector will make a final interpretation on all welds.

Radiation Safety Whenever X-ray equipment or radioactive sources are in use, adequate precautions should be taken to protect the radiographer and others in the vicinity. Radiation hazards should be minimized by adherence to requirements cited in relevant standards.

Random Radiography Where 10% radiography is specified, the following rules should be applied:

1. At least 10% of the number of welds within the specified line class should be radiographed around their total circumference.
2. At least 10% of the number of welds made by each welder should be radiographed around their total circumference. If a welder makes fewer than 10 welds, one of these should be fully radiographed.
3. Radiographs should sample the entire range of pipe sizes that have been welded, where practicable.

4. Since the intent of 10% examination is to evaluate the quality of welder performance, radiographs should be made as soon after weld completion as practicable.

Radiography of Pipe Material With P-No = 1 Radiographic examination of a field weld of these pipes are not mandatory. However, if the process condition (i.e., operating pressure and temperature and also service fluid) warrants, in the opinion of the engineer, radiographic examination should be performed at the rate of 10%.

Radiography of Pipe Material With P-No > 1 Field welds of these pipes should be examined by radiography of the rate of 100%.

11.4.2.4.7 Brinell Hardness Test

Submerged arc-welding procedure qualifications for P-1, P-3, and P-4 group materials should have a hardness not exceeding 225 Brinell in the weld deposit.

On production welds for P-5 and P-6 group material, a Brinell hardness test should be taken on 10% of the butt welds in each P-groups material for all air-hardening filler metal. If fewer than 10 welds are made per P-group, one weld should be tested. Brinell hardness should not exceed 225 BHN.

The Brinell hardness testing should be limited to piping and tubing greater than DN 100 (NPS 4) and a wall thickness over 6.35 mm.

Carbon steel welds (sour water service) should have a maximum Brinell hardness of 200.

Carbon steel welds (sour water service) should have a maximum Brinell hardness of 185.

Hardness test results and locations should be recorded. The engineer should be permitted to witness hardness testing and should have access to the test results.

11.4.3 Inspection by Destructive Testing

When the qualified welders perform welding, the engineer should have the right to cut out one weld made by each welder on the works to prove the quality of his workmanship. This weld will be selected at random by the engineer. The executor should bear all expenses in connection with cutting out and replacing this initial test weld for each welder, even if the weld is found on test to be entirely satisfactory.

Twelve coupons may be cut for testing, and of these coupons four may be used for tensile test, four for a nick-break test, and two each for root-bend and face-bend tests. The standard of the tests will be in accordance with ASME, Section IX, Part A.

11.5 WELD DEFECTS AND ACCEPTANCE CRITERIA

Welds which are deposited by procedures differing from those properly qualified and approved should be rejected and completely removed from the piping.

Weld metal should be properly fused with the parent metal without significant undercutting or overlapping at the toes of the weld; slight intermittent undercut should be permitted provided that it does not form a sharp notch and that it meets the following requirements.

The stop and start of each run of weld should merge smoothly and should show no pronounced hump or crater in the weld surface.

Acceptance criteria should be as stated in the engineering design and should at least meet the limits stated in Table 341-3.2A of ANSI/ASME B.31.3.

For ultrasonic examination of welds, supplementary acceptance criteria cited in Clause 344.6.2 of ANSI/ASME B.31.3 should be considered.

11.6 WELD REPAIR

When a defective weld is detected, either visually or by any other method in accordance with this chapter, it should be removed to sound metal and repaired. The repair weld should be made using qualified welding procedure as well as qualified welders/welding operators. Preheating and heat treatment should be as required for the original welding.

External undercuts should be repaired by grinding off the weld cap in the undercut location and recapping the location.

No weld containing cracks, regardless of size or location, is acceptable. Cracked welds should be cut, removed, and rewelded.

On completion of repair, the weld should be radiographed, whether the defect in the original weld was detected by radiographic examination or not.

When a defective weld has been detected, the next two welds produced by the same welder should be radiographed.

Should two or more welders participate in making a defective weld, the executor and engineer should decide together which welder is responsible for the defective work. The engineer should have the right to cut out welds for further tests. The test welds (except the initial free test referred to in Clause 9.3) that meet the specified requirements and specifications when properly tested, should be replaced by a satisfactory tie-in at company's expense. In the event any test weld cut from the line does not prove satisfactory to the engineer when properly tested, it should be replaced at the executor's expense. Test welds should be cut out as soon as practicable after completion to avoid unnecessary delay and expense. When welding the line together at places where the test weld has been cut, one weld will be used if it is practicable to pull the line back into position; otherwise, two welds will be made by setting in a pipe with a minimum length of 2½× the pipe diameter, or 1.25 m, whichever is the longer.

11.7 PRE- AND POSTWELD HEAT TREATMENT

Heat treatments may be carried out either full body or locally, depending on:

- Type of heat treatment.
- Material composition of pipe.
- Number and sizes of pipe.
- Availability and cost of energy.
- Required accuracy of heat treatment.
- Welding process and welding consumable.
- Code requirement.

Heat treatment should be carried out in accordance with a qualified heat treatment procedure specification, which is to be submitted by the executor for approval of the inspector.

During heating up and cooling down, no temperature gradient should exceed:

100°C/m in axial direction, nor 40°C/m in tangential direction, to be checked by temperature recorder.

For wall thicknesses of pipe up to and including 20 mm, the rate of heating should not exceed 200–250°C/h.

For wall thicknesses of pipe over 20 mm, the rate of heating should not exceed:

5500/*t*° C/h (*t* = maximum pipe wall thickness)
or
55° C/h, whichever is greater.

The workpiece should be cooled to 400°C whereby the cooling rate is limited as follows:

- For wall thickness of pipe <20 mm 275°C/h
- For wall thicknesses ≥ 20 mm 6875/*t*°C/h (*t* = maximum pipe wall thickness) or 55°C/h, whichever is greater.

11.7.1 Preheat Requirements

Preheating of the parent metal prior to any welding, tack welding, and thermal cutting may be necessary to avoid cold cracking of certain ferritic steels in the weld and HAZ. Preheating could also be required for welding of nonferrous materials to remove moisture or to prevent hot cracking.

For preheating temperatures below 200°C for fuel gas/air burner systems, high-velocity gas/oil burners or infrared elements may be employed either locally or in a furnace.

For preheating temperatures above 200°C, electric resistance or induction heating is preferred although infrared radiators are acceptable.

An even temperature distribution is required.

Temperature control may be carried out with a Tempilstick, digital pyrometers, or a contact thermometer.

For piping shop welds, electrical heating is preferred, but ring torches are allowed when burning sulfur-free fuel.

When required for field welds of piping, the following methods of preheating should be applied:

a. For DN <250 (NPS <10)	Heating by appropriate torches is allowed
b. For DN ≥250 (NPS ≥10)	Electrical heating or heating by means of infrared or ring burners is required

The following requirements should be adhered to for the preheating zone:

1. Width of the heated zone is $2t$ (t = wall thickness) with a minimum of 100 mm on each side of the weld. Width of the insulated zone = width of zone heated +150 mm.
2. For pipe butt welds, the width of the heated band on each side of the weld is 2.5 t, with a minimum of 75 mm.
3. Special attention should be paid to the extent of heated bands in order not to aggravate the problems related to residual stress distribution, such as cracking, buckling, and distortion.

Where weld preheating is specified, welding should continue without interruption.

For Cr-Mo steel with a wall thickness of 25 mm, the above preheat should always be applied.

For other ferritic steels, intermediate lowering of preheat temperature is permitted only when at least 50% of the weld is completed. The joint should be cooled under insulation. Preheating should be restored to the specified temperature and maintained for 30 minutes, before the welding is recommenced.

Minimum preheat temperature required and recommended for materials of various P-numbers are given in Table 330.1.1 of ASME/ANSI B.31.3.

Preheat requirements for an unlisted material should be specified in the WPS.

11.7.2 Postweld Heating Treatments

This section covers basic practices that are suitable for most welding operations, but not necessarily appropriate for all service conditions.

11.7.2.1 Methods of Heating

Heat treatment should be carried out by one of the following methods, ensuring that the minimum stipulated temperature is achieved through the thickness of the pipes.

1. Heating in a stationary industrial furnace.
2. Local heating:
   a. Portable muffle furnace.
   b. Induction coils.
   c. Resistance heaters. The method of securing resistance-heating elements around the joint should be capable of holding the elements securely in contact with the pipe work throughout the heat treatment cycle. Any fixing, e.g., galvanized wire, likely to be injurious to the joint, should not be used.

   Selection of the method for heat treatment is subject to prior approval of the engineer. Manually-operated gas torches should not be used.

11.7.2.2 PWHT Requirements

Any PWHT should conform to the requirements stipulated in one of the following documents provided that approval of the engineer is obtained.

1. table 331.1.1 in ASME/ANSI B31.3;
2. approved welding procedure.

The upper limit of the PWHT temperature range given in Table 331.1.1 in ASME/ANSI B31.3 is a recommended value that may be exceeded provided the actual temperature does not exceed the lower critical temperature of material, which is given in table below.

Approximate Lower Critical Temperature

Material		**Approximate Lower Critical Temperature (°C)**
Carbon steel		725
Carbon molybdenum steel		730
1	%Cr–½%Mo, 1¼%Cr–½%Mo	775
2¼	%Cr–1%Mo, 3%Cr–1%Mo	805
5	%Cr–½%Mo	820
7	%Cr–½%Mo	825
9	%Cr–½%Mo	810

When parts of two different P-Numbers are joined by welding, the PWHT should be that specified for the material requiring the higher PWHT temperature.

When a nonpressure part is welded to a pressure part and PWHT is required for either part, the maximum PWHT temperature should not exceed the maximum temperature acceptable for the pressure retaining part.

Caution is necessary to preclude metallurgical damage to some materials or welds not intended or qualified to withstand the PWHT temperatures required.

It is preferred that PWHT be carried out in a stationary industrial furnace, but when it is necessary to apply a local heat treatment, the temperature gradient should be such that:

1. For the butt joint welds the length of material on each side of weld is at least $2.5\sqrt{rt}$, where r is the bore radius and t is the pipe thickness at the weld, and minimum insulation width should be $10\sqrt{rt}$.

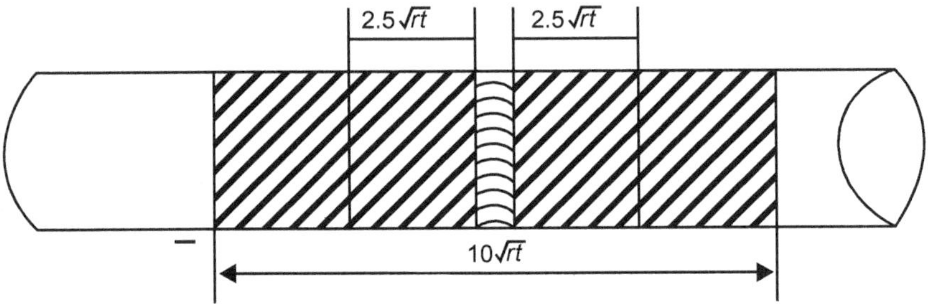

Minimum insulation width

2. For branch connection welds, the length of material from each crotch is at least:

    a. $2.5\sqrt{r_m t_m}$ along the main pipe where r_m is the bore radius and t_m is the thickness of the main pipe;
    b. $2.5\ \sqrt{r_b t_b}$ along the branch pipe where r_b is the bore radius and t_b is the thickness of the branch pipe.

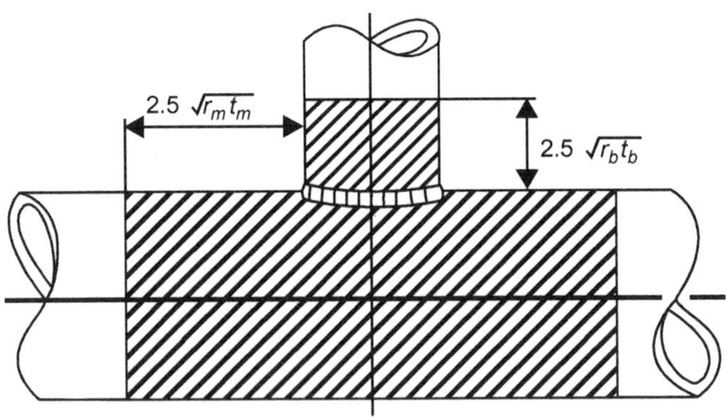

Notes:

r_m is the bore radius of main pipe;
t_m is the thickness of main pipe;
r_b is the bore radius of branch pipe;
t_b is the thickness of branch pipe.

Area (shaded) to be heated for the local treatment of branch connections

PWHT is not required for nonferrous material.

All lines in caustic service with an operating temperature more than 60°C or with caustic concentration exceeding 25% by weight should be postweld heat treated according to Table 331.1.1 of ASME/ANSI B31.3 P-No. 1.

All lines in sour water service should be postweld heat treated according to Table 331.1.1 of ASME/ANSI B31.3 P-No. 1.

All lines operating in main service above 65°C should be postweld heat treated according to Table 331.1.1 of ASME/ANSI B31.3 P-No. 1.

After PWHT, the hardness of weld deposit of lines, described in clause nos. 12.3.2.7 and 12.3.2.9, should not exceed 200 BHN.

After PWHT, the hardness of the weld deposit of sour water service should not exceed those specified in Clause nos. 9.2.6.4 and 9.2.6.5.

11.7.2.3 Alternative Heat Treatment

Alternative heat treatment should be in accordance with ANSI/ASME B31.3 Clause 331.2.1.

11.7.2.4 Hardness Test

Hardness tests of production welds are intended to verify satisfactory heat treatment. Where a hardness limit is specified in Table 331.1.1 in ASME/ANSI B31.3, at least 10% of welds in each furnace, heat-treated batch and 100% of those locally heat treated should be tested. The hardness limit applied to the weld and to the heat-affected zone (tested as close as practicable to the edge of the weld).

Chapter 12

Plant Piping Systems Pressure Testing

12.1 INTRODUCTION

This chapter covers the minimum requirements of pressure testing to be carried out on plant piping systems. Upon completion of piping systems and before commissioning, it should be pressure tested in order to prove the strength of the system, its tightness (absence of leaks), and the integrity of weldments and materials.

Before commencing a pressure test, the executor should prepare and submit for the company's approval a detailed test procedure together with the pressure test flow sheets and hydrotest diagram.

Atmospheric lines such as flare, vent, and nonhydrocarbon drain lines do not require pressure testing unless specifically called for in a project's specification. These lines should be inspected to ensure completeness of fabrication and installation.

Care should be taken to ensure that the following equipment is not subjected to field test pressures:

1. Pumps, turbines, and compressors.
2. Rupture disks, safety valves, flame arrestors, filter elements.
3. Instruments, including gage glasses, and pressure gages.
4. Equipment where the internal lining may be damaged by the medium.
5. Equipment such as filters and driers where the contents could be damaged or contaminated.
6. Vessels and exchangers which have hydrostatic test pressures less than the line test pressure.

Vents should be provided at all high points in the piping system and drains should be provided at all low points.

Fabrication and welding should be completed prior to system testing.

Welding cleanup, nondestructive examination, stress-relieving, and other heat treatments should be completed before pressure testing is performed.

Insulation and paint should not be applied over welded, screwed, and mechanical joints before pressure testing.

Oil and Gas Pipelines and Piping Systems. DOI: http://dx.doi.org/10.1016/B978-0-12-803777-5.00012-5

Testing should be performed by the executor's qualified personnel, who should be thoroughly familiar with all equipment and test procedures.

Pressure testing should be performed in the presence of the company inspector.

Pressure testing should be performed when weather conditions are suitable for inspection.

12.2 HYDROSTATIC PRESSURE TESTING

1. Where the test pressure of piping attached to a vessel is the same as or less than the test pressure for the vessel, the piping may be tested with the vessel at the piping test pressure.
2. Where the test pressure of the piping exceeds the vessel test pressure, and it is not considered practicable to isolate the piping from the vessel; the piping and the vessel may be tested together at the vessel test pressure, provided the owner approves and the vessel test pressure is not less than 77% of the piping test pressure calculated by Eq. (12.1) (see Section 12.2.4.1).

If a portion of a piping system cannot be hydrostatically tested because the presence of water is objectionable, a notation should be made on the pressure test flow sheets indicating that hydrostatic testing by water is prohibited.

Underground lines inside plot limits should be completely tested, inspected, drained, and approved before coating of field welds and backfilling are carried out.

Cast iron piping with spigot and socket joints should be intermittently covered, leaving the joints clear and all the bends securely anchored before testing.

12.2.1 Testing Area

If possible, test equipment including pumps, gages, recorders, and other items that should be located in the same area.

The area should be continuously monitored by test personnel while testing is being performed so that any change in test procedures or conditions will be noted immediately and corrective action taken as required.

12.2.2 Test Equipment

The engineer approval should be obtained prior to use of equipment intended for the execution of the pressure test.

12.2.2.1 Instruments

The required instruments for each test should be as follows:

1. Pressure gage, accurate to within 1% of its range. The range selected should be such that the pressure reading occurs between 25% and 90% of the full range of instrument.
2. Dead weight tester and temperature indicator, if specified. Instruments should be suitably protected from all elements. Prior to testing, they should be calibrated (preferably on site), and a record of the instrument calibration and identification should be kept as part of the inspection records. The calibration date and identification should be affixed to the instruments. Calibration should be done according to standards.

12.2.2.2 Pumps

The executor should be responsible for the supply of suitable pumps for testing. At a minimum, a high-volume centrifugal pump for filling and a variable, speed-positive, displacement-reciprocating pump with a sufficient discharge head and capacity for pressurizing should be provided.

The executor should provide a chemical injection pump and the means necessary for measuring the inhibitor that's added to test the water.

12.2.3 Test Medium

The test fluid normally should be fresh water and should not contain suspended solids, which may plug small lines. If it is found necessary, a water filter should be used equipped with 100-mesh screen to remove 99% of particles that are 140 μm or more in size.

The water should have a corrosion inhibitor that meets the approval of the engineer.

If the water temperature is likely to fall to 0°C or below, glycol or another antifreeze approved by the engineer should be added.

12.2.3.1 Carbon and Low-Alloy Steel

Piping manufactured from carbon steel or low-alloy steel may be hydrostatically tested with fresh water. Brackish or sea water may be used provided that prior approval of the engineer is obtained.

12.2.3.2 Stainless Steels

The chloride content in the water used for the hydrostatic testing of piping manufactured from or containing parts manufactured from austenitic stainless steel (e.g., 18Cr-8Ni or 18Cr-10Ni-2Mo) should be controlled as follows:

1. If the piping metal temperature does not exceed 50°C during commissioning, operation or nonoperation, water containing up to 150 mg/kg (150 mass PPM) chlorides may be used for the hydrostatic test.
2. If the piping metal temperature exceeds 50°C during commissioning, operation or nonoperation, the piping should be tested using condensate water, demineralized water, or oil with a minimum flash point of 50°C.

12.2.3.3 Duplex Stainless Steels

The preferred method is to use potable water (up to 150 mg/kg chlorides) and to carefully rinse the equipment immediately after the pressure test with condensate or demineralized water (up to 2 mg/kg chlorides), followed by through drying or dewatering the equipment and filling with nitrogen in such a way that all oxygen is removed from the system.

12.2.3.4 9% Nickel Alloy Steel

The test water should be of a neutral pH value, clear and free from sulfides that may otherwise precipitate stress cracking of the equipment.

In systems where residual moisture cannot be tolerated, e.g., in SO_2, acid, ammonia, and LPG service and where certain catalysts are used, other test mediums should be selected. If flammable liquid is used, consideration should be given to the test environment.

12.2.4 Test Pressures

12.2.4.1 General Process Piping-Test Pressure

At any point in a metallic piping system, the hydrostatic test pressure should be as follows:

1. Not less than 1.5× the design pressure.
2. For design temperature above the test temperature, the minimum test pressure should be calculated by the following formula, except that the value of S_T/S should not exceed 6.5:

$$P_T = \frac{1.5PS_T}{S} \tag{12.1}$$

where:
P_T = Minimum hydrostatic test gage pressure
P = Internal design gage pressure
S_T = Allowable stress at test temperature
S = Allowable stress at design temperature (see Table A-1 ANSI B 31.3)

3. If the test pressure as defined above produces a stress in excess of the yield strength at test temperature, the test pressure may be reduced to the maximum pressure so that it will not exceed the yield strength at test temperature.

12.2.4.2 Refrigeration Service-Test Pressure

Refrigerant piping should be tested in accordance with the provisions of ANSI B 31.5 ("Refrigeration Piping") and the aforementioned requirements should be successfully dried out accordingly, or other measures approved by the engineer should be applied against freezing.

Note:

However no oxygen or any combustible gas or combustible mixture of gases should be used for testing within the system. Water should not, if possible, be used for testing of refrigerant piping, but if used, it must be completely removed.

12.2.4.3 Vacuum Services-Test Pressure

Vacuum service lines should be tested at a minimum internal pressure of 200 kPa (29 psi) unless limited to a lower pressure by design and/or construction.

12.2.4.4 Utility Services-Test Pressure

The test pressure for service air and water piping should be 1½× the design pressure or 700 kPa (100 psi) for steel piping, whichever is higher. Flanged cast iron piping should be tested to 1½× the design pressure + maximum static head of pipe content during operation, but in no case more than the following test pressures:

Size (Dia.)	Class 125 Flange	Class 250 Flange
DN 25-300 (NPS 1–12)	1700 kPa (250 psi)	3500 kPa (500 psi)
DN 350-600 (NPS 14–24)	1400 kPa (200 psi)	2800 kPa (400 psi)
DN 750 (NPS 30 & LARGER)	700 kPa (100 psi)	1400 kPa (200 psi)

12.2.4.5 Jacketed Lines-Test Pressure

In jacketed lines the jacket should be tested not less than 1½× the jacket design pressure. The internal line should be tested on the basis of internal or external pressure, whichever (in the opinion of the engineer) is critical. When the pressure in the jacket is high, it might be advisable or essential to pressurize the inner pipe during hydraulic test of the jacket.

12.2.4.6 Nonmetallic Piping-Test Pressure

The hydrostatic test pressure at any point in a nonmetallic piping system should be not less than 1.5 × the design pressure, but should not exceed 1.5× the maximum-rated pressure of the lowest-rated component in the system.

12.2.5 Test Duration

The test pressure should be maintained for an adequate time (not less than 1 hour) to detect small and slow seepage leaks and to permit a thorough inspection.

12.2.6 Test Preparation

For hydrostatic testing, all pipe supports should be in position and completed before testing is undertaken.

All piping intended for other than liquid service should be adequately supported by temporary supports, if necessary, particularly on lines using spring or counterweight supports. Large adjacent lines should not be tested simultaneously where the weight of the combined test water load may overload the supports structure.

If blind flanges are necessary, the flanges should be indicated on the pressure test flow diagrams.

All lines should be cleared of debris by flushing with water or blowing with steam or air. Precautions should be taken to ensure that debris is not flushed into associated vessels, control and soft seated valves, equipment, or "dead ends."

The executor of the test should consider temporary gaskets when making connections that will be broken for testing or reassembly after testing.

Ball valves and gate valves should be completely open during the test.

The executor should take care to avoid contaminating valve seats with foreign particles.

Lines containing check valves should have the pressure source located on the upstream side. If impossible, the check valves should be removed from the line or blocked open.

Lines which have spring hangers or are counterweight-supported should be temporarily blocked up during testing in order to sustain the weight of the test fluid. Sometimes spring hangers are provided with stops for carrying the test load and need not be blocked up.

All orifice plates that interfere with filling, venting, and draining should be removed for the test.

Equipment that is not to be included in the test should be isolated by blinds or should be disconnected from the piping.

Control valves should remain in the line for testing. Those that are normally closed should be opened by clean dry air. The valves should not be jacked or forced open.

Pressure control valves with internal passages between the process fluid and the diaphragm should be isolated from the test. External connections should be disconnected or blocked during the test. The diaphragm pressure should be bled off.

Expansion joints of the sliding sleeve or bellows type should be provided with temporary means to limit lateral movement.

Expansion joints, instruments, filters and similar equipment for which the maximum permissible cold test pressure is considerably lower than the maximum hydrostatic test pressure permissible for the other components of the system should be removed or blanked off from the line before testing. Such equipment should be inspected during the commissioning with due regard to the pressure limitation.

Plugs should be removed from tell-tale holes in compensating rings and reinforcement pads around branches (nozzles) and the holes left open during hydrostatic testing and observed for any leakage.

Prior to commencement of the test, a thorough check should be made to ensure all fittings, flanges, plugs, etc., are in place. All flanges and flanged fittings should be bolted and bolts properly torqued.

Testing of more than one individual section simultaneously may be carried out by connecting the sections with suitable jumper lines. The materials used in the fabrication of these jumpers should be of the same quality as the system under test, as a minimum. Threaded piping should be minimized on jumper lines.

Instrument piping at orifice flanges should be removed, or a drain or vent should be opened to ensure that the instrument does not become pressurized because of a leaking valve.

Mounted instrumentation should be isolated from the system during testing.

The supply connection size should ensure filling of the system within a reasonable time, and it should have a flanged valve.

A valve should be used for depressurizing, for which a globe valve is preferred.

12.2.7 Procedure for Hydrostatic Pressure Test

Pressure test sections should be chosen by the executor, based on testing as much piping as possible at one time without exceeding the allowable test pressure of the weakest element in the system.

Vents and other connections that can serve as vents should be open during filling so that all air is vented prior to the application of test pressure to the system.

For convenience, exchangers and/or vessels may be tested simultaneously with connected piping, provided the test pressure of both the connected piping material and equipment does not exceed the maximum allowable pressure. Vessels should not be tested with the piping when the following conditions apply:

1. Large vessel that would overload the foundation or support if filled with water.
2. Large vessels requiring too much water to fill.

3. Vessels with internals that would be damaged by water.
4. Vessels that are extremely difficult to drain or vent.

Note:

When vessels or tanks are drained, vents should be opened to avoid pulling any vacuum.

Test pressures should be taken at the lowest point of a line or test system.

During hydrostatic testing, care must be exercised to limit the applied pressure to the particular portion of the system designated on the field pressure test flow diagram.

Care must also be taken to avoid overloading any parts of supporting structures.

Where conditions require a test pressure to be maintained for a period of time, during which the test medium in the system might be subject to thermal expansion, provision should be made for the relief of excess pressure due thereto (by either installation of a proper relief valve or discharging excess pressure).

Test pressure should not be applied against any closed valve unless the maximum allowable working pressure (MAWP) of the valve exceeds the test pressure.

If the test pressure is the same upstream and downstream of a control valve, the block and bypass valves should be left open, with the control valve open or closed (whichever is most convenient).

If the test pressure is different at upstream and downstream of a control valve, the test on the upstream portion should be made with the upstream block valve and control valve open, and with the downstream block valve and bypass valve open and blinded off.

All plain test blanks required for field testing should be provided and/or made in the field according to standards.

12.2.8 Measures to be Taken After Hydrostatic Testing

After hydrostatic testing is completed and approved by the engineer, the system should be depressurized.

After depressurizing, lines and equipment should be completely drained. The equipment should be cleaned, if necessary (in the opinion of the engineer).

While draining, the system should be vented to avoid a vacuum.

Special attention should be given to points where water may be trapped, such as valve bodies or low points.

Systems should be dried after draining. The drying process may be accomplished by blowing warm compressed air through the system or by another method approved by the engineer.

All temporary blanks and blinds should be removed. Valves, orifice plates, expansion joints, and short pieces of piping that have been removed should be reinstalled with proper and undamaged gaskets in place. Valves that were closed solely for hydrostatic testing should be opened. Temporary piping supports should be removed so that insulation and painting may be completed.

12.2.9 Leaking Welds

Welds or portions of welds that leak during hydrostatic test should be cut out, rewelded, radiographed, and the line should be hydrostatically tested again at the executor's expense.

12.3 PNEUMATIC PRESSURE TESTING

With prior approval by the engineer, pneumatic testing may be substituted for hydrostatic testing if the following conditions apply:

1. When special supports or other special arrangements would be required on large, low-pressure lines.
2. When hydrostatic tests are uneconomical or impractical.
3. When the possibility exists that piping, insulation, refractory material, or attached equipment might be damaged by water.
4. When even a small amount of water left in a system, such as a refrigerant system, could be injurious. Pneumatic test should be specified in pressure test flow sheets when it is substituted for hydrostatic testing.

The following lines are usually excluded from hydrostatic testing and are usually tested with compressed air and soap suds:

1. Instrument air lines (test with dry air if possible).
2. Air lines to air-operated valves (test with dry air only).
3. Very large (usually over DN 600-NPS 24) gas or steam overhead line.
4. Pressure parts of instruments in gas or vapor service.

If an investigation determines that testing is required for relief or blow down header systems or individual pipe stacks more than 15.2 m in height that are open to atmosphere, they need only be tested for tightness with soap suds at 0.4 bars (5 psi) air pressure.

Piping with linings subject to damage by water, should be subject to pneumatic pressure test.

Lines over DN 200 (NPS 8) from tank tops will be air tested in order to minimize excessive loads, and/or the need for special designed supports or foundations. This also includes flare lines if testing is required.

12.3.1 Test Equipment

12.3.1.1 Compressors

1. The executor should be responsible for the supply of suitable compressors for the testing.
2. Air compressors with oil-lubricated construction should not be used for purging and testing instrument air lines.

12.3.2 Test Medium

Unless otherwise specified, pneumatic testing should be performed with clean dry air or nitrogen.

Air for instrument air lines should be oil-free.

If other gases are used as a test medium, it should be nonflammable and nontoxic.

12.3.3 Pneumatic Test Pressure

The pneumatic test pressure of any piping system should be equal to 110% of the design pressure of the system. If the test pressure exceeds 150 psig, written approval of the engineer should be obtained.

12.3.4 Precautions to Be Taken in Pneumatic Pressure Testing

Due to the large energy stored in compressed gas and, hence, the potential hazard of a sudden release of this energy, pneumatic testing should be carried out in small sections to minimize the risk involved.

Materials involved in testing in which resistance to brittle fracture at low temperatures has not been enhanced, test temperatures above 16°C may be considered to reduce the risk of brittle fracture during the test.

Flange joints should be masked with tape. One small hole should be punched in the tape to indicate leaks.

Welds of piping subjected to pneumatic strength test above 150 psig should be 100% radiographed.

In view of its hazardous nature, pneumatic pressure test should only be undertaken with the presence of the engineer and inspector.

All nonessential personnel and members of the public, when necessary, should be evacuated from the test area.

12.3.5 Pneumatic Testing Procedure

The following steps should be taken in performing a pneumatic test:

1. Gradually bring the piping system pressure up to 100 kPa (ga), 14.5 psig maximum, and make a preliminary inspection. Carry out a leak test using a sonic detector or soapy water. Hold pressure for at least 10 minutes.

2. The test pressure should be increased gradually in increments of 100 kPa (14.5 psig) to provide time for stress in the piping to equalize. The gradual increase also provides sufficient time to allow the company inspector to check for leaks.
3. When the full test pressure has been reached, the pressure should be held for a sufficient time to permit inspection of piping welds and connections. The time should never be less than 1 hour.
4. The pressure should be gradually reduced to atmospheric at the end of the test.

12.4 DOCUMENTATION REQUIREMENTS

Test procedures, reports, and test results should be submitted to the engineer on completion of tests and prior to start-up of unit. The documents include following:

12.4.1 Diagrams

Pressure test flow diagrams should be prepared by the executor.

Pressure test flow diagrams should show the systems to be tested, the valves to be opened, and the equipment to be blanked with blind flanges.

Pressure test flow diagrams should provide a record of the systems tested.

A reproduced copy of the mechanical flow sheets may be marked to produce the required pressure test flow diagrams.

12.4.2 Recorder Charts (if Required)

Charts are required on all tests performed by executor.

Each chart should depict only one system test.

Charts should be signed by the executor and the engineer on completion of the test to certify that the test was performed.

12.4.3 Test Record

Actual line and equipment test pressures should be accurately recorded by the executor. These records should be made of each piping installation during the testing and should include the following information:

1. Date of test.
2. Identification of piping or equipment tested (equipment included in the test of piping system).
3. Design pressure and temperature.
4. Test pressure.
5. Test medium used (if not water).

6. Water temperature and chloride content (if applicable).
7. Test duration.
8. Applicable remarks concerning defects.
9. Inspector's approval signature. The approval signature certifies that the piping system has been tested as required.

The executor should collect, check, and furnish all information required to enable the engineer to maintain a complete test record.

Chapter 13

Engineering Aspects for Plant Piping Systems

INTRODUCTION

This covers the minimum requirements for Design of Plant Piping Systems to be used in oil, gas, and petrochemical industries. The Standard is based on ASME B31.3. The chapter consists of the following parts:

Part 1—General Design Requirements
Part 2—Above-Ground Piping Systems
Part 3—Underground Piping Systems

This chapter covers minimum requirement(s) for general aspects to be considered in design of piping for oil, gas, and petrochemical plants to be designed in accordance with ASME B31.3. These requirements include, but are not limited to, the following:

1. Loading and unloading terminals.
2. Crude oil- and gas-gathering central facilities.
3. Process units.
4. Package equipment, in accordance with ASME B31.3.
5. Pump house and compressor stations (booster stations).
6. Tank farms and oil/gas depots.

This chapter is not intended to be applicable to the following systems:

1. Heating, ventilation, and domestic water system within building (HVAC).
2. Steam piping system: within the steam generation unit and power station plant designed in accordance with ASME B31.1.
3. Nonmetallic piping systems.
4. Hydraulic systems.
5. Offshore facilities.

Oil and Gas Pipelines and Piping Systems. DOI: http://dx.doi.org/10.1016/B978-0-12-803777-5.00013-7

DESIGN REQUIREMENTS

13.1 DESIGN PROCEDURE

Piping design is characterized by two successive phases as follows:

13.1.1 Basic Design

The following documents are minimum requirements for piping design in this stage:

- Plot plan and/or equipment layout.
- Piping and instruments diagrams.
- Piping specifications relating to individual project.
- Line identification list.

13.1.2 Detail Engineering Design

Layout for erection and detailed piping drawings for construction should be produced during this stage.

Detail design of piping should include but not be limited to the following:

1. Final (detailed) P&ID (piping and instrument diagram).
2. General plot plan.
3. Unit plot plan or equipment layout.
4. Above-ground piping layout.
5. Underground piping and foundation layout.
6. Piping plans (erection drawings).
7. Isometric drawings.
8. Line identification list.
9. M.T.O. (Material Take Off list).
10. Piping material specification.
11. Pipe support schedule.
12. Stress analysis calculation.
13. Design model (optional).
14. Pressure testing P&ID.
15. Tie-in diagram.

13.1.3 Piping and Instrumentation Diagram

The P&ID should be completed in accordance with standards. The following items should be considered and shown in the P&ID:

1. Data and information of equipment.
2. Line identification.
3. Nozzle's position and size for vessels and towers.

4. Type of valves.
5. Vents, drains, and relief systems for lines and equipment.
6. Insulation and tracing on lines.
7. Pipe class (wall thickness and material).
8. Control systems and loops (instrumentation).

The Utility Flow Diagrams is a type of P&ID that represents the utility systems within a plant and shows all equipment and piping in respect of utilities (water, air, steam, . . .).

13.1.4 General Plot Plan

The space and arrangement of units should be designed in accordance with the standard requirements. The general plot plan should give the layout of the whole plant. It should be prepared to one of the following scales: 1:500, 1:1000, 1:2000. The following items should be shown in the plot plan:

1. Battery limits of complex (area boundary).
2. Geographic and conventional or plant north.
3. Elevation, with regard to the nominal plant: 0 elevation.
4. Coordinates of main roads, process units, utility units, buildings, storage tanks, and main pipe rack.
5. Location of flares and burn pit.
6. Direction of prevailing wind.

Note: Plant coordinates may be started from point N = O and E = O.

The arrangement of units areas, storage areas, buildings, and devices for shipment to be provided within the plant should be decided upon on the basis of the following factors:

1. Soil characteristics.
2. Main road or rail access ways.
3. Location of pipelines to and from plant.
4. Direction of prevailing wind.
5. Local law and regulation that may affect the location of units and storage facilities.
6. Natural elevation for location of units and equipment (such as storage tanks, waste water unit, oil/water separator, etc.).

The units should be separated by roads. Major roads should have a minimum width of 6 m, with a maximum length of 400 m. The minor roads should have a minimum width of 4 m (minor roads should not be in an area classified as zone 0 or 1).

A plant may contain one or several process units. Where any unit processes flammable fluids and may be operated independently (i.e., one unit may be shut down with others in the operation), the minimum spacing between equipment on the two adjacent units should be at least 20 m.

For units processing flammable fluids, the central control building should be adjacent to a road. It should not be located in any area classified as zone 0 to 1 (based on standards).

13.1.4.1 Security Fence

1. All sites (plants or complex) should be within a security fence.
2. Any public building, such as administration office, restaurant, clinic, etc., should be located outside of the process area boundary.
3. Except for case 4 in this section the minimum space between the security fence and units' boundary should be 20 m, and between the security fence and equipment it should be 30 m.
4. In case of special units, such as flammable material storage with vapor release and toxic materials, the minimum space should be at least 60 m from site boundaries adjacent to centers of population (domestic, work, or leisure).

Except where they are an integral part of a process unit, site utility units should be grouped together in an area classified as nonhazardous.

Fire water pumps and equipment should be sufficiently remote from the processing, storage, and loading areas, where a major fire could occur. A firefighting system should be designed in accordance with standards.

The wastewater treatment facilities should be located at the lowest points of the plant.

Loading/unloading areas for road transport should have adequate space to provide access for filling, parking, and maneuvering. A drive-through rack arrangement is preferred. The loading and unloading facilities should be downwind or crosswind from process units and sources of ignition, based on the direction of prevailing wind.

13.1.4.2 Flares

The location, spacing, orientation, and general design consideration should be in accordance with standards.

13.1.4.3 Access Requirements

1. Access ways within the plant should be provided for maintenance, emergency cases, and for firefighting from the road around the plant. The piping system should be laid in such a way to make possible passage of mobile equipment.
2. Minimum widths of access way should be as follows:

a. Vehicular access ways within units:	4.0 m
b. Pedestrian access ways and elevated walk way:	1.2 m
c. Stairways and platforms:	0.8 m
d. Footpaths in tankage areas:	0.6 m
e. Maintenance access around equipment:	1 m

3. Minimum headroom clearance for access ways should be as follows:

a. Over railways or main road:	6.8 m
b. Over access roads for heavy trucks:	6 m
c. For passage of trucks:	4 m
d. For passage of personnel:	2.1 m
e. Over fork-lift truck access:	2.7 m

13.1.5 Unit Plot Plan

The drawing should be prepared in one of the following scales: 1:200, 1:500.
The drawing should show the following items:

1. Conventional north.
2. Coordinates of battery limits and roads.
3. Symbols for equipment and coordinates of their center lines.
4. Finished floor elevation.
5. Equipment index list.

The area of any unit should not exceed 20,000 m^2, and the length of each side should not exceed 200 m.

13.1.6 Layout

Layout of equipment and piping should be designed in accordance with standards.

The piping layout should minimize piping runs on very high pressure and corrosive/toxic services such as acidic gases, and should consider economy, accessibility for operation, maintenance, construction, and safety.

Layout of equipment are as follows:

1. Compressors
   a. Generally, compressors should be installed outdoors. In case a shelter is required, the ventilation of rooms should be taken into consideration.
   b. Insofar as it is practical, all compressors should be positioned under one shelter. This arrangement makes work easier for operators and maintenance crews; in addition, one crane may serve all compressors if its deployment becomes necessary.
   c. Minimum spacing between gas compressor and open flames should be 30 m.
2. Pumps
   a. Pumps should generally be located in the open area, at or near the grade level. Adequately ventilated shelters should be provided for

large machines requiring in situ maintenance. The pumps should also be located under the pipe rack.

b. All pumps should be accessible for operation and maintenance. Adequate space for lifting and handling facilities for maintenance should be provided.

c. Pumps should be located and specified so that an acceptable NPSH can be obtained without undue elevation of suction vessels or columns. Pumps on flammable or toxic duties should not be located in pits to meet this requirement.

d. In flammable-fluid service, the horizontal distance between the related pump and adjacent heat source of 650°C or more should be 30 m minimum.

3. Fired heaters

a. A heater, or group of heaters, should be located on the periphery of a plot and immediately adjacent to an unrestricted road. There should be adequate access for firefighting from all sides of a heater.

b. The layout and design of heaters should normally be such that the tube removal can be effected by mobile lifting equipment for which there should be proper access.

4. Air-cooled heat exchangers (fin fan)

a. The location of air-cooled heat exchangers should be specifically considered with respect to any areas of special fire risk. Such consideration should include:

i. The effect of the exchanger on air movement and increased fire spread.

ii. The possibility of failure of exchanger tubes releasing more combustible fluid to the fire.

b. Air-cooled heat exchangers may be located above pipe racks, where practicable and economical.

c. Air coolers should not be located within 7.5 m horizontally from pumps on hydrocarbon service, and where practicable be at least 20 m horizontally from fired heaters, to minimize the possibility of circulating hot air.

5. Shell-and-tube heat exchangers

a. Heat exchangers should be located so that when their tube bundles are withdrawn they do not project into an emergency escape route or any road with unrestricted vehicle access. They should be so arranged that can be readily dismantled for cleaning and maintenance.

b. Heat exchangers should be located collectively and at one point as far as possible, and their tube bundle pulling area should be provided (tube bundle length + minimum 2 m).

6. Cooling towers

The direction of the prevailing wind should be considered in selecting the location of cooling towers. The towers should be located to minimize any nuisance, both within and outside the site, from the water blowout, evaporation, drift, and ice formation. The requirements of BS-4485:1988 should be met.

7. Air intakes and discharges

a. Air intakes, including intakes to heating and ventilating system, air compressors for process, instrument, plant and breathing air, and to gas turbines, should be located as far as is practicable away from areas where air contamination by dust or by flammable or toxic material can occur. They should not be located in any area classified as zone 0, 1, or 2 (except for gas turbine air intakes, which should be in accordance with the manufacturer's requirement), nor located above or below an area classified as zone 0, or 1.

Note: Intakes and discharges should be separated to prevent cross-contamination by recirculation, taking into account natural wind effects. The distance between intakes and discharges should be not less than 6 m.

8. Storage tanks (liquids)

a. Storage tanks in tank farms should be laid out in a separate area (unit) and should be completely surrounded by a bund or dyke as specified in NFPA 30:2003 (for minimum tank spacing) and IP marketing code, part 4, section 7 (for bund).

b. For tanks with diameter less than 48 m, individual bounded compounds are not required, but for each crude oil tank with a diameter of 48 m or greater, a separate bounded compound should be provided. In no case should the number of tanks in any bounded compound exceeds 6; nor the total capacity should exceed 60,000 m^3. Intermediate walls of lesser height than the main bunds may be provided to divide tankage into groups of a convenient size to contain small spillage and act as firebreaks.

c. Tanks should be laid out to provide access for firefighting. There should be no more than two rows of tanks between adjacent access roads.

d. Pumps associated with tankage operation should not be located inside a bounded tank compound.

e. For distance and spacing in respect of storage tanks reference should be made to standards.

9. Pressurized LPG storage

a. LPG storage should be laid out in accordance with standards.

b. Any site boundary to third-party property should have such a distance that the radiation at ground level, in the event of ignition of the leakage from a single relief valve and/or from a fire in a spill contaminated area, should not exceed 4.7 kW/m^2.

The ground-level radiation should be calculated using the method in API Recommended Practice 521, Appendix A.

10. Sour NGL storage

In sour NGL storage tanks, in addition to heat radiation, a safe distance with regard to an H_2S-contaminated area should be considered.

13.1.7 Isometric Drawings

All lines DN 50 (NPS 2) and larger in the process and utility areas should have isometric (spool) drawings; utility and instrument piping DN 50 (NPS 2) are exempted, unless otherwise specified.

Isometric drawings should be prepared for construction of each pipe (prefabrication or site fabrication), as per piping plan drawings.

Drawings should be designed without using scale and should include graphic art, dimensional tables, list of materials, and plant north, design data, insulation, tests, etc.

13.1.8 Line Identification List

Line identification list (line list) should include, but not limited to the following information:

1. Start point and end of line (connected to equipment or other lines).
2. Medium service.
3. Phase of flow (liquid, vapor, etc.).
4. Pressure and temperature (design, operating).
5. P&ID and reference drawing.
6. Line number (in accordance with standards).
7. Piping specification code (line class).
8. Type of insulation.
9. Pipe size.
10. Corrosion allowance.
11. Heat treatment.
12. Branch reinforcement.
13. Special information, if required.

13.1.9 Pipe Supports

Pipe supports schedule should be prepared with the following data:

1. Type of support.
2. Reference drawing for fabrication and installation.
3. Line number.
4. Location of installation (unit, area, coordinates).
5. Piping plan and civil drawing number.

The location and identification of all pipe supports should be shown on the piping plan and isometric drawings.

13.1.10 Model

Model or three-dimensional software (if required) should be made during detail engineering. The model of software should be reviewed by company with designer for any necessary modification.

The scale for construction of model components should be 1:33⅓ (or 1:30, if approved by the company).

13.1.11 Pressure Testing Diagram

This drawing should be prepared based on final P&ID(s) with the following considerations:

1. Position of valves (closed or open).
2. Isolation of equipment nozzles and limit of test section with the spectacle blind or similar facilities.
3. Installation of vent and drain connections for test.
4. Isolation or removing of all instruments.
5. Test pressure and test medium.
6. Test procedure.

13.1.12 Tie-In Diagram

1. In case of a modification or expansion of existing plants, the tie-in diagram(s) should be prepared to clarify piping connection points and their tie-ins between the existing plant and its expansion parts.
2. This diagram should be as detailed as P&ID.
3. The tie-in diagram should show the location points and procedure of tie-in.

ABOVE-GROUND PIPING SYSTEMS

This section sets forth minimum engineering requirements for safe design of above-ground piping within the property limits of plants handling oil, gas, and petrochemical products.

13.2 PIPING DESIGN

13.2.1 Design Pressure

Design pressure for piping systems should be determined in accordance with ASME B31.3 with the following additions:

1. Where the pressure is limited by a relieving device, the design pressure should not be less than the pressure that will exist in the piping systems when the pressure-relieving device starts to relieve or when the set pressure of the pressure-relieving device, whichever is the greater.

The maximum differences in pressure between the inside and outside of any piping component or between the chambers of a combination unit, e.g., a jacketed pipe, should be considered, including the loss of external or internal pressure.

Piping subject to a vacuum should be designed for a negative pressure of 100 kPa (1 bar) unless a vacuum breaker or similar device is provided, in which case a higher design pressure may be approved.

2. The value of the design pressure to be used should include the static head, where applicable, unless this is taken into account separately.

   Design pressure of a piping system subject to internal pressure should be defined as one of the following:

   1. Design pressure of the equipment to which the piping is connected.
   2. Set pressure of relief valve of the piping equipment system (if lower than (1)).
   3. A pressure not lower than the shut-off pressure or that resulting from the sum of the maximum suction pressure plus 1.2× the design differential pressure, for discharge lines of pumps and/or compressors not protected by a relief valve.
   4. The maximum differences in pressure between inside and outside of any piping component or between chambers of a combination unit, e.g., a jacketed pipe, should be considered, including the loss of external or internal pressure.

13.2.1.1 Vacuum Piping

The piping subject to vacuum should be designed for a negative pressure of 100 kPa (15 psi) unless a vacuum breaker or similar device is provided.

13.2.2 Design Temperature

Design temperatures should be determined in accordance with ASME Code B31.3 with following additions:

1. Design temperatures should include an adequate margin to cover uncertainty in temperature prediction.
2. Design maximum temperature should not be less than the actual metal temperature expected in service and should be used to determine the appropriate design stress "S" for the selected material.

In case the exterior of components are thermally insulated, the lowest metal temperature should be taken to be the minimum temperature of the contents of the pipe.

13.2.3 Operating Temperature

The operating temperature of a piping should be determined as the temperature corresponding to that of the fluid in normal operating conditions.

In case of steam-traced piping, the operating temperature should be assumed as equal to one of the following:

1. Temperature equal to 70% of steam operating temperature, if conventional tracing is employed without the use of heat-conductor cement, and when the steam temperature is higher than the operating temperature of process fluid.
2. Steam operating temperature, in case of tracing with the use of heat-conductor cement.
3. Steam operating temperature, in case of jacketed piping.

13.2.4 Piping Components

The selection of type and material of pipe should be in accordance with standards. For sour services, requirements of NACE.MR-01-75 should be considered.

13.2.4.1 Pipe

13.2.4.1.1 Pipe Material

Pipe material as cited in ASME B31.3. The selection of type and material of pipe should be in accordance with standards. For sour services, requirements of NACE.MR-01-75 should be considered.

13.2.4.1.2 Pipe Size Requirement

Pipe smaller than DN 15 (NPS ½″) should not normally be used.

The use of steel pipe in sizes: DN 32 (NPS 1¼), DN 65 (NPS 2½), DN 90 (NPS 3½), DN 125 (NPS 5), DN 175 (NPS 7), DN 225 (NPS 9) and DN 550 (NPS 22) should be avoided.

13.2.4.1.3 Pipe Wall Thickness

The required thickness of pipes should be determined in accordance with ASME B31.3:2004.

The selection of standard wall thicknesses of pipes should be in accordance with ASME B36.10M:2004.

13.2.4.2 Fittings, Bends Miters, and Branch connections

Branch connections should be calculated in accordance with ASME B31.3:2004

13.2.4.2.1 Pipe Bending

- The bending radius should be given on the isometric drawing, but in principle should not be less than 5× the nominal pipe diameter.

- All bending of stainless steel and nickel alloy pipe should be done cold. Where the size and schedule of pipe is such that cold bending becomes impracticable, a hot bending and subsequent solution heat treatment procedure should be prepared for the engineer's review and approval.
- Stress relieving is not normally considered necessary for cold-bended stainless steel and high nickel alloys but, in the case of austenitic stainless steels, it may be necessary to obtain some reduction in the residual stress level after cold bending, e.g., in case where the bend is subject to chloride or polythionic acid attack. Where stress relief is specified by the engineer, the bend should be stress relieved by heating rapidly to a temperature of 900–950°C and holding for a period of 1 hour per 25 mm thickness for a minimum period of 1 hour followed by cooling in still air. Heating should be either by local electric heating blankets or by the use of a furnace. In the latter case, the furnace gas should have a controlled sulfur content.
- For ferritic steels, with the exception of the quenched and tempered grades, a normalizing heat treatment should be applied if the cold deformation is more than 15% or when the hardness increase in Vickers or Brinell is more than 100.
- For quenched and tempered ferritic steels an appropriate stress relieving heat treatment should be applied, if the cold deformation is more than 15% or when the hardness increase in Vickers or Brinell is more than 100. Stress relieving should be at least 10°C below the tempering temperature.

Pull Bends Pull bends with a center line bend radius not less than 5× nominal pipe size (NPS) may be made in pipe sizes up to and including DN 40 (NPS 1½).

13.2.4.2.2 Miter Bends (Addition to ASME B31.3)

- Miter bends should not be used where the pressure exceeds 500 kPag (72.5 psig) or where the stress range reduction factor "f" in the case of thermal or pressure cycling, would be less than 1.
- In any case, the use of miter bends should be restricted, according to ASME B31.3. Fabrication details should be in accordance with the standard drawing.

13.2.4.3 Valves

For economy and interchangeability, types of standard valves to be selected should be kept at a minimum. Valves material and design should comply with standards.

13.2.4.3.1 Selection of Valves

Appendix A may be used as guidance for valve selection.

All valves for steam services in piping designation PN100 (Classes 600) and higher should be of the butt weld end type, except for instrument isolation valves, and blowdown valves.

Low-Temperature Services

- The following types of valves may be used for services down to −20°C:
  - gate, globe, and check valves, with metal seats or with soft insert;
  - ball valves, floating, and trunnion mounted ball with soft and metal seats;
  - butterfly valves, off-set type, soft and metal seated.
- Gate and ball valves may have cavities where cryogenic liquid could be trapped and cause excessive pressure build up during warm-up of the valve body. For such valves, in services of −20°C and lower, a cavity relief should be provided by drilling a hole of 3–5 mm in the upstream side of the wedge or ball outside of the seat facing area. These valves should be clearly marked to indicate the cavity relief side.

Plug Valves-Lubricated These valves should not be used for general purposes and should only be used when the product allows the use of a plug lubricant.

Plug valves should not be used as a single item, since they require periodical maintenance by trained staff. Alternatively, in high-pressure gas systems the use of pressure balanced (nonlubricated) plug valves may be considered.

Valves of Special Types Many special valves have been developed and proved suitable for process requirements and special services. Care should be taken to select the correct valves, with a view to the design, materials, fabrication and testing. Reference should be made to standards.

1. Flush bottom valve

   Drain valve on piping or equipment without dead nozzle end, for viscous or solidifying products.
2. Multiport valves

   A multiport ball valve or plug valve with sleeve can be selected to divert flows. Use of two normal valves is, however, preferred.
3. Iris-type valve with flexible diaphragm sleeve

   For pneumatic or gravity feed of solids and powders.
4. Steam stop valves and hydrogen valves

   Flexible wedge gate valves should be used in main steam and hydrogen lines, DN 150 (NPS 6) and above. The double-disc parallel-seat type is an acceptable alternative.

13.2.4.4 Location of Valves

1. Piping layout should ensure that valves are readily accessible to allow operation and maintenance at site.

2. Valves for emergency isolation of equipment, and valves that must be frequently operated or adjusted during operation should be accessible from grade, platform, or permanent ladder.

13.2.4.4.1 Elevated Valves

1. Generally chain-operated valves should not be used. Where this in not practical, elevated valves with the center line elevation more than 2.2 m should be chain operated.
2. Hand wheels and stems of valves should be kept out of operation aisles. Where this is not practical the elevation of valves should be 2 m from grade to bottom of hand wheel.
3. Valves above roads and in overhead pipe rack should be avoided.

13.2.4.4.2 Control Valves

1. Control valves should be located so as to facilitate their maintenance and manual operation from the operating floor. Unless due to a process reason, the valves should be located at grade level.
2. Control valve should normally be installed in horizontal piping with the valve stem vertically upward.
3. Sufficient clearance should be provided above the diaphragm and below the bottom of the body for easy maintenance.
4. Control valves should not be bolted up against stop valves on both sides.

13.2.4.4.3 Check Valves

All check valves should preferably be located in a horizontal position. The valves may be mounted in the vertical position with upward flow.

13.2.4.4.4 Safety and Pressure Relief System

Location and arrangement of safety and pressure relief valves and relief systems should be in accordance with standards. Consideration should be given to the followings:

1. Safety valves should be located as close as possible to the equipment they are protecting and be accessible for checking and maintenance purposes.
2. Discharges of relief/safety valves to the atmosphere should be arranged as follows:
   a. Minimum height from grade: 15 m.
   b. Height higher by 3 m with respect to any equipment or service platform that are located within a radius of 15 m (7.5 m for steam).
   c. Minimum distance from open flames 30 m (such as from furnace burners).

d. Each pipe discharging to atmosphere should have a 6-mm (¼″) diameter weep hole at the lowest point.

3. Safety valves for thermal relief should always be installed on those lines that may be intercepted at ends, when the line internal fluid, at max. ambient temperature, may reach conditions higher than the rating of the line itself.

   When discharge of a safety valve is connected to a close circuit (blow down), such valve should be higher than the header, so as to create a natural drainage.
4. The pressure drop between vessel and safety valve should not exceed 3% of operating pressure of the vessels.
5. Installation of block valves or spades in any location where they would isolate a vessel or system from its pressure or vacuum relief device should be in accordance with ASME B31.3:2004.
6. Discharge piping and overpressure protection system should be in accordance with ASME B31.3: 2004.

13.2.4.4.5 Block and Bypass Valves

1. Unless otherwise required by process, block and bypass valves should be provided for control valve installation as per standard.
2. Block and bypass assemblies should have a means of depressurizing and draining the associated valve and pipework.
3. A valved drain connection should be provided upstream of each control valve between the control valve and the block valve.

13.2.4.5 Flanges

13.2.4.5.1 Flange Types

Flange type should be in accordance with standards with following considerations:

1. Flanges should normally be a welding neck type.
2. Slip-on flanges should not be welded directly on to elbows or other fittings and should be double welded for all services.
3. PN 68 (class 400) flanges should not be used.
4. Where flanges in accordance with other standards such as BS 3293 are required, because of adjacent equipment, a check calculation on the suitability of the flange design for hydrostatic test conditions should be made.

13.2.4.5.2 Limitation on Flange Facings

1. The facings of PN 20 (class 150), PN 50 (class 300), PN 100 (class 600), and PN 150 (class 900) flanges should be raised face except where ring type joints are required.

2. Unless otherwise specified, the facings of flanges in PN 250 (class 1500) and PN 420 (class 2500) should be of a ring-joint type. Class 900 flanges, when used on hydrogen services, should also be of ring-joint type.
3. A steel flange should be plain (flat) face at a joint with a cast iron or nonferrous flange having a plain (flat) face.
4. For flanges, finishing standards should be considered.

13.2.4.5.3 Blind Flanges

1. Thickness of blinds for test should be in accordance with standards.
2. Blind's size DN 300 (NPS 12) and larger should be supplied with jack screws.

(All heavy flanges such as DN 300 (NPS 12) and larger should be equipped with facilities for jack bolting.)

13.2.4.5.4 Limitations on Gaskets

1. Selection of gasket should be made in accordance with standards.
2. The use of asbestos should be avoided. However, if its use is inevitable then the following should be considered.
   a. To avoid galvanic corrosion, graphite compressed asbestos fiber (caf) gasket should not be used in austenitic stainless steel piping on corrosive aqueous duties.
   b. Spiral wound gasket for use with class 900 RF flanges should be provided with both inner and outer guide rings (for class 600 and lower, outer rings would suffice).
   c. Unless required with special properties, caf gasket should be specified oil resistant to be suitable for oil refinery and chemical plant duty.
   d. Caf flat gaskets should be specified 1.5 mm thick for flanges up to DN 600 (NPS 24).

13.2.4.6 Piping Joints

Piping joints not covered by standards should be designed in accordance with ASME B31.3.

13.2.4.6.1 Threaded Joints (Addition to ASME B31.3 Para.314)

Threaded joints may be used for normal fluid service and when:

1. The fluid handled is nonflammable and nontoxic, nonhazardous, nonerosive, and the duty is noncyclic.
2. The design pressure does not exceed 1000 kPag (150 psig).
3. The design temperature is between −29°C (−20°F) and 186°C (366°F), except that steam is not included in this category.
4. The connection is provided for the pressure test.

Where seal welding of threaded joints is used, the material should be weldable.

Threaded joints and fittings should not be used for:

1. General chemical service.
2. Corrosive fluid.
3. Steam service.

To reduce the incidence of leakage, the use of threaded joints and unions where permitted should be minimized and be consistent with the needs of pipe work fabrication. Sufficient threaded joints or unions, where permitted, should be provided to facilitate dismantling of pipe work for all operational, maintenance, and inspection purposes, including requirements for shutdown and gas freeing.

With the exception of connections to instruments and instrument valve manifolds, threaded joints should not be used in stainless steel, alloy steel, or aluminum piping systems.

No threaded joints or fittings should be used between a pressure vessel or main pipes DN 50 (NPS 2) or above and the first block valve isolating a piping system. This valve should be flanged or may be socket-welded with a flanged joint immediately downstream.

Layout of piping employing threaded joints should minimize stress on joints insofar as possible.

13.2.4.6.2 Socket Weld Joints (Additional to ASME B31.3)

1. Socket welding connections should be used wherever possible up to the limiting size of DN 40 (NPS 1½) except for water service which could be used up to DN 80 (NPS 3).
2. Where permitted, socket welding joints are preferred to threaded joints, except for nonhazardous service.
3. Socket welding rather than threaded fittings should be used for searching fluid service (e.g., for glycol service).

13.2.4.6.3 Flanged Joints (Additional to ASME B31.3)

The use of flanged joints should be kept to a minimum particularly on hazardous services. However, sufficient break flanges should be provided to allow removal and replacement of piping where:

1. Process duties may be fouling.
2. Deterioration of piping or valves is anticipated in service due to corrosion, erosion, etc.

13.2.4.6.4 Expansion Joints

Guidance on the selection and application of expansion joints is contained in BS 6129, ASME B31.3 and EJMA (Expansion Joint Manufacturers Association).

Vibration Attention should be given to any expected vibration, e.g., from associated machinery, when specifying the design requirements of the bellows.

External Shrouds External shrouds should be provided for mechanical protection; the shrouds should be designed to minimize the ingress and prevent the retention of water within the convolutions.

Internal Sleeves Internal sleeves are required when:

1. Flow-induced vibrations may occur.
2. It is possible for solid debris to collect in the convolutions. This debris may be due to insufficient flushing of lines or equipment after initial construction or due to internals coming loose because of corrosion, vibration, or erosion.
3. The duty is fouling, corrosive, or contains solids. In this case purging behind the sleeve should be specified.

When an internal sleeve is provided, the bellows should be installed in the vertical position with the sleeve pointing downwards and the convolutions should be self-draining.

Spool Pieces Bellows are to be welded directly into a line; they should be purchased with spool pieces of the same material as that of the line and welded to the ends of the bellows.

Lubricants The use of molybdenum disulfide lubricants should be avoided on external tie bars, etc., if the bellows operate at a high temperature.

Support The supporting and guiding of piping systems containing bellows expansion joints should be fully assessed at the design stage, preferably also by the bellows expansion joint manufacturer, to ensure that the bellows expansion joint deflects in the manner for which it was designed, and that the system retains structural stability.

13.2.5 Instrument Piping

Piping specification should apply up to and including the first block valve(s) from the process lines or equipment.

13.2.5.1 Requirements

Limit and type of piping for instrument connections should be in accordance with ASME B31.3.

Pressure points should be as short as possible. Long connections, if unavoidable, and connections to vibrating lines should be properly braced.

13.2.6 Instrument Connections

13.2.6.1 Level and Contents (Level Gages)

1. Connection for external float and displacer chambers should be in accordance with standards. Branch connection should be DN 50 (NPS 2).
2. Level gages should be connected with block valves to the equipment. Level gages in PN 150 (pressure rating 900) and higher should have a double block valve.
3. Stand pipes should be applied if more than two pair of level gage connections are required. The minimum diameter of a stand pipe should be DN 80 (NPS 3), and the equipment connection should be DN 50 (NPS 2).
4. If the required level gage is too large for a single gage, multiple level gages should be used. The connection of stages should have a minimum overlap of 25 mm.
5. Loads on equipment nozzles, from the weight of a long stand pipe with level gages and/or thermal expansion forces, should be considered.
6. There are two alternatives for connecting level gages (case II is the preferred design):
   a. with block valves between the level gage and the stand pipe;
   b. with block valves between the stand pipe and the equipment.
7. Connections for internal level instruments and flange mounted differential pressure instruments should be sized to permit float removal.
8. Drainage from gage glasses, external float and displacer chambers should be led via a tundish to a suitable drain at grade.

A closed disposal system should be provided for drainage of vaporizing liquids that evolve H_2S, other toxic gas or large volumes of flammable vapor.

13.2.6.2 Temperature Measurement

1. Connection for thermowells should be in accordance with standards.
2. Branches for flanged thermowells should be DN 40 (NPS 1½) minimum and should conform to the process duty line specification.
3. Thermowells should be located at least 10 times the pipe inside diameter downstream from the mixing point of two different temperature streams.

13.2.6.3 Pressure Measurement

1. Connections for pressure instruments should be in accordance with standards.

2. Isolation valves should be full bore when rodding is specified.
3. Instrument connections should be orientated horizontally in vertical lines and, on or the top of horizontal lines. Tappings at side or 45 degrees from horizontal will be permitted only where necessary to prevent fouling from adjacent pipework, structure, or equipment. The selected orientation should permit remote mounted instruments to be self-venting or self-draining. Connections for instruments on immiscible fluids should be horizontal.

13.2.6.4 Flow Measurement

1. Orifice flange connections should be in accordance with standards.
2. For inline flow measurement devices, the manufacturer's installation requirements should be followed.
3. Piping and primary valving at the orifice fitting or other flow measurement device should conform to the line specification.
4. Line-tap connections should conform to the requirements for pressure tapping.
5. Orifice flanges should be installed in horizontal pipes, as far as possible. When it is impossible to find a sufficient meter run in a horizontal piping, orifices may be installed in a vertical piping with upwards flow for liquids and downward flow for gas and steam.
6. The minimum meter run required for both upstream and downstream of the orifice flanges should be in accordance with standards.

13.2.6.5 Analyzers

1. Process line connections for analyzer sampling systems should be either:
   a. A flanged connection to the line specification, to accommodate a probe.
   b. A line connection as detailed for pressure instruments on standards.
2. Fast loops up to and including the main sample filter or pressure reducing valve should be to line specification.
3. Sample lines from the process connection or offtake from the fast loop should be under instrument impulse lines.
4. For inline analyzers, manufacturer's installation requirements should be followed.

13.2.6.6 Sample Systems

13.2.6.6.1 Manual Sampling Systems

1. The system should be designed to minimize the possibility of loss of containment, taking particular account of the process conditions and hazardous nature of the material to be sampled.

2. A representative sample should be obtained by sitting the sample connection at a suitable point on the process line in accordance with standards.
3. The effect on area classification should be considered in positioning sample points.
4. Sample piping should be as short as possible.
5. All operational parts of sampling systems should be easily accessible, including the primary isolation valve.
6. Sample points should either be approximately one meter above ground or 1 m above an accessible level.

13.2.6.6.2 Sample Connections

1. Liquid sample connection should not be located at the dead ends of piping.
2. Gas sample connections should be located at the top of a horizontal run piping, or at the side of vertical piping.
3. Minimum size DN 20 (NPS ¾) sample connection in accordance with standards Appendix A should be provided for nonhazardous duties on feed and product lines, and elsewhere specified by process requirement in P&ID diagrams.
4. Sample points with two valves should have one valve at the take-off point of the process line with the size as standard drain valve, and another one at the sampling point with the size max. DN 15 (NPS ½).
5. Unless otherwise specified, sample coolers should be provided for sample connections where the fluid is above 70°C (160°F). Requirements for sample cooler should conform to ASME B31.3.

13.2.7 Vents and Drains

All piping systems should have adequate numbers of vent and drain connections to ensure effective venting and draining of the system.

The sizes of vent and drain valves should be selected as a function of the characteristics of the liquid to be drained, but they should not be smaller than those cited in the following:

1. For lines DN 40 (NPS 1½) and smaller: DN 15 (NPS ½).
2. For lines DN 50 (NPS 2) and larger: DN 20 (NPS ¾) (for nonhazardous service).
3. Small connections from hazardous service lines, e.g., nipples for vents, drains, pressure tappings, sample points, (excluding connections from orifice flanges, carriers and other instrumentation) should be DN 25 (NPS 1) minimum for strength.

If vent and drain connections are not provided on the equipment, blanked branches should be provided on the attached piping providing that effective venting and draining of the equipment through the piping system is practicable. Minimum sizes of vent and drain connections are given in the table below:

Equipment Capacity		Vent		Drain	
m^3	ft^3	DN	NPS	DN	NPS
Up to 1.5	Up to 50	25	1	25	1
Up to 6	Up to 200	25	1	40	1½
Over 6 to 17	Over 200 to 600	25	1	50	2
Over 17 to 71	Over 600 to 2500	40	1½	80	3
Over 71	Over 2500	50	2	80	3

Operational vents, sample points, and drains that may discharge flammable fluids should be minimized and should be not less than 15 m (50 ft) minimum from possible sources of ignition, e.g., furnaces, hot lines, and rotating machinery.

Drainage lines for hazardous or valuable process chemicals should be accessible for inspection and maintenance.

Drainage from such equipment should be permanently piped to a sump tank provided with means for emptying.

For other services, the drainage from sample points, gage glasses, and level controllers should be led via tundishes into an adjacent drain. Sample points and drains, where necessary, should be protected against static-induced ignition of flammable materials.

Where a significant release of H_2S, other toxic gases or large volumes of flammable vapor could occur, vents and drains should be piped to a closed disposal system.

Separate arrangements should be made for the drainage and the tracing of lines containing materials likely to solidify. Provisions should be made for rodding and high-pressure water cleaning.

Vents and drains for pressure test purposes should be sized as follows:

Pipe Size	DN	NPS	Vent Size		Drain Size	
			DN	NPS	DN	NPS
Up to	150	6	20	¾	20	¾
Between	200–350	8–14	20	¾	25	1
Above	350	14	20	¾	40	1½

Vents and drains for custody metering systems should include double isolation with one lockable valve, and drip type sight glass for leakage detection.

In cases where the vent connection is provided only for a pressure test, the valve should be replaced with a plug.

On process piping (conveying hazardous fluids), vents and drains should be designed as follows:

- For piping DN 40 (NPS 1½) and smaller: valve and cap (plug).
- For piping DN 50 (NPS 2) and larger: valve with blind flange.

13.2.8 Blow Down

Blowdown headers should run with a minimum slope of 1:500 toward the separator, avoiding intermediate piping components restricting flow.

Branching off a discharge pipe to blowdown header should always be at the top of pipe with a configuration of 45 degrees in the direction of flow.

13.2.9 Utility Piping

13.2.9.1 Utility Services (Utility Stations Standard Drawing)

In process areas and where necessary for maintenance and cleaning, the valved hose connections should be installed for compressed air, low-pressure steam, service water, and inert gas (when required).

Diameter of service connections should be DN 20 (NPS ¾) and hose length should be 15 m.

All branch piping from utility headers should be taken from the top of the header to prevent plugging.

Hose connections should be located with respect to the operator, air on left, steam at middle, water on right.

Emergency showers and eye wash fountains should be provided for process plants handling dangerous chemicals as cited in standards.

13.2.9.2 Cooling Water

1. Cooling water lines should have block valves at the unit limit.
2. Restriction orifices should be installed in open cooling water outlet lines to maintain a slight overpressure, in order to avoid vapor locks at the channel side of coolers and condensers in elevated positions.
3. In freezing areas, a closed cooling water system should have a bypass with globe valve upstream of the supply and downstream of the return block valve for each unit for winterization purposes.
4. Cooling water system with a cooling tower should have block valves at the inlet to each cooling tower cell.

13.2.9.3 Cooling Water Pipe Work to and From Heat Exchangers

1. Where sea water or aggressive water is used as cooling medium, the water velocity in the exchanger tubes should be controlled given the design limits to prevent scaling, erosion, corrosion, or the formation of

deposits. A control or butterfly valve should be used in the outlet piping to achieve this purpose.
2. Where further means of controlling water velocity in the tubes is necessary, some means of a measuring device should be used. If the orifice plate is used, it should be of type 316 stainless steel, Monel, or Incoloy 825.
3. Where the water pipe work is associated with elevated condensers or coolers, irrespective of the type of water used, the control should be positioned on the outlet side of the equipment, to ensure that the equipment and pipe work always runs full of water to avoid vacuum conditions leading to boiling, impingement, and scale formation.

13.2.9.4 Fuel Oils and Fuel Gas

1. For low flashpoint fuel, a separate all-welded supply system should be provided with steam connections for purging.
2. Fuel gas pipe work should be provided with a condensate knockout drum with means of disposal of the condensate to a closed system.
3. If necessary fuel gas lines should be steam traced and insulated downstream of the condensate knockout drum.
4. Burner piping below fired heaters should not restrict access or exit from the area, for reasons of safety and for the removal and cleaning of gas and oil burners.
5. Fuel oil piping loops to burners should be fully steam traced, with the steam supply being kept separate from the atomizing steam. It is not permissible to insulate the fuel oil and atomizing steam lines in a common jacket in lieu of providing steam tracing. Pressure tap pings to fuel oil pressure controllers should also be adequately traced.

13.2.9.5 Hydrogen Service

1. Leakage should be minimized by the use of welded joints and the exclusion of threaded connections. Vents and drains should be kept to a minimum and their valves blanked.
2. Purging connections permanently piped to a supply of nitrogen should be provided on all units on hydrogen service and on separate pieces of equipment which may have to be isolated during operation of the unit. Such connections should incorporate double block valves with a valved bleed between.
3. Flange facing in hydrogen service should be a smooth finish to 100–150 AARH (see MSS-SP-6).

13.2.9.6 Instrument Air

1. The main instrument for air supply should be an independent self-contained system in accordance with standards.

2. Headers or manifolds should be fitted with isolation valves and removable caps, plugs, or flanges to allow for blow down.
3. The main header isolating valves should be socket weld or flanged gate type. Isolating valves local to each individual instrument off take may be threaded and of brass construction.
4. All low points on main and subheaders should be provided with an accessible drain valve DN 25 (NPS 1) or header size, whichever is less. Individual supply lines will not normally require drain valves.
5. Distribution off takes should be taken from the top of horizontal headers.
6. The specification of tubing for use downstream of the air header isolation valve and for control lines is detailed in standards.

13.2.9.7 Process Air

1. Process air should be supplied by permanently installed compressors that also supply service air.
2. Process air headers should be taken from the upstream of an air dryer with the separately piping from instrument air system.

13.2.9.8 Service Air

1. Service air may be supplied by means of portable compressors. Where a permanent system is specified, materials should comply with the requirements of standards.

13.2.9.9 Breathing Air Systems

Breathing air systems used for the prevention of vacuums in nonpressurized vessels should be of high integrity.

13.2.9.10 Steam System

1. Main steam distribution headers should have a block valve at the main header off take. The line-up should allow for spading.
2. Main steam distribution headers and entering process units should have a double block and bleed valve on the easily accessible location at the unit battery limit.
3. Steam distribution header(s) should have a block valve with a spectacle spade at the off take of the main steam distribution header. The off take should be located on the top of the line. Instruments and recorder connections for flow, pressure, and temperature should be installed downstream of the block valves to the plant or unit.
4. Steam required for smothering, snuffing, tracing, and similar services should be supplied through separate distribution header(s).
5. Block valves should be of parallel slide type on the main steam distribution system serving the refinery or works together with those at the battery limit of any other area, e.g., a tank farm, administration area.

Within any process unit where any section can be taken out of service for maintenance with normal operation continuing on the remaining sections, the section isolating valves should also be of the parallel slide type.

6. If it is necessary to discharge large quantities of steam, noise suppressors should be provided.
7. The draining facilities of a steam supply line should not discharge into sewer systems. They should run to a safe location such as collecting condensate pits, contaminated water rundown systems, gravel pits, gullies, etc., and be combined as far as practical. Situations jeopardizing personnel and goods should be avoided. In cold areas, icing-up of personnel access surfaces should be avoided.
8. Stagnant and reverse flow conditions should be avoided in steam distribution systems.
9. For steam services, valves DN 150 (NPS 6) and larger with pressure rating PN100 (class 600) and higher should have a bypass valve for preheating and pressure-balancing. The bypass size should be:

Nominal Size of Main Valve		Nominal Size of Bypass Valve for Warming-Up of Pipe and Pressure-Balancing of Lines With Limited Volumes		Nominal Size of Bypass Valve for Pressure-Balancing	
IN		IN		IN	
DN	NPS	DN	NPS	DN	NPS
150	6	20	¾	25	1
200	8	20	¾	40	1½
250	10	25	1	40	1½
300	12	25	1	50	2
350	14	25	1	50	2
400	16	25	1	80	3
450	18	25	1	80	3
500	20	25	1	80	3
600	24	25	1	100	4

10. Steam lines connected to process lines should be fitted with a block valve. A check valve should be installed upstream of the block valve, with a bleeder in between. Block valves and check valves should be close together and close to the process line.
11. Inline silencers should be fitted with a small drain line at the bottom of the silencer to prevent accumulation of condensate.
12. Vent facilities should be installed to permit warming-up of the lines prior to commissioning.
13. All steam supply lines should have drain facilities at the low points and at the end to remove condensate (e.g., during commissioning).

13.2.9.10.1 Steam Connections

1. Branch connections for steam systems at 4500 kPa (gage) (650 psig) and above should be DN 25 (NPS 1) minimum and, should be taken off the top of main steam distribution system. A block valve between branch and steam mains should be provided.
2. Process steam connections to fired heaters should be provided with a check valve and block valve in series; the check valve should be located between the block valve and the fired heater.
3. Process steam connections to fractionating columns and similar process equipment should be provided with a check valve and a block valve in series, the block valve being located at the column.
4. Utility connection up to DN 50 (NPS 2) should not be connected permanently to the steam header.
5. Steam lines to groups of pumps should have individual block valves for independent shut-off.

13.2.9.10.2 Steam-Out Connections

1. Piping installed for steaming-out fractionating columns and similar process equipment should be provided with two valves (one gate valve and one check valve). The gate valve may be required, and a drain valve should be installed between them.
2. The steaming-out connection, should be independent of the drain from the vessel and should be provided with a spade for positive isolation.

Steam-out connection sizes should be as follows:

Vessel Capacity		Connection	
m^3	(ft^3)	DN	NPS
Up to 28	Up to 1000	25	1
Over 28 to 57	Over 1000 to 2000	40	1½
Over 57 to 1400	Over 2000 to 50,000	50	2
Over 1400	Over 50,000	80	3

Process steam and steam-out connections should be provided with drainage arrangements.

13.2.9.10.3 Exhaust Lines

1. Exhaust lines from steam machinery discharging to the atmosphere should be fitted with an exhaust head suitably drained.
2. Exhaust steam lines should enter in the top of the exhaust steam collecting header.

13.2.9.10.4 Steam Trapping

1. Steam traps should not be installed in superheated main steam headers or superheated main steam distribution headers. For saturated steam

service steam, traps should be fitted to drain pockets at low points of main steam headers and main steam distribution headers.

2. Sections of steam distribution headers, heating elements, coils, tracers, etc., should each have a steam trap.
3. Steam traps should be as near as possible to the condensate outlet of the unit to be drained, unless a cooling leg is required. Traps should be at all low points or at natural drainage points, e.g., in front of risers, expansion loops, changes of direction, valves, and regulators.
4. Steam traps should have a bypass arrangement if the system cannot accommodate replacement and/or repair time without causing a process problem.
5. Steam traps should be easy to maintain and replace. The connecting piping up to and including the first downstream block valve should be designed for the full steam pressure and temperature. Steam traps inside buildings should have a bypass and should not discharge into an open drain inside the building.
6. Open steam trap discharges should be located away from doors, windows, air intakes, ignition sources, stairs, and access ways.
7. Steam trapping arrangements should conform to standards. All trap pipework should be designed to provide flexibility to allow for thermal movement between the main, trap, and condensate return main.
8. Trap size should be based on the maximum quantity to be discharged at the minimum pressure difference between inlet and outlet.
9. Traps should be fitted with a strainer on the inlet, unless it is an integral part of the trap and should be of cast or forged steel, according to the duty.
10. Socket weld trap assemblies should be provided with flanges to allow for maintenance. The flanges should be so arranged that the upstream atmospheric blow down will be effective whilst the trap assembly is removed for maintenance.
11. No steam trap should be connected to more than one steam line nor to more than one section of the same steam line.
12. Open tail pipes should terminate 75 mm (3 in) above ground level and should not discharge onto stanchions, pipe supports, or directly into salt glazed drains, etc., which might be adversely affected by the discharge. They should be directed in such a way as not to present any hazard, and in paved areas the discharge should be directed in a manner such that the condensate does not run across the paving.
13. Traps discharging to atmosphere should be mounted to be self-draining to avoid frost damage.
14. Traps operating on different steam pressures may discharge into the same header, providing the condensate line is adequately sized to accommodate the flash steam.
15. Traps should be located adjacent to the equipment they serve, and should be accessible for maintenance and firmly supported.

16. Multiple traps should be grouped together and installed in enclosures so that the operation of each trap can be checked and prevent frost damage to traps not in use. Tail pipe discharges should be arranged to allow maintenance on any one trap whilst all others are operating
17. Trapping systems not detailed on the pipe work drawings should be site-run, ensuring that steam and condensate lines do not interfere with normal operation and maintenance, and in particular with access to valves and other equipment.
18. Drainage from large steam consumers such as heaters, condensers, and reboilers, should get use of level-controlled collection pots.
19. Condensate pots should be sized as follows:

Main Size	Pot Size
DN 100 (NPS 4) and below	Main size
DN 150 (NPS 6)	DN 100 (NPS 4)
DN 200 (NPS 8) and above	DN 150 (NPS 6)

20. Valves on condensate pots should be DN 25 (NPS 1) in nominal bore minimum, they may be of gate, parallel slide, or globe type. For high pressure, superheated steam, large or important steam mains, globe or parallel slide types should be used. Globe valves should be used where it may be necessary to control flow.
21. All globe valves should be capable of passing the full-rated flow of condensate.

13.2.9.10.5 Steam Tracing

1. Steam tracing of piping should be installed generally in accordance with standards.
2. Materials for steam tracers should comply with the appropriate line specification.
3. Flattening or crimping of the tracer line should be avoided.
4. Fittings between steam supply pipes or condensate drain pipes and the copper tracer should be carbon steel adaptors socket-welded to the carbon steel pipe and brazed to the copper tube. The fittings should be separately insulated from the traced line.
5. Fittings should be used only where necessary to join the longest possible length of tracer and not only for ease of installation. Essential joints should be located at the pipe flanges. Loops should be provided adjacent to pipe flanges to allow for future use of compression fittings.
6. Piping DN 40 (NPS 1½) and smaller may be grouped together with a single tracer. Piping DN 50 (NPS 2) and larger should be individually traced.
7. Piping on corrosive services and piping liable to blockage due to deposition of solids or to the formation of solid polymers should have

individual steam tracing, irrespective of the pipe size, and not be grouped together with other pipes.

8. External tracing should consist of a single steam line, run at the bottom of the line to be traced, and the pipe and tracer insulated with the standard insulation for the next larger size pipe. Where heat requirements dictate, however, multiple tracers should be provided. Tracers for vertical lines may be coiled around the lines.
9. Expansion loops should be installed where necessary in tracers, and should coincide with flanged joints in traced lines. Loops coinciding with flanges should be such as to allow flanges to be sprung apart, and at spaded flanges should allow the spades to be swung.
10. Expansion loops should be installed in the horizontal plane and pockets should be avoided.
11. Each steam distribution or supply point should be located above the highest point of the piping system being traced. Each condensate collection header should be located at an elevation low enough to permit gravity flow of condensate from all connected lines.
12. Each individual tracing line should be provided with a block valve located at the steam header or subheader. Valves should be steel socket welding type. Valves should be readily accessible from ground or platform level and positioned on the subheader for ease of maintenance. Each tracer or leg of parallel tracers should be provided with its own trap except that groups of tracers that are self-draining may be drained to a level-controlled condensate pot or a collection header. Tracing on control valves and bypasses should allow control valve removal without interfering with the tracing of the bypass.
13. Tracers should be attached to lines by strapping or binding wires. Heat transfer cement may be used to improve the transmission of heat from tracer to traced line.
14. Where degradation of a product or metallurgical deterioration of the pipe may occur due to local hot spots pipe or where an internal lining may be damaged, direct contact with an external tracing line should be prevented by a suitable insulating strip between the pipe and the tracing line.
15. For pipe work on which moisture may form due to low-temperature operating conditions or intermittent service that would allow the pipe work to cool down during idle periods, corrosion due to a galvanic couple between the pipe and/or support clamps and the tracing line should be avoided by use of an approved insulating strip.
16. Steam-heated pumps and other equipment should have their own individual steam supply independent of the line tracing. On pumps, valves, or other equipment requiring removal for maintenance, steam tracing and condensate connections should be flanged.
17. Each tracing circuit should be labeled clearly and permanently upstream of the supply isolating valve and immediately before the steam trap.

The label should be in stainless steel and should have the following information:

a. Steam supply identification code.
b. Line designation.
c. Steam trap designation.

18. A steam or condensate line should not be attached to, or supported from, any line other than the one it is tracing.

13.2.10 Jacketed Piping

This part of the standard should be read in conjunction with standards.

Jacketed piping is classified as "partly jacketed" and "fully jacketed."

General points common to all jacketed piping:

1. The design of the piping and flanges should consider differential expansion between the inner pipe and jacket during start-up, shutdown, normal operation, or any abnormal conditions. The design should ensure no buckling of the inner pipe due to external pressure or differential expansion. Jacket, connections, and inner pipe should all be of the same material of construction to avoid problems due to thermal stresses or welding of dissimilar metals.
2. Line/jacket sizes should be as follows:

Line Size		Jacket Size	
DN	NPS	DN	NPS
20	¾	40	1½
25	1	50	2
40	1½	80	3
50	2	80	3
80	3	100	4
100	4	150	6
150	6	200	8
200	8	250	10
250	10	300	12

3. Spacers should be used to ensure that the inner pipe is concentric with the jacket.
4. The radius of pulled bends should not normally be less than five times the jacket nominal I.D.
5. Forged elbows may also be used for the fabrication of bends in certain instances where the radius of the bends coincide and where the tangent lengths allow assembly.
6. Where the main process line is sectionalized by block valves, the supply and trapping arrangement should be similarly arranged to facilitate maintenance on isolated sections.

7. Transfer of steam from one jacket to another should be arranged as follows:

   A single jump-over connected into the jackets by radial branches should be used on vertical jacketed pipe. These should be positioned as low as possible on the upper jacket and as high as possible in the lower one. where the jacketed pipe run is not vertical, one of the following alternatives may be used:

   a. Single jump-over from lowest part at upstream to the highest part of downstream of steam.
   b. Single jump-over connections branched into the highest part of the jackets where absence of condensate is ensured.
   c. Double jump-over connections one for steam at highest point and one for condensate at lowest point.
8. Jump-over connections at main line flanges should include break flanges. Jump-over at other locations should be all welded.
9. Main line flange bolt holes positioned at the center to allow bolt access may be necessary at tangentially branched jump-over.
10. Heat transfer cement may be used on any unjacketed parts between tracer and pipe to eliminate cold spots.
11. Steam jacket supply valves and traps should be identified by labels as required for steam tracing circuits.

13.2.10.1 Partly Jacketed Piping

1. Partial jacketing is suitable for lines carrying materials where there is no risk of blockage at cold spots and may be specified where contamination of the process fluid with water cannot be tolerated and all butt welds on the jacketed line are required to remain uncovered by the jacket.
2. Jackets should be swaged on to the inner pipe adjacent to flanges and at inner line butt welds where these are required to remain uncovered. Alternatively, butt welded caps bored to suit the outside diameter of the inner line may be used. Where branch welds on the inner line remain uncovered, the main jacket should be locally swaged to the inner line and the branch separately jacketed.
3. Main line flanges should be standard raised face slip-on flanges to ASME B16.5 (inch dimensions) sized to suit the inner line.

13.2.10.2 Fully Jacketed Piping

Full jacketing should be used for duties where it is essential that cold spots are eliminated and butt welds on the inner line may be covered by the jacket.

The jacket should be welded to the back of the main line flange. Main line flanges should be raised face slip-on weld flanges to ASME B16.5 (inch dimensions) sized to suit the jacket but with a bore to suit the inner line.

Reducing orifice-type flanges complete with a tell-tale hole drilled radially through the rim may be used where leak detection at flange welds is necessary.

The use of blinds or blanks bored out for this duty is not acceptable.

Split forged tees should be used for the jackets of branch connections.

13.2.11 Piping Adjacent to Equipment

13.2.11.1 Isolation of Equipment

In this chapter, isolation is subdivided into two categories:

1. Positive isolation. Where no leakage can be tolerated, e.g., for safety or contamination reasons.
2. General isolation. Where the requirements are less critical than (1) above.

In certain circumstances, it may be necessary for operational reasons or additional security to provide a combination of (1) and (2) above.

13.2.11.1.1 Positive Isolation

1. Positive isolation should be by one of the following:
   a. The removal of a flanged spool piece or valve and the fitting of blind flanges to the open-ended pipes.
   b. Line blind.
   c. A spade in accordance with standards. The arrangements of spading points, together with venting, draining, and purging facilities, should enable a section of line containing a spade to be checked as free from pressure before spade insertion or removal.
2. Positive isolation methods should be provided:
   a. To permit isolation of major items of equipment or group of items for testing, gas freeing, making safe, etc. A group of items containing no block valves in the interconnecting pipework may be considered as one item of equipment.
   b. To isolate a section of plant for overhaul.
   c. To isolate utility services, e.g., fuel gas, fuel oil, atomizing, snuffing, or purge steam to individual fired heaters.
   d. To prevent contamination of utility supplies, e.g., steam, water, air, and nitrogen where permanently connected to a process unit.
   e. Steam and air connections for regeneration and steam/air decoking should be positively isolated from the steam and air systems, preferably by spool pieces or swing bends.

13.2.11.1.2 General Isolation

1. General isolation should be done by one of the following:
   a. A bidirectional block valve.

- **b.** A unidirectional block valve, where isolation in only one direction is required for all conditions and where internal relief of the valve cavity is required.
- **c.** Double block valves with a bleed valve mounted on the pipe work between them in following cases and point (4) in this section:
  - **i.** Drain connection in gas services when ice formation or freezing can occur.
  - **ii.** All steam piping entering or leaving process units.
  - **iii.** High-pressure steam supply lines 4137 kPag (600 psig) to turbines over 300 kW (400 HP).
  - **iv.** All manifolds in the tank age area should have a double block and bleed valve in cases where there is a possibility of product contamination.
- **d.** Double valves without a bleed valve mounted on the pipe work between them should be provided for sample connection, drains, and vents:
  - **i.** In hazardous fluid service.
  - **ii.** In PN 150(class 900) piping or higher.
  - **iii.** An unmanned installation where vibration is anticipated.
  - **iv.** Where freezing due to cooling on expansion may occur.

The block valves should be far enough apart to allow safe access to the upstream valve with material discharging from the point of emission.

For fluids such as LPG, the valves should be separated by at least 1 m to reduce the risk of simultaneous obstruction of both valves by ice or hydrate formation.

2. Block valves should be provided:
- **a.** In all lines at recognized separation of process units or limits of operating areas.
- **b.** At vessel branches, excluding:
  - **i.** Connections for overhead vapor lines, transfer lines, reboiler lines, and side-stream vapor return lines.
  - **ii.** Pump suction and reflux-line connections on vessels that have a block valve located within 10 m (30 ft) in a horizontal direction from the vessel branch. This exception should not apply where there is a particular process requirement to isolate a vessel inventory.
  - **iii.** Vents on tanks and vessels open to atmosphere.
  - **iv.** Relief device connections and/or connection to other vessels or piping systems fitted with relief devices that protect the vessel in question.
  - **v.** Overflow connections.
- **c.** At suction and discharge of pumps and compressors, but not at suctions of air compressors that take suction from atmosphere.
- **d.** For isolating equipment, e.g., individual or groups of heat exchangers as requiring servicing during plant operation.
- **e.** For isolation of instruments.

f. Where required to prevent flow at vents and at drains, sample points, steam-out points, and for diverting flow through alternative routes and bypasses.
g. At inlets, outlets, and drains of storage tanks.
h. To isolate utility services, i.e., fuel oil, fuel gas, atomizing, snuffing, and purge steam to individual fired heaters.
i. On supply lines to road and rail car filling stations handling toxic or flammable materials; the block valve should be located remote from the filling point for emergency shut off.
j. In all utility services at:
   i. Each branch on the refinery or main process plant.
   ii. The inlet at each individual user.
   iii. The outlet of each cooling water user.
k. In steam and condensate systems at each branch on the header in each plant or unit if the length of pipe between header and user is 5 m (16 ft) or greater.
l. At vents placed in piping for operational purposes. These vents should be spaded or blanked in accordance with the line specification. Vents for testing purposes only should not be valved.
m. At drain connections in the low point of piping systems, where the arrangements should be as in point (l).
n. At suitable points in ring, main, or distribution systems to allow sectionalizing.
o. At the outlet from each noncondensing steam user, where the user discharges to a pressurized system.

3. Process area limit block valves for relief and blowdown piping should be provided with a purpose-built locking device, for locking valves open. Gate valves fitted on this duty, should be installed in the horizontal or inverted position so that the valves tend to fail in the open position.
4. Double block valves with an intermediate valve to vent the space between the block valves should be provided:
   a. Where frequent isolation is required and temporary, not positive isolation is acceptable for operational safety, e.g., in segregating products.
   b. For temporary isolation between permanently connected utility supplies and process units where the utility requirement is in frequent use and there is a need to make the utility readily available for injection into the process, e.g., stripping steam where the system has to be made condensate free right up to the point of injection. (Nonreturn valves will also be required in all utility services connected directly with process equipment.)
   c. Where equipment may be taken out of service while the unit remains in operation, e.g., compressors or where there is a requirement to isolate equipment yet hold it readily available for use with the operating unit, e.g., hydrogen storage vessels.

13.2.11.1.3 Pump Emergency Isolation

1. Except for glandless type pumps, an emergency isolation valve should be provided in the suction line between a vessel and a pump when any of the following apply:
   a. The suction inventory at normal operating level is 30 m^3 (1060 ft^3) or greater and the pumping temperature is greater than the auto-ignition temperature.
   b. The liquid is toxic.
2. Where a pump is paired or spared, the common line should be fitted with the valve.
3. The valve should be located as close to the vessel as possible outside the vessel supports and not less than 3 m (10 ft) from the pump.
4. The valve should be remotely operable when it is less than 15 m (50 ft) horizontally from the pump, unless a fire wall is installed between the valve and the pump.
5. For remotely operated valves, the valve control station should be located not less than 15 m (50 ft) from the pump and, if practicable, within sight of the valve. The control station should not be located in a special fire-risk area, and should be accessible through a low-fire-risk area.

13.2.11.2 Provision of Strainers and Filters

Provision of strainer should be in accordance with standards with following considerations:

1. Permanent strainers or filters should be fitted in the following instances if they are not already an integral part of the equipment:
   a. In a fuel oil supply to burners and for each set of gas pilots.
   b. Hydraulic systems for remote control of valves.
   c. At the inlet of steam turbines, jet ejectors, and trapping systems.
   d. Loading installations, when necessary, to maintain product quality.
   e. In the suction lines of pumps, where the liquid may contain solids liable to damage pumps.
   f. In compressor suction lines.
   g. In any lubricating system, i.e., sealing oil, gland oil, and gear coupling lubricators.
   h. Any flushing oil system.
2. Permanent strainers in compressor suction lines should be provided with isolation to permit easy removal. Maximum design pressure drops and flow direction should be indicated by a nameplate permanently attached to the strainer.
3. Temporary strainers should be fitted between the suction valve and the equipment and should ensure that the debris is completely removed from the system when the strainer is cleaned, and be cleaned easily without disturbing the main pipe work. Pipe work should be designed to incorporate

the strainer, and no pipe springing should be allowed for retroactive installation. A suitable spacer should be provided for use on removal of the strainer.

4. If permanent strainers are fitted, the temporary ones should normally be located between the pump or compressor and its associated suction side block valves. On main crude oil charge pumps, twin strainers arranged in parallel and each having its own block valve should be installed upstream of the suction header.
5. Twin parallel filters should be provided on vital process pumps that do not have a standby and the shutdown of which for strainer cleaning could lead to a shutdown of the refinery or works, or of an important unit.
6. Pressure tappings should be provided for measuring pressure drop at all permanent and temporary strainers.

13.2.11.3 Piping to Equipment

13.2.11.3.1 Pumps

1. Pump suction piping should cause minimum flow turbulence at the pump nozzle. Suction piping should not have pockets where gas can accumulate. However, if this is unavoidable, venting facilities should be provided.
2. If the suction nozzle of a pump is smaller in size than the connecting piping and a reducer is required in a horizontal line, it should be eccentric, installed with the belly down (top flat). This may require an additional drain.
3. If the discharge line size differs from the pump discharge nozzle, a concentric reducer should be applied.
4. A block valve should be installed in the suction line of each pump upstream of the strainer. The discharge line should also have a block valve. A nonreturn valve should be installed upstream of the discharge block valve.
5. The discharge valve, suction strainer, and suction valve may be of the same size as the pump nozzles for economic reasons and also to avoid comparatively heavy attachments, unless the pressure drop is too high.
6. For spared pumps which have common suction and discharge lines, a DN 20 (NPS ¾) bypass with throttling valve should be installed around the discharge nonreturn valve in the following cases:
   - **a.** if discharge and suction line working temperatures are above 230°C;
   - **b.** if process fluid can solidify at ambient temperature, e.g., water lines in frost areas;
   - **c.** if discharge/suction line working temperature is below −100°C;
   - **d.** if draining of the space upstream of the nonreturn valve is required.
7. When the discharge and suction lines are working at ambient or below ambient temperatures, a 3- to 5-mm hole in the closing member of

the nonreturn valve may be considered instead of a bypass around the nonreturn valve. Valves with such a hole in the closing member require marking on the valve body and on the process engineering flow schemes and isometric drawings.

8. Permanent strainers should be installed in all pump suction lines. Y or T type strainers should be used for permanent installation in vertical suction lines, in services of high content of impurities use of Y-type is preferred. In horizontal suction lines, Y-type or bucket-type strainers may be used. For suction lines ≥ DN 450, (NPS 18), bucket-type strainers should be used. The installation of the Y-type strainer of double suction pumps should not disturb an even flow to the two suction nozzles of the pump. In a vertical suction line the Y-type strainer should be installed pointing away from the pump. In a horizontal suction line the Y-type strainer should be installed pointing downwards.
9. Warming-through connections should be supplied for pumps as per standards.
10. For multistage pump fitted with pressure relief device for pump casing protection, the suction side design pressure need not be greater (but should not be less) than the relief set pressure.
11. Pump vents should be connected to the vapor space of the suction vessel for operation under vacuum. This allows filling of the pump before start-up. The vent line should have two valves, one at the pump and one at the vessel. Pump vent and drain nozzles should be fitted with valves; if not connected to a drain system, the valves should be fitted with plugs. Pump handling butane or lighter process fluids should have a vent line to the flare system. The vent line should have a spectacle or spade blind.
12. In order to avoid a fire hazard, lubricating oil, control oil, and seal oil lines should not be routed in the vicinity of hot process and hot utility lines.
13. Cooling water lines to pumps and compressors should not be less than DN 20 (NPS ¾). Lines DN 25 (NPS 1) or less should have the take-off connection from the top of the water main line so as to prevent plugging during operation.
14. Pumps for vacuum service require a sealing liquid on the stuffing boxes and a vent line to the process system to prevent dry-running.
15. Reciprocating, positive displacement pumps and also centrifugal pumps (if required) should be safeguarded against a blocked outlet with a pressure relief device. This should not be an integrated part of the pump. The relief valve should be installed in a bypass between the discharge line upstream of the check valve and the suction vessel. Alternatively, the relief valve may be installed in a bypass between the discharge line upstream of the check valve and the suction line downstream of the

block valve. However, it should be assured that this will not create an overpressure of the suction system.

16. Spools should be provided between pump suction strainer/pump suction connection and pump discharge nozzle/nonreturn valve to facilitate easy removal of pump for maintenance.

13.2.11.3.2 Compressors

1. To prevent fatigue failure of compressor piping, the effect of vibrations and pressure surge should be considered. Piping should have a minimum of overhung weight.
2. Suction line should be designed with special consideration for straight and minimum length. Interstage and discharge piping should be sufficiently flexible to allow expansion due to the heat of compression.
3. Block valves should be in the suction and discharge lines, except for air compressors, which should have block valves in the discharge lines only. Discharge lines should have a check valve between block valve and discharge nozzle. In case of reciprocating compressors, check valve may not be required.
4. In each compressor suction line, a suction strainer should be installed downstream of the block valve of the compressor and as close as possible to the compressor suction nozzle. Screens and filters should be reinforced to prevent failure and subsequent entry into the compressor. Provision should be made to measure the pressure difference across filter.
5. Reciprocating compressors should be safeguarded against a blocked outlet with a pressure-relieving device installed in a bypass between the discharge line upstream of the block valve and the suction vessel. Interstage sections should also be protected by relief valves.
6. The suction line between a knockout drum and the compressor should be as short as practicable, without pockets, and slope toward the knockout drum.
7. In the single stage compressor, the pressure rating of the suction valve and piping between this valve and the suction nozzle should be equal to the rating of the discharge line. The pressure rating of the suction piping of a reciprocating compressor should have the same rating as the discharge of that stage, including valves and suction pulsation dampeners.
8. In case of multistage compression, the suction design pressure should be equal to the highest design pressure of the equipment from which it takes suction. If the design pressure turns out to be lower than the maximum shut-in pressure and/or the discharge pressure, then the suction piping must be relief protected.
9. The two-design pressure system is not preferred for less than three-stage stations.

10. Suction lines should be connected to the top of the header, except for suction lines at least one pipe size smaller than the header, which may be connected concentrically at the side of the header.
11. Compressors in hydrocarbon or very toxic services should have purge facilities. Possibility of spading should be provided by spectacle blinds, removable spool pieces, or elbows.

13.2.11.3.3 Steam Turbines

The set pressure of the relief valve in the turbine exhaust system should not exceed either the turbine design pressure or the pressure of the exhaust piping. The calculation for the relief valve orifice should be based on the turbine inlet nozzle.

1. Warming-up facilities for the turbine should be provided.
2. Piping should be designed to permit steam-blowing up to the inlet and outlet flanges of the turbine before start-up.
3. Steam vents should be routed to a safe location and should not be combined with any lubricating oil, seal oil, or process vent.

13.2.11.3.4 Heat Exchanger

1. The nozzle positions of heat exchangers should allow an optimum piping layout.
2. Sufficient space should be kept between an adjacent heat exchanger inlet and outlet (control) valve manifolds as per standards.
3. Heat exchanger piping should not be supported on the shell and should not hamper the removal of the tube bundle and shell/channel covers. A removable pipe spool may be required.
4. When shell-and-tube exchangers can be blocked in by valves, causing trapped liquid, attention should be paid to:
   a. Preventing exposure of the low-pressure side piping to the maximum pressure of the high-pressure side, irrespective of whether caused by internal failure or otherwise;
   b. potential increase of pressure due to thermal expansion of the trapped liquid on the cold side or due to solar radiation;
5. The equipment and the connected piping should be protected by thermal expansion relief valves, if pressures and/or pressure differences can increase beyond the design limits.
6. Steam heat exchangers should have a nonreturn valve in the steam inlet if the normal steam pressure is less than 110% of the process relief valve set pressure or, without relief valve, 110% of the process design pressure.
7. Where contamination is critical in heat exchangers, a check valve should be installed in the inlet of low pressure side if the normal pressure of low pressure side is less than 110% of the design pressure of the high-pressure side.

13.2.11.3.5 Pressure Vessel

1. Piping to columns should drop or rise immediately after the nozzle and run parallel and close to the column. For ease of support a number of lines can be routed together, parallel in one plane.
2. If a tall slender vessel ($L/D \geq 10$) is susceptible to aerodynamic oscillations, the piping platforms and ladders of the top of the vessel should be located such that the platform projected area against wind is kept at minimum.
3. Pressure vessels that are grouped together should have platforms and interconnecting walkways at the same elevation. The number of stairways and ladders to the platforms should be sufficient to meet safety requirements.
4. If not controlled in another way, process steam lines to pressure vessels should have a regulating globe valve direct to the pressure vessel nozzle. A check valve to prevent the product from entering the steam line should be installed close to and upstream of the regulating valve with a valved low point drain between them. A gate valve upstream of the check valve should isolate the line from the main steam header.
5. The steaming-out pressure for columns should be 350 kPag (50.8 psig) for tall columns, a higher pressure may be considered if the design permits.
6. Pressure vessel drain valve should be located outside of skirt.

13.2.11.3.6 Fired Heaters (Furnaces)

1. Burner utilities headers (fuel oil, fuel gas, atomization steam) should be arranged with a vertical bundle along furnace walls.
2. The set of valves controlling the feeding of smothering steam should be located in a safe location at least at a distance of 15 m from the furnace.
3. A throttling balance valve should be provided in the inlet to each coil of a set of parallel coils.
4. Outlet lines should be provided with the following:
   a. A check valve should be installed in the outlet from each heater with the check valve nearer to the furnace.
   b. Drain valves should be provided to drain each coil.

13.2.12 Piping Layout Design

The elevation of overhead pipe rack should be such to provide minimum free height access.

Generally all process and utility piping should be installed above ground.

Inside plants (process units and utilities areas) piping should be routed on overhead pipe bridges (pipe rack).

Equipment that is a potential source of fire, should not be located under pipe rack.

Firefighting water lines if installed above ground, should not run along pipe bridges or pipe racks.

Outside plants and in the interconnection areas (manifold, tank farms, flares, etc.) piping should preferably be installed on the ground on concrete sleepers in pipe racks.

Pipe trenches close to process equipment should be avoided. Where it is not practicable to run pipe over rack, and trenches below paving level are unavoidable, such trenches should be divided into sections about 10 m minimum length by fire break.

13.2.12.1 Slope in Piping

Except for branch and equipment connections, all lines should run in horizontal direction. Where lines need to be drained completely, the piping should be sloped and provided with drainage points:

Minimum slope of lines should be as follows:

Minimum slope of lines should be as follows:

1. Process lines on sleeper	1:120
2. Process lines on pipe racks	1:240
3. Service lines	1:200 to 1:240
4. Drain lines	1:100

The slope of lines should be indicated on P&ID.

13.2.12.2 Pipe Racks

Overhead racks may contain more than one level. For steel pipe racks, the height of levels should have one of the following elevation.

1. Main pipe racks: 4.60, 6.20, 7.80 m
2. Individual or secondary pipe rack: 3.80, 5.40, 7.00 m.

Arrangement of pipe rack should be made in accordance with standards.

In special case for large size pipes or concrete pipe racks, the distance between the various floors may be increased.

Except for special cases, minimum width of pipe rack should be 6 m. The width of pipe rack should be designed to accommodate all pipes involved plus 20% space for future expansion or modification. Where the pipe rack support air coolers, the preferred width should be the width of air coolers.

In multilevel pipe racks, pipe carrying corrosive fluids should be on the lower level, and utility lines should be at the upper floor. Large size or heavy weight pipes should be located at the lower level and on extreme sides.

13.2.12.3 Pipes Space

The space between the axis of two adjacent pipes should be at least equal to the sum of ½ O.D of flanges (with higher rating) or ½ O.D. of each pipe plus 25 mm. This dimension (25 mm) should be increased to 50 mm for insulated pipes.

Sufficient space should be allowed between adjacent lines at points of change in direction to prevent damage of one line by another, due to expansion or contraction (see relevant standards).

At crossings, lines should have a minimum clearance of 25 mm, after allowing for insulation and deflection.

Hot lines in pipe racks should be grouped together and consideration should be given to the expansion loops.

13.2.12.4 Pipe Branches

For gases or vapors, the branches should be taken from the top of the main lines.

Where main trunk lines for steam, water, and other common systems run through a number of process units, off takes to the users in any single unit should not be taken directly from the main trunk lines. Off takes should be from a header or headers supplying each unit and should be provided with a valve at their junction with the main trunk lines.

Where the main trunk lines run in offsite pipe racks, off takes into each unit should be fitted with a valve at the plot limit of each unit.

Type of branches should be in accordance with piping specification conforming to relevant standards.

Take-offs from pipe rack to process areas should rise from rack level, run low, and rise at the battery limit to the elevation of the piping within the process area.

13.2.12.5 Battery Limits

All process and utilities piping at the battery limits of one unit or groups of units linked should be equipped with block valves and a spectacle blind as specified in relevant standards.

Battery limit valves, spades, and blinds should preferably be located in the vertical riser.

13.2.13 Piping Flexibility (Additional to ANSI B 31.3)

13.2.13.1 Expansion and Contraction

Piping system should be designed for thermal expansion or contraction in accordance with relevant standards taking the following into consideration as minimum:

1. Depressurizing temperature.
2. Drying-out temperature.
3. Minimum/maximum working temperature.
4. Defrosting temperature.

13.2.13.2 Flexibility Design

Consideration and calculation of stress analysis should be made in accordance with relevant standards with following requirements:

1. Start-up, shutdown, steam-out, where applicable and upset conditions, including short-term excursions to higher temperatures or pressure as well as normal operating conditions, should be considered in flexibility analysis.
2. Sufficient flexibility should be provided in the piping to enable spade, line blinds, or bursting discs to be changed.
3. Vessels or tanks at which piping terminates should be considered inflexible for the initial piping analysis.
4. Only where it is impractical to increase flexibility sufficiently, to reduce the stress range or anchor loads to acceptable levels, use of bellows or expansion joints should be considered.
5. Specific attention should be given in the design and flexibility analysis of piping connecting to machinery to ensure that piping loads transmitted to the machine are within the acceptable limit under all operating conditions. Hanger supports should be used on these lines wherever practicable. Where impractical, low-friction pads (e.g., PTFE) may be used.
6. A flexibility analysis for pipe work connecting to machinery should include mismatch to the allowed tolerance between pipe work and the machine to ensure that the calculated loads represent the worst loads which might be generated. A number of calculations may be necessary to determine the conditions of mismatch that will generate the maximum loads. Specific attention should be given to mismatch in flexibility analysis of smaller lines where such lines are unusually rigid.

13.2.14 Piping Supports

1. Piping should not be supported off other pipes, particularly if either or both pipes are subject to thermal expansion or vibration. Nor should it be supported from vessels or other equipment, except where brackets have been specifically provided. Piping should not be placed in direct contact with concrete, nor be supported from removable flooring, flexible flooring, deck plating, or hand rails.
2. Stainless steel piping should be protected at pipe supports against galvanic and crevice corrosion.
3. The line should be designed to be self-supporting when pressure relief valves are removed. In case of spades, line blinds, bursting discs, and others, temporary support may be used.
4. Pipes in a pipe rack or pipe track (sleeper way) should be grouped according to size to permit longer spans for the larger pipes.
5. Local stresses in piping due to pipe support should be considered.

13.2.14.1 Design and Selection

In general, the location and design of pipe supporting elements may be based on simple calculations and engineering judgment. However, when a more refined analysis is required and a piping analysis, which may include support stiffness, is made, the stresses, moments, and reactions determined thereby should be used in the design of supporting elements.

Piping should be supported, anchored, or guided to prevent line deflection, vibration, or expansion/contraction that could result in stresses in excess of those permitted by ASME B31.3 in the piping or inline connected to equipment.

In design of supporting element the following considerations should apply:

Each support assembly, including the spring supports, should be designed to sustain the hydrostatic test load.

Field welding of pipe supports to piping should be kept to a minimum. Structural provisions to connect pipe supports should be made as much as possible during the civil and mechanical engineering phase (e.g., concrete plinths, inserts, cleats and brackets to steel structure, clips to vessels). Welded guides and other welded support elements in hot dip galvanized steel structures (e.g., pipe racks/bridges) should be attached to steel members before they are hot dip galvanized.

Pipes should be supported in groups at a common support elevation of the supporting structure. Inserts should be poured in vertical and horizontal concrete beams, allowing supports and hangers to be bolted. Since civil design is often well ahead of the pipe support design, provisions should be made to incorporate these inserts at standard locations.

To prevent galvanic corrosion, carbon steel clamps on pipes of other metallic materials should be separated from the pipe by using synthetic rubber, glass fiber paper tape, or other insulating material between the clamp and the pipe.

Individual lines may be suspended by hanger supports only when no other methods of support are practical. Suspending of one line from another should be avoided.

Special attention should be given to locations with potentially high load concentrations, such as valves, strainers, inline instruments, and equipment. The supports should be suitable for these high loads and should facilitate maintenance exchange of the heavy valves/equipment.

For piping containing gas or vapor, weight calculations need not include the weight of liquid if the designer has taken specific precautions against entrance of liquid into the piping, and if the piping is not to be subjected to hydrostatic testing at initial construction or subsequent inspections.

In addition to the weight effects of piping components, consideration should be given in the design of pipe supports to other load effects introduced by service such as pressure and temperatures, vibration, deflection, shock, wind, earthquake, and displacement strain.

The layout and design of pipe supporting elements should be directed toward preventing the following:

1. Excessive thrusts and moments on connected equipment (such as pumps and turbines);
2. Excessive stresses in the supporting (or straining) elements;
3. Resonance with imposed or fluid-induced vibrations;
4. Excessive interference with thermal expansion and contraction in a piping system that is otherwise adequately flexible;
5. Unintentional disengagement of piping from its supports;
6. Excessive heat flow, exposing supporting element to temperature extremes outside their design limits.

Design of the elements for supporting or restraining piping systems, or components thereof, should be based on all the concurrently acting loads transmitted into the supporting elements.

In load calculations, where required, consideration should be given to the following:

1. Weights of pipe, valves, fittings, insulating materials, suspended hanger components, and normal fluid contents.
2. Weights of hydrostatic test fluid or cleaning fluid, if normal operating fluid contents are lighter.
3. Additional loading that may occur during erection.
4. Intentional use of restraints against normal thermal expansion.
5. The effects of anchors and restraints to provide for the intended operation and protection of expansion joints.
6. Reaction forces due to operation of safety or relief valves.
7. Wind, snow, or ice loading on outdoor piping.
8. Additional loadings due to seismic forces.

13.2.14.1.1 Additional Requirements for Lines Connected to Equipment

Lines connected to columns and other vertical vessels should have a resting support as close as possible to the column or vessel nozzle, and be guided at regular intervals to safeguard the line against wind load and/or buckling.

Maximum guide distance should be 6 m for lines smaller than DN 200 (NPS 8) and 10 m for lines DN 200 (NPS 8) and larger.

Pipe supports on equipment should be bolted to cleats welded to the equipment. The cleats should be supplied by the equipment manufacturer. The executor should use standard cleats for the connection of pipe supports, ladders and platforms.

To support piping systems connected to equipment, maximum use should be made of platforms, fire decks, etc.

To allow adequate clearance for the removal of covers, heads, channels, bundles, and shells, lines should not be supported on heat exchanger shells and heads.

Onshore reciprocating compressors and integral piping should be supported on a common slab.

Piping connected to rotating equipment should have adjustable supports to facilitate alignment, spading, and equipment exchange. The supports should allow for thermal expansion and vibration and should be modeled in the pipe stress analysis.

To prevent damage to lines and tank connections caused by settlement of the tank, the first pipe support should be located sufficiently far away from the tank. The following distances should be adhered to:

Nominal Pipe Size DN	Distance Between Tank and First Support (m)
100 and smaller	5
150	6
200	7
250	8
300	9
350	10
400	10
450	10
500 and larger	12

13.2.14.1.2 Allowable Stresses, Load Ratings, and Temperatures

This section is a supplement to reference standard MSS-SP-58 1993, section 4. For ease of reference, the section numbering of the reference standard has been used for this supplement.

Allowable stresses for materials commonly used in the design of pipe supporting elements are listed in Table 13.1.

Materials may not be used above the highest temperature for which a stress value appears.

Allowable stresses for materials not listed in Table 13.1 with known physical properties should be determined as the lower of the following values:

1. 1/4 of minimum tensile strength at service temperature.
2. 5/8 of minimum yield strength at service temperature.

13.2.14.2 Seismic Loadings

When the region is designated as susceptible to earthquakes, the anticipated earthquake loadings should be established.

A properly designed piping system will have sufficient inherent flexibility to absorb large movements without leading to excessive strains or failure.

TABLE 13.1 Steel Materials Mainly Used for Pipe Supports Basic Allowable Stresses in kPa (ksi) at Metal Temperature °C (°F)

Material	Min Temp, °C (°F)	Tensile Strength, kPa (ksi)	Yield Strength, kPa (ksi)	Min Temp, T_o						
				37.8°C (100°F)	93.3°C (200°F)	148.9°C (300°F)	204.4°C (400°F)	260°C (500°F)	315.6°C (600°F)	371.1°C (700°F)
A-36	−28.89	399,900	248,200	1227	1165	1165	1165	1165	1165	1165
	(−20)	(58)	(36)	(17.8)	(16.9)	(16.9)	(16.9)	(16.9)	(16.9)	(16.9)
A-53 Gr.B	−28.89	413,700	206,900	1379	1379	1379	1379	1303	1193	1138
	(−20)	(60)	(30)	(20)	(20)	(20)	(20)	(18.9)	(17.3)	(16.5)
A-105	−28.89	482700	248200	1607	1510	1469	1420	1338	1227	1193
	(−20)	(70)	(36)	(23.3)	(21.9)	(21.3)	(20.6)	(19.4)	(17.8)	(17.3)
A-106Gr.A	−28.89	331,000	206,900	11.3	1103	1103	1103	1103	1020	993
	(−20)	(48)	(30)	(16)	(16)	(16)	(16)	(16)	(17.8)	(14.4)
A-106Gr.B	−28.89	413,700	241,300	1379	1379	1379	1379	1303	1193	1138
	(−20)	(60)	(35)	(20)	(20)	(20)	(20)	(18.9)	(17.3)	(16.5)
A-106Gr.C	−28.89	482,700	275,800	1607	1607	1607	1579	1489	1358	1324
	(−20)	(70)	(40)	(23.3)	(23.3)	(23.3)	(22.9)	(21.6)	(19.7)	(19.2)
A-283Gr.B	−28.89	344,800	186,200	1055	1007	965	917	862	814	765
	(−20)	(50)	(27)	(15.3)	(14.6)	(14)	(13.3)	(12.5)	(11.8)	(11.1)
A-307Gr.B	−28.89	413,700	–	945	945	945	945	945	–	–
	(−20)	(60)		(13.7)	(13.7)	(13.7)	(13.7)	(13.7)		

A-515Gr.60	−28.89	413,700	220,600	1379	1345	1303	1262	1193	1089	1062
	(−20)	(60)	(32)	(20)	(19.5)	(18.9)	(18.3)	(17.3)	(15.8)	(15.4)
A-516 Gr.60	−28.89	413,700	220,600	1379	1345	1303	1262	1193	1089	1062
	(−20)	(60)	(32)	(20)	(19.5)	(18.9)	(18.3)	(17.3)	(15.8)	(15.4)
A-387 Gr.11	−28.89	413,700	241,300	1379	1379	12379	1358	1303	1262	1214
	(−20)	(60)	(35)	(20)	(20)	(20)	(19.7)	(18.9)	(18.3)	(17.6)
A-387 Gr.22	−28.89	413,700	206,900	1379	1276	1241	1234	1234	1234	1234
	(−20)	(60)	(30)	(20)	(18.5)	(18)	(17.9)	(17.9)	(17.9)	(17.9)
A-387 Gr.5	−28.89	413,700	206,900	1379	1218	1200	1186	1179	1158	1124
	(−20)	(60)	(30)	(20)	(18.1)	(17.4)	(17.2)	(17.1)	(16.8)	(16.3)
A-210TP 304	−28.89	517,100	206,900	1379	1379	1379	1289	1207	1131	1103
	(−425)	(75)	(30)	(20)	(20)	(20)	(18.7)	(17.5)	(16.4)	(16)
A-240 TP 347	−28.89	517,100	206,900	1379	1379	1379	1379	1372	1331	1282
	(−425)	(75)	(30)	(20)	(20)	(20)	(20)	(19.9)	(19.3)	(18.6)
A-312 TP 304	−28.89	517,100	206,900	1379	1379	1379	1289	1207	1131	1103
	(−425)	(75)	(30)	(20)	(20)	(20)	(18.7)	(17.5)	(16.4)	(16.0)

The following aspects, however, should be carefully examined and, where necessary, adequate measures should be taken.

Piping should be provided with sufficient flexibility between two anchor points, taking into account that the two anchor points might respond in different modes during an earthquake.

Piping offsets, expansion loops, etc., are normally only provided for absorbing thermal movements. Suitable limit stops should be provided to restrict this movement in case of a seismic shock.

Supports for branch-off lines and supports for vital control equipment should be determined by careful scrutiny.

Instrument lead lines should have sufficient flexibility to absorb seismic movements of the columns, pipe rack, and/or structures to which the instrumentation lines are attached.

Piping going through bund walls, building walls, and floors should be provided with sleeves large enough to allow for the anticipated differential movements due to seismic loadings. Dampening and sealing material should be provided where it is required to maintain a liquid tight connection.

In earthquake areas the following pipe supporting aspects require further scrutiny:

1. Providing additional limit stops for both horizontal and vertical lines with horizontal thermal displacements, thus preventing further movement in case of a seismic shock force;
2. Providing restraints for risers (vertical piping, usually with wind-load guides) in the longitudinal pipe direction, thus restraining the pipe against jumping in case of a seismic shock force;
3. Providing additional guides at column resting supports;
4. Providing sway braces or sway struts;
5. Providing snubbers;
6. Providing jump-restraining pipe clamps or clips, thus preventing the lines from jumping off their support member (especially for horizontal lines, running along pipe racks or pipe tracks in places where no branch-off lines are holding the line in place).

13.2.14.2.1 Material Selection

Permanent supports and restraints should be of material suitable for the service conditions. If steel is cold formed to a center line radius less than twice its thickness, it should be annealed or normalized after forming.

Cast, ductile, and malleable iron may be used for rollers, roller bases, anchor bases, and other supporting elements subject chiefly to compressive loading. Cast iron is not recommended if the piping may be subject to impact-type loading resulting from pulsation or vibration. Ductile and malleable iron may be used for pipe and beam clamps, hanger flanges, clips, brackets, and swivel rings.

Steel of an unknown specification may be used for pipe supporting elements that are not welded directly to pressure containing piping components. (Compatible intermediate materials of known specification may be welded directly to such components.). Basic allowable stress in tension or compression should not exceed 82 MPa, and the support temperature should be within the range of −29°C to 343°C.

Attachments welded or bonded to the piping should be of a material compatible with the piping and service.

Materials commonly used in the design of pipe supporting elements should be selected from Table 13.1.

13.2.14.3 Pipe Hangers (Selection and Application)

Where negligible movement of pipe occurs at hanger locations, simple rod hangers should be used for suspended lines.

Pipe, straps, or bars of strength and effective area equal to the equivalent hanger rod may be used instead of hanger rods.

Hanger rods, straps, etc., should be designed to permit the free movement of piping caused by thermal expansion and contraction.

Welded link chain of 5.0 mm or larger diameter stock, or equivalent area, may be used for pipe hangers with a design stress of 62 MPa maximum.

Adjustable supports should be used where differential settlement between equipment and piping may occur.

13.2.14.3.1 Counterweight Supports

Counterweights should be provided with stops to limit travel. Weights should be positively secured. Chains, cables, hangers, rocker arms, or other devices used to attach the counterweight load to the piping should be subject to the requirements of standards.

13.2.14.3.2 Spring Hangers

Where significant vertical movement of the pipe occurs at the hanger location, a resilient support should be used. Selection of resilient supports should be based on permissible load variation, as illustrated in Table 13.2. Load and movement calculations should be made for the proper selection of spring hangers other than spring cushion types. The effect of vertical movement transfer from the top of risers along horizontal runs should be taken into consideration when applying spring hangers.

13.2.14.4 Spring Supports and Sway Braces

This section should be in accordance with section 10 of MSS-SP-58 1993.

TABLE 13.2 Spring Supports

Vertical Expansion	Allowable Load Change	Single Spring Hanger	Double Spring Hanger	Stanchion Support
	Note (1)	Note (2)		Note (3)
Max. 6 mm	Nominal—25%	48, 51 SS	49, 51 SS, 53 SS	49, 52 SS
	Medium—15%	48, 51 SS	49, 51 SS, 53 SS	49, 52 SS
	Critical—6%	51 S	51 S, 53 S	52 S
Max. 25 mm	Nominal—25%	51 S	51 S, 53 S	52 S
	Medium—15%	51 S	51 S, 53 S	52 S
	Critical—6%	54, 55	54, 55, 56	54, 55
Max. 50 mm	Nominal—25%	51 LS	51 LS, 53 LS	52 LS
	Medium—15%	51 LS	51 LS, 53 LS	52 LS
	Critical—6%	54, 55	54, 55, 56	54, 55
Max. 75 mm	Nominal—25%	51 LS	51 LS, 53 LS	52 LS
	Medium—15%	54, 55	54, 55, 56	54, 55
	Critical—6%	54, 55	54, 55, 56	54, 55
Over 75 mm	Nominal—25%	54, 55	54, 55, 56	54, 55
	Medium—15%	54, 55	54, 55, 56	54, 55
	Critical—6%	54, 55	54, 55, 56	54, 55

Notes:
1. Load change at maximum spring working capacity not to exceed percentages given herein.
2. Numbers in these columns are type numbers from figure 1 of MSS-SP-58:1993.
3. Variable spring types 51, 52, and 53, i.e., standard spring, short spring, and long spring models are identified above as S, SS, and LS, respectively.

13.2.14.5 Anchors and Guides

13.2.14.5.1 Anchors and Guides for Pipework

A supporting element used as an anchor should be designed to maintain an essentially fixed position.

To protect terminal equipment or other (weaker) portions of the system, restraints (such as anchors and guides) should be provided where necessary to control movement or to direct expansion into those portions of the system which are designed to absorb them.

The design, arrangement and location of restraints should ensure that expansion joint movements occur in the directions for which the joint is designed. In addition to the other thermal forces and moments, the effects of friction in other supports of the system should be considered in the design of such anchors and guides.

Where corrugated or slip-type expansion joints, or flexible metal hose assemblies are used, anchors and guides should be provided where necessary to direct the expansion into the joint or hose assembly. Such anchors should be designed to withstand the force specified by the manufacturer for the design conditions at which the joint or hose assembly is to be used. If this force is otherwise unknown, it should be taken as the sum of the product of the maximum internal area times the design pressure plus the force required to deflect the joint or hose assembly.

Where expansion joints or flexible metal hose assemblies are subjected to a combination of longitudinal and transverse movements, both movements should be considered in the design and application of the joint or hose assembly.

Flexible metal hose assemblies should be supported in such a manner as to be free from any effects due to torsion and undue strain as recommended by the manufacturer.

13.2.14.6 Bearing Type Supports

Bearing type supports should permit free movement of the piping, or the piping should be designed to include the imposed load and frictional resistance of these types of supports, and dimensions should provide for the expected movement of the supported piping.

13.2.14.6.1 Support Feet

Selection, design, and application of supports feet should be according to standards.

To ensure unrestricted movement of sliding supports, bearing surfaces should be clean.

13.2.14.7 Structural Attachments

External and internal attachments to piping should be designed so that they will not cause undue flattening of the pipe, excessive localized bending stresses, or harmful thermal gradients in the pipe wall. It is important that attachments be designed to minimize stress concentration, particularly in cyclic services.

13.2.14.7.1 Nonintegral Attachments

Nonintegral attachments, in which the reaction between the piping and the attachment is by contact, include clamps, slings, cradles, U-bolts, saddles, straps, clevises, and pick-up supports. If the weight of a vertical pipe is supported by a clamp, it is recommended to prevent slippage that the clamp be located below a flange, fitting, or support lugs welded to the pipe.

In addition, riser clamps to support vertical lines should be designed to support the total load on either arm in the event the load shifts due to pipe and/or hanger movement.

13.2.14.7.2 Integral Attachments

Integral attachments include ears, shoes, lugs, dummy supports, rings, and skirts that are fabricated so that the attachment is an integral part of the piping component. Integral attachments should be used in conjunction with restraints or braces where multiaxial restraint in a single member is to be maintained.

Consideration should be given to the localized stresses induced into the piping component by the integral attachments.

The design of hanger lugs for attachment to piping for high temperature service should be such as to provide for differential expansion between the pipe and the attached lug.

To prevent lines subjected to thermal expansion/contraction moving off their supports, consideration should be given to the actual length of the cradle or pipe shoes to be used.

Pipe stanchions, pipe dummies, and trunnions should have welded end plates.

Weld-on support attachments, such as cradles or pipe shoes, pipe stanchions, pipe dummies, trunnions, and lugs, should not be attached to tees, reducers, and elbows. When stress analysis permits, pipe stanchions, pipe dummies, and lugs may be attached to elbows.

Field welding to pipes for pipe supporting purposes should be limited as far as possible. Field welding for pipe support purposes should not be performed on the following pipe materials:

- materials requiring post weld heat treatment;
- lined carbon steel (glass, PTFE, rubber, cement, etc.);
- nonferrous materials.

For pipes requiring postweld heat treatment, attachments required for supporting purposes should be indicated on the piping isometric drawings, and welding should be executed at the pipe shop before postweld heat treatment.

All welds of support elements and of supports to piping should be continuous. The fabricated and supplied supports should conform the "Bill of Material for Supports" drawings and standards should be able to withstand the allowable loads.

Welds should be proportioned so that the shear stresses do not exceed either $0.8\times$ the applicable S values for the pipe material shown in the allowable stress tables, or the allowable stress values determined in accordance with standards.

If materials for attachments should have different allowable stress values than the pipe, the lower allowable stress value of the two should be used.

13.2.14.8 Supports for Insulated Pipes and Attachments

Insulated lines running in pipe trenches should be supported high enough to assure the insulation will remain above the highest expected storm water levels.

Clamped cradles or pipe shoes should be used on the following insulated lines:

- piping lined with glass, rubber, plastics, etc.;
- piping requiring post weld heat treatment;
- piping requiring approval of the engineer;
- expensive materials such as titanium, hastealloy, monel, etc.;
- piping with corrosion resistant coating (e.g., galvanized piping).

For all other insulated lines welded cradles or pipe shoes should be used.

13.2.14.8.1 Insulated Lines for Hot Service

Reference should be made to relevant standards.

Piping DN 40 (NPS 1½) and smaller should be supported directly from the insulation by addition of a loadbearing metal sleeve or saddle outside of the insulation.

Piping DN 50 (NPS 2) and larger should be supported on pipe shoes or saddles that allow for the full insulation thickness.

The lines should be set above the supporting structure by cradles or pipe shoes to provide adequate clearance for painting and insulation. The clearance between the insulation and the supporting structure should be at least 50 mm.

To maintain common support levels the back or underside of pipes (or underside of supports) should be on the same plane or level, irrespective of pipe size or insulation thickness.

Supports for lines with heat tracing should be shaped such that they will not obstruct with the tracing or impede dismantling of supports and tracing.

To limit water ingress into insulation, welded, rather than clamped, cradles/pipe shoes should be used where the latter could pierce through the insulation.

Cradles and pipe shoes of lines operating at temperatures above 400°C should be isolated from the supporting structure by incombustible insulating blocks of sufficient load bearing and insulation capabilities.

Alternatively, clamped cradles or pipe shoes can be installed around the insulation. At the location of these supports the insulation should have sufficient load-bearing capabilities.

13.2.14.9 Uninsulated Lines

Uninsulated lines should rest directly on the supporting structure. However, cradles or pipe shoes should be considered if:

- The operating temperature of the line is below ambient and therefore the line will often have surface condensation.
- The line is a permanently operating transport line, operating at ambient temperatures, without switch-over possibilities.

- The line requires a slope.

  Note: This is only for small slope corrections. The height of cradles or pipe shoes measured from underside of pipe should be maximum 400 mm.
- The line may operate (even temporarily) at such a low temperature that this may cause embrittlement of the supporting member.
- They are needed to avoid unacceptable pipe corrosion in high corrosion areas (e.g., due to coating damage caused by movement and water collection on top of the supporting structure).

  Note: The application of cradles or pipe shoes in these situations does not alleviate corrosion problems of the supporting members themselves. In very corrosive environments saddles may be considered instead of cradles or pipe shoes.
- If the span between the supports (e.g., where the economic span is dictated by a majority of bigger lines) is too big for a pipe, the size of that pipe may be increased to meet an acceptable deflection and stresses, provided an economic evaluation justifies such an increase versus the cost of intermediate supports. Such decision is subject to the purchaser's approval.

13.2.14.10 Spacing (Span)

1. Maximum allowable span should be in accordance with standards.
2. Where line self-draining is essential, the deflection of the pipe due to self-weight at the point of maximum deflection should not exceed the vertical fall over the span due to the set slope.

13.2.14.11 Threads

Where supports have vertical adjustment, any bolt projection below the foot plate should not exceed two bolt diameters. The use of metric threads on pipe supporting elements is permitted (additional to ASME B31.3).

13.2.14.12 Fixtures

Inextensible supports other than anchors and guides.

1. Pipe support clips for carbon steel lines should be made from round bar (i.e., U-bolts) rather than flat strip to reduce corrosion (addition to ASME B31.3).
2. Shoes should be provided at supports, on insulated lines where the normal operating fluid temperature exceeds 100°C (212°F). Shoes should be sufficiently large to prevent disengagement from the supporting structure under abnormal temperature cycle or hydraulic shock conditions (additional to ASME B31.3).

13.2.15 Insulation

1. Insulation should be calculated as a function of piping operating temperature and ambient conditions. Ambient reference temperature should be deduced as the average of the minimum values occurred in the coldest month of the year for hot insulation and average of the maximum values occurred in the hottest month of the year for cold insulation.
2. Design for thickness and type of various insulation should be made in accordance with the
3. To reduce the possibility of condensation within the bellows, thermal insulation or screening may be used, but the bellows material should not be subjected to a continuous high temperature which would lead to an unacceptable fatigue life.

13.2.15.1 Winterization

Equipment and piping should be winterized if any of the following conditions applies in a stagnant system for the fluid being handled:

1. The lowest ambient temperature is below the pour point or freezing point.
2. Undesirable phase separation, deposition of crystals, or hydrate formation will occur at any ambient temperature.
3. In gas systems where condensation, hydrate or ice formation can occur at any ambient temperature (or due to cooling caused by expansion of the gas).
4. Viscosity at low ambient temperature is so high that an inadequate flow rate is obtained with the pressure available for starting circulation.
5. Lines that are normally dry, e.g., flare lines or instrument air lines, but which may carry moisture during an operating upset.

13.2.16 Painting

1. Painting should be applied to uninsulated piping in order to protect them from the corrosion due to environmental agents.
2. Any piping component should be furnished completely painted by the manufacturer as specified by purchase order.
3. Painting should not be applied to galvanized and stainless steel pipes.
4. Painting system should be designed in accordances with relevant standards.

13.2.17 Piping Connections to Existing Plant (Additional to ASME B31.3)

1. Cut-ins and tie-ins should be designed so that:
   a. Fabrication and field work is minimized and consistent with plant operating and shutdown limitation requirements.
   b. "Hot work" (field welds, etc.) should be minimized in restricted areas.
2. The method and the location of cut-ins and tie-ins should be approved by the operating management at the design stage.

UNDERGROUND PIPING SYSTEMS

This chapter covers the minimum requirements for design of underground piping in oil, gas, and petrochemical plants.

13.3 UNDERGROUND PIPING DESIGN

13.3.1 Design Condition

Design pressure and temperature of underground piping should be determined in accordance with ASME B31.3.

13.3.2 Underground Piping Drawings

The following items should be shown in underground piping drawings:

1. All buried sewers.
2. Underground process and water lines.
3. Equipments and structures foundations.
4. Cables trenches of electrical and instruments.
5. Coordinates of roads, buildings, storage tanks, and pipe racks.
6. Natural elevation.

Underground piping plan should be prepared to the scales: 1:100 or 1:200 or 1:500.

13.3.3 Underground Piping Layout Design

Minimum clear space between underground piping and sewer lines should be 300 mm. Except on cooling water lines where heat transfer might occur, there should be a minimum clearance space of 460 mm. Minimum cover should be 460 mm (or greater where frost or loading may govern below high point of finished surface). Minimum cover for water lines should be 1 m.

Embedded lines should be arranged as horizontally and plainly as possible.

The external surface of embedded piping should be spaced at a distance of 0.3 m minimum from other installations.

Where underground pipings cross each other, about 150–300 mm differential elevation (higher or lower) must be provided at the crossing point during installation.

As a rule, the underground piping should not be installed on two levels on the same route.

As a rule, branch off from the underground piping should be taken out from the upper side of main header; however, water pipe can be taken out from the edge side as possible.

The underground piping should be shaped so as not to permit air pocket or drain pocket, as much as possible.

Road crossings should be properly constructed concrete under passes, allowing ample space for maintenance purposes and 20% allowance for installation of future lines. These crossings should be specified on the drawings.

Where piping is embedded below the surface of subterranean water, the study of floating force to piping should be performed.

13.3.4 Valve Installation

Block valves should be provided near the header of each branch.

All valves should be installed in valve pits according to standard drawings.

13.3.5 Water Lines

The cooling lines should be arranged at the equipment side.

Fire hydrant lines should be arranged in the outer of pipe rack route and near to access road.

13.3.6 Protection of Underground Piping

Underground water lines outside process unit boundaries should also be cathodically protected.

APPENDIX A

Types of Valves

Type	Size	Service
Ball valves	DN 15 and larger	General service
Gate valves	DN 15 and larger	General service
Butterfly valves, lined	DN 100 and larger	General (water) service, class 150#
Butterfly valves, lined	DN 100 and larger	150# and 300# corrosive service
Butterfly valves, soft or metal seated (high performance)	DN 80 and larger	General service and special applications, e.g., cryogenic, high temperature
Diaphragm valves, lined	DN 15 to DN 300	150#, corrosive service
Ball valves, lined	DN 15 to DN 150	150# and 300# corrosive service
Plug valves (pressure balanced)	DN 15 and larger	High-pressure gas systems (e.g., hydrogen)
Needle valves	DN 15 to DN 40	Accurate control
Globe valves	DN 15 to DN 200	General service
Diaphragm valves, lined	DN 15 to DN 300	Low-pressure corrosive service
Butterfly valves, lined	DN 50 and larger	Moderate pressure corrosive service
Choke valves	DN 50 to DN 200	For high-pressure difference and/or erosive service
Butterfly valves (high performance)	DN 80 and larger	General service
Check valves	DN 15 to DN 40	Piston type, horizontal flow
	DN 15 to DN 40	Ball type, horizontal flow
	DN 50 and larger	Swing type
	DN 50 and larger	Dual plate type, spring energized

Chapter 14

Strainers and Filters

14.1 INTRODUCTION

This chapter covers general requirements for design material, fabrication, inspection, and test of different types of strainers and filters mainly used in suction lines of pumps, compressors, and turbines in oil, gas, and petrochemical industries.

This chapter is intended to supplement purchase orders placed for strainers and filters. If requirements of this chapter differ from, or are in conflict with, purchase documents, the following will take precedence in the order of priority as indicated hereunder:

1. Purchase order,
2. Data sheets and drawings.

Filters should be a horizontal or vertical vessel with full-end, quick-opening closures and should be designed, constructed, and inspected in accordance with ASME Requirement for Pressure Vessels, Section VIII, Div. 1.

The supplier should complete all performance data as indicated on the filters and strainers data sheet.

Differential pressure-protection devices, such as differential pressure indicators, alarm and trip, should be provided for each filter and strainer as specified in the data sheet, and should be positioned so that the pressure drop across the filter or strainer element can be determined. This pressure drop should not exceed 0.5% of the flowing line pressure, if it is not specified on the data sheet.

Pressure-tight covers or flanged-access openings should be provided to permit internal access for inspection, maintenance, and filter or strainer element replacement.

Flanges or other type of connections should match the line class and rating for which it is intended.

Maintenance servicing should be possible without the necessity for breaking pipe connections. Internals should be able to be changed out within 10 minutes once the vessel has been drained and the head cover is removed.

The filter element of a cartridge filter should be designed as a disposable component, to be replaced with a new cartridge when clogged. Some

Oil and Gas Pipelines and Piping Systems. DOI: http://dx.doi.org/10.1016/B978-0-12-803777-5.00014-9

cartridge elements of robust construction may be specified as cleanable and reusable.

Cartridge filters should be compact, reliable, and easy to operate.

Cartridge filters should be used in systems where the contaminant level is less than 0.01% by weight (100 ppm).

Different materials may be used for cartridge elements.

All air filters should be designed to be installed upright directly onto the compressor, blower, or engine inlet or on the remote inlet to air intake piping.

Filters and strainers should be designed so that standard elements available from various sources may be used.

Filter inlet design should include provision to prevent direct impingement of the incoming fluid onto the filter element (Fig. 14.1).

Air dryer filters should be a cartridge type: cleanable and suitable for oil and vapor removal. When desiccants are used, other filters should be suitable for removal of desiccant particles.

For compressor suction lines bayonet filters as shown in relevant standards should be used.

14.2 STRAINER AND FILTER DESIGN, FABRICATION, AND ASSEMBLY

Permanent strainers should normally be of a design that has a steel body incorporating a basket that can be removed without dismantling pipework.

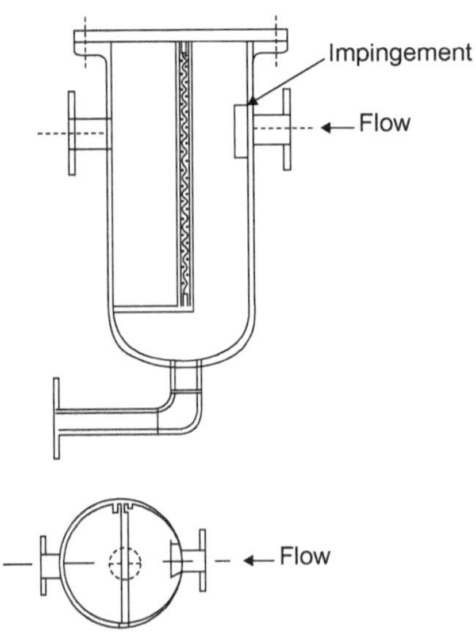

FIGURE 14.1 Pump suction filter.

Strainers should be so designed and fabricated to prevent their damage due to vibration, differential pressure, pulsating flow, or impact of objects.

Screen of mesh size 20 or finer should be reinforced with perforated plate, or heavier screen and steel bars.

Strainer open areas should be not less than 3× the inlet pipe cross-section area. For suction lines to compressors in air service, this criteria should be followed unless otherwise specified.

Cone or basket type strainers should be provided with identification tabs that protrude from the holding pipe flanges.

In "TEE" type strainers, guide rods and shelf rods supporting the strainer element should be sized as per standard.

End flanges of strainers should be integral with the body. For steel strainers, only flanges may be added by full penetration butt-welding.

Strainer bodies should have an arrow raised on the body itself, indicating the direction of flow.

The construction of a strainer or filter should be such that parts can be reassembled in the intended manner after being dismantled to the extent needed for servicing.

A strainer or filter should be constructed so that, when in its intended operating position, any air trapped within will not reduce the rate of liquid flow or the effective strainer element capacity.

A unit element should be constructed to hold in its intended position to ensure that joints or seals required to prevent fluid bypass of the element will be maintained.

A strainer should be such that, when the screen or filter element is removed for cleaning, all foreign matter (sediment and dirt) will be removed or can be removed without the probability of any foreign matter being deposited in the outlet side of the strainer.

In each strainer or filter, clean-out and drain openings should be closed by a standard pipe plug or a threaded shouldered plug. The plugs should not create a galvanic cell with the housing that will accelerate corrosion.

Both external and internal parts of the assembly should be free of rough or sharp edges that are likely to cause injury to persons servicing the unit.

The assembly should be capable of disassembly and reassembly with ordinary tools.

The assembly should be constructed to withstand the stresses and strains likely to be encountered in service.

An opening thread for connection of pipe should be threaded in accordance with the Standard for Pipe Threads, ISO 7-1.1982.

Strainers provided with screwed covers should employ either ground joints gaskets, or "O" rings suitable for the purpose. If a gasket or "O" ring is used, it should be retained by the body, cover, or cap when the part is removed and should not be damaged when the cover or cap is screwed in place.

Cementing or retaining of the gasket is not necessary provided that a complete set of new gaskets is furnished with each replacement cartridge for those filters employing a cartridge type filtering element.

Body and cover joint should be clamped or flanged as specified in the fabrication design of strainer. This joint should be suitable to meet the rating conditions required.

14.3 MATERIAL AND DIMENSIONS

The material for strainer or filter body (including bolting) should be equal to the materials of valves in the same service.

In general the screen for factory fabricated strainers should be a basket type made of the same (or better) materials as the valve trim of the line classes, e.g., 11–13% chrome or Type 316 stainless steel.

If corrosion of a ferrous part will interfere with the intended function of a strainer or filter, the part should be provided with a corrosion-resistant protective coating.

A protective coating should provide resistance against corrosion to a degree not less than that provided by the protective plating.

Cadmium plating should have a thickness of not less than 0.008 mm (0.0003 in.) and zinc plating should have a thickness of not less than 0.013 mm (0.0005 in.), except on parts where threads constitute the major portion of the area, in which case the thickness of the cadmium or zinc plating should be not less than 0.0038 mm (0.00015 in.).

Wire cloth type elements, if finer than 60 mesh, should be resistant to corrosion. A 60 mesh or coarser element should be resistant to the fluid it may normally contact.

A part made of drawn brass or machined from a brass rod should be capable of withstanding, without cracking, a mercurous nitrate test for copper and copper alloys.

The mesh size and material of suction pump strainers should be in accordance with Table 14.1.

The mesh size and material for strainer elements used in compressor systems should be in accordance with Table 14.2.

The mesh size and material for strainer elements used in turbine systems should be in accordance with Table 14.3.

For the dimensions of screen and perforated plate opening, refer to Table 14.4.

Many different types of filter media should be available for separating solid matter from liquids and gases; the range may include paper, natural and synthetic fibers felt, plastic sheet, ceramic, carbon, cotton, yarn, cloth, woven wire, woven fabric, organic and inorganic membranes, perforated metal, sintered metals, and many other materials.

TABLE 14.1 Mesh Size and Element Material for Pump Suction Strainers

Pumps	Strainer Type	Pipe or Flange Size DN (NPS)	Mesh(1) Size or Opening 25.4 mm	Material for or Strainer Element
Centrifugal: horizontal single stage vertical in-line	Temporary	80 (3) and under	5′5	–
		100–150 (4–6)	3′3	
		Over 150 (6)	13 mm (½″)	
Horizontal multistage vertical deep well	Temporary	150 (6) and under	20′20	–
		Over 150 (6)	3 mm (1/8″)	
Reciprocating	Temporary	All sizes	5′5	–
Controlled volume	Permanent	All sizes	Note (2)	Type 304 stainless Steel
Rotary and turbine pumps	Temporary	All sizes	20′20 Note (3)	–
External flush oil systems for mechanical seals	Permanent used in main header	All sizes	20′20	Type 304 stainless steel

Notes:
1. Mesh size (or opening) as listed, are the usual maximums for normal operation.
2. Mesh size for controlled volume pumps should be selected on the basis of the pumped fluid characteristics. Chemical or mechanical cleaning of suction lines may be substituted for strainer where the pumped fluid can be expected to be free of sediment.
3. Mesh size listed for rotary and turbine pumps assumes preliminary operation on low-viscosity fluid.

14.4 TYPES OF STRAINERS

Different types of suction strainers are as follows:

14.4.1 Flat Type Strainer

Flat disc strainers should be in the form of a perforated plate and should be used as a temporary strainer in suction lines. Its size and characteristics are shown in relevant standards.

TABLE 14.2 Mesh Size and Element Material for Compressor Suction Strainers

Compressor	Service	Pipe or Flange Size DN (NPS)	Strainer Type	Mesh Size or Opening 25.4 mm	Material for Strainer Element
Centrifugal	Air suction	All sizes	Permanent	3′3	Type 304 stainless Steel
	Note (1) Gas suction		Temporary	5′5 (2)	Type 304 stainless Steel
	Buffer seal		Permanent	100′100	Type 304 stainless steel
	Flushing		Permanent	80′80	Monel
Reciprocating	Air suction	All sizes	Perm. Filter	–	–
	Gas suction		Temporary	20′20	Type 304 stainless Steel
	Flushing		Permanent	80′80	Monel
Rotary Screw	Air suction	All sizes	Permanent	20′20	Type 304 stainless
	Gas suction		Temporary	20′20	
	Buffer seal		Permanent	100′100	
Axial	Air suction	All sizes	Permanent screen on dry Filter	5′5	Type 304 stainless steel
	Gas suction		Temporary	5′5 (2)	
	Buffer seal		Permanent	100′100	
	Washing		Permanent	80′80	

Notes:
1. Suction service for centrifugal compressors includes side stream suctions.
2. Wire mesh screens for centrifugal compressors, gas suction service, should be reinforced with perforated plate or heavier mesh and steel bars.

TABLE 14.3 Mesh Size and Element Material for Turbine Suction Strainers

Turbines	Service	Pipe or Flange Size DN (NPS)	Strainer Type	Mesh Size or Opening 25.4 mm	Material for Strainer Elements
Gas turbines	Inlet air	All sizes	Permanent screen on dry filter	5′5	Type 304 stainless steel
	Washing		Permanent	80′80	
Steam turbines(1)	Inlet	All sizes	Permanent	3 mm (1/8″)	Monel

Note:
An additional strainer element should be provided for special purpose and generator drive steam turbines for use during start up. Opening size should be 1.5 mm (1/16″)

TABLE 14.4 Dimensions for Screen and Perforated Plate Opening

	Screen				Perforated Plate Opening	
Meshes	Wire Diam		Arrange Opening Width		mm	inch
PER 25 mm (Linear Inch)	mm	inch	mm	inch		
100′100	0.102	0.004	0.15	0.006	1.5	1/16
80′80	0.140	0.0055	0.18	0.007	3.0	1/8
20′20	0.406	0.016	0.85	0.033	13	½
5′5	1.6	0.063	3.5	0.137		
3′3	1.6	0.063	6.9	0.272		

14.4.2 Conical Type Strainer

This kind of strainers should be used as temporary strainers in suction lines and is shown in relevant standards.

14.4.3 "T" Type Strainer

This type of strainer, with a bathtub screen, should be used as temporary or permanent strainers and is shown in detail in relevant standards (Figs. 14.2 and 14.3).

FIGURE 14.2 'T' line strainer.

14.4.4 "Y" Type Strainer

This kind of strainer should be required for permanent installation in vertical or horizontal suction lines, and is shown with detail in relevant standards.

14.4.5 Basket Type Strainer

On large pipelines basket type strainers should be used (Fig. 14.4). They should provide greater dirt-holding capacity and easier access for removal of the strainer for cleaning. They normally have a higher pressure drop than a simple Y type strainer.

14.4.6 Dual Strainer

Where continuous operation is required in a pipeline service, dual strainers should be used in an integral unit with provisions to isolate them at any time for cleaning (Fig. 14.5).

14.4.7 Self-Cleaning Strainer

This type of strainer should be used as an alternative to dual or multiple strainers where continuous supply is critical in a process system, and should be divided into a brush model and a screw model, as illustrated in (Fig. 14.6).

Twin strainers or self-cleaning strainers, as appropriate, should be used in unspared equipment services.

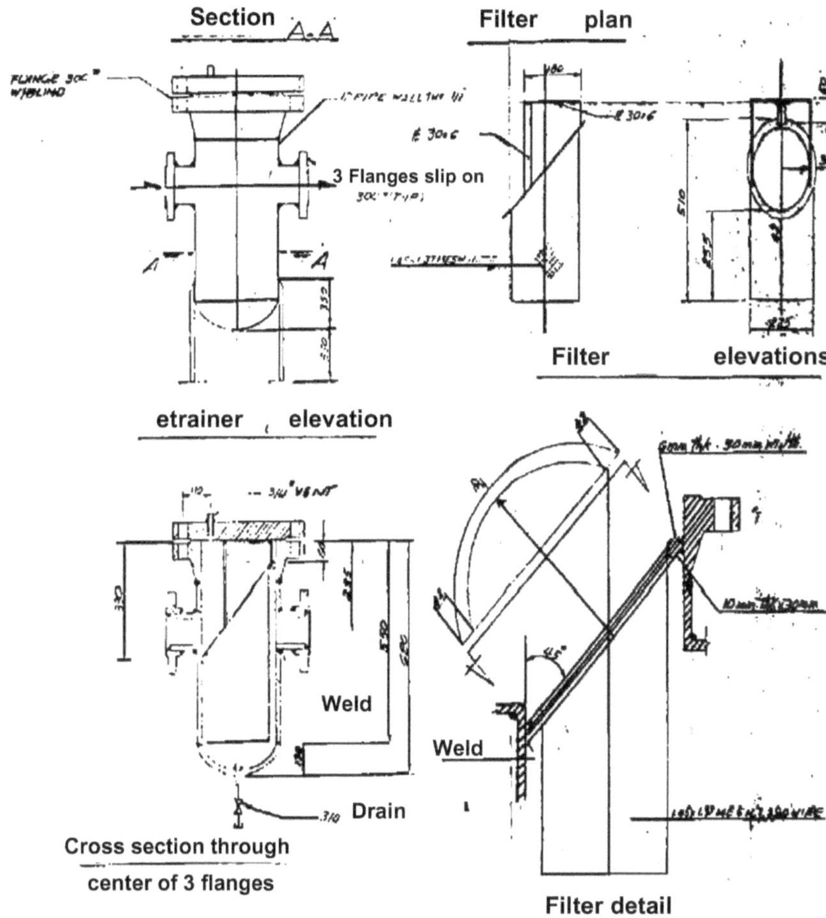

FIGURE 14.3 Filter detail.

FIGURE 14.4 Basket type strainer.

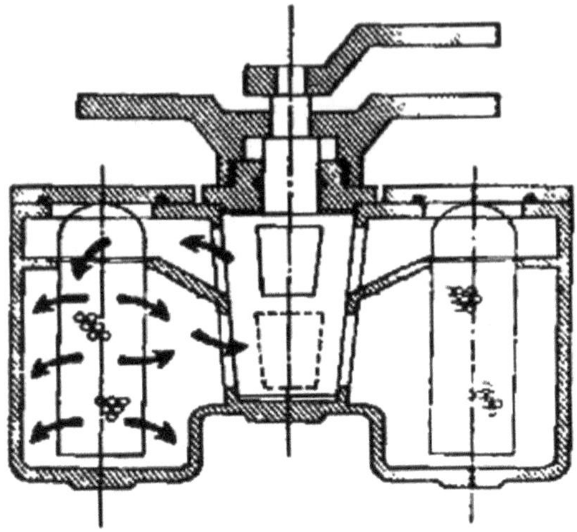

FIGURE 14.5 Dual in-line strainer (schematic).

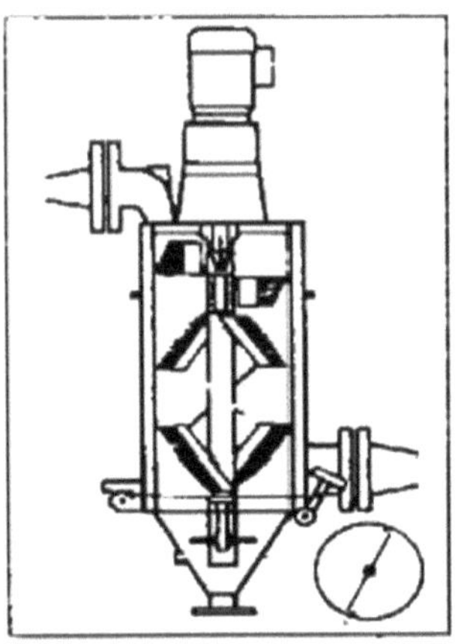

Brush model

Suitable for low-viscosity products.

A special feature is the two bars in the perforated cylinder for cleaning the brushes.

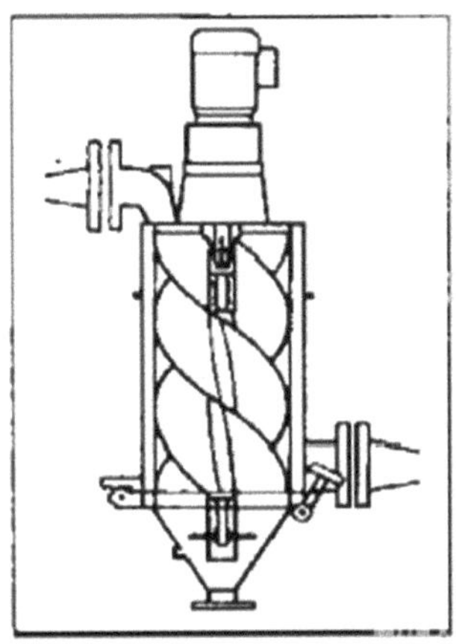

Screw model

Suitable for heavy-viscosity products such as animal fat and wax and to enable cleaning-in-place (CIP).

FIGURE 14.6 Self-cleaning strainers.

Self-cleaning strainers should be provided with valved, blowdown connections.

If considerable clogging of strainers is anticipated, due to coke or similar conditions (as indicated on the data sheet), a strainer should be of the automatic self-cleaning type to permit continuous flow of clean liquid (Fig. 14.7).

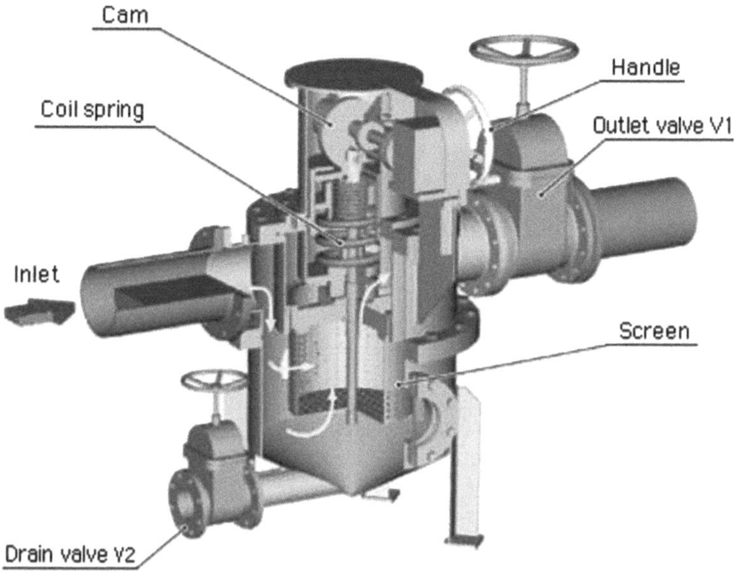

FIGURE 14.7 Automatic self-cleaning strainer.

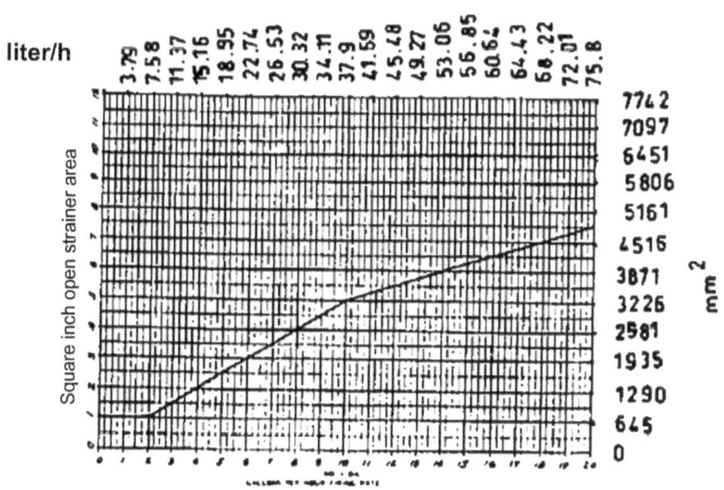

FIGURE 14.8 Strainer area for * no. 1 fuel oil. *As designated by ASTM D396-1997 I square inch 6.45 cm^2, 1 Gallon = 3.79 L.

14.4.8 Oil Burner Strainer

Each strainer assembly intended for use with oil burning equipment is to be rated for capacity in terms of the maximum firing rate of the burner equipment as expressed in terms of liter of fuel oil per hour (0.264 gallons/h). Figs. 14.8–14.10 provide for strainers employing wire cloth or perforated

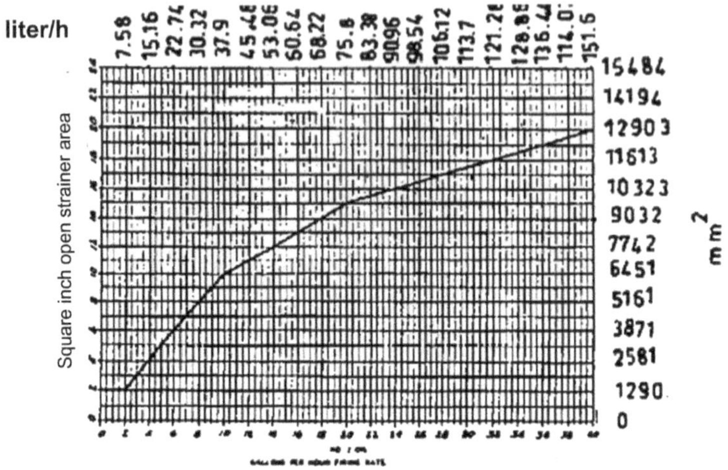

FIGURE 14.9 Strainer area for * no. 2 fuel oil. *As designated by ASTM D396-1976 *Note:* 1 square inch = 6.45 cm^2, 1 gallon = 3.79 L.

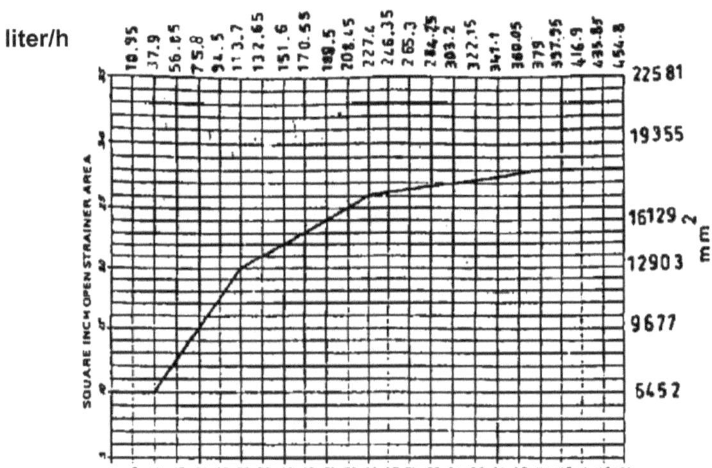

FIGURE 14.10 Strainer area for * nos. 4, 5, 6 fuel oil. *As designated by ASTM D396-1976. *Note:* 1 $in.^2$ = 6.45 cm^2.

screens, minimum areas of screen opening based upon the grade of fuel oil used, and the size of the equipment to be served.

Strainer capacity ratings for oil burner strainers are expressed in terms of maximum burner firing capacity rather than actual flow capacity. The curves as shown in Figs. 14.8–14.10 have been predicated on experience and are designed to provide freedom from excessive strainers for a reasonable period of time.

The selection of capacity ratings based on firing rates versus open or free area for various grades of fuel oil also provides for low initial pressure losses or drops through the strainer assembly when clean.

The capacity of each size and type of element is to be determined. When possible, this is to be accomplished by calculation for wire cloth and perforated types. Elements other than these, such as filter elements of felt, cotton waste, ceramics, and the like, are to be subjected to the tests to establish ratings by comparison to the recognized properties of wire cloth or perforated screens.

A wire cloth type element has an open area equal to the total area of the cloth minus the area covered by wrap woof, and supports are multiplied by the screen factor, which is the percentage of open area of the cloth to the whole area. If the screen factor is unknown, it may be calculated as follows:

$$\text{Screen factor} = (1 - \text{ND}) \times (1 - \text{nd})$$

where:

N = Number of wires in warp per unit length
n = Number of wires in the woof per unit length
D = Diameter of wire in warp (mm)
d = Diameter of wire in the woof (mm)

A perforated type element has an open area equal to the total number of openings multiplied by the area of each opening minus the area covered by seams, ribs, and supports.

The effective area of a perforated type element that includes a wire cloth insert is considered to be the smaller of the two areas determined by Paragraphs 8.8.5 and 8.8.6.

A filter–type element when clean, other than the wire cloth or perforated type (i.e., filter elements such as felt, cotton waste, or ceramics), should not cause a pressure differential between inlet and outlet openings in excess of 51.8 mm of mercury (approximately 6.9 kPa or 1 psi) when passing the intended grade of fuel oil at a rate not exceeding the rated capacity of the strainer.

14.5 INSPECTION

14.5.1 Normal Inspection

The manufacturer should be prepared for the inspection of strainers/filters. The inspection should include the following:

1. Check of chemical and mechanical properties of the materials used.
2. Visual inspection of parts and casing.
3. Dimensional inspection.
4. Assembly inspection.

14.5.2 Additional Inspection

Additional inspection should be carried out unless otherwise specified in the purchase order. These additional inspections are listed below.

- Radiographic examination of welding of flanges to the body of steel strainers or filters:
  - Weld quality should conform to requirements of ANSI B.31.3 with regards to the butt-welded flanges.
- Radiographic examination of body of strainers:
  - Radiographic examination should be made in areas shown in Fig. 14.11 (i.e., zones 1, 2, and 3).
  - The radiographic procedure and criteria for acceptability of defects indicated in radiographs should be in accordance with ANSI B.16.34, Annex B.
  - The types and degrees of discontinuities in radiographs and also the use of film and recording media can be compared with those cited in ASTM E- 142, E-186 and E-446.
- Magnetic-particle examination of the whole surface of body of strainers and filters:
  - Inspection procedure and degree of acceptability should comply with requirements laid down in ANSI B.16.34, Annex C. Type and classification of defects should be compared with reference photographs of ASTM E-125.

14.5.2.1 Liquid Penetrant Inspection

This type of inspection should be used in lieu of magnetic examination where this examination is not applicable. The inspection method should conform to ASTM E-165. The procedure and acceptance standard should comply with requirements outlined in ANSI B.16-34, Annex D.

14.5.2.2 Charpy Impact Test on Strainers and Filters (V-Notch)

Charpy impact test should be performed for toughness indication. The test should be performed in accordance with ASTM A 370 and ISO 148.

14.6 TESTS

Except as otherwise indicated, representative samples of a filter or strainer are to be subjected to the tests described in this chapter.

If a series of strainers or filters are to be investigated in which the bodies differ in size only, three representative samples are to be chosen to include the largest, smallest, and one intermediate size. If a strainer or filter having a single body size is being investigated, one sample is sufficient.

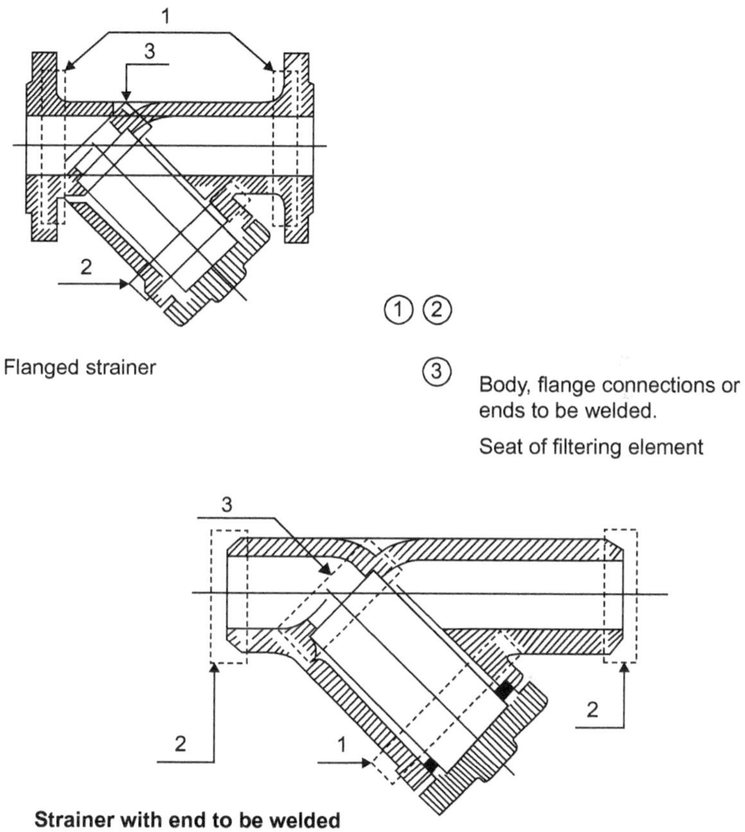

FIGURE 14.11 Typical parts of "Y" type strainer.

The strainer or filter is to be investigated for a specific fluid or fluids and for the service conditions for which it is to be recommended, such as fluid temperature and fluid pressure.

14.6.1 Deformation and External Leakage (Tightening Test)

This test should be performed according to Clause 11 of ANSI/UL 331 (1977).

14.6.2 Pressure Drop Test of Strainers or Filters for Oil Burners

This test should be carried out according to Clause 14 of ANSI/UL 331 (1977).

14.6.3 Clogging Test of Strainers for Oil Burners

Clogging test should be performed according to Clause 15 of ANSI/UL 331 (1977).

14.6.4 Mercurous Nitrate Immersion Test

Mercurous nitrate immersion test should be carried out according to Clause 16 of ANSI/UL 331 (1977).

14.6.5 Fire and Thermal Shock Test

This test should be done in accordance with Clause 16 of ANSI/UL 1105 (1977).

14.6.6 Vibration Test

Vibration test according to Clause 12 of ANSI/UL 1105 should be applied if specified by the engineer.

14.6.7 Shock Test

Shock test should be applied to strainers if specified by the engineer according to ANSI/UL 1105 (1977).

14.6.8 Cold-Temperature Test

A cold-temperature test, according to ANSI/UL 1193 (1985), should be performed on strainers and filters, if they are used in areas in which the temperature falls down to $-30°C$. After this test, a shock test, according to relevant standards, should be performed.

14.6.9 Hydrostatic-Strength Test

This test should be done in accordance with Clause 12 of UL 331 (1977).

Testing performed in the presence of purchaser's representative should not relieve the manufacturers of their own responsibilities and guarantees and of any further contractual obligations.

14.7 PAINTING AND PROTECTION

On completion of tests all strainers and filters should be painted with two layers of an antirust undercoat and one final layer of paint suitable for the specified environment following surface preparation. The color of final layer should be as per relevant standards.

Stainless steel or bronze strainers or filters should not be painted.

All unpainted surfaces (inside or outside) should be adequately protected with suitable antirust compound, easily removable by hydrocarbon solvent.

Flanged or butt-welded ends should be protected with wooden covers of a diameter not less than the outside diameter of the ends. Screwed and socket-welding ends should be protected with plastic or cardboard plugs.

14.8 IDENTIFICATION AND MARKING

Corrosion-resistant stainless steel nameplates should be securely fastened by screws or rivets to each identifiable piece of equipment.

The following information should be embossed on nameplates:

1. Manufacturer's name;
2. Manufacturer's serial number;
3. Date of manufacture;
4. Equipment item or tag number;
5. Size (nominal diameter of connecting pipe and face-to-face dimension);
6. Weight, including internals;
7. Whether or not radiographed and/or stress relieved;
8. Pressure rating;
9. Body and element material;
10. Hydrostatic Test Pressure

A strainer or filter should be also marked with the following information:

1. The fluid service or services for which the strainer or filter is intended.
2. The direction of flow.
3. In oil burner strainers or filters, the rated effective area or the catalog designation of the element, or equivalent, if more than one size element is available for a particular strainer or filter. This marking should be on the element.
4. Marking should be legible and reasonably permanent, such as afforded by a metal nameplate, decalcomania transfer, or waterproof marking ink.
5. If a manufacturer produces strainers or filters at more than one factory, each strainer or filter should have a distinctive marking to identify it as the product of a particular factory.

14.9 PACKING AND SHIPMENT

Strainers or filters should be suitably packed for export and protected against all damages or defects which may occur during handling, sea shipment to the port and rough road haulage to site and extended tropical open air storage, generally as per the purchaser's general condition of purchase.

Spare parts should be packed for long-time storage under site atmospheric conditions, as cited in the data sheet.

All equipment and component parts should be guaranteed by vendors against defective material, design, and workmanship when operated under normal condition for 12 months starting from the completion of seven days continuous test *in situ* at full load, but not exceeding 18 months after date of shipment. If any malperformance or defects occur during the guarantee period, the vendor should make available repaired, altered, or replacement parts free of any charges, whatsoever, direct on the purchaser's job site. The vendor should make available free of charge qualified representatives as he deems necessary to supervise the removal, repair, and replacement of the defective parts in such a manner that the guarantee be maintained.

The guarantee period for repaired or replaced parts should be 12 months after startup of repaired equipment but not more than 18 months after the repaired parts and/or equipment are shipped.

The guarantee period for the remaining equipment, whose operation is dependent upon the proper performance of the repaired part, should be extended by the number of days or fraction thereof that the equipment have been inoperative because of defects. Field labor charges for work during the guarantee period should be subject to negotiation between the purchaser and the vendor.

If defects are found, and the vendor is not in position to take necessary action and perform the repairs within the time required by the purchaser and agreed upon every time according to the purchaser's requirements, the purchaser should have such modification and repairs made and the relevant expense will be charged to the vendor. It is understood that in this instance the vendor should not be relieved of his guarantee and contract obligations.

Furthermore, the vendor should guarantee the provision of spare parts for a minimum period of 15 years from the late date of dispatch of the materials and/or equipment.

Suppliers should be notified of any insurance facilities and rates in cases where the safety of the strainers and filters to be shipped or boarded are deemed essential.

The manufacturer/supplier's proposal should include the following information:

- Preliminary outline drawings showing dimensions and weights of equipment.
- Description of any special tools furnished for installation or maintenance of the equipment.
- Completed specification (data sheet).
- Complete list including make, model, and size of all equipment to be supplied including auxiliary equipment.
- Spare parts for 2 years of operation and commissioning.

Chapter 15

Corrosion in Pipelines and Piping Systems

15.1 INTRODUCTION

This chapter covers an outline for the corrosion consideration in selection of materials used in oil, gas, and petrochemical industries.

The designers have to apply provisions of this chapter during the design stage of a system in order to avoid or minimize corrosion hazards technically, economically, and safely during the designed life of such systems.

Today, there is a great deal of refinement in the availability of materials of construction, varying from metallic to nonmetallic. Also there are a large number of factors to be taken into consideration when selecting a material for a given application; these factors can be put into the following major groups:

15.1.1 Environment

Corrosion consideration arises when the environment exploits the chemical or physical properties of a material. Therefore, an understanding of the various type of environments is necessary to select a stable and resistance material.

15.1.2 Cost

The selection of a material for use in a corrosive situation must be based on sound economics. Both the cost of the material and the ongoing cost of preventive measures must be included. A sound judgment on materials must recognize that the relative corrosion resistance of materials and the cost of supplementary protective measures can change significantly from one corrosive media to another.

15.1.3 Safety

Most environments in the oil, gas, and petrochemical industries involve flammable hydrocarbon streams, highly toxic and explosive gases, and strong

Oil and Gas Pipelines and Piping Systems. DOI: http://dx.doi.org/10.1016/B978-0-12-803777-5.00015-0

acids or caustics that are often at elevated temperatures and pressures. Therefore, engineering standards vis-à-vis corrosion in material selection that address the safety concerns for such areas are considered.

Factors that influence corrosion consideration in material selection are distinguished from those that interact in a more complex fashion. For example, "application" influences selection because the type of process, variables during operation, etc., will define whether a material can be used for the intended purpose or not. On the other hand, mechanical and metallurgical properties are not uniquely defined for all environments. If the material is to be used at low temperatures, then embrittlement can be a serious problem.

Therefore, these considerations involve the kind of information on those factors that have a direct influence on corrosion consideration in material selection. However, when there is discrepancy among sections of this chapter or between this and other disciplines for selection of materials, the following priorities should be regarded:

- client discretion and in-house experience,
- specific industry standards,
- general material selection procedure (Section 15.5).

15.2 CORROSIVE ENVIRONMENTS

The first step in the material selection is a thorough review of the corrosive environments and equipment operating conditions.

This section briefly classifies the corrosive environments.

15.2.1 Atmospheric Environments

Atmospheric environments are defined under categories of dry, damp, humid, rural, industrial, coastal, municipal, etc.

Generally increase in humidity, temperature, and the percentage of acidic gases such as CO_2, H_2S, SO_2, CO, Cl_2 will increase its corrosivity.

15.2.2 Soil

Most of the industrial equipment in contact with soil or embedded underground will suffer corrosion. Increase in water content and decrease in pH and resistivity will enhance corrosivity of soil.

15.2.3 Seawater

The greatest attack on offshore structures occurs in the splash zone due to alternate wetting and drying and also aeration. In quite stagnant conditions the effect of bacteria and pitting–type corrosion are predominant. The rapid growth of marine fouling in the tropics may provide a protective shield that counteracts the effect of the greater activity of the hotter water.

15.2.4 Natural Waters

The corrosivity of natural waters depends on their constituents such as dissolved solids, gases, and sometimes colloidal or suspended matters. The effect may be either one of stimulation or one of suppression of the corrosion reactions.

Constituents or impurities of water are dissolved gases such as oxygen, CO_2, SO_2 , NH_3, H_2S, and are the results of bacterial activity.

Dissolved mineral salts are mostly calcium, magnesium, sodium, bicarbonate, sulfate, chloride, and nitrate. The effect of each of these ions on corrosion rates (CRs) is different, but briefly, the chlorides have received the most study in this regard.

Organic contaminants of water directly and indirectly can affect the CR of metals and alloys. Bacteria, under favorite conditions, can be doubled in 10–60 minutes. This characteristic is typical of wide spread biodeterioration caused by microbes in all industries of which corrosion is a special case. With a few exceptions, such as synthetic polymers, all materials can be attacked by bacteria.

15.2.5 Chemicals

A chemical may be defined as a substance containing over 95% of the principal chemical.

To classify all the chemical environments and suitable materials is impossible because of the enormity of collating such a large amount of data. For example, if some 400 chemicals are identified as being handled and processed on a large scale and there are 10 suitable materials, then 4000 systems would have to be considered. Since temperature, concentration, and solution velocity are important in determining CR, and if only five levels of each of the three variables are considered, then the number of environments to be considered would be $4000 \times 5^3 = 600{,}000$. Therefore in this chapter only those chemical environments that are corrosive and have a detrimental effect on material selection in oil, gas, and petrochemical industries are briefly discussed.

15.3 THE MATERIALS OF ENGINEERING

Today's engineers have a vast range that comprises several thousand materials available to them. Also parallel to the invention of new and improved materials there have been equally important developments in materials processing, including vacuum melting and casting, new molding techniques for polymers, ceramics, and composites, and new joining technology.

In addition to the need for an increased knowledge of materials and technology, other challenges are having to be met by material engineers. In earlier times, with a much smaller number of materials available, engineers often made their design selections by a process of trial and error, in many cases using more material than was really necessary. Today there is a requirement to have more effective and efficient knowledge about materials in order to minimize cost.

15.3.1 The Range of Materials

The complete range of materials can be classified into the four categories:

- Metals
- Polymers
- Ceramics and inorganic glasses
- Composites

The classification composites contains materials with constituents from any two of the first three categories.

15.3.2 Properties of Engineering Materials

A broad comparison of the properties of metals, ceramics, and polymers is given in Table 15.1.

Very many properties, or qualities, of materials have to be considered when choosing a material to meet a design requirement (see Table 15.2). These include a wide range of physical, chemical, and mechanical properties together with a forming, or manufacturing characteristic, cost and availability data, and in addition, more subjective esthetic qualities, such as appearance and texture. Some of these values for different materials are given in Table 15.3.

15.4 CORROSION PREVENTION MEASURES

To select required materials for a given process, necessarily and economically, feasible protective measures should also be considered.

Basically, protection comprises those protective measures providing separation of metal surfaces from corrosive environments or those that cater for adjustment or altering the environments.

Since there are specific relevant standards on these fields, there is no necessity to further enlarge this section. For this reason, only an outline of corrosion preventive measures is discussed in this section.

TABLE 15.1 Comparison of Properties of Metals, Ceramics, and Polymers

Property	Metals	Ceramics	Polymers
Density ($kg/m^3 \times 10^{-3}$)	2–16 (average 8)	2–17 (average 5)	1–2
Melting points	Low to high	High, up to 4000°C	Low
	Sn 232°C, W 3400°C		
Hardness	Medium	High	Low
Machinability	Good	Poor	Good
Tensile strength (MPa)	Up to 2500	Up to 400	Up to 120
Compressive strength (MPa)	Up to 2500	Up to 5000	Up to 350
Young's Modulus (GPa)	40–400	150–450	0.001–3.5
High-temperature creep resistance	Poor	Excellent	—
Thermal expansion	Medium to high	Low to medium	Very high
Thermal conductivity	Medium	Medium but often decreases rapidly with temperature	Very low
Thermal shock resistance	Good	Generally poor	—
Electrical properties	Conductors	Insulators	Insulators
Chemical resistance	Low to medium	Excellent	Generally
Oxidation resistance at high temperatures	Poor, except for rare metals	Oxides excellent	good
		SiC and Si_3N_4 good	—

15.4.1 Cathodic Protection

Cathodic protection is possible only when the structure to be protected and the auxiliary anode are in both electronic and electrolytic contact.

A reduction in metal to electrolyte potential of −0.850 V (reference to saturated copper sulfate electrode) is specified as the necessary potential that must be obtained for either optimum or absolute protection of ferrous structures in soil or water.

Cathodic protection is applied by one of two methods: power impressed current or sacrificial anodes.

TABLE 15.2 Material Properties and Qualities

Physical properties	Density, melting point, hardness, elastic moduli, damping capacity
Mechanical properties	Yield, tensile, compressive and torsional strengths ductility fatigue strength, creep strength, fracture toughness
Manufacturing properties	Ability to be shaped by: molding and casting, plastic deformation, powder processing, machining. Ability to be joined by adhesives, welding, etc.
Chemical properties	Resistance to oxidation, corrosion, solvents and environmental factors
Other nonmechanical properties	Electrical, magnetic, optical and thermal properties
Economic properties	Raw material and processing costs, availability
Esthetic properties	Appearance, texture, and ability to accept special finishes

15.4.2 Coating, Painting, and Lining Materials

More metal surfaces are protected by coating, painting, and lining than by all other methods combined. Coating, painting, and linings that act as a protective film to isolate the substrate from the environment exist in a number of different forms. Therefore the selection of a proper corrosion resistance system depends on a number of factors.

15.4.3 Inhibitors

Altering the environments provide a versatile means for reducing corrosion. Typical changes in medium that are often employed in petroleum industries are:

- lowering temperature,
- decreasing velocity,
- removing oxygen or oxidizer,
- filtration,
- changing concentration of corrosives,
- use of corrosion inhibitors.

An inhibitor is a substance that, when added in small concentrations to an environment, decreases the CR considerably. To be fully effective, all inhibitors are required to be present above a certain minimum concentration.

Corrosion inhibitors are chemicals that may be divided in different categories. Among these, the most used class in oil industries is the film-forming

TABLE 15.3 Properties (at 25°C) of Some Groups of Materials

Material	*E* (GPa)	Yield Strength (MPa)	Tensile Strength (MPa)	Fracture Toughness (MPa $M^{1/2}$)	Density ($kg/m^3 \times 10^{-3}$)
Steels	200–220	200–1800	350–2300	80–170	7.8–7.9
Cast irons	150–180	100–500	300–1000	6–20	7.2–7.6
Aluminum alloys	70	25–500	70–600	5–70	2.7–2.8
Copper alloys	90–130	70–1000	220–1400	30–120	8.4–8.9
Magnesium alloys	40–50	30–250	60–300		1.7–1.8
Nickel alloys	180–220	60–1200	200–1400	>100	7.9–8.9
Titanium alloys	100–120	180–1400	350–1500	50–100	4.4–4.5
Zinc alloys	70–90	50–300	150–350		6.7–7.1
Polyethylene (LDPE)	0.12–0.25		1–16	1–2	0.91–0.94
Polyethylene (HDPE)	0.45–1.4		20–38	2–5	0.95–0.97
Polypropylene (PP)	0.5–1.9		20–40	3.5	0.90–0.91
PTFE	0.35–0.6		17–28		2.1–2.25
Polystyrene (PS)	2.8–3.5		35–85	2	1.0–1.1
Rigid PVC	2.4–4.0		24–60	2.4	1.4–1.5
Acrylic (PMMA)	2.7–3.5		50–80	1.6	1.2
Nylons (PA)	2.0–3.5		60–100	3–5	1.05–1.15
PF resins	5–8		35–55		1.25
Polyester resins	1.3–4.5		45–85	0.5	1.1–1.4
Epoxy resins	2.1–5.5		40–85	0.3–0.5	1.2–1.4
GFRP	10–45		100–300	20–60	1.55–2.0
CFRP	70–200		70–650	30–45	1.40–1.75
Soda glass	74		50[a]	0.7	2.5
Alumina	380		300–400[a]	3–5	3.9
Silicon carbide	410		200–500[a]		3.2
Silicon nitride	310		300–850[a]	4	3.2
Concrete	30–50		7[a]	0.2	2.4–2.5

[a]*Modulus of rupture value.*

ones. The effect of a film-forming inhibitor is to establish a molecular layer just on the steel surface and a second layer of aliphatic tail in hydrocarbon. Therefore water cannot reach the steel surface and promote corrosion. The efficiency of an inhibitor in reducing corrosion depends on concentration, rate of dispersion, film persistency, velocity, temperature, pH, flow regime, presence of disturbances able to perturbate the flow, and fluid composition.

15.5 MATERIAL SELECTION PROCEDURE

This section of analysis deals with the procedure and the order of appreciation, evaluation and selection of materials both for their functional suitability and for their ability to sustain this positive function for the required length of time at a reasonable cost.

Considering the profusion of materials and material-oriented literature, this procedure is, of necessity, schematic.

It attempts only to indicate the system of parallel evaluation of corrosivity of different petroleum environments, the effect of process parameters, estimation of CRs, determination of corrosion allowances (CAs) for a given life, and a few guidelines in selection of materials in conjunction with corrosion control measures and economic principals. However, the knowledge of materials is so vast and far reaching that close cooperation of the designer with metallurgists, material engineers, corrosion engineers, and other materials specialists is underscored and emphasized.

15.5.1 Guidelines on Material Selection

Materials should be selected with due consideration to their functional suitability and ability to maintain their function safely for an economical period of time at a reasonable cost. The particular material selected should be accurately specified.

The whole material complex should be considered as an integrated entity, rather than each material separately. The more highly resistant materials should be chosen for the critical components and where relatively high fabrication costs are involved. It may be necessary to compromise and sacrifice some mechanically advantageous properties to satisfy corrosion requirements and vice versa.

Where a CR is either very low or very high, the choice of materials is simple; where it is moderately low, a thorough analysis of all aspects is required.

In dry environments and carefully controlled fluids, many materials can be used—and these often may be left unprotected. Under atmospheric conditions, even polluted atmospheres, including such metals as stainless steels and aluminum alloys, may be left unprotected. Also, copper and lead have a long life. In a more severe wet environment, e.g., in marine conditions, it is generally more economic to use relatively cheap structural materials (mild steel) and apply additional protection, rather than use the more expensive ones. For the most severe corrosive conditions, it is preferable in most cases to use materials resistant to corrosion than to use cheaper materials with an expensive protection.

Materials more expensive than absolutely necessary should not be chosen unless it is economical in the long run and necessary for safety of personnel or product or for other important reasons. Using fully corrosion-resistant materials is not always the correct choice, a balance between first cost and cost of subsequent maintenance should be found over the full estimated life of the designed utility.

Certain combinations of metal and corrosive are a natural choice:

1. aluminum-nonstaining atmospheric exposure;
2. chromium-containing alloys-oxidizing solutions;
3. copper and alloys-reducing and nonoxidizing environments;
4. hastelloys (chlorimets)-hot hydrochloric acid;
5. lead-dilute sulfuric acid;
6. monel-hydrofluoric acid;
7. nickel and alloys-caustic, reducing, and nonoxidizing environments;
8. stainless steel-nitric acid;
9. steel-concentrated sulfuric acid;
10. tin-distilled water;
11. titanium-hot strong oxidizing solutions;
12. tantalum-ultimate resistance.

Composition of alloy alone does not ensure quality of the product. Evaluation of resistance to corrosion in a given environment, adverse effect of corrosion products on utility or contents, susceptibility to a specific type of corrosion and fouling, and tendency to corrosion failure due to fabrication and assembly processes such as welding, forming, machining, heat treatment, etc., are of prime importance for selection of material.

Due consideration should be given to special treatments required to improve resistance to corrosion, e.g., special welding techniques, stress relieving, blast peening, metallizing, sealing of welds, etc., as well as to any fabrication or assembly methods that would aggravate any tendency of the material to corrosion failure.

Alloys in a highly alloyed condition as necessary should be used when the cost of fabrication is higher than the cost of basic material. Proportional cost of material in some multishaped or complicated components is much less than in the simple ones.

An alloy or temper should be selected which is free of susceptibility to the localized corrosion under the respective general and local environmental conditions in the utility and that meets the strength and fabrication properties required for the job. It is sometimes better to use a somewhat weaker but less sensitive alloy than to use the one that does not lend itself to reliable heat treatment and, due to this, whose resistance to a particular form of corrosion is poor.

If heat treatment after fabrication is not feasible, materials and method of fabrication chosen should give optimum corrosion resistance in the as-fabricated condition. Materials prone to stress corrosion cracking should be avoided in environments conducive to failure, observing that stress relieving alone is not always a reliable cure.

When corrosion or erosion is expected, an increase in wall thickness of the structure or piping should be provided over that which is required by other functional design requirements. This allowance in the judgment of the designer should be consistent with the expected life of the structure or piping. The allowance should secure that various types of corrosion or erosion (including pitting) do not reduce the thickness of structure or piping below the thickness that is required for mechanical stability of the product. Where no thickening can be allowed or where lightening of product is contemplated, a proportionally more corrosion-resistant alloy or better protection measure should be used.

Short-life materials should not be mixed with long-life materials in non-repairable subassemblies. Materials forming thick scales should not be used where heat transfer is important.

Where materials could be exposed to atomic radiation it is necessary to consider whether the effect will be harmful or beneficial, observing that some controlled radiation may enhance the property of a metal.

Not only the structural materials themselves but also their basic treatment should be evaluated for suitability (e.g., chromate passivation, cadmium plating, etc.) at the same time.

Nonmetallic materials complying with the following requirements are preferred: low-moisture absorption, resistance to fungi and microbes, stability through temperature range, compatibility with other materials, resistance to flame and arc, freedom from out-gassing, and an ability to withstand weathering.

Flammable materials should not be used in critical places. Their heat could affect the corrosion stability of structural materials. Toxic materials producing dangerous volumes of toxic or corrosive gases when under fire or high-temperature conditions should not be used.

Fragile or brittle materials that are not, by design, protected against fracture should not be used in corrosion-prone spaces.

Materials that produce corrosion products and can have an adverse effect on the quality of contents should not be used, especially when the cost of wasted contents exceeds the cost of containment.

All efforts should be made to obtain from the equipment suppliers an accurate detailed description of materials used within their products.

The following should be noted with regard to electrical equipment. The use of hygroscopic materials and of desiccants should be avoided. The latter, when their use is necessary, should not be in contact with an unprotected metallic part.

- Fasteners should be of a well-selected corrosion-resistant material, or materials better protected than the parts that they join together.
- Materials selected should be suitable for the purpose and be either inherently resistant to deterioration or adequately protected against deterioration by compatible coatings, especially in problem areas where corrosion can cause low conductivity, noise, short circuits or broken leads, thus leading to a degradation of performance.
- Insulation materials used should not be susceptible to moisture. Stainless steels or precipitation hardening stainless steel should be passivated.

Appendices A and B briefly discuss the corrosion behavior and materials in oil and gas industries and refinery & petrochemical plants.

15.5.2 Procedure

The first step in material selection is a thorough review of the corrosive environment, process parameters, and equipment operating conditions, including temperature, pressure, flow rates, liquid versus gaseous phase, aqueous versus anhydrous phase, continuous versus intermittent operation, media used for cooling or heating, etc.

Second step: the CR should be predicted, and the CAs should be determined for the service life of the equipment and its various components in carbon steel material (see Section 15.6).

Third step: when the determined total wall thickness (Mechanical + CA) is not acceptable by other means, great effort should be exercised to select and evaluate a suitable corrosion preventive measure to lower the required CA. Some of these measures cannot withstand process conditions, i.e., temperatures too high for polymer lining, no facilities for chemical injection, no continuous electrolyte for cathodic protection, and so on.

Fourth step: if the preventive measures cannot reduce the CRs of carbon steel to an acceptable level, then a more corrosion-resistant material should be considered. Many materials should be immediately excluded because of service conditions.

Fifth step: if required, a conduction of standard test methods for evaluation of nominated materials in simulated environment. These tests give the assurance to the right selection (Section 15.6.3).

Sixth step: the evaluation of the cost of material for service life of plant. Also, a comparison between the selection of an expensive material and a cheaper one along with a protective measure (see Section 15.6.6).

When materials of construction are selected, the preparation of a clear and concise specification to ensure that the material is fabricated and obtained as ordered meets the requirements of the specific material standards is mandatory.

Any problem associated in the fabrication of equipment with selected material or rejection during inspection should be reported for reanalysis and ratification in selection and specification.

It should be realized that specified material may fail owing to an undesirable or unknown properties induced during fabrication or installation, such as metallurgical changes, inclusion, and chemical composition changes. Measures that may be required to prevent or limit such factors (e.g., special heat treatment) are outside the scope of this chapter.

15.5.3 Process Parameters

The major factors controlling corrosion in oil and gas industries are:

- the CO_2 partial pressure,
- the H_2S partial pressure,
- the fluid temperature,
- the water cut,
- the water salinity,
- the flow dynamics,
- the pH of solution.

It must be emphasized that corrosion is likely to occur only in the water phase. Vaporized water in streams at temperatures above the dew point are considered noncorrosive. For more data on the main corrosion processes in oil, gas, and petrochemical industries, refer to Appendices A and B.

15.6 CR AND CAs

CR means a uniform decrease in thickness of a material per year.

CA, expressed in terms of thickness, is a measure of extra thickness with which a material can survive its design life.

Therefore to determine a suitable material with enough CA for design life of a plant, the CR should be predicted. This is possible by the following:

- calculation,
- corrosion abstracts and data survey handbooks,
- experiences and in-house data,
- materials vendors data,
- equipment fabricators,
- testing of materials.

15.6.1 Calculation

Many different formula, nomographs, curves, and software based on laboratory or field experiences exist to evaluate the general CR. Results obtained following these approaches should be considered "worst case CR" or R_{max}.

Note:

Due to the complexity of the corrosion phenomenon and the diversity of the operating conditions in oil, gas, and petrochemical plants, the selected calculation model should be presented to the company for review and approval.

This rate may be adjusted by considering the influence of the rest of the environment. The final CR may be thus expressed as following:

$$R_{(corr)} = R_{max.} \times F_{(s)} \times F_{(i)} \times F_{(c)} \times F_{(o)} \times F_{(w)} \times F_{(pH)}$$

where:

$F_{(s)}$ = Scale factor
$F_{(i)}$ = Inhibition factor
$F_{(c)}$ = Condensation factor
$F_{(w)}$ = The percentage of water
$F_{(o)}$ = The percentage of oil
$F_{(pH)}$ = (pH) factor

15.6.2 Corrosion Study by Literature Survey

In corrosion studies undertaken for the purpose of finding suitable material to withstand a particular service, it is best to first take advantage of the vast amount of published literature in the field of corrosion.

Such a study will in general give a very good clue as to the general types of metals or alloys that should prove most satisfactory for the particular job.

NACE's corrosion data survey handbook has two sections:

- Metals,
- nonmetals.

There are more than 50,000 points of data (Nelson method) on the performance of metallic and nonmetallic materials of construction in corrosive environments. Also, NACE's corrosion survey and KO DAB software is highly recommended.

Note:

The method Nelson has used for many years to present corrosion information. This example shows the corrosion of lead by sulfuric acid as a function of temperature and concentration (Fig. 15.1).

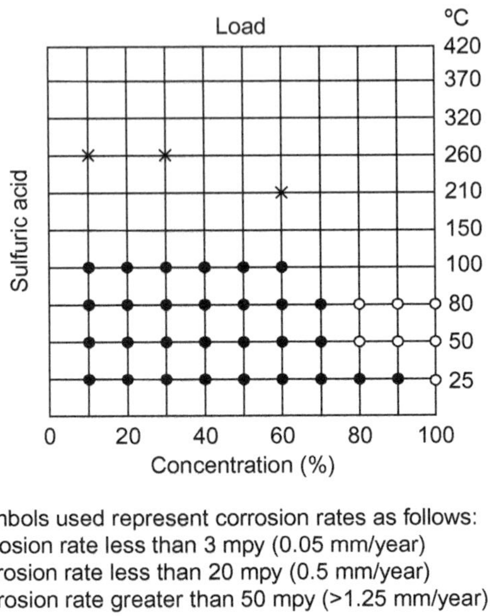

FIGURE 15.1 Nelson's method for summarizing corrosion data.

15.6.3 Corrosion Tests

Corrosion tests are the best appropriate technique for acquisition of data for material selection.

In most corrosion data survey handbooks, CRs are evaluated just for a few factors (e.g., temperature and concentration). However, there are many other factors besides concentration and temperature that influence CRs. While they are often extremely important, it is impossible to list them all in a survey of this type.

15.6.3.1 Test Methods

Corrosion tests are primarily aimed at the acquisition of data in relatively short times, compared to service lifetimes, to predict service behavior. Corrosion test methods may be divided into three categories:

- Laboratory tests,
- pilot plant tests,
- full-size equipment tests.

The complexities of service situations consider many factors: as velocity, temperature, pressure, aeration, heat flux, the presence of oxidizing agents, partial pressure of corrosive gases, inhibitor concentration, etc., all of which can either increase or decrease the CR and may cause pilot plant exposure tests to be preferred to laboratory tests.

15.6.3.2 The Need for Testing

The necessity for corrosion testing depends on:

- degree of uncertainty after available information has been considered,
- the consequences of making a less-than-optimum selection,
- the time available for evaluation.

Therefore, the company's material/corrosion engineer should determine when and which type of tests are required.

15.6.3.3 Pilot Plant Tests

These tests give more information for a primary selection of materials than most laboratory tests. Test conditions are more like the final application; therefore, the results are more reliable.

15.6.3.4 Full-Size Tests

Reliability is further enhanced when it is possible to test full-size components fabricated from candidate material.

15.6.3.5 Laboratory Tests

In some cases laboratory testing is the only means for final material selection. The primary laboratory tests on the selected materials should be as simple as possible.

Depend on the nature of the environment in which the material is to be used, at least the more important corrosion controlling factors should be simulated in tests.

Briefly, test for metallic material should at least cover the following:

- Actual fluids should be used or mixtures simulating the actual.
- Test coupons should be provided from the selected material.
- Generally, experimental time is approximately 1 week.
- Microscopic examination is essential to look for local attack.

  and for nonmetals, test should cover
- weight, volume, hardness, strength, and appearance changes before and after exposure,
- generally, the test period is 1–3 months.

In any laboratory test, great care should be taken in the interpretation of the data. At best the results can only be qualitative and a great deal of common sense and experience has to be applied to such results before they become useful to the engineer.

Although it is not the intention here to catalog the various standard test methods in details, those listed below may be helpful where and when required.

15.6.3.5.1 Standard Corrosion Test Methods

Such test methods can be used to compare the many variables in material composition, manufacturing, and field performance.

The tests also can be used to compare different type of materials, or the same material grade from various manufacturers using a different manufacturing process, and to evaluate different corrosion control measures. However, the objective set of variables for the test dictates the selection of test methods. Therefore a careful study of both processes and material properties (physical, chemical, and mechanical) and the type of protective measure (e.g., inhibition) is usually required to make the proper selection of test method.

15.6.3.5.2 Reporting Test Results

Test results should be tabulated to indicate at least the followings:

- chemical composition of material,
- heat treatment,
- pretest metallurgy,
- posttest metallurgy,
- estimated CR,
- type of localized attack (if any),
- pH of test solution,
- other information pertinent to the evaluation of material such as pressure, temperature, additives to the test solution, etc.,
- method of CR calculation.

15.6.4 Corrosion Allowance

The minimum CA to be considered for an equipment depends on the required service life of the equipment multiplied by the expected CR under process conditions.

$$CA_{(\text{in mm})} \geq \text{Life}_{(\text{year})} \times CR_{(\text{mm/year})}$$

where:

CA = corrosion allowance
CR = corrosion rate

TABLE 15.4 Corrosion Allowances Versus Corrosion Rate for 20 Years Service Life

Class	Average Corrosion Rate (mm/year)	Corrosion Allowance (mm) For a Design Life of 20 Years
A–mild corrosion	<0.05	1.0
B–medium	0.05–0.15	3
C–severe	0.15–0.30	6.0

According to the above equation, the following classes should be considered for an equipment with a design life of 20 years (Table 15.4).

Where the CR is more than 0.3 mm/year or the total corrosion over the design life exceed 6 mm then other alternatives should be evaluated. These alternatives may include the following:

1. Replacement at intervals (e.g., every 10 years where the CR is 0.6 mm/year),
2. corrosion resistance linings,
3. alternative solid corrosion resistance materials.

15.6.5 Selection of Corrosion-Resistant Alloys

Alloy selection, from a corrosion standpoint, can be considered to be a three-step process. First, resistance to general corrosion must be ensured. This is primarily a function of the chromium content of the alloy. Second, resistance to a localized attack also must be ensured. This is primarily a function of molybdenum content. Finally, resistance to environmental stress cracking is sought at the highest feasible strength level. Nickel content plays a principal role in this instance, particularly in providing resistance to anodic cracking.

The close correlation between pitting resistance and resistance to anodic cracking should be noted. This apparently results from the ease of crack initiation under the low-pH, high-chloride conditions found in pits. Therefore, higher molybdenum can also increase resistance to anodic cracking.

With the procedures given below, regions of alloy applicability can be shown schematically as a qualitative function of environmental severity. This has been attempted in which an aqueous, CO_2-containing environment (hence low pH) has been assumed and the effects of temperature, chloride, and H_2S concentration are illustrated. The effect of yield strength is not shown, but if environmental cracking is the limiting factor, reducing the yield strength should extend applicability to more severe environments.

The reader should be cautioned that a diagram is really more of a guide to alloy qualification than a direct selection for a particular application. Therefore, it may aid in developing a more efficient approach to alloy testing.

Note:

Where the corrosion problem is not general (uniform), and is localized to include stress corrosion cracking, pitting, crevice, sulfide stress cracking (SSC), etc., the material selection should be based on the specific corrosion problem. In these cases the selection procedure is to follow all parts of the aforementioned section except for the CR calculation and CA.

15.6.6 Economics in Material Selection

Corrosion is basically an economic problem. Thus the corrosion behavior of materials is an important consideration in the economic evaluation of any project.

The two extremes for selecting materials on an economic basis without consideration of other factors are:

Minimum cost: Selection of the least expensive material, followed by scheduled periodic replacement or correction of problems as they arise.
Minimum corrosion: Selection of the most corrosion resistance material regardless of installed cost or life of equipment.

15.6.6.1 Cost-Effective Selection

This generally falls somewhere between these extremes and includes consideration of many other factors.

In most instances, there will be different alternative materials that may be considered for a specific application. Calculation of true long-term costs requires estimation of the following:

Total cost of fabricated equipment and piping (see Note).

Note:

It should be realized that the costs of processed products, such as sheet, plate, sections, and forgings will be much higher than ingot. Every process and every heat treatment will give added value and increase the final material cost. Also the process of alloying will mean that, generally, the cost of alloys will be higher than those for unalloyed metals. (see Tables 15.5 and 15.6)

15.6.6.2 Economic Evaluation Techniques

Several different techniques exist for economic appraisal of different materials and alternative corrosion control measures. Among these are the concept of:

- internal rate of return,
- discounted pay back,

TABLE 15.5 Estimated Costs of Some Materials (See Note 1), by Mass and by Volume (See Note 2)

Material	Cost ($/kg)	Material	Cost ($/100 cm^3)
Germanium	365.75	Germanium	213.50
Silver	163.98	Silver	172.20
Cobalt	31.54	Cobalt	27.42
PTFE	12.25	Nickel	7.95
Nickel	8.93	Chromium	5.90
Chromium	8.31	Tin	4.43
Tin	6.06	Brass (sheet)	4.17
Titanium	5.41	Beryllium-copper	3.45
Brass (sheet)	5.02	Cadmium	2.94
Al/Cu alloy sheet	4.38	Phosphor bronze (ingot)	2.85
Beryllium-copper	3.90	18/8 stainless (sheet)	2.71
Nylon 66 (PA 66)	3.85	PTFE	2.63
18/8 stainless (sheet)	3.50	Copper (tubing)	2.45
Cadmium	3.40	Titanium	2.43
Phosphor bronze (ingot)	3.24	Copper (grade A ingot)	2.12
Magnesium (ingot)	2.89	Manganese	1.87
Acrylic (PMMA)	2.80	Brass (ingot)	1.84
Copper (tubing)	2.75	Al/Cu alloy sheet	1.30
ABS	2.63	Zinc (ingot)	0.81
Manganese	2.52	Lead (ingot)	0.67
Copper (grade A ingot)	2.38	Magnesium (ingot)	0.51
Brass (ingot)	2.18	Mild steel (sheet)	0.47
Amino resin thermoset	1.49	Nylon 66 (PA 66)	0.44
Aluminum (ingot)	1.37	Aluminum (ingot)	0.37
P-F thermoset	1.31	Acrylic (PMMA)	0.33
Silicon	1.24	Silicon	0.30
Polystyrene	1.12	ABS	0.28
Zinc (ingot)	1.12	Mild steel (ingot)	0.25
Polyethylene(HDPE)	1.09	Amino resin thermoset	0.23

(*Continued*)

TABLE 15.5 (Continued)

Material	Cost ($/kg)	Material	Cost ($/100 cm^3)
Polypropylene (PP)	1.00	Cast iron	0.19
Natural rubber	0.98	P-F thermoset	0.16
Polyethylene (LDPE)	0.74	Polystyrene	0.12
Rigid PVC	0.72	Natural rubber	0.12
Mild steel (sheet)	0.60	Polyethylene (HDPE)	0.11
Lead (ingot)	0.60	Rigid PVC	0.11
Mild steel (ingot)	0.32	Polypropylene (PP)	0.09
Cast iron	0.26	Polyethylene (LDPE)	0.07
Portland cement	0.09	Portland cement	0.03
Common brick	0.07	Common brick	0.01
Concrete (ready mixed)	0.04	Concrete (ready mixed)	0.01

Notes:
1. The costs are based on bulk quantities quoted in July 1991.
2. It is usual to see the cost of materials quoted per unit mass. This may give a misleading picture, as it is often the volume of material that is more important than its mass.

TABLE 15.6 Cost Buildup (Steel Products)[a]

Material	Cost ($ per Tonne)
Iron from blast furnace	210
Mild steel (ingot)	315
Mild steel (black bar)	490
Mild steel (cold drawn bright bar)	665
Mild steel (hot-rolled sections)	498.75
Mild steel (hot-rolled strip coil)	476
Mild steel (cold-rolled strip coil)	593.25
Mild steel (galvanized sheet)	689.50
Austenitic stainless steel (cold-rolled sheet)	3500

[a]*These cost figures applied in July 1991.*

- present worth, also referred to as net present value,
- present worth of future revenue requirements,
- benefit–cost ratios.

Some of these techniques lack adequate sophistication; others are unduly complex and do not lend themselves readily to comprehension and use especially as methods of calculation.

Generally, the applied method should embodied accepted economic terminology at the accounting and managerial levels, so that a material/corrosion engineer's judgment can be properly expressed and understood by project management.

Therefore, for economic evaluation reference is made to the NACE Standard method "RP-02-72, Direct Calculation of Economic Appraisal of Corrosion Control Measures."

15.7 MATERIALS APPRECIATION AND OPTIMIZATION

To indicate, approximately, the general trend of parallel appreciation of materials, selective check-off lists are given in this section. These can, of course, vary for different materials or designs and a selective adjustment will be required.

It is obvious from the contents of these selection lists that a thorough expert knowledge is required, both in engineering and in corrosion control, to complete and evaluate the required data. Only very seldom, and then mostly in simple or repetitive projects, can this task be left to an individual; normally, close cooperation of designer and corrosion or material engineer is needed, and each will have to bring into play his overall and specialized expertise.

The data obtained in such a selection list, after appropriate evaluation and comparative appreciation, should serve as a base for a decision as to whether the appreciated conglomerate of material and its fabrication methods are suitable for the considered purpose. Although in some cases a clear-cut confirmation of suitability may be secured, in many more cases several materials and methods may be evaluated before the optimal one is found. Even then such materials will not always satisfy all required properties and under such circumstances the most satisfactory compromise should be accepted.

15.7.1 Material Appreciation

15.7.1.1 Metals: Selective Check-off List

15.7.1.1.1 Physical Character of Material

1 General

Ballistic properties
Chemical composition (%)
Contamination of contents by corrosion products

Corrosion characteristics in:
- atmosphere
- waters
- soil
- chemicals
- gases
- molten
- metals
- other environments

Creep characteristics versus temperature
Crystal structure
Damping coefficient
Density (g/cm^3)
Effect of cold working
Effect of high temperature on corrosion resistance
Effect on strength after exposure to hydrogen
Effect on strength after exposure to higher temperature
Electrical conductivity (mho/cm)
Electrical resistivity (ohm/cm)
Fire resistance
Hardenability
Hydraulic permeability
Magnetic properties (Curie point)
Maximum temperature not affecting strength (°C)
Melting point (°C)
Position in electromotive series
Rapidity of corrosion—corrosion factor
Susceptibility to various types of corrosion:
- general
- hydrogen damage
- pitting
- galvanic
- stress corrosion cracking
- corrosion fatigue
- fretting
- concentration cell/crevice
- corrosion/erosion
- cavitation damage
- intergranular
- selective attack
- high temperature
- others

Thermal coefficient of expansion ($°C^{-1}$)
Thermal conductivity (W/(m °C))

Toxicity
Wearing quality:
- inherent
- given by heat treatment
- given by plating

2 Strength or mechanical

Above strength properties:
- at elevated temperatures and for long holding times at temperature
- at room temperature after exposure to elevated temperatures

Bearing ultimate* (N/mm^2 or kN/m^2 or kg/mm^2)
Complete stress–strain curve for tension and compression
Compression modulus of elasticity (kg/mm^2)
Compression yield* (N/mm^2 or kg/mm^2)
Fatigue properties:
- S–N curve
- endurance limit

Hardness (Vickers)
Impact properties (Charpy kg/cm^2 at 20°C):
- notch sensitivity
- effect of low temperature
- maximum transition temperature (°C)

Poisson's ratio
Response to stress-relieving methods
Shear modulus of elasticity (kg/mm^2)
Shear ultimate* (Pa)
Tangent modulus curves in compression—with and across grain
Tension modulus of elasticity (Pa)
Tension—notch sensitivity
Tension yield*

*Typical and design values, variability—with and across grain.

15.7.1.1.2 Design Limitations

Restrictions

Size and thickness
Velocity
Temperature
Contents
Bimetallic attachment
Geometric form
Static and cyclic loading
Surface configuration and texture
Protection methods and techniques
Maintainability

15.7.1.1.3 Fabrication Character of Materials

1. **Brazing and soldering**
    - Compatibility
    - Corrosion effect
    - Flux and rod
2. **Formability at elevated and room temperatures**
    - Aging characteristics
    - Annealing procedure
    - Corrosion effect of forming
    - Heat-treating characteristics
    - Quenching procedure
    - Sensitivity to variation
    - Tempering procedure
    - Effect of heat on prefabrication treatment
3. **Formability in annealed and tempered states**
    - Apparent stress × local strain curve
    - Characteristics in:
        - bending
        - dimpling
        - drawing
        - joggling
        - shrinking
        - stretching
    - Corrosion effect of forming
    - Elongation × gage length
    - Standard hydropress specimen test
    - True stress–strain curve
    - Uniformity of characteristics
4. **Machinability**
    - Best cutting speed
    - Corrosion effect of:
        - drilling
        - milling
        - routing
        - sawing
        - shearing
        - turning
    - Fire hazard
    - Lubricant or coolant
    - Material and shape of cutting tool
    - Qualitative suitability for:
        - drilling
        - milling

- routing
- sawing
- shearing
- turning

5. **Protective coating**
   - Anodizing
   - Cladding
   - Ecology
   - Galvanizing
   - Hard surfacing
   - Metallizing
   - Necessity of application for:
     - storage
     - processing
     - service
   - Paint adhesion and compatibility
   - Plating
   - Prefabrication treatment
   - Sensitivity to contamination
   - Suitability
   - Type of surface preparation optimal
6. **Quality of finish**
   - Appearance
   - Cleanliness
   - Grade
   - Honing
   - Polishing
   - Surface effect
7. **Weldability**
   - Arc welding
   - Atomic hydrogen welding
   - Corrosion effect of welding
   - Cracking tendency
   - Effect of prefabrication treatment
   - Electric flash welding
   - Flux
   - Friction welding
   - Heat zone effect
   - Heli-arc welding
   - Pressure welding
   - Spot welding
   - Torch welding
   - Welding rod

8. **Torch cutting**
   Cutting speed

15.7.1.1.4 Economic Factors

1. **Availability**
   In required quantities:
      single
      multiple
      limited
      unlimited
   In different forms:
      bar
      casting (sand, centrifugal, die, permanent mold)
      extrusion
      forging
      impact extrusion
      pressing
      sintered
      powder pressing
   In metallized and pretreated forms:
      galvanized
      plastic coated
      plated
      prefabrication treated
   In cladded forms
   Uniformity of material
   Freedom from defects
   Time of delivery:
      present
      future
2. **Cost in different forms**
   Bar, shape, or plate
   Casting (sand, centrifugal, die, permanent mold)
   Extrusion
   Forging
   Impact extrusion
   Pressing
   Sintered
   Powder pressing
3. **Size limitation in different forms**
   Gage
   Length
   Weight
   Width

15.7.1.2 Selective Check-Off List, to Be Adjusted for Generic Groups

15.7.1.2.1 Physical Character of Material

1. General

Anisotropy characteristics-main and cross-direction
Area factor
Burning rate
Bursting strength (1 mm thick, Mullen points)
Change in linear dimensions at 100°C for 30 min (%)
Clarity
Color
Contamination of contents (decomposition and extract)
Creep characteristics versus temperature-creep-apparent-modulus (10° kg/mm^2)
Crystal structure
  Crystalline melting point
  Damping coefficient
Decay characteristics in:
  atmosphere
  alcohol
  chemicals
  gases
  high relative humidity
  hydraulic oils
  hydrocarbons
  lubricants and fats
  solvents
  sunlight
  water
  other environments
Deflection temperature (°C)
Density (g/cm^3)
Dielectric constant (60, 10^3, 10^4 c/s)
Dielectric strength—V to laminations (v/mm):
  short time
  step by step
Dissipation factor (1 MΩ)
Effect of high temperature on decay
Effect of low temperature on decay
Effect on strength after exposure to temperature
Electrical conductivity (mho/cm)
Electrical properties at room temperature per thickness
Electrical loss factor (1 MΩ)
Electrical power factor (60. 103, 104 c/s)

Electrical resistivity:
- arc/s
- insulation (96 h at 90% RH and 35°C) MΩ
- volume (Ω/cm—50% RH and 25°C)

Evolvement of combustion products
Evolvement of corrosives
Fillers
Fire resistance
Flammability (ignited by flame)
Folding endurance
Fuming on application
Gas permeability (cm^3/M^2/mm thick/24 h/atm at 25°C):
- CO_2, H_2, N_2, O_2

Generic group and composition
Gloss
Haze (%)
Heat distortion temperature
Heat expansion (coefficient of linear thermal expansion)
Heat degradation
Heat insulation (thermal conductivity)
Heat-sealing temperature range
Hydraulic permeability
Intensity of color
Light transmission, total white (%)
Maximum service temperature (°C)
Maximum temperature not affecting strength (°C)
Melt index (dg/min)
Migration of plasticizers
Minimum temperature not affecting strength (°C)
Opacity
Rate of water vapor transmission at 37.8°C
Reflectance
Refractive index
Reinforcement
Softening temperature (°C)
Specific gravity (25°C)
Specific heat (kcal/(°C g))
Stability of color
Stiffness—Young's modulus
Susceptibility to various types of deterioration:
- general
- cavitation/erosion
- erosion
- fatigue

fouling
galvanic (metal-filled plastics)
impingement
stress cracking and crazing
others
Taste
Thermal conductivity (W/(m °C))
Toxicity
Transmittance (%)
Unit weight (m^3/kg)
Water absorption (24 h/l cm thick/%)
Wearing quality:
inherent
given by treatment

2. **Strength or mechanical**
Abrasion resistance
Average yield (cm^2/kg/mm)
Bearing ultimate* (N/mm^2)
Bonding strength (kg/thickness)
Brittleness
Bursting pressure (kg/mm^2)
Complete stress–strain curve for tension and compression
Compressive strength:
flatwise (N/mm^2)
axial (N/mm^2)
at 10% deflection (N/mm^2)
Deformation under load
Elongation (%)
Elongation at break (%) (24°C)
Fatigue properties
S–N curve
endurance limit
Flexibility and flex life
Flexural strength* (N/mm^2)
Hardness (Rockwell)
Impact strength, Izod
Inherent rigidity
Modulus of elasticity (kg/mm^2)
in compression (kg/mm^2)
in flexure (kg/mm^2)
in tension (kg/mm^2)
in shear (kg/mm^2)
Resistance to fatigue

Safe operating temperature (°C)
Shear, ultimate* (Pa)
Tear strength:
- propagating (g/mm)
- initial (kg/cm)

Tensile strength (kg/mm^2)
Vacuum collapse temperature
*Typical and design values, variability—with and across grain.

15.7.1.2.2 Design Limitations

Size and thickness
Velocity
Temperature
Compatibility with adjacent materials:
- at ambient temperature
- at elevated temperature

Geometric form
Static and cyclic loading
Surface configuration and texture
Protection methods and techniques
Maintenance

15.7.1.2.3 Fabrication Character of Materials

1. **Molding and injection**
   - Compression ratio
   - Compression molding pressure (kg/cm^2)
   - Compression molding temp. (°C)
   - Injection molding pressure (kg/cm^2)
   - Injection molding temp. (°C)
   - Molding qualities
     - Mold (linear) shrinkage (cm/cm)
     - Specific volume, etc. (cm^3/g)
2. **Lamination**
   - Laminating pressure (kg/cm^2)
   - Laminating temperature (°C), etc.
3. **Formation at elevated temperatures**
4. **Machinability**
   - Adverse effect of:
     - drilling
     - milling
     - sawing
     - shearing
     - turning

Best cutting speed
Fire hazard
Machining qualities
Material and shape of cutting tool, etc.

5. **Protective coating**
Cladding
Painting
Plating
Sensitivity to contamination
Suitability
Type of surface preparation optimal

6. **Quality of finish**
Appearance
Cleanliness
Grade
Polishing
Surface and effect

7. **Joining**
Adhesive joining
Bonding
Cracking tendency
Heat zone effect
Welding, etc.

15.7.1.2.4 Economic Factors

1. **Availability**
Cladded forms
Forms available
Maximum width (cm)
Thickness range
Treated forms
Uniformity of material
Freedom from defects

15.7.1.3 Natural and Synthetic Elastomers: Selective Check-Off List

15.7.1.3.1 Physical Character of Material

1. **Uncured properties**
Color
Consistency
Shelf life
Viscosity

2. **Cured properties**
   - Chemical composition
   - Compression set
   - Contamination of contents by contact
   - Creep
   - Drift (room temperature)
   - Elasticity
   - Elastic memory
   - Flexing
   - Hardness range
   - Linear shrinkage (%)
   - Permeability to gases
   - Permanent set
   - Resilience
   - Self-damping
   - Specific gravity
   - Swelling resistance (in various environments)
3. **Thermal properties**
   - Coefficient of thermal expansion $\times 10^{-3}$/°C
   - Drift at elevated temperature
   - Elongation % at elevated temperature
   - Flame resistance
   - Heat aging
     - Low-temperature brittle point (°C)
     - Low-temperature stiffening (°C)
   - Maximum service temperature range (°C)
   - Tensile strength-elevated temperature (Pa)
   - Thermal conductivity (kcal/(h m °C m))
4. **Electrical properties**
   - Dielectric constant
   - Dielectric strength
   - Dissipation factor
   - Electrical resistivity (Ω/cm)
   - Volume resistivity

15.7.1.3.2 Mechanical Properties

Crack resistance
Cut growth
Elongation % plain
Elongation % reinforced
Resistance to abrasion
Resistance to impact and shock
Resistance to wear
Tear resistance

Tensile strength–reinforced (mPa)
Tensile strength–unreinforced (mPa)

15.7.1.3.3 Environmental Properties

Chemical resistance in:
- water
- acid
- alkali
- aliphatic hydrocarbons
- aromatic hydrocarbons
- halogenated hydrocarbons
- alcohol
- oil and greases
- synthetic lubricants (e.g., diester)
- hydraulic fluids-silicates, phosphates

Light
Oxidation
Radiation
Weather

15.7.1.3.4 Subjective Properties

Odor
Staining
Taste

15.7.1.3.5 Fabrication Character of Materials

Bonding to rigid materials
Limits of geometric form
Molding limits
Processing characteristics
Sagging

15.7.1.3.6 Corrosion Effect on Substrate

Fumes in combustion
Physical contact
Vapor on setting and set

15.7.1.4 Adhesives: Selective Check-Off List

15.7.1.4.1 Physical Character of Material

1. **General**
   - Chemical composition
   - Color
   - Contamination of contents by adhesive

Characteristics in:
- atmosphere
- waters and humidity
- soil
- chemicals
- lubricating oil
- hydraulic fluid
- degreasing solvent
- gases
- other environments

Creep characteristics and temperature
Density
Effect of forming on substrate
Effect of organic vapors on substrate
Effect of high temperatures
Effect of low temperatures
Effect on strength after exposure to higher temperature
Electrical conductivity
Electrical resistivity
Fire resistance
Flammability

2. Strength or mechanical

Load on glue line:
- cleavage (kg/cm of width)
- peel (kg/cm of width)
- shear (kg/cm^2)
- tension (kg/cm^2)

15.7.1.4.2 Design Limitations

Utility:
- development
- field repair
- production
- prototype

Application:
- bonding
- reinforced joint
- restriction on joined area
- restriction on pressure
- scaling
- type of drying

Surface types:
- base metal
- finish

Type of joint:
- edge
- filled
- lap

Hydraulic permeability
Maximum temperature not affecting strength
Melting point
Odor
Smoke development on fire
Surface effect (corrosion):
- metals
- nonmetals
- coatings

Susceptibility to environments:
- outdoor sheltered
- outdoor direct exposure
- outdoor sunlight only
- indoor room condition
- indoor controlled
- stored sealed
- stored sealed (opened for test)
- stored outdoor sheltered

15.8 CORROSION IN OIL AND GAS PRODUCTS

Corrosion occurrence has been widely experienced in the oil and gas industry. In the following, the main corrosion processes in oil and gas phases are discussed.

First of all it must be emphasized that corrosion is likely to occur only in the water phase, as the oil phase is considered noncorrosive. Consequently, the presence of free water is necessary for corrosion to occur, i.e., vaporized water in streams at temperatures above the dew point are considered noncorrosive.

In addition, it is necessary, especially for mixed phases streams (oil + gas + water) to verify the water wetting of materials; in fact, if water is confined in the middle of the stream, or trapped by oil, no corrosion attack may develop.

The principle factors controlling corrosion are:

- the CO_2 partial pressure
- the H_2S partial pressure
- the fluid temperature
- the water salinity
- the water cut
- the fluodynamics
- the pH

Additional factors influencing CRs in petroleum refineries and petrochemical plants including off-site facilities and pollution-control facilities are:

- organic acids (naphthenic acids)
- hydrogen (atomic)
- amine solution
- sulfur
- sodium hydroxide
- ammonia
- hydrofluoric acid
- glycol
- cyanide
- sulfuric acid
- galvanic couple
- stress (plus chlorides, caustic, ammonia, amines, polythionic acid)
- bacteria
- concentration of corrosives
- aeration
- heat flux
- welding defects
- high-temperature oxidation and corrosion

15.8.1 Effect of CO_2

Dry CO_2 is noncorrosive until about 400°C, while it is corrosive when dissolved in a water phase.

CO_2 corrosion in the presence of free water is known to cause sweet corrosion. CO_2 dissolves in the water phase forming carbonic acid, which decreases the water acidity; the final pH of the solution will depend on temperature and CO_2 partial pressure.

The corrosivity of CO_2-saturated solutions is much higher than other acid solutions at the same pH, because of a direct action of CO_2 in the corrosion phenomena.

15.8.1.1 Effect of Temperature

Same studies in laboratories (Fig. 15.2) show that the CR increases up to 70°C, probably due to the increase of mass transfer and charge transfer rates. Above these temperatures, the CR starts to decrease. This fact is attributed to the formation of a more protective scale, due to a decrease in iron carbonate solubility, and consequently a diffusion process becomes the rate-determining step.

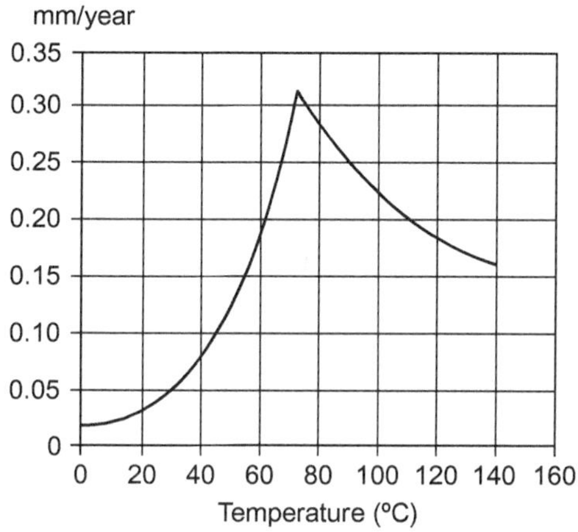

FIGURE 15.2 Corrosion rate evaluation according to De Waard and Milliams.

15.8.1.2 Effect of Pressure

The partial pressure of CO_2 has a detrimental effect on CR. Since the partial pressure of CO_2 is proportional to the total pressure by $P_{CO_2} = P \times \text{Mole\%}\ CO_2$, therefore, the CR will increase by increasing pressure.

15.8.1.3 Prediction of CO_2 Corrosion Rate

The evaluation of a CR of carbon steels in CO_2-saturated waters may be done* according to De Waard and Milliams formula, where the CR is an exponential function of CO_2 partial pressure and temperature. Results obtained following this approach should be considered worst case corrosion.

The formula is easily usable in the form of a curve (Fig. 15.2) and nomogram (Fig. 15.3):

**Note*:

Several corrosion rate prediction models have been developed by oil companies and research institutes, due to the complexity of the phenomenon and the diversity of the operation conditions in oil and gas production and treatment networks. Therefore, the model itself, and all modifications, in order to use it for calculation of CO_2 corrosion in a specific regime, should be presented to the company for review and approval.

$$\text{Log}(R_{\max}) = 5.8 - \frac{1710}{T} + 0.67\,\text{Log}(P_{CO_2})$$

where:

$R_{\max}$ is the corrosion rate (mm/year)
T is the temperature (°K)
$P_{CO_2} = P \times m(CO_2)$

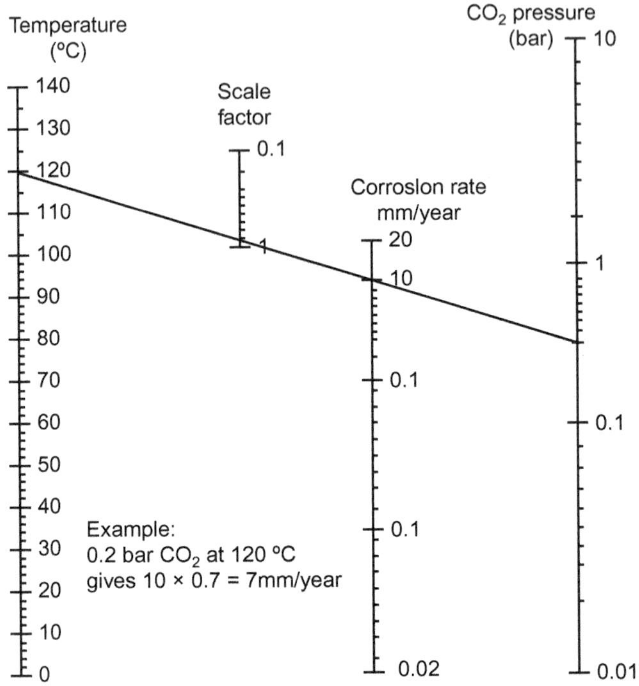

FIGURE 15.3 Nomogram for calculation of corrosion rates as a function of CO_2 partial pressure and temperature. Example shown: at 0.2 bar CO_2 and 120°C, corrosion rate is 10 mm/year. Scale factor = 0.7, thus expected corrosion rate is 7 mm/year.

being P the total pressure (bar) and m (CO_2) the CO_2 molar fraction in the gas phase.

For high pressures, it is recommended to substitute the partial pressure with the fugacity, defined as:

$$f_{CO_2} = a.P_{CO_2}$$

being a the activity coefficient, given by:

$$\text{Log}(a) = \left(0.0031 - \frac{1.4}{T}\right)P$$

The validity limits under which the above mentioned formula was originated are:

- P (CO_2) < 2 bar
- temperature < 70°C
- distilled water
- flowrate = 1 m/s

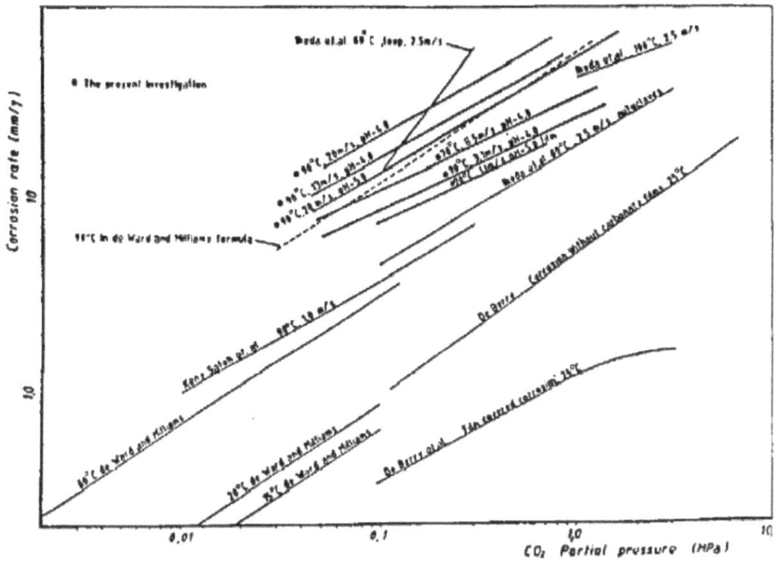

FIGURE 15.4 Corrosion rates as a function of CO_2 partial pressure. *Data from various sources.*

The fitness of this formula has been confirmed also for CO_2 partial pressures much higher than the original experimental work and for high flowrate (Fig. 15.4).

This rate, called R_{max}, may be adjusted by considering the influence of the rest of the environment other than $P(CO_2)$ and T. The final CR may be thus expressed as:

$$R_{corr} = R_{max} \times F(g) \times F(s) \times F(w) \times F(i) \times F(c) \times F(pH)$$

where these reduction factors are applicable.

15.8.1.4 Effect of Glycol—F(g)

In wet CO_2-containing pipelines and flowlines, glycol is often injected to prevent hydrate formation. Glycols have a significant inhibitive effect on corrosion. The reduction of the expected CR due to the presence of triethylene glycol can be conservatively expressed by F(g):

$$\text{Log}(F(g)) = 1.6\ \text{Log}\ W(g) - 3.2$$

For mono and di-ethylene glycol the available data are more limited, but the results are also covered by the above F(g) factor.

15.8.1.5 *Effect of Scaling—F(s)*

At temperatures higher than about 70°C, the steel may be protected by its corrosion products (iron carbonate, $FeCO_3$), and consequently the CR may be depressed by a coefficient *F* scale representing the scaling factor:

$$\text{Log}\,(\text{F}(s)) = \frac{2400}{T} - 0.6\,\text{Log}(f_{CO_2}) - 6.7 \leq 1$$

In the range of temperature between 70°C and 150°C, carbon steels are more prone to localized attack in case of high turbulence, as a consequence of the failure of the $FeCO_3$ film. In this case, the CR may be much higher, as predicted.

At temperatures higher than 150°C and CO_2 partial pressure below 50 bar, the steel is protected by a strong film of $FeCO_3$, which is not removed even by high turbulent streams, and the corrosion rate becomes negligible.

A summary of temperature range and corrosion morphology is drawn in Fig. 15.5.

Corrosion morphology may be either uniform or localized (mesa or pitting), according to the process parameters (temperature, CO_2 partial pressure, water phase composition, flowrate).

15.8.1.6 *Effect of Water Cut—F(w)*

Oil presence is generally considered beneficial, as far as oil exerts a kind of inhibition effect. Oil, in fact, may form on a steel surface a film enough thick and adherent to inhibit water wetting. On the contrary, gas and condensates do not generally exert any beneficial effect, as they have no inhibition property.

Hydrocarbon condensates are assumed not to influence corrosion significantly. Field experience has shown that, as opposite to oil, the hydrophobic behavior of condensates is negligible.

As far as vertical tubing is concerned, an oil film on the steel surfaces is stable up to about 20–40% water cuts. For higher water quantities, the CR can be correctly predicted by De Waard and Milliams, as the steel may be considered water wet.

As far as horizontal pipes is concerned, the amount of water is not an important factor. In fact, as water is generally heavier than oil, gas, and condensed products, in case of stratified flow, it may separate on the lowest surfaces, generally at the 6 o'clock position. In this case, the expected corrosion will occur only on the water-wetted surfaces.

In case of stratified flow, corrosion is also likely to occur in the top of the line, due to the condensation of water droplets from the wet gas. The effect of inhibition is poor in this case, and experience shows that the CR in the top of the line can be assumed as 10% of the predicted rate in fully immersed conditions, with a maximum of about 0.3 mm/year, irrespective of

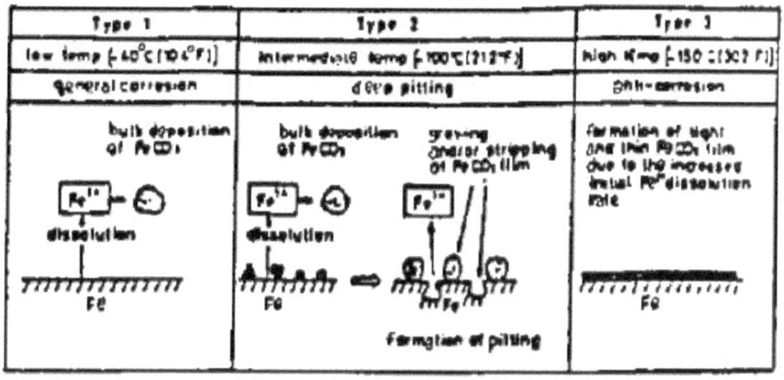

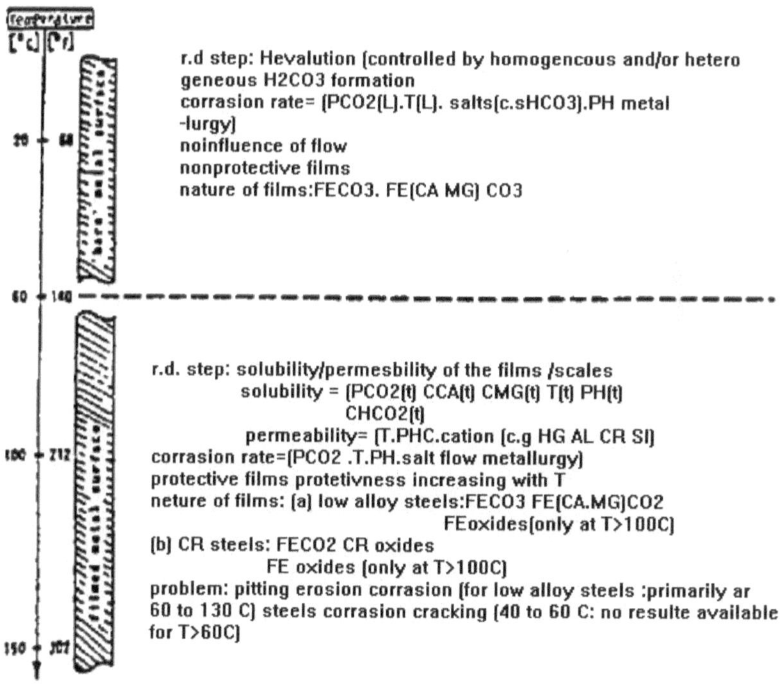

FIGURE 15.5 Schematic summary of nature and morphology of corrosion in the different temperature ranges.

the CO_2 content. To increase the inhibition efficiency in case of stratified flow, a periodical pig launch should be made to allow the inhibitor film to be formed in the top of the line.

Along flowlines and pipelines, minimum areas are expected, as shown in Fig. 15.6, where free water may become stagnant, particularly if the flow is not able to remove it (low velocity). In these cases, corrosion attack will be localized at the 6 o'clock position, as schematically shown in Fig. 15.7B,

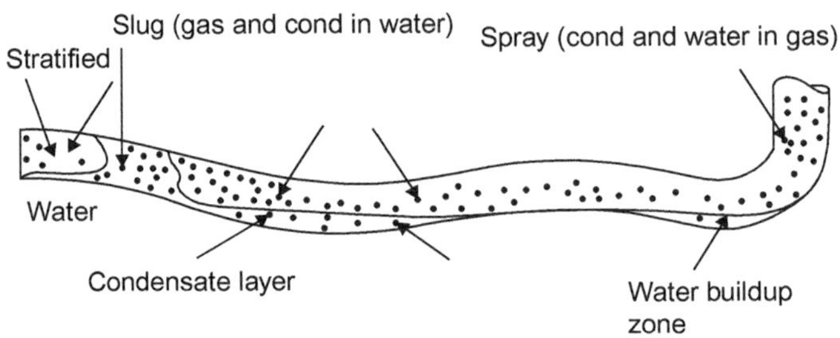

FIGURE 15.6 Schematic summary of flow regime in gas flowlines.

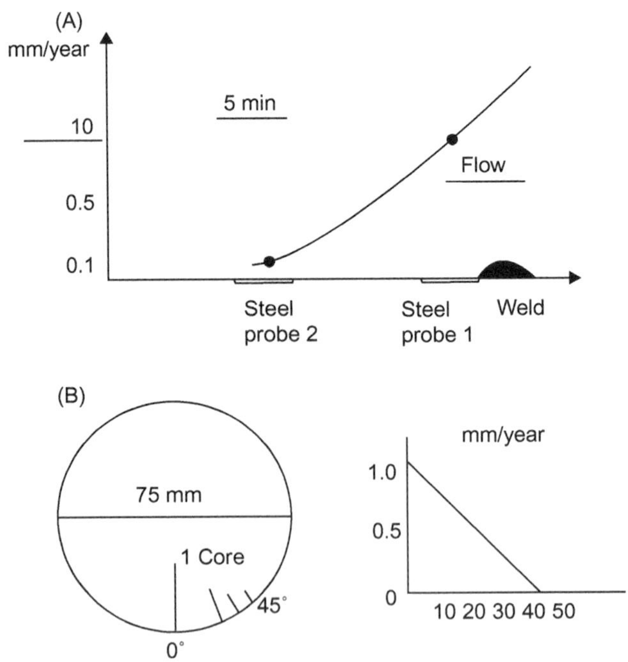

FIGURE 15.7 (A) Corrosion rate at two distances from the simulated weld in flowing oil/water mixture. liquid flow rate about 1 m/s. (B) Corrosion rate in the pipe bottom and at various angles from the pipe bottom in a flowing oil/water mixture at 20°C. Five percent water (4% NaCl) in oil, liquid flow rate about 1 m/s.

and the rate may be correctly predicted anyway with the approach discussed in this paragraph.

Pigging of the line is generally useful to remove the water remaining stagnant on the pipe bottom at the 6 o'clock position. This will also allow a better inhibition of the pipe surfaces; in fact, being inhibitors oil-soluble products, they are transported by the oil phase, and stagnant water may result more difficult to be inhibited. When corrosion inhibitors protection is utmost important, it is a common practice to pig the lines regularly (daily or weekly).

In case of higher gas flowrate, the flow pattern may become annular, as shown for horizontal pipes in Fig. 15.8B. In this case, a continuous liquid film (which varies in thickness around the circumference of the pipe) exists over the full pipe circumference, while gas is flowing in the middle of the pipe. Since the steel surface is completely wetted, corrosion is equally likely to occur at any point around the circumference. As far as the use of the flow pattern diagrams, superficial velocity may be defined as the velocity the (liquid/gas) phase would exhibit if it flowed through the total cross section of the pipe alone.

Protection in this case may be achieved through continuous injection of film-forming corrosion inhibitors, as they can be transported by the water phase and film on the full pipe surface. Attention should be paid in this case to the flow velocity, as highly turbulent flow may produce high shear stresses on the pipe wall and remove the inhibitor film. Another important factor in this case is the avoidance (or reduction) of disturbances, like small radius bends, over penetrated root welds, as shown schematically in Fig. 15.7A. A sudden change of diameter or direction could create turbulence and impingement after the discontinuities and remove the inhibitor film and promote a high rate corrosion.

To conclude, for light hydrocarbon condensate water wetting may occur at any velocity; thus, F(w) is always set equal to 1.

15.8.1.7 Effect of Corrosion Inhibitors—F(i)

It has been common practice for many years to inject corrosion inhibitors into CO_2-containing production tubing and process streams carried by carbon steels. In some cases inhibitors have been injected into nominally dry gas lines as a second defense to back up the drying process in the event of misoperation. In some other cases, inhibitors are applied as the first line of defense against corrosion in carbon steel lines carrying wet gas from satellite to central gathering facilities where bulk drying can be carried out. Temperature drops can be considerable over such intrafield lines, so that condensation of water and hence corrosion will take place around the full internal pipe surface.

Corrosion inhibitors are chemicals that may be divided into a few categories. Among these, the most used class in horizontal flowlines/pipelines is the film-forming amine type.

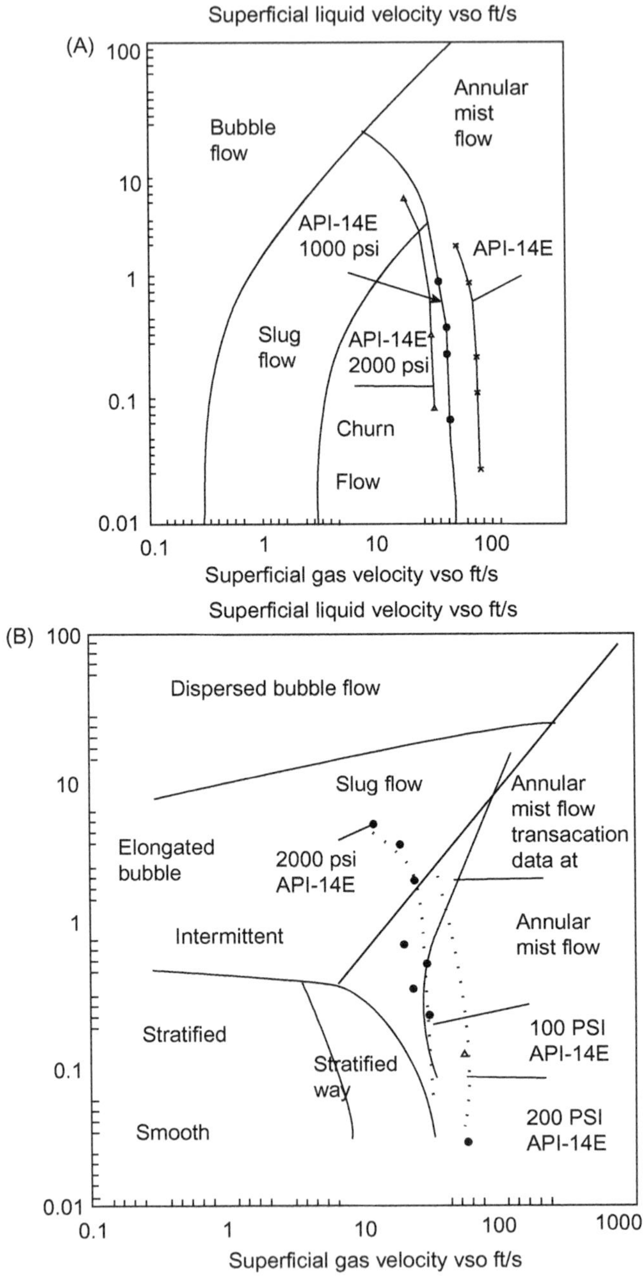

FIGURE 15.8 (A) Flow pattern in vertical multiphase flow.[15] (B) Flow pattern in horizontal multiphase flow.[15]

In this case, the inhibitor is composed of a flat aromatic molecule (amine), which is polar and is attracted by the steel surface and thus able to establish some absorption link; the molecule has also an aliphatic long tail, which is oil soluble. The effect of a film-forming inhibitor is thus to establish a first layer of flat molecules just on the steel surface, a second layer of aliphatic tails, and a third layer of oil/condensates. Water cannot thus reach the steel surface and promote corrosion.

The stability of this film is dependent on the chemistry and fluodynamics of the transported fluids, as they can remove the inhibitors, because of chemical affinity between oil products and the aliphatic tail, from the steel surface, or to promote breaking of this film by impingement of water droplets.

The effect of velocity on corrosion inhibitor performance is to reduce the film life and increase the concentration of inhibitors required to maintain protection.

It is generally accepted to define the capacity of inhibitors to protect against corrosion using a parameter, called efficiency, defined as:

$$\text{Inh.Eff} = \frac{R_{\text{Corr}}(\text{With inhibitors})}{R_{\text{Corr}}(\text{Without inhibitors})}$$

Often used as a percentage. To give an example, if a system would exhibit 2 mm/year CR without inhibitors and 0.2 mm/year with inhibitors injection, the calculated efficiency is 90%.

Under ideal conditions for inhibitors application, an inhibitor efficiency of above 85% can be achieved when comparing the CR seen and that predicted by the De Waard–Milliams nomogram. However, an efficiency of 85% is dependent upon even distribution of inhibitors over the whole circumference of the pipe wall, something unlikely to be achieved in flowlines transporting a mixture of fluid and gas. In addition, areas of extreme turbulence can appear in connection with disturbances, which reduce the level of protection that an inhibitor can provide. Such disturbances have been seen at flanges and at over penetrations at welds and may also occur in areas of growing corrosion damage.

Some recent data have been published by Tulsa University, where all inhibitors were tested under two flow conditions: CRs increased as superficial gas velocity increased; inhibition efficiency, above 95% in a single-phase flow decreased significantly as shown in Fig. 15.9 in the range 40–95%, highlighting the necessity of qualifying the chemicals before injection through properly designed corrosion testing.

The inhibitors film persistency on steel surfaces depends on the inhibitor type (Fig. 15.10) and dosage (Fig. 15.11). Moreover, it was found that the capacity of an inhibitor to produce resistant films is dependent also on the environment pH. Moreover, it was shown that increasing inhibitor concentration is usually required when a high flowrate (i.e., high shear stresses) is expected, as shown in Fig. 15.12.

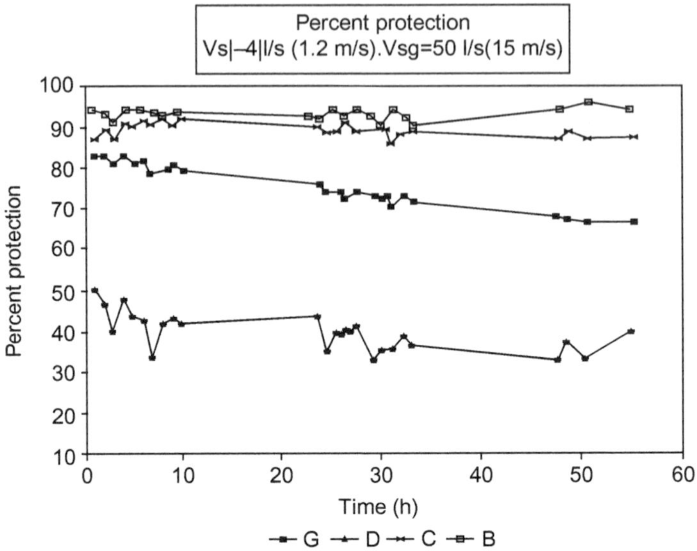

FIGURE 15.9 Percent protection for four commercial inhibitors under two-phase flow conditions.

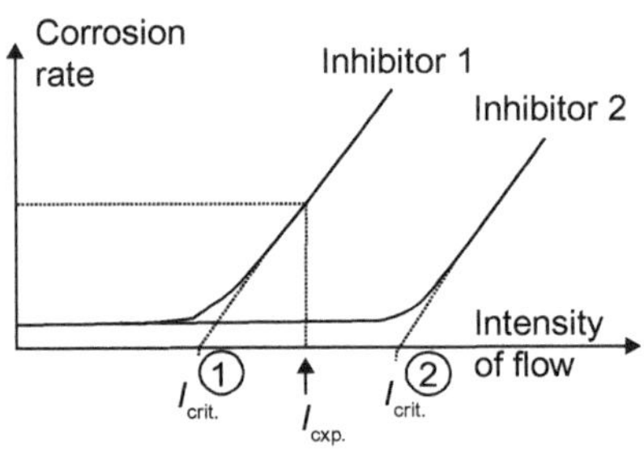

FIGURE 15.10 Effect of the nature of inhibitors on critical flow velocity.

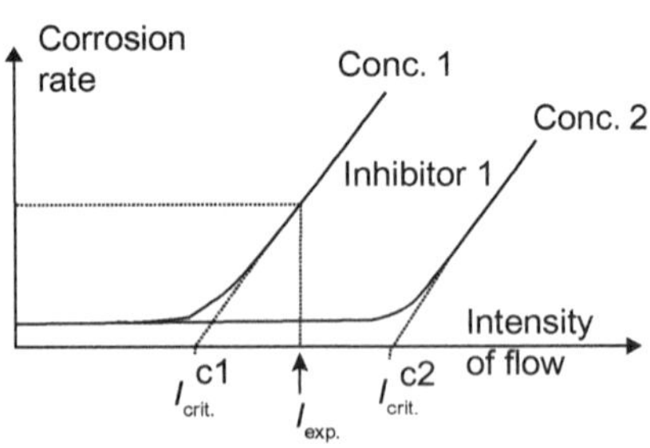

FIGURE 15.11 Effect of inhibitor concentration on critical flow intensity.[8]

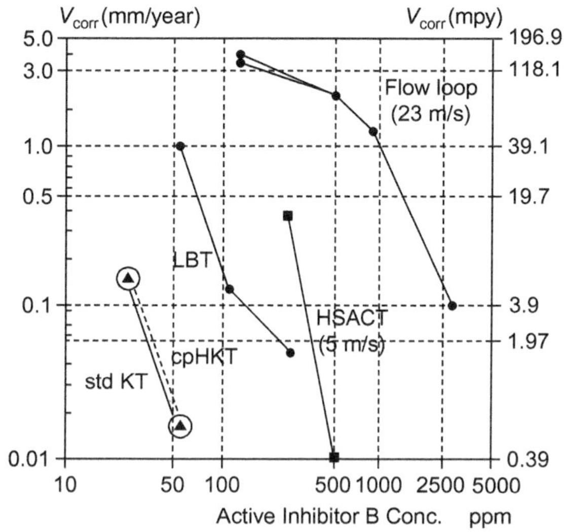

FIGURE 15.12 Performance/concentration curves for one inhibitor type for conditions of increasing severity.

Most inhibitors exhibit a maximum temperature, above which they do not function properly. Generally, it is believed that inhibitors in pipelines can work up to about 90°C.

For a given oil-soluble inhibitor the parameters of primary importance that control CRs in inhibited systems include inhibitor concentration, dispersion of inhibitor in water, film persistency, and velocity. The parameters of secondary importance in predicting corrosion in inhibited wet gas pipeline include partial pressure of CO_2, temperature (if below a critical level), and composition of the aqueous environment (including pH). In other words, if the local concentration of an appropriate inhibitor is sufficiently high, corrosion may be controlled regardless of CO_2 partial pressure, fluid composition, or temperature in the range normally found in pipelines.

From this last approach, instead of calculating the expected CR or inhibitors efficiency, the reliability of corrosion inhibitors is the most important task to define.

To conclude, the inhibitors efficiency is a very difficult task to establish, being dependent on:

- proper inhibitors selection and dosage,
- fluid velocity and flow regime,
- presence of disturbances able to perturbate the flow,
- operating temperature, in order to ensure persistency of film inhibitors.

It is general practice, in the design stage, to assume 0.9* inhibitors efficiency, thus $F(\text{i}) = 0.1$. However, lower figures should be considered where high flow velocities are expected to produce erosion corrosion attacks in the presence of disturbances. In fact, it is possible to have erosive liquid flow at local flow disturbances such as weld beads, pin ends in connections, bends, size reduction, flanges, even where the bulk liquid flow rate is not high.

For all these cases, the capacity of field repairing and repair costs should become the effective design criteria.

*Note:

More precisely this figure should be 0.85 for condensate, 0.9 for gas, and 0.95 for oil stream.

15.8.1.8 Effect of Condensing Phase—F(c)

In case of water condensation from saturated vapor as a consequence of the stream cooling along the route of piping, corrosion is likely to occur under the condensed droplets. These conditions are very likely to produce a protective film, as scale deposition and adherence are favored because of the quiescent conditions. In fact, experiments and experience demonstrated that, despite what the nomogram predicts, the same value in immersed and condensing conditions, the CR drops at max. 0.3 mm/year, irrespective of the partial pressure of CO_2.

Similar reduction in the CR may be applied to reduce the expected CR for the top surface of vessels and separators, etc. It is also applicable to the top of pipelines where the flow regime is stratified, or nominally dry gas lines where cooling below the dew point occurs. No reduction is applicable to immersed conditions.

Such a reduction factor may be evaluated as:

$$F_{\text{cond}} = 0.4(\text{cond. rate, } g/(\text{m}^2\ \text{s})) \leq 1$$

In most cases, the adoption of $F(\text{c}) = 0.1$ is suggested, where applicable.

15.8.1.9 Effect of Water Salinity—F(pH)

Corrosion in production fluids is mainly controlled by the presence of free water, which may come from the reservoir itself (formation water) and/or condense along the route (condensation water).

These two kinds of waters differ very much with regard to composition and the effects on the corrosion phenomena.

In fact, whereas condensate being free of salts can achieve very low pH, the salts dissolved in formation waters may have a high buffering effect, which leads to higher pH at the same CO_2 partial pressures. The solution pH will depend, finally, on the amount and kind of dissolved solids as well as the dissolved gases.

In addition, the presence of ions (like Ca^{2+}, Mg^{2+}) can increase the resistance of the corrosion product film, where the corrosion resistance may be influenced by this phenomena.

To evaluate $F(\mathrm{pH})$, the saturation pH must be first calculated as the lowest of the two following:

$$\mathrm{pH}_{sat} = 1.36 + \frac{1307}{T} - 0.17\,\mathrm{Log}(f_{CO_2})$$

which refers to the formation of Fe_3O_4 and

$$\mathrm{pH}_{sat} = 5.4 - 0.66\,\mathrm{Log}(f_{CO_2})$$

which refers to the formation of $FeCO_3$.

Once known the real environment pH, called pH_{act}, the corrective factor $F(\mathrm{pH})$ may be calculated as:

if: $\mathrm{pH}_{sat} \geq \mathrm{pH}_{act}$

$$\mathrm{Log}(F_{pH}) = 0.32(\mathrm{pH}_{sat} - \mathrm{pH}_{act})$$

if: $\mathrm{pH}_{sat} \leq \mathrm{pH}_{act}$

$$\mathrm{Log}(F_{pH}) = -0.13(\mathrm{pH}_{act} - \mathrm{pH}_{sat})1\hat{.}6$$

The adoption of such a reduction factor is not allowed if also F_{scale} is used, thus F_{pH} is set to 1 if F_{scale} is lower than 1.

If the real environment pH is unknown, $F(\mathrm{pH})$ is set at 1.

15.8.2 Effect of H_2S

H_2S in the presence of water is known to cause sour corrosion. This name includes different mechanisms, known as uniform corrosion, SSC, and hydrogen-induced cracking (HIC).

15.8.2.1 Uniform Corrosion

H_2S is a weak acid, so it causes a small decrease in the pH of the water solution. Nevertheless, it may corrode also in neutral solutions, with a uniform CR generally quite low, as shown in Fig. 15.13.

Furthermore, H_2S may play an important role in the mechanical resistance of corrosion products film, increasing or decreasing their strength, depending on the relative amount, as shown in Figs. 15.14 and 15.15.

15.8.2.2 Hydrogen-Induced Cracking

This form of attack, also known as Stepwise Cracking, is typical of carbon steels, showing ferritic structures, when in the presence of MnS (Type II) elongated inclusions as a consequence of a rolling manufacturing process.

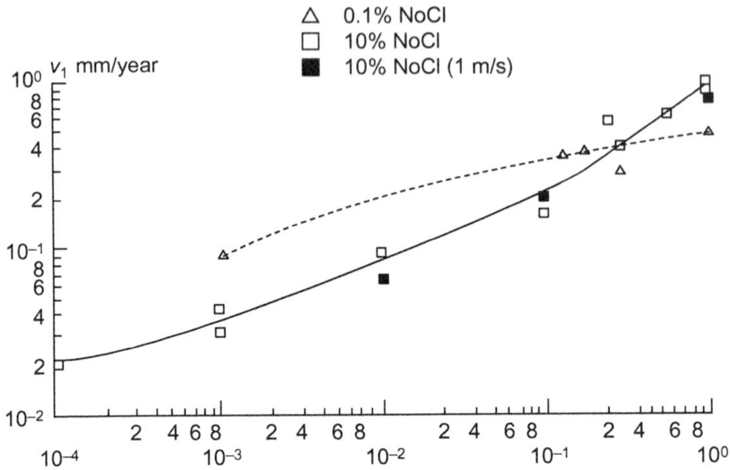

FIGURE 15.13 Influence of $P(H_2S)$ on corrosion rate of steel.

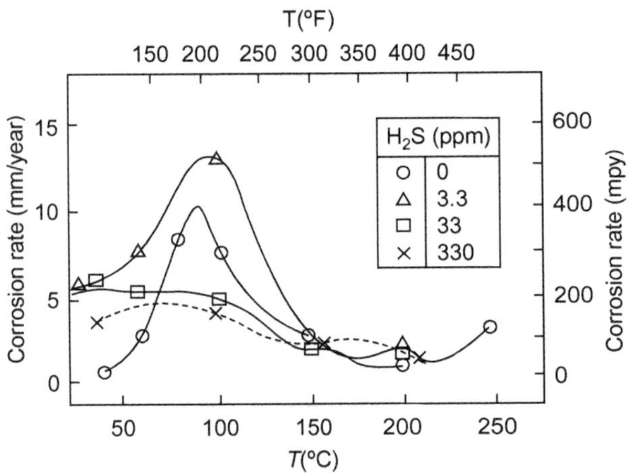

FIGURE 15.14 Effect of the addition of a small quantity of H_2S and of the temperature on the corrosion rate of pure iron (5% NaCl, 3.0 MPa $CO_2 + H_2S$ @ 25°C; Test duration 96 hours; flowrate 2.5 m/s; specific volume 25 cc/cm^2).[12]

This kind of attack may develop if the operating conditions fall above the limits as shown in Fig. 15.16 for sour gas systems and Fig. 15.17 for sour multiphase systems.

At operating temperatures higher than 65°C, the risk of HIC is anyway very low, and precautions against HIC damage are generally unnecessary.

This kind of attack must be avoided through proper selection of carbon steel chemical analysis and corrosion resistance verified through testing during manufacturing.

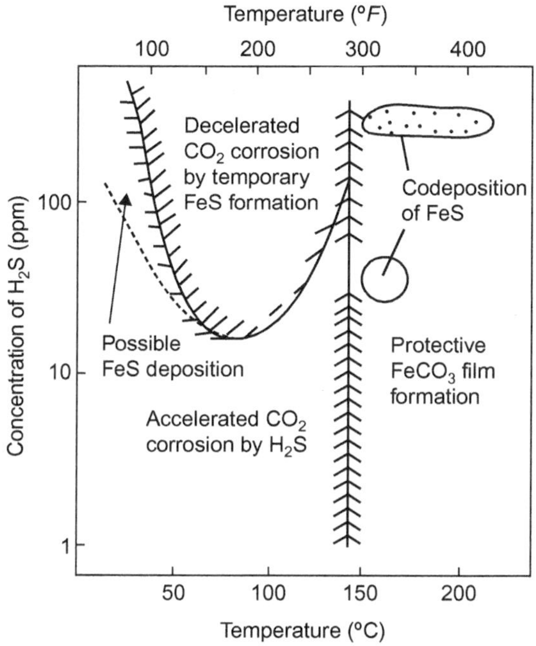

FIGURE 15.15 Effect of the H_2S contamination on the CO_2 corrosion rate.

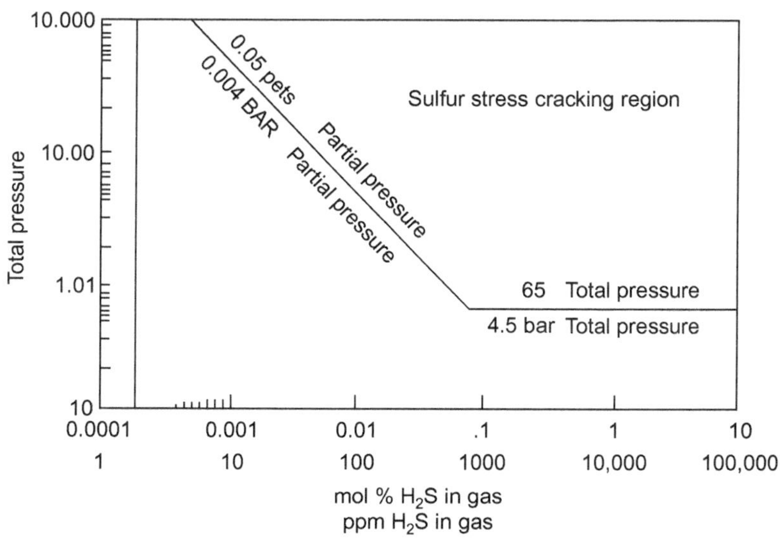

FIGURE 15.16 Sour gas systems.

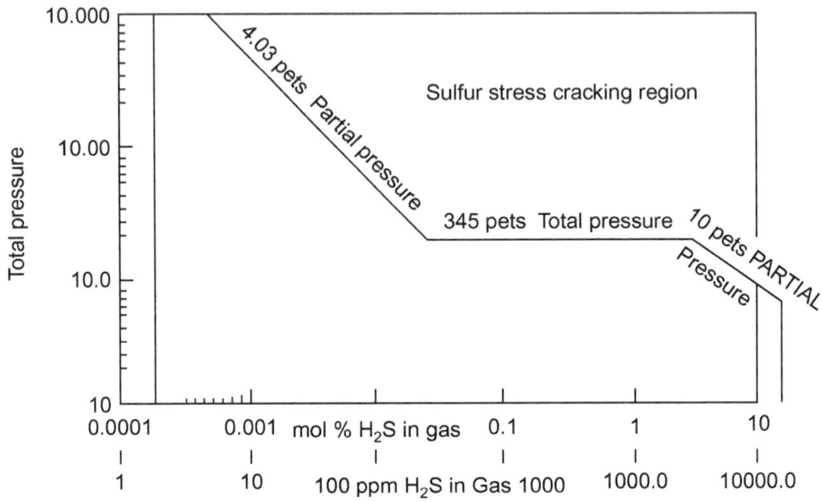

FIGURE 15.17 Sour multiphase systems.

15.8.2.3 Sulfide Stress Cracking

This kind of attack occurs under the combined action of tension stresses, aggressive environment (H_2S) when in the presence of a susceptible material.

Standards define the conditions for sour service (Figs. 15.16 and 15.17) and specify acceptable limits for the materials to be used in these conditions. The requirements imposed by the standard are a restriction to the list of approved materials and a limitation in materials hardness.

15.8.3 Effect of Fluodynamics

Fluodynamics exerts an important influence on the CR.

When increasing the flowrate, the primary effect is a higher mass transfer from the bulk of the solution to the near metal surface, which enhances both the corroding species and the corrosion products mobility. This effect is shown in Fig. 15.18, where the rate calculated using De Waard and Milliams approach is 4.6 mm/year (approach valid at 1 m/s flowrate): higher flowrate produce higher CRs.

The dependence of the CR on liquid flow velocity decreases with increased pH, as demonstrated in Fig. 15.19. This is important for practical situations, where dissolved $FeCO_3$ can increase the pH significantly.

The effect of velocity on CR of steels is also depended on steel composition and microstructures. Steels with more homogeneously distributed carbides like in tempered martensite and in bainitic structures are not expected to form lamellar cementite, which can act as a cathodic depolarizer and stimulator of the CO_2 corrosion. Also the effect of Cr appears to be a partial blockage of the surface, probably by Cr-oxide which interferes with the

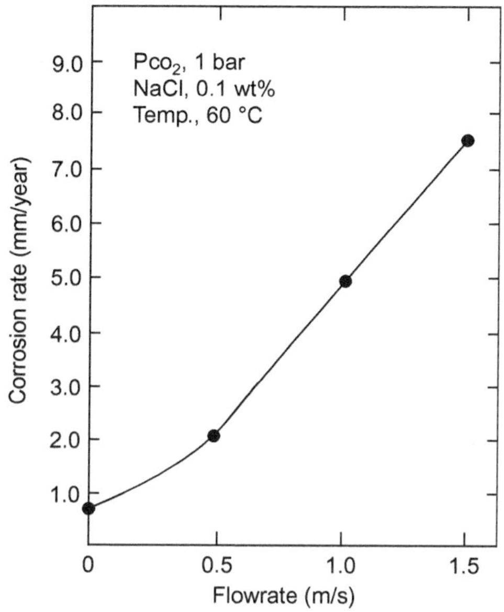

FIGURE 15.18 The effect of flow rate on the corrosion rate of carbon steel in CO_2 saturated water.

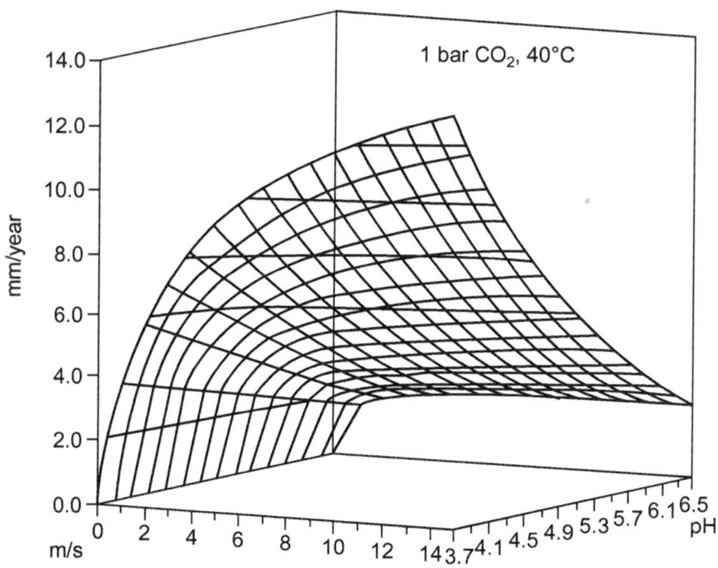

FIGURE 15.19 Effect of flowrate and pH at 40°C, 1 bar CO_2.

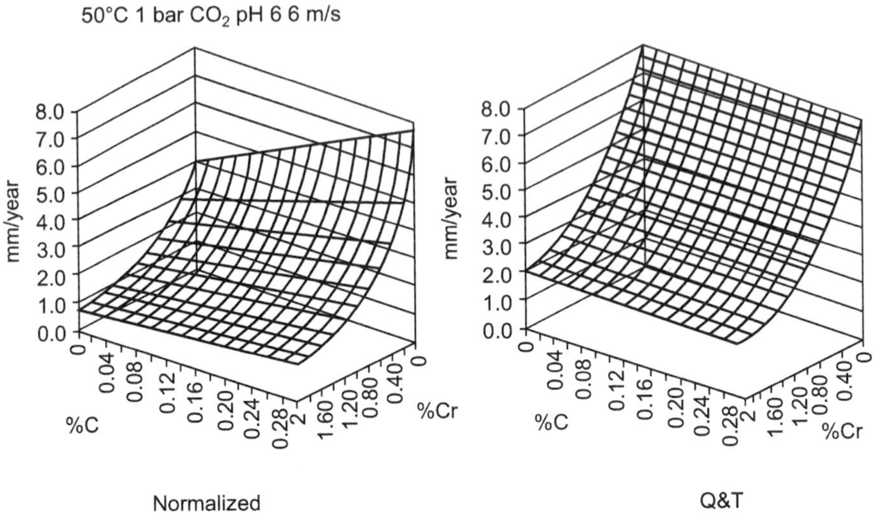

FIGURE 15.20 Example of calculated effect of composition of low-alloy steels on CO_2 corrosion rates.

corrosion reaction. Fig. 15.20 gives an example to demonstrate the differences obtained for the two groups of steels. The differences in the baselines with Cr% = 0 or C% = 0 indicates that martensitic steels could corrode somewhat faster than normalized ones.

When flowrate produces friction stresses on the corrosion products (or inhibitors) film, it may break them, leading to a much higher CR.

On the contrary, as with horizontal pipes, low flowrates are generally not able to remove the free water which is stagnant at the 6 o'clock position.

When the flowrate is high, it is possible that corrosion products protective or inhibitors films are removed, giving the chance of a high CR to occur. This kind of attack is also called erosion corrosion. API RP-14E contains a simple formula for estimating the velocity beyond which accelerated corrosion due to erosion corrosion may occur.

The formula is empirical and derived from field experiences and is meant to describe the velocity for the possible onset of erosion corrosion in uninhibited corrosive oil and gas well surface-production equipment fabricated from carbon steel in the absence of sand:

$$V = \frac{122}{\sqrt{P}}$$

where:

V = that velocity beyond which accelerated corrosion due to high velocity may occur (m/s)
P = fluid density (kg/m^3)

Several authors have observed that erosion corrosion happens in an annular mist regime. It is also indicated that the increase of the CR with velocity at the Khuff Gas sour production system (Fig. 15.21) was associated with the onset of annular mist regime in multiphase flow, as indicated also in Fig. 15.22. Here, without inhibitors injection, CRs dramatically increase at about 5 m/s flowrate, with inhibitor injection; on straight sections, this change happens at about 7–8 m/s. Of course these figures are strictly valid

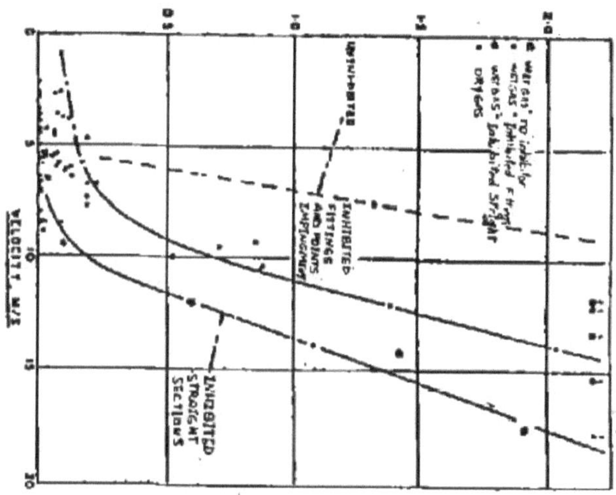

FIGURE 15.21 Effect of velocity on corrosion in Khuff gas.

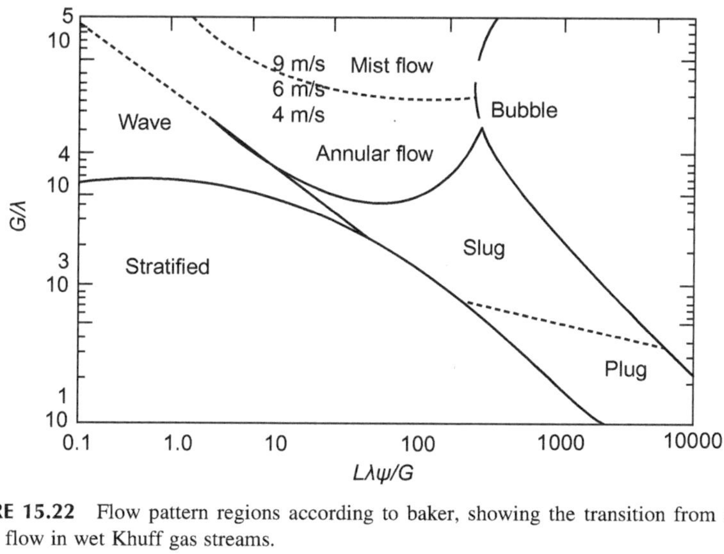

FIGURE 15.22 Flow pattern regions according to baker, showing the transition from annular to mist flow in wet Khuff gas streams.

for the Khuff Gas experience only, and cannot be extrapolated to other conditions.

From this experience, the following general rules can be derived:

- the calculated API erosional velocity is associated with the onset of annular mist flow;
- the onset of erosion corrosion in uninhibited systems is associated with the onset of annular mist flow regime in multiphase flow in the surface piping;
- the onset of erosion corrosion in inhibited systems (straight portions) occurs at a velocity of about 1.5× the calculated API erosional velocity.

15.9 SPECIAL CONSIDERATION IN REFINERIES AND PETROCHEMICAL PLANTS

15.9.1 Corrosives and Corrosion Problems

The following are the main corrosives and corrosion problems that require special material consideration in petroleum refineries and petrochemical plants.

15.9.1.1 Sulfur Content

Organic sulfur-bearing fluids ($\geq$0.1 wt%) corrode steel at high temperature. Based on accumulated experience in actual plant design, chromium-molybdenum steel is normally employed instead of carbon steel for the parts, where the operating temperature requires it.

15.9.1.2 Erosion

A serious erosion problem may arise around H.D.S piping and effluent air coolers downstream of the water injection point. Materials and fluid velocity should be determined by R.L. Piehls Method.

15.9.1.3 Naphthenic Acid

Crude oil with an acidic value exceeding 0.5 mg KOH/g may pose serious corrosion problems for heater tubes, piping, and rotary machines. Austenitic stainless steel of 316 grade should be selected for parts handling high-temperature (over 260°C) crude oil having high acidity.

15.9.1.4 Hydrogen

High temperature and pressure sections are subject to hydrogen attack. Materials for such services should be selected using the curves in the latest edition of API. This should also be considered for equipment located downstream of the waste heat boiler.

15.9.1.5 Polythionic Stress Cracking

Normal 304 stainless steel becomes subject to polythionic stress corrosion cracking because it sensitizes after welding and/or long-term exposure under high-temperature service. To avoid sensitization, stabilized austenitic stainless steel such as Type 321 or 347 may be selected.

15.9.1.6 Caustic Embrittlement by Amine Solution

Piping and equipment in direct contact by amine solution should be stress relieved to avoid caustic embrittlement, provided that the operating temperature exceeds 90°C.

15.9.1.7 Salts

If the salt concentration of the crude oil is more than 1 lb/1000 bbls of crude, consideration needs to be given to the problem of chloride fouling and corrosion caused by hydrochloric acid resulting from the hydrolysis of $MgCl_2$ and $CaCl_2$ in distillation tower, crude preheating exchangers, as indicated in the following:

1. In cases where the salt concentration is between 1 and 10 ptb, the corrosion prevention measures should be provided.
2. If the salt concentration is over 10 ptb, a desalter must be necessarily provided to reduce it to under 1 ptb, or to reduce it to around 1 ptb and then corrosion preventive measures should be provided.

15.9.1.8 Condensate

To prevent condensation of water and corrosion, the operating temperature at the top section of the atmospheric distillation tower is raised. Therefore, more economical materials can be used here.

15.9.1.9 High Temperature

Regarding material selection for high-temperature piping and reformer tubes used in reforming furnace, the selection should be made with consideration given to economics because of the very expensive materials used.

15.9.1.10 CO_2 Corrosion

Carbon steel has no corrosion resistance in wet-reforming gas services at temperatures above 40°C.

15.9.1.11 Amine Solution

The use of copper and copper alloys should be avoided. The corrosiveness of rich-amine solution is the highest in the case of hydrogen units and town gas units where only CO_2 is handled, and moderate in the case of F.C.C units where $CO_2 + H_2S$ ($H_2S/(CO_2 + H_2S) = 0.01$) is handled and lowest in the case of hydrodesulfurization, where only H_2S is handled. Equipment that undergoes maximum corrosion are lean/rich heat exchangers, regenerator, reboiler, reclaimer, and overhead condensers.

15.9.1.12 H_2S

Parts and components to come in contact with wet hydrogen sulfide should be provided with sulfide stress corrosion cracking prevention measures in accordance with NACE MR 01–75 latest edition.

15.9.1.13 H_2SO_4

Sulfuric acid in concentrations above 85% by weight are usually not corrosive to carbon steel if temperatures are below 40°C. Cold-worked metal (usually bends) should be stress relieved. Flow velocities above 1.2 m/s can destroy protective iron sulfate film. Also localized attack immediately downstream of piping welds has been attributed to a spherodized structure; a normalizing postweld heat treatment at 870°C is required to minimize corrosion. All valves and pumps require corrosion resistance internals or trim. In addition to sulfuric acid, reactor effluent contains traces of alkyl and dialkyl sulfates from secondary alkylation reaction. These esters decompose in reboilers to form sulfur dioxide and polymeric compounds and finally sulfurous acids that can cause severe corrosion in overhead condensers (particularly deisobutanizer tower).

Neutralizers or filming amine corrosion inhibitors can be injected into the overhead vapor lines of various towers to prevent corrosion.

15.9.1.14 Hydrogen Flouride

In general, hydrofluoric acid is less corrosive than hydrochloric acid because it passivates most metals. However, if these films are destroyed by dilution or else, severe corrosion in the form of hydrogen blistering of carbon steel and stress cracking of hardened bolts will occur.

Specific areas where corrosion is likely to occur include the bottom of the acid rerun tower, depropanizer towers, the overhead condensers of these towers, the reboiler of the propane stripper, and piping around the acid rerun tower.

By proper design practices to keep the feed stocks dry, and prescribed maintenance procedures to keep the equipment dry during shutdowns, there will be few corrosion problems by this catalyst.

15.9.1.15 Acetic Acid

Corrosion by acetic acid can be a problem in petrochemical process units. As a rule even a tenth of a percent of water in acetic acid can have a significant influence on CR of this acid. Temperature increases the CR; bromide and chloride contamination causes pitting, and S.C.C., while an addition of oxidizing agents, including air, can reduce the CR.

Type 304 stainless steel can be used for temperatures below 90°C and Type 316 and 317 for hot acetic acid applications.

15.9.1.16 Ammonia

Ammonia causes two types of SCC in petrochemical plants. The first is cracking of carbon steel in anhydrous ammonia service, and the second type is cracking of copper alloys. Use of low-strength steels, postweld heat treatment of welds and regular inspection are some actions to be taken to minimize cracking. Cracking of copper alloys tube bundles during shutdowns should be prevented by neutralizing the residual ammonia by acid.

15.9.1.17 Fuel Ash

Corrosion by fuel ash deposits can be one of the most serious operating problems with boiler and preheat furnaces.

All fuels except natural gas contain certain inorganic contaminants that leave the furnace with products of combustion. In particular, vanadium pentoxide vapor (V_2O_5) reacts with sodium sulfate (Na_2SO_4) to form sodiumvanadate ($Na_2O \cdot 6V_2O_5$). This compound will react with steel, forming a molten slag that runs off and exposes fresh metal to attack.

In general, alloys with high chromium and nickel contents provide the best resistance to this type of attack. Also, an addition of magnesium type compound can raise the melting points of fuel ash deposits and prevent the formation of highly corrosive films. These additives also offer additional benefits with regard to cold-end corrosion in boilers by condensation (150–170°C) of sulfuric acid produced by sulfur content of fuel, by forming magnesium sulfate.

15.9.1.18 Microorganism

The corrosion action of the sulfate reducing bacteria is well known in oil industries, especially in cooling water systems, fire water loops, after hydrotesting of tanks and vessels, and in mothballed or water flooded systems.

15.9.2 Special Material Requirements for Refinery's Equipment

15.9.2.1 Austenitic Stainless Steel

The use of austenitic stainless steel should be kept to a minimum. When the use of such a material cannot be avoided, and where there is danger of transgranular stress corrosion cracking, the use of higher alloy materials such as stabilized Incoloys or ferritic stainless steel such as Type 444 (18 Cr-2 Mo), should be considered.

15.9.2.2 *Parts to Be Welded*

For parts to be welded, including tank plates and structural steel, no bottom or side, air or enriched air-blown converter steels should be used. Oxygen-blown converter steel may be used only below the creep temperature range.

15.9.2.3 *Copper Base Alloy*

The use of copper base alloys in direct contact streams in which ammonia acetylene or its homologs may be present is prohibited.

15.9.2.4 *Carbon Content*

Carbon content and a carbon equivalent, based on C + Mn/6 for any plain carbon or carbon-manganese steel that is to be joined by welding, should not exceed 0.25% and 0.41%, respectively.

In cases where the above requirements are not met, welding procedure qualification tests in accordance with applicable codes should be performed, and welding procedures for the portions to be subjected to the tests should be submitted for agreement separately.

15.9.2.5 *Ferritic Stainless Steel*

The use of ferritic stainless steel should be considered on the basis of the following characteristics:

1. Weldability
2. 474°C (885°F) embrittlement
3. Pitting corrosion
4. Caustic embrittlement

15.9.3 Special Equipment Requirements

15.9.3.1 *Pressure Vessels (Including Exchanger Shells, Channels, Etc.)*

15.9.3.1.1 Low-Temperature Vessels

When pressure vessels are subjected to low temperatures, i.e., below 0°C, materials and fabrication practices, e.g., postweld heat treatment, should be selected to minimize the risk of brittle fracture. Requirements for material selection depend upon the minimum design temperature. However, where this minimum design temperature is not a normal continuous operating condition, e.g., and if it arises as a result of autorefrigeration due to rapid depressurization, the full range of temperatures and coincident pressures should be evaluated in order to determine the appropriate conditions for material selection.

The use of postweld heat treatment can extend the use of carbon steel to temperatures lower than would be acceptable for as-welded vessels.

However, unless postweld heat treatment is required for process reasons, it should be specified only when the requirement cannot be met by using carbon steel in the as-welded condition.

15.9.3.1.2 Corrosion Resistance

Carbon steel should normally be selected for pressure vessels, and an appropriate CA applied where total corrosion is not expected to exceed 6 mm over the design life of the vessel.

Where the CR is predicted to exceed 6 mm over the design life, the various alternatives should be evaluated. These alternatives may include, but should not be limited to, the following:

1. replacement at intervals, e.g., every 12 years,
2. increased CA,
3. corrosion-resistant internal linings,
4. alternative solid corrosion-resistant material.

Where pressure vessels are relatively thin in the absence of any CA, the use of solid corrosion-resistant alloys such as stainless steel and nickel-based alloys may be more suitable than CAs or the use of internal cladding. However, for thicker vessels it is likely that internal corrosion-resistant alloy cladding will provide the most economic solution. In certain circumstances, the use of coal tar epoxy, glass flake–reinforced resins, or other nonmetallic coatings may be appropriate. The following should be considered where there is a choice between clad or lining and solid corrosion-resistant material:

The alloy material of vessel shells and heads required for corrosion resistance may be provided as alloy clad plate or as alloy cladding, provided that the backing material of the clad plate is suitably resistant to the other conditions of the designated service.

The choice between alloy plate and alloy clad plate should be made and be based on economic considerations. However, when austenitic stainless steel is the material required for corrosion resistance, alloy cladding should be used.

The use of alloy sheet lining, instead of cladding or deposit lining for vessel shells and heads, should be subjected to separate approval and should be considered only for localized areas where the use of lining may be desirable from an economic standpoint.

For heavy shells and heads, alloy deposit lining may be used in lieu of cladding. The automatic strip-arc deposit welding process is acceptable. In all overlay weld metal, the ferrite content should be between 4% and 5%.

Where the anticipated erosion corrosion rates on carbon steel wear plates exceed 1.5 mm/year, alloy steel wear plates should be employed to prevent this CR being exceeded.

Cast iron pressure-retaining parts should not be used in process fluid services, but may be used in fresh cooling water services for heat exchanger channels and cover sections.

Strip cladded plate should not be used if postweld heat treatment is required.

15.9.3.2 Storage Tanks

The decision on steel used for storage tanks should be made from an economic viewpoint, between either normal or high-tensile steel; however, the yield strength of the plate, weld metal, and heat-affected zone should be 60 kg/mm^2 maximum.

Steels containing deliberately added chromium, nickel, or molybdenum should not normally be used for tankage.

Austenitic stainless steels should not be used for swing arm cables.

15.9.3.3 Heat Exchanger Tube Bundles

Materials for heat exchanger tubes and tubesheets should be selected for resistance against both shell and tube side fluids.

Allowances for corrosion should be made on both sides of single tubesheets.

For water-cooled heat exchangers, the following considerations should be made.

15.9.3.3.1 Seawater-Cooled Heat Exchanger

For heat exchangers on seawater duty where long life is required, titanium may be used. Alternatively, Cu/Ni alloys may be selected provided that the fluid velocities are kept within the range specified in the specifications.

Normally, seawater is permitted only on the tube side of a heat exchanger. On some high-pressure gas coolers, however, this is not possible because of the risk of tubes collapsing under external pressure. In such cases titanium tubes and shell are necessary to allow seawater to be used on the shell side.

Where the materials of interconnecting seawater piping and the mating surfaces of the heat exchanger are dissimilar, rubber-lined couplings will be required if galvanic corrosion would otherwise occur. This is particularly important in the case of titanium and Cu/Ni dissimilar metal joints. An alternative solution which may be considered is the use of sacrificial spool pieces of austenitic spheroidal graphitic cast iron between the titanium and Cu/Ni components.

Titanium plate exchangers should be used in closed circuit systems where treated fresh water is exchanged with seawater.

Whether tubes should be either inhibited aluminum brass or aluminum bronze, tubesheets should be of the same material as tubes.

Where corrosion of copper base alloys by sulfides in the hydrocarbon stream are excessive, consideration should be given to either providing a greater CA than normal or to the use of materials such as nickel-based alloys (such as Monel, Incoloy 801), aluminum alloys (such as Alclad), and titanium alloys.

15.9.3.3.2 Freshwater-Cooled Heat Exchanger

Carbon steel tubes may be used only where the water has a low dissolved solid content and where a water recirculation system is employed.

Copper-based alloys subject to stress corrosion cracking in hydrocarbon streams containing free ammonia (with pH exceeding 7.2 even for short periods) should not be employed in heat exchangers. Cu-Ni (70–30) alloy may be considered satisfactory for such applications.

For hydrocarbon/hydrocarbon heat exchangers, where one or both hydrocarbon streams have a high H_2S content, consideration should be given to the use of double dip-aluminized material for the material in contact with liquids in order to prevent corrosion due to sulfide scale fouling.

Unstabilized austenitic stainless steel should not be used for U-tubes. Low carbon unstabilized stainless steels such as Type 304L or Type 316L are acceptable for U-tubes.

Normally air-cooled heat exchanger tubes should be carbon steel. For corrosive or heavy fouling services, the application of internal coatings will be considered, if required.

15.9.3.4 Furnaces

Material for furnace tubes should be selected from an economic viewpoint. However, high-temperature strength, corrosion resistance, and scaling-resistance factors must be satisfied.

Regarding furnace tube wall thickness, material selection and calculation should be based on 100,000 hours operation in accordance with API RP 530.

CRs in excess of 0.5 mm/year are not normally acceptable. The design CR of furnace tubes should be determined on the basis of available information from corrosion experience in similar applications.

In the absence of any suitable information, a minimum CA of 3.2 mm should be provided for furnace tubing in hydrocarbon services, and 1.6 mm in steam services.

Material selected for furnace tubes and other parts of furnace coils exposed to firebox conditions should be such that free scaling temperatures will not be exceeded under normal operation.

The composition and physical and mechanical properties of materials for headers and return bends, irrespective of whether they are cast or wrought, rolled or welded in, should be compatible with those tubes to which they will be connected and should be of a weldable quality. The use of cast alloy steel parts should require approval.

Carbon steel tubes for steam generating units should be seamless.

15.9.3.5 Piping

Materials for piping should be selected from an economic viewpoint; however, the strength based on the pressure–temperature rating against corrosion resistance should be satisfied.

Where high alloy or nonferrous material is employed, special consideration should be given to decide the economical limits of the pipe size for which a clad or solid design is to be adopted.

15.9.3.5.1 Low-Temperature Piping

Where piping systems are subjected to low temperatures, i.e., below 0°C, materials and fabrication practices, e.g., postweld heat treatment, should be selected to minimize the risk of brittle fracture. However, where this minimum design temperature is not a normal continuous operating condition, e.g., if it arises as a result of autorefrigeration due to rapid depressurization, the full range of temperatures and coincident pressures should be evaluated in order to determine the appropriate conditions for material selection.

The use of postweld heat treatment can extend the use of carbon steel to temperatures lower than would be acceptable for as-welded pipework. However, unless postweld heat treatment is required for process reasons, it should be specified only where the requirement cannot be met by using carbon steel in the as-welded condition.

15.9.3.5.2 Corrosion Resistance Piping

Where carbon steel is the selected material for piping, and the total corrosion is not expected to exceed 6 mm over the design life of the piping, an appropriate CA should be applied. Where the CR is predicted to exceed 6 mm over the design life, the various alternatives should be evaluated. These alternatives may include, but should not be limited to, the following:

1. replacement at intervals, e.g., every 10 years,
2. increased CA,
3. application of internal corrosion-resistant cladding,
4. alternative solid corrosion-resistant material,
5. injection of corrosion inhibitor, or other treatments of the process stream.

Where the choice is between options (**3**) or (**4**), the selection will generally be governed by costs. However, austenitic stainless steel in solid form should be used in a marine environment only where the external skin temperature does not exceed 50°C, and should normally be AISI Type 316, because of the risk of chloride stress corrosion cracking.

Where the external skin temperature exceeds 50°C, carbon steel lines internally clad with austenitic stainless steel may be used. Alternatively, solid pipe of one of the duplex stainless steels may be used, as they exhibit much greater resistance to chloride stress corrosion cracking. They also possess much higher yield strengths, and their use therefore results in weight saving.

Where carbon steel vessels or pipework are connected to pipework that is either internally clad or of solid corrosion-resistant alloy, and where galvanic corrosion of the carbon steel is likely, such corrosion should be prevented by installing electrically isolating joints, insulating flanges, or pipe spools coated internally with a nonmetallic lining, whichever is appropriate for the conditions. It should be noted that, in many locations, insulating flanges and joints will be rendered ineffective by electrical short circuiting through connections to the supporting steelwork.

Piping for seawater duty should normally be of 90/10 Cu/Ni conforming to piping specification. High-molybdenum austenitic and 25% Cr duplex stainless steels may offer cost and weight advantages over Cu/Ni, and should be evaluated for specific projects. These stainless steels have, in addition to much higher strength, excellent resistance to pitting corrosion, chloride stress corrosion cracking, and flow-induced erosion and, therefore, may permit the use of higher flow velocities, which in turn may permit the use of smaller bore piping, and thinner pipe walls, thus saving weight and cost.

Recent limited laboratory testing indicates that the stainless steels may be susceptible to chloride crevice corrosion in seawater at temperatures above about 30°C. This should be taken into consideration for piping downstream of heat exchangers.

Nonmetallic materials, such as glass-reinforced plastics, may offer advantages for seawater pipework, particularly with respect to corrosion resistance and weight saving. However, specialist expertise in design, fabrication, and installation techniques will be required for the evaluation of factors such as cost, susceptibility to mechanical damage, and safety implications, before such materials are selected. Any proposal to use nonmetallic materials should be subject to approval by company. Also, it may be necessary to obtain waivers from the statutory authorities regarding the use of combustible materials.

15.9.3.6 Corrosion Resistance Valves

In general, valve bodies and bonnets should be manufactured in a material similar to that used for the piping or vessel to which they are attached. Valve trims should be manufactured in a more resistant material to prevent erosion/corrosion; the choice of materials being dependent upon the process conditions.

Where electroless nickel plating of valve internals is approved by the company (e.g., for ball valves or parallel slide gate valves), the following requirements should be met:

The substrate should be stainless alloy for high integrity valves, e.g., those for subsea applications, although carbon steel may be permitted for other applications.

Phosphorous content of the coating should be within the range 8–11% by weight.

Plating thickness should be not less than 0.075 mm.

Components should not be baked after plating.

Plated components should be subjected to a ferroxyl test to ASTM B 733 for evidence of porosity or cracking.

Plated test pieces should be sectioned and checked for coating thickness using a micrometer.

The adhesion of the nickel coating should be evaluated using coated test pieces subject to testing in accordance with ASTM B733 and ASTM B571.

15.9.3.7 Flare Systems

For high-pressure flares of the Indair type, alloy 800H should be used for the bowl and stool. For Mardair type flares, Incoloy DS should be used for the trumpet. The gas-filled parts and base may be fabricated in alloy 800H; however, AISI Type 316 may be used if there is a heat shield in Incoloy DS with ceramic fiber insulation.

Indair and Mardair are trade names of Kaldair Limited. Incoloy is a trade name of Inco Limited.

15.9.3.8 Rotating Machinery

Generally, pump casings are fabricated in a material that matches that used for the piping system. Pump internals are usually fabricated in corrosion-resistant materials with additional resistance to erosion; the choice of materials being dependent upon the process conditions. Pumps handling seawater or brine above 40°C contaminated with oil and H_2S should be fabricated in one of the super duplex stainless steels or a more corrosion-resistant material.

When a wet gas stream is corrosive, the first stage of wet gas compressors should be fabricated in a corrosion-resistant material such as 13% Cr steel.

Exhaust stacks for gas turbines should be fabricated in corrosion-resisting carbon steel. Also, stacks should be protected externally by aluminum metal spray with an aluminum silicone sealer. Care should be taken to ensure that the design eliminates thermal fatigue.

15.9.4 Special Material Requirement in Petrochemical Plants

In selecting materials for petrochemical plants, considerable effort should be paid to fluid composition, sizing of lines, valve and pump details, and processing temperature and pressure. Most environments in petrochemical processes involve flammable hydrocarbon systems, highly toxic chemicals, explosive gases, and strong acids and caustics. Therefore corrosion could be mysterious and a costly enemy to the safety of personnel and the community. Material and CAs should be selected on the basis of corrosion tests and procedures outlined in Section 15.5 of this chapter.

15.9.5 Supplemental Requirements for Equipment in Sour Service

Equipment in sour service, as set forth in the process data sheet or in the mechanical drawings, should strictly comply with the requirements of NACE standard MR 0175, as supplemented by the following articles.

15.9.5.1 Carbon Steel

15.9.5.1.1 Process of Manufacture

All carbon steel products should be fully killed and fine grain treated and should be supplied in the normalized, normalized and tempered, or quenched and tempered condition.

Production should be by a low-sulfur and low-phosphorus refining process (e.g., electric furnace with double deslagging or in the basic oxygen converter).

The heat should be vacuum degassed and inclusion shape control treated, preferably by calcium.

15.9.5.1.2 Chemical Analysis

Chemical analysis should be restricted as follows:

Check Analysis			Heat Analysis	
Carbon	0.020	% max	0.19	% max
Sulfur	0.003	% max	0.002	% max
Phosphorus	0.020	% max	0.020	% max
Manganese	1.20	% max	1.20	% max
Silicon	0.45	% max	0.45	% max
Residuals				
Chromium	0.25	% max		
Copper	0.25	% max		
Molybdenum	0.10	% max		
Nickel	0.30	% max		
Vanadium	0.05	% max		
Niobium	0.04	% max		
Ca + O + N to be reported				

15.9.5.1.3 Carbon Equivalent

Carbon equivalent (CE) should not exceed 0.42%.

$$CE(\%) = C + \frac{Mn}{6} + \frac{(Cr + Mo + V)}{5} + \frac{(Ni + Cu)}{15}$$

15.9.5.1.4 Through-Thickness Tension Testing

Plates 25 mm and greater in thickness, directly exposed to sour environments (see NACE MR 0175 par. 1.3) should have a minimum reduction of area of tension test specimens (*Z* value) of 35%.

Testing should be conducted according to ASTM A770.

15.9.5.1.5 Ultrasonic Testing

Ultrasonic testing, in accordance with ASTM A435, should be carried out on all plates having a thickness greater than 12 mm.

15.9.5.1.6 Hardness

Hardness across the width and thickness of each product/weld should not exceed 200 HB.

Specimen for hardness surveys should be taken in the same area where the coupons are to be removed for mechanical tests.

15.9.5.1.7 Weldability Tests

It should be demonstrated that the proposed plates are suitable for welding and subsequent postweld heat treatment by carrying out weldability tests on representative plates. The hardness value in heat-affected zones should not exceed 200 HB. Details of these tests should be provided for the company's approval.

15.9.5.1.8 Stainless Steel

Stainless steel products should be supplied in the fully solution-treated condition.

Cold works resulting in a material deformation degree of more than 5% should be followed by a solution annealing heat treatment of the parts involved.

15.9.5.1.9 Fabrication Requirements

All carbon steel vessels should be postweld heat treated after completion of all welding. Minimum temperature should be 595°C as stated in NACE MR 0175. This condition will lead to specify a PWHT target temperature of 615 ± 20°C.

Internal and external fittings as well as attachments welded to pressure parts should be fully penetrated.

Nozzles should be self-reinforced type with integral or welding-neck flanges.

Hardness testing should be conducted on base metal, weld metal, and heat-affected zones, as follows:

- on macro-examination specimens taken from production test coupons
- on internal welds of the equipment (one set reading for each longitudinal and/or circumferential shell weld, at least).

Hardness value should not exceed 200 HB.

Because of the significantly greater risk of crevice corrosion in sour/chloride service, the use of screwed couplings and some types of weld details, which could result in a crevice on the process side, is not permitted.

C-½% Mo welding consumables and those having more than 1% Ni should not be used for welding carbon-manganese steel.

Weld repair of plate surface defects will not be permitted without the company's approval, and should be subjected to an agreed-upon repair procedure prior to the work being carried out.

ERW pipes should not be used for sour service.

15.10 ENGINEERING MATERIALS

15.10.1 Ferrous Alloys

Some 94% of the total world consumption of metallic materials are in the form of steels and cast irons. This is also true in the oil industries, with a figure around 98%.

Therefore, the primary choice in any material selection is steel or cast iron unless they cannot afford the design requirements.

15.10.1.1 Carbon Steels

The strength and hardness of steels vary very considerably with both carbon content and type of heat treatment. Certain names, which relate to the carbon content, are used in connection with steels.

- Mild or low carbon steels: those containing up to 0.3% of carbon.
- Medium carbon steels: steels containing between 0.3% and 0.6% of carbon. These may be hardened and tempered.
- High carbon steels or tool steels: steels containing 0.6% of carbon and always used in the hardened and tempered condition. Table 15.7 gives some typical uses of carbon steels.

TABLE 15.7 Compositions and Typical Applications of Steels

C (%)	Name	Applications
0.05	Dead mild steel	Sheet and strip for presswork, car bodies, tin-plate; wire, rod, and tubing
0.08–0.15	Mild steel	Sheet and strip for presswork; wire and rod for nails, screws, concrete reinforcement bar
0.15	Mild steel	Case carburizing quality
0.1–0.3	Mild steel	Steel plate and sections, for structural work
0.25–0.4	Medium carbon steel	Bright drawn bar
0.3–0.45	Medium carbon steel	Shafts and high-tensile tubing
0.4–0.5	Medium carbon steel	Shafts, gears, railway tyres
0.55–0.65	High carbon steel	Forging dies, railway rails, springs
0.65–0.75	High carbon steel	Hammers, saws, cylinder linings
0.75–0.85	High carbon steel	Cold chisels, forging die blocks
0.85–0.95	High carbon steel	Punches, shear blades, high-tensile wire
0.95–1.1	High carbon steel	Knives, axes, picks, screwing dies and taps, milling cutters
1.1–1.4	High carbon steel	Ball bearings, drills, wood-cutting and metal-cutting tools, razors

15.10.1.1.1 Surface Hardening

Generally, the toughness of a material decreases as the hardness increases. There are very many service conditions where the requirement is for a tough material of very high surface hardness, such as shafts and gears. Table 15.8 shows different methods to accomplish surface hardening.

15.10.1.2 Alloy Steels

The main effects conferred by specific alloying elements are given in Table 15.9. Major categories of steels are as followings:

15.10.1.2.1 Low-Alloy Steel

Contains up to 3–4% of one or more alloying elements and is characterized by possessing similar microstructures to, and requiring similar heat treatment to, plain carbon steels. But this has improved strength and toughness over the plain carbon steels with the same carbon content.

TABLE 15.8 Surface Hardening Methods

	Carburizing	Nitriding	Cyaniding
Effect	A high carbon surface is produced on a low carbon steel and is hardened by quenching	A very hard nitride-containing surface is produced on the surface of a strong tough steel	A carbon- and nitride-containing surface is produced on a low carbon steel and is hardened by quenching
Suitability	Suitable for plain carbon or alloy steels containing about 0.15 percent C	Nitralloy steels containing aluminum; a typical nitriding steel contains 0.3% C, 1.6% Cr, 0.2% Mo, 1.1% Al. This steel is hardened by oil quenching from 900°C and tempered at 600–700°C before being nitride	Suitable for plain carbon or alloy steels containing about 0.15% C
Method	Low carbon steel is heated at 850–930°C in contact with gaseous, liquid, or solid carbon-containing substances for several hours. The high carbon steel surface produced is then hardened by quenching	The steel is heated at 500–540°C in an atmosphere of ammonia gas for 50–100 hours. No further heat treatment is necessary	Low carbon steel is heated at 870°C in a molten percent sodium cyanide bath for about 1 hour quenching in oil or water from this bath hardens surface of the steel
Result	Case depth is about 1.25 mm. Hardness after heat treatment is HRC 65 (HD 870). Negligible dimension change caused by carburizing. Distortion may occur during heat treatment	Case depth is about 0.38 mm. Extreme hardness (HD 1100). Growth of 0.025–0.05 mm occurs during nitriding. Case is not softened by heating for long times up to 420°C. Case has improved corrosion resistance	Case depth is about 0.25 mm. Hardness is about HRC 65. Negligible dimension change is caused cyaniding. Distortion may occur during heat treatment
Applications	Typical uses are for gears, camshafts, and bearings	Typical uses are for valve guides and seatings, and for gears	Typical uses are for small gears, chain links, nut bolts, and screws

	Flame Hardening	Induction Hardening	Siliconizing (Ihrigizing)
Effect	The surface of a hardenable steel or iron is heated by a gas torch and quenched	The surface of a hardenable steel or iron is heated by a high-frequency electromagnetic field and quenched	A moderately hard corrosion-resistant surface containing 14% silicon is produced on low carbon steels
Suitability	Steel containing 0.4–0.5% carbon or cast iron containing 0.4–0.8% combined carbon may be hardened by this method	Steel containing 0.4–0.5% carbon or cast iron containing 0.4–0.8% combined carbon may be hardened by this method	Suitable for plain carbon steels containing 0.1–0.2% carbon
Method	A gas flame quickly heats the surface layer of the steel and a water spray or other type of quench hardens the surface	The section of steel to be hardened is placed inside an induction coil. A heavy induced current heats the steel surface in a few seconds. A water spray or other type of quench hardens the surface	The steel parts are heated at 930–1000°C in contact with silicon carbide and chlorine gas for 2 hours. No further heat treatment is required
Result	The hardened layer is about 3 mm thick. Hardness is HRC 50–60 (HD 500–700). Distortion can often be minimized	The hardened layer is about 3 mm thick. Hardness is HRC 50–60 (HD 500–700). Distortion can often be minimized. Surface remains clean	Case depth is about 0.63 mm. Hardness is about HD 200. Case has good corrosion resistance. Growth of 0.025–0.05 mm occurs during siliconizing
Applications	Used for gear teeth, sliding ways, bearing surfaces, axles, and shafts	Used for gear teeth, sliding ways, bearing surfaces, axles, and shafts	Typical uses are for valves, tubing, and shafts

TABLE 15.9 Effects of Alloying Elements in Steels

Alloying Element	General Effects	Typical Steels
Manganese	Increases the strength and hardness and forms a carbide; increases hardenability; lowers the critical temperature range, and when in sufficient quantity, produces austenitic steel; always present in a steel to some extent because it is used as a deoxidizer	Pearlitic steels (up to 2% Mn) with high hardenability used for shafts, gears, and connecting rods; 13% Mn in Hadfield's steel, a tough austenitic steel
Silicon	Strengthens ferrite; raises the critical temperatures; has a strong graphitizing tendency; always present to some extent because it is used, with manganese, as a deoxidizer	Silicon steel (0.07% C; 4% Si) used for transformer cores; used with chromium (3.5% Si; 8% Cr) for its high-temperature oxidation resistance in internal combustion engine valves
Chromium	Increases strength and hardness; forms hard and stable carbides; raises the critical temperatures; increases hardenability; amounts in excess of 12% render steel stainless	1.0–1.5% Cr in medium and high carbon steels for gears, axles, shafts, and springs, ball bearings and metal-working rolls; 12–30% Cr in martensitic and ferritic stainless steels; also used in conjunction with nickel (see below)
Nickel	Marked strengthening effect lowers the critical temperature range; increases hardenability; improves resistance to fatigue; strong graphite-forming tendency; stabilizes austenite when in sufficient quantity	0.3–0.4% C with up to 5% Ni used for crankshafts and axles as well as other parts subject to fatigue
Nickel and chromium	Frequently used together in the ratio Ni/Cr = 3/1 in pearlitic steels; good effects of each element are additive, each element counteracts disadvantages of the other; also used together for austenitic stainless steels	0.15% C with Ni and Cr used for case carburizing; 0.3% C with Ni and Cr used for gears, shafts, axles and connecting rods; 18%, or more, of chromium and 8%, or more, of nickel give austenitic stainless steels
Tungsten	Forms hard and stable carbides; raises the critical temperature range, and tempering temperatures; hardened tungsten steels resist tempering up to 600°C	Major constituent in high-speed tool steels; also used in some permanent magnet steels

(Continued)

TABLE 15.9 (Continued)

Alloying Element	General Effects	Typical Steels
Molybdenum	Strong carbide-forming element, and also improves high-temperature creep resistance; reduces temper-brittleness in Ni-Cr steels	Not normally used alone; a constituent of high-speed tool steels, creep-resistant steels and up to 0.5 percent Mo often added to pearlitic Ni-Cr steels to reduce temper-brittleness
Vanadium	Strong carbide-forming element; has a scavenging action and produces clean, inclusion-free steels	Not used on its own, but is added to high-speed steels and to some pearlitic chromium steels
Titanium	Strong carbide-forming element	Not used on its own, but added as a carbide stabilizer to some austenitic stainless steels
Aluminum	Soluble in ferrite; also forms nitrides	Added to nitriding steels to restrict nitride formation to surface layers
Cobalt	Strengthens but decreases hardenability	Used in Stellite type alloys, Magnet steels, and as a binder in cemented carbides
Niobium	Strong carbide former, increases creep resistance	Added for improved creep resistance and as a stabilizer in some austenitic stainless steels
Copper	Increases strength and corrosion resistance. >0.7% Cu permits precipitation hardening	Added to cast steels to improve fluidity, castability, and strength. Used in corrosion-resistant architectural seals
Lead	Insoluble in iron	Added to low carbon steels to give free-machining properties

High-Strength, Low-Alloy Steel This is a group of low-alloy steels with a very fine grain size with tensile yield strengths between 350 and 360 MPa. This is achieved by addition of small controlled amounts of Nb, Ti, or Va.

15.10.1.2.2 High-Alloy Steels

High-alloy steels are those that possess structures and require heat treatment that differ considerably from those of plain carbon steels. Generally they contain more than 5% of alloying element. A few examples of some high-alloy steels are given in following paragraphs.

High-Speed Tool Steels High carbon steels rich in tungsten and chromium provide wearing metal-cutting tools, which retain their high hardness at temperatures up to 600°C. An example is 18/4/1 steel containing 18% tungsten, 4%of chromium, 1% vanadium, and 0.8% carbon.

Stainless Steels When chromium is present in amounts in excess of 12%, the steel becomes highly resistance to corrosion. Several types of stainless steel are summarized below.

Ferritic Stainless Steels Ferritic stainless steels contain between 12% and 25% of chromium and less than 0.1% of carbon. This type of steel cannot be heat treated but may be strengthened by work hardening.

Martensitic Stainless Steels These class of steels contain between 12% and 18% of chromium, together with carbon content ranging from 0.1% to 1.5%. These steels can be hardened by quenching from the austenite range temperatures.

Austenitic Stainless Steels These are nonmagnetic and contain 18% chromium, 8% nickel, and less than 0.15% carbon. Carbides may form in these steels if they are allowed to cool slowly from high temperature, or if they are reheated in the range 500–700°C (heat-affected zones adjacent to welds). Small stabilizing additions of titanium, or niobium, prevent intercrystalline corrosion, weld decay. They are widely used in chemical engineering plants.

Maraging Steels These are very high-strength materials. They can be hardened to give tensile strengths of up to 1900 MPa. They contain 18% nickel, 7% cobalt, and a small amount of other elements such as titanium. The carbon content is less than 0.05%. A major advantage of maraging steels is that before the hardening process they are soft enough to be worked and machined and precipitation hardening treatment is at a fairly low temperature when distortion of machined parts is negligible. Although the basic material cost is very high, the final cost of a complex component is less than other high strength because of the much lower machining costs.

Manganese Steels This is a high-alloy steel that contains 12–14% of manganese and 1% of carbon. They are nonmagnetic and are very resistant to abrasion coupled with the fact that the core of material remains comparatively soft and tough. They are used for drill bits, rock crusher jaws, excavator bucket teeth, etc.

15.10.1.3 Cast Iron

The carbon content of cast irons is generally between 2%and 4%. They are generally cheap, with an ease of melting and casting along with a high-damping capacity and very good machinability. Cast irons are classified as

TABLE 15.10 Composition and Properties of Some Cast Irons

Approximate Composition	Tensile Strength (MN/m^2)	Type and Uses
3.2% C, 1.9% Si	250	Pearlite and graphite. Motor brake drums
3.25% C, 2.25% Si	220	Pearlite and graphite. Engine cylinder blocks
3.25% C, 2.25% Si, 0.35% P	185	Ferrite, pearlite and graphite. Light machine castings
3.25% C, 1.75% Si, 0.35% P	200	Ferrite, pearlite and graphite. Medium machine castings
3.25% C, 1.25% Si, 0.35% P	250	Pearlite and graphite. Heavy machine castings
3.6% C, 2.8% Si, 0.5% P	370	Wear resistant. Piston rings
3.6% C, 1.7% Si	540	Pearlitic S.G
3.6% C, 2.2 Si	415	Ferritic S.G
2.8% C, 0.9% Si	310	Blackheart malleable
3.3% C, 0.6% Si	340	Whiteheart malleable
2.9% C, 2.1% Si, 1.75% Ni, 0.8% Mo	450	Shock resistant. Crankshafts for petrol and diesel engines
2.9% C, 2.1% Si, 15% Ni,	220	Ni-resist. Corrosion-resistant austenitic iron
2% Cr, 6% Cu		Used in chemical plant
2.5% C, 5% Si	170	Silal. A growth-resistant iron for high-temperature service

SG, shaft and gear.

either white or gray. These terms arise from the appearance of a freshly fractured surface. The structure of cast irons is affected by the following factors:

- the rate of solidification,
- carbon content,
- the presence of other elements,
- the effect of heat treatment.

Table 15.10 indicates the composition and properties of some cast irons.

15.10.1.3.1 Malleable Irons

These cast irons are produced by heat treatment of certain white cast irons. There are two processes used to give rise to black heart and white heart irons. The names arise from the appearance of the fracture surface of the treated iron. The white heart structure is composed of ferrite at the surface of casting and ferrite, pearlite, and some graphite nodules at the center.

15.10.1.3.2 Alloy Cast Iron

Alloy cast irons are high strength, hard, and abrasion- and corrosion-resistant materials and are suitable for high-temperature services.

Addition of about 5% nickel causes the formation of martensitic structure at the surface of casting that is very hard constituents.

About 15–25% of nickel plus some chromium and copper give the best cast iron for corrosion resistance.

For prolonged service at elevated temperatures, the carbon content of alloy is kept down to about 2%. The alloy contains 5% silicon or silicon and nickel. The presence of these reduces the oxide scale formation at high temperatures.

15.10.2 Nonferrous metals

All the other metallic elements (some 70 in number) and their alloys are classified as nonferrous. Out of all the nonferrous metals, only a few—aluminum, copper, lead, magnesium, nickel, tin, titanium, and zinc—are produced in moderately large quantities. Only a brief description of nonferrous metals and their alloys are mentioned in the following paragraphs; a detailed coverage of the metallurgy of these metals is outside the scope of this chapter.

15.10.2.1 Aluminum

Aluminum possess a number of properties that makes it an extremely useful engineering material. It has good corrosion resistance, low density, and good electrical conductivity. The corrosion resistance of aluminum is due to the presence of a thin oxide layer which is only a few atoms in thickness, but it is permeable to oxygen and protects the surface from further attack. The corrosion resistance may be improved by anodizing. High-purity aluminum is too weak to be used for many purposes. The material commonly termed pure aluminum is an aluminum iron alloy, by adding up to 0.5% of iron. This small iron addition gives a considerable increase in strength, although there is some reduction in ductility and corrosion resistance. Commercial-purity aluminum is used extensively, and accounts for about nineteenths of aluminum product sales.

15.10.2.2 Aluminum alloys

Aluminum may be alloyed with a number of elements to produce a series of useful engineering materials. For properties of some aluminum alloys and their use, see Table 15.11.

TABLE 15.11 Properties of Some Aluminum Alloys

Alloy Number		Approximate Composition	Condition[a]	Tensile Strength	Type of Product	Uses
British	American					
		99.99% A1	O	45	Sheet, strip	Linings for vessels in food and chemical plants
1080	1060	99.8% A1	O	75	Sheet, strip	
1200	1200	99.0% A1	O	90	Sheet, strip, wire, extruded sections	Lightly stressed and decorative paneling, wire and bus bars, foil for packaging, kitchen and other hollow-ware
			H14	120		
			H18	150		
3103	3103	A1 + 1.75% Mn	O	110	Sheet, strip, extruded sections	Hollow-ware, roofing, paneling, scaffolding tubing
			H14	160		
			H18	210		
5251	5052	A1 + 2% Mg	O	180	Sheet, plate, tubes and extrusions	Stronger deep-drawn articles; ship and small boat construction and other marine applications
			H24	250		
5154 A	5454	A1 + 3.5% Mg	O	240		
			H24	300		
5056 A	5056 A	A1 + 5% Mg	O	280		
			H24	335		

LM6	S12C	A1 + 12% Si	M	180[b]	Sand and die castings	Excellent casting alloy
			M	210[c]		
6082	6082	A1 + 0.9% Mg;	T4	220	Sheet, forgings, extrusions	Structural components for road and rail transport vehicles
		1% Si; 0.7% Mn	T6	320		
2014	2014	A1 + 4.5% Cu	T4	440	Sheet, forgings, extrusions, tubing	Highly stressed parts in aircraft construction and general engineering
		0.5% Mg; 0.8% Mn	T6	480		
2024	2024	A1 + 4.5% Cu;	T3	480		
		1.5% Mg; 0.6% Mn				
2L95		A1 + 5.6% Zn;	T6	500	Plate Rod and bar	Aircraft construction
L160		1.6% Cu; 2.5% Mg				
7075	7075	A1 + 7% Zn;	T6	620	Sheet and extrusions	
		1.75% Cu; 2% Mg				
2090	2090	A1 + 2.2% Li;	T6	580	Sheet plate	Aircraft construction
		2.7 Cu; 0.12% Zr				
8090	8090	A1 + 2.5% Li;	T6	495		
		1.3% Cu; 0.7% Mg				

[a]*Key: The symbols for condition are: O, annealed; M, as cast; H14, partly work hardened; H18, fully work hardened; H24, hardened and partially annealed; T3, solution-treated, cold-worked, and aged; T4, solution-treated and naturally aged; T6, solution-treated, precipitation-hardened.*
[b]*Sand cast.*
[c]*Chill or die cast.*

15.10.2.3 Copper

Copper is one of the oldest metals known to man and one of its alloys, bronze, has been worked for over 5000 years. Some applications of the various grades of pure copper are: wire, for electrical windings and wiring; sheet for architectural cladding, tanks, and vessels; and tubing for heat exchangers in chemical industries. There are very many useful applications of copper alloys in industries, but due to its high price it has been replaced by cheaper material.

15.10.2.4 Copper Alloys

Copper may be alloyed with a number of elements to provide a range of useful alloys. The important alloy systems are:

- copper-zinc (brasses),
- copper-tin (zinc) (bronzes and gun metals),
- copper-aluminum (aluminum bronzes),
- copper-nickel (capronickels).

Small addition of beryllium or chromium to copper give high-strength alloys, a small addition of cadmium gives a significant increase in strength with little loss of electrical conductivity, while an addition of tellurium to copper gives an alloy with very good machinability. Properties of copper and some copper alloys are classified in Table 15.12.

15.10.2.5 Lead and Its Alloys

Lead is soft and malleable, and possess an excellent resistance to corrosion. It has been used for water pipeworks and waste disposal systems, but nowadays it's replaced by other materials. A major application for lead is in manufacture of lead-acid storage batteries, which account for almost 30% of the annual world consumption of lead. Cable sheathing, soft solders, and fusible plugs in sprinklers of firefighting systems are other application of lead alloys (see Table 15.13).

15.10.2.6 Nickel

Pure nickel possess an excellent resistance to corrosion by alkalis and many acids and, consequently, is used in chemical engineering plant. For cheapness, nickel is frequently used as a cladding of thin sheet on a mild steel base. Nickel may also be electroplated on a number of materials, and an intermediate layer of electrodeposited nickel is essential in the production of chromium-plated mild steel.

15.10.2.7 Nickel Alloys

The principal nickel base alloys used industrially are Monel, Inconel, Incoloy, and the Nimonic series of alloys. Table 15.14 indicates composition and use of some nickel alloys.

TABLE 15.12 Properties of Copper and Some Copper Alloys

Alloy	Approximate Composition	Condition[a]	Tensile Strength (MPa)	Type of Product	Uses
Pure copper	99.95% Cu	O	220	Sheet, strip, wire	High-conductivity electrical applications
		H	350		
Arsenical copper	99.85% Cu	O	220	All wrought forms	Chemical plant, deep drawn and spun articles
		H	360		
Brasses	99.25% Cu; 0.5% As	O	220	All wrought forms	Retains strength at elevated temperatures
		H	360		
Gilding metal	90% Cu; 10% Zn	O	280	Sheet, strip	Heat exchange steam pipes
		H	Wire		
Cartridge brass	70% Cu; 30% Zn	O	325	Sheet, strip	Imitation jewelry and decorative work
		H	700		
General cold-working brass	65% Cu; 35% Zn	O	340	Sheet, strip, extrusions	High-ductility brass for deep drawing decorative work
		H	700		
Muntz metal	60% Cu; 40% Zn	M	375	Hot-rolled plate and extrusions	General purpose cold-working alloy
High-tensile brass	35% Zn; 2% Mn; 2% A1;	M	600	Cast and hot worked forms	Condenser and heat exchanger plates
	2% Fe; balance Cu				

(*Continued*)

TABLE 15.12 (Continued)

Alloy	Approximate Composition	Condition[a]	Tensile Strength (MPa)	Type of Product	Uses
Bronzes	95.5% Cu; 3% Sn;	O	325	Strip	Ships screws, rudders, and high-tensile applications
	1.5% Zn	H	725		
	5.5% Sn; 0.1% P;	O	360	Sheet, strip, wire	British copper coinage
	Balance Cu	H	700		
	10% Sn; 0.5 P;	M	280	Castings	Springs and steam turbine b lades
	Balance Cu				
Gunmetal	10% Sn; 2% Zn;	M	300	Castings	General purpose castings and bearings
	Balance Cu				
Aluminum bronze	95% Cu; 5% A1	O	400	Strip, tubing	Pressure-tight castings, pump and valve bodies
		H	770		
	10% A1; 2.5% Fe;	M	700	Hot worked and cast products	Imitation jewelry and condenser tubes
	2–5% Ni; bal. Cu				
Cupronickle	75% Cu; 25% Ni	O	360	Strip	High-strength castings and forgings
		H	600		
	70% Cu; 30% Ni	O	375	Sheet, tubing	British silver coinage
		H	650		

Monel	29% Cu; 68% Ni;	O	550	All forms	Condenser tubing, excellent corrosion resistance
	1.25% Fe; 1.25% Mn	H	725		
Beryllium-copper	1¾–2½% Be; ½% Co;	WP	1300	Sheet, strip	Excellent corrosion resistance, used in chemical plant
	balance Cu				
Cadmium-copper	99% Cu; 1% Cd	O	285	Wire, rod	Springs, non-spark tools
		H	500		
Chromium-copper	0.4–0.8% Cr; bal. Cu	WP	450	Wrought forms and castings	Overhead electrical wire, spot-welding electrodes
Tellurium-copper	0.3–0.7% Te;	O	225	Wrought forms	Welding electrodes, commutal segments
	bal. Cu	H	300		Free-machining properties

[a]*Key: O, annealed; H, work hardened; M, as manufactured (cast or hot worked); WP, solution heat-treated and precipitation–hardened.*

TABLE 15.13 Some Lead Alloys

Alloy	Composition (%)					Applications
	Pb	Sb	Sn	Bi	Hg	
Antimonial lead	99	1	—	—	—	Cable sheathing
Hard lead	94	6	—	—	—	Lead-acid batteries, lead shot
Plumbers solder	60	2.5	37.5	—	—	Soft solders
Common solder	50	—	50	—	—	
Tinman's solder	38	—	62	—	—	
Type metal	62	24	14	—	—	For casting into printing type
Linotype metal	81	14	5	—	—	
Woods alloy	24	—	14	50	—	(+12% Cd) Alloy with melting point of 71°C used for fusible plugs in sprinkler systems
Roses alloy	28	—	22	50	—	Low melting point alloy with melting point of 100°C
Dental alloy	17.5	—	19	53	10.5	Dental cavity filling (M.Pt. 60°C)

15.10.2.8 Titanium

The useful properties of titanium are its relatively high strength coupled with a low density, and its excellent corrosion resistance. However, it does possess some characteristics that make processing both difficult and costly.

The main use of titanium alloys are where an excellent corrosion resistance is required.

15.10.2.9 Polymers

The group of materials known as polymers (or plastics) can be subdivided into thermoplastic, elastomers, and thermosetting materials. The use of plastic materials is continually increasing with a rate of 7% per annum. The major increase in the use of plastics are due to low cost, low densities, high resistance to chemical attack, good thermal and electrical insulation properties, and ease of fabrication. The main disadvantages are the low strength and elastic modules values, low softening, and thermal degradation temperatures and their comparatively high thermal expansion coefficients. A brief general definitions of subdivisions of plastics are discussed in this section.

TABLE 15.14 Composition and Uses of Some Nickel Alloys

Name	Composition (%)										Uses
	Ni	Cu	Cr	Fe	Mo	W	Ti	A1	Co	C	
Monel	68	30	–	2	–	–	–	–	–	–	Chemical engineering plant
											Steam turbine blades
Inconel	80	–	14	6	–	–	–	–	–	–	Chemical engineering plant
											Electric cooker heating elements
											Exhaust manifolds
Brightray	80	–	20	–	–	–	–	–	–	–	Heating elements for kettles,
											toasters, electric furnaces
Hastelloy C	55	–	15	5	17	5	–	–	–	–	Chemical engineering plant
Hastelloy X	47	–	22	18	9	1	–	–	–	–	Furnace and jet engine components
Nimonic 75	77	–	20	2.5	–	–	0.4	–	–	0.1	Thermocouples sheaths, furnace components, nitriding boxes
Nimonic 90	56.6	–	20	1.5	–	–	2.4	1.4	18	0.06	Gas turbine discs and blades
Nimonic 115	56.5	–	15	0.5	4	–	4	5	15	0.1	Gas turbine discs and blades
Incoloy 825	45	3	22	25	3		1.2	0.2	–	0.05	Chemical engineering plant

15.10.2.10 Thermoplastics

There are several varieties of thermoplastics, but generally they have the property of softening by heating and hardening by cooling within a temperature range (ASTM D883).

Thermoplastics are categorized as following: (see also Table C.9).

- Polyethylene (PE) consists of: High-density polyethylene (HDPE), low-density polyethylene (LDPE), linear-low-density polyethylene (LLDPE), ultrahigh-molecular–weight polyethylene (UHMWPE), cross-linked polyethylene (XLPE) and polyethylene foam.
- Ethylene copolymers consist of: ethylene-vinyl acetate, ionomer
- Polypropylene (PP)
- Polyvinyl chloride (PVC) consists of: unplastisized (UPVC), plasticized (PPVC), copolymers (CPVC and PVDC).
- Polytetrafluoro ethylene (PTFE) consists of: ECTFE
- Polystyrene (PS) consist of: SBR, SAN, ABS
- Acrylic materials consist of: PMMA, PAN
- Polyamides (nylon) (PA) consist of: PA 6, PA 6.6, PA 6.10, PA 6.12, PA 11, PA (R1M)
- Polycarbonate (PC)
- Acetal (POM) polyoxymethylene
- Saturated polyesters: PET, PETP, PCDT, PBT
- Cellulosics : (CN), CA, CAB, CP, EC
- Polyether ether ketone (PEEK)
- Polyphenylenes: ppo, pps
- Polysulfones and polyarylates: PSU, PES, PPSU, PPS
- Polyimides (PI): PEI, PAI

15.10.2.11 Elastomers

Elastomers are materials that have a low elastic module that shows a great extensibility and flexibility when stressed but which return to their original dimensions, or almost so, when the deforming stress is removed.

There are several classes of elastomers, these being (see also Table C.9).

- natural rubber (NR),
- synthetic R class elastomers (unsaturated carbon chains),
- m class elastomers (saturated),
- class elastomers (heterochain with oxygen),
- u class elastomers (heterochain with O,N),
- q class elastomers (heterochain with Si),
- thermoplastic elastomers.

15.10.2.12 Thermosetting Materials

Thermosetting material undergoes chemical changes when first heated and is converted from a plastic mass into a hard and rigid material. There are also a number of materials that will set hard and rigid at ordinary temperatures. The commercially available thermosetting materials are as following:

- phenolic materials, phenol formaldehyde,
- amino-formaldehyde materials, urea (UF), melamine (MF),
- polyester materials,
- epoxies,
- polyurethanes,
- allyl resins, diallyl phthalate (DAP).

15.10.3 Composite Materials

There are very many situations in engineering where no single material will be suitable to meet a particular design requirement. However, two materials in combination may possess the desired properties and provide a feasible solution to the material selection problem. In this section some of the composites in current use just will be mentioned.

- Timber and plywood, which is built up of thin layers of wood bonded with a water-resistant glue or a thermosetting resin, with the grain of successive layers at right angles to each other.
- Fiber-reinforced materials: in these composite materials, high-strength fibers are encased within a tough matrix. Fibers include glass, carbon, polymer, ceramics, wire filaments; matrices include thermoplastic and thermosetting resins, glasses, ceramics, and metals. Many different matrix/fiber combinations have been developed with different properties for different applications.
- Sandwich structures are structures generally composed of two skins of high strength with a lightweight core. This arrangement provides a material with low-density and high-specific stiffness, whereas the maximum tensile and compressive stresses are carried by the skin. These combinations also provide useful thermal and sound insulation. Skin material includes sheet metal, plywood, plastics (including GRP), concrete and plasterboard, with cores that may be metal or paper honeycomb structures, rigid plastic foam, chip board or low-density porous masses of glass fiber or rockwool bonded with a plastic resin.

Chapter 16

Control Valves

16.1 INTRODUCTION

This engineering and material chapter covers the minimum requirements for control valve bodies, actuators and accessories, designed, constructed, and materially tested in accordance with the references outlined herein. This practice is intended to be used in the petroleum industries.

16.2 CONTROL VALVE BODY DESIGN REQUIREMENTS

The following description is primarily intended to indicate the general and minimum requirement of the control valves body design criteria to be used in the petroleum industries. The supplier's design calculations, drawings, and material selection should be approved by the purchaser before an order is placed.

Control valves have inherent operation characteristics that hinder precise positioning under varying operating conditions. Factors such as pressure differential across the valve seat, overtightening of packing, and viscous or fouling service can create additional forces preventing the valve from assuming the position called for by the controller.

Valve types should be selected by taking into account such factors as operating and design conditions, fluids being handled, rangeability required, cost, allowable leakage, noise, and other special requirements.

Pneumatic control valves in electronic control loops should be equipped with a 24-V. D.C. device (convertor or transducer) to convert 4–20 mA, D.C. electronic instrument signal to a 0.2–1 barg pneumatic signal.Control valves should be furnished with pneumatic type (or DC electric-pneumatic) positioners in the following applications:

1. Temperature control valve other than minor applications.
2. Valves 4 in. body size and over.
3. More than one valve on a single controller.
4. Critical pressure drop service for valve trim 1 iin. and larger.
5. Line pressure exceeding 20 bar.
6. Extension bonnets (includes radiation fins and belows seals).
7. Butterfly valves (rotary action).

Oil and Gas Pipelines and Piping Systems. DOI: http://dx.doi.org/10.1016/B978-0-12-803777-5.00016-2

8. Sounders patent bodies.
9. Three-way valves.
10. Slow system, such as mixing, thermal, or level process.
11. For fast systems, a detailed stability analysis should be presented and if required, boosters will be used instead.
12. On duties where the controlled liquid will vaporize across the ports.

Filter regulators and positioners or boosters should be factory mounted and tubbed. All connecting tubing in instrument air service should be plastic-coated copper with brass compression type fittings, unless otherwise specified.

Single-seated bodied valves should be top guided, double-seated bodied valves should be a top- and bottom-guided construction, unless otherwise as specified in the data sheet.

Control valves should have removable trims, and sufficient clearance should be allowed for access and removal.

Butt welding valves should not be used. However, if the line specification calls for butt welding, consideration should be given to the welding of control valves.

For flashing conditions, the type and size and, additionally, the flashing condition of the control valve should be specified in data sheet and/or agreed upon with the user.

For control valves intended for operating high temperatures, particular attention should be paid to the clearance between plug and guide bushing to avoid valve sticking when the valve is hot.

In gas-pressure let-down stations when a high differential pressure is present across the valve, a special low noise valve should be used. In this case the noise level should not exceed certain limits as specified in relevant standards (noise and vibration control standard) or through the requirement of data sheets.

The action of valves on failure of the operating medium should be determined by process requirements with regard to safe operation and emergency shutdown requirements.

Where cage-guided control valves are specified, a balanced trim should be considered for large sized valves.

For control valves on vacuum services, special provisions should be considered for the prevention and detection of leakage.

Where temperature of control fluid is below 0°C, a bonnet extension should be used.

An extension bonnet or finned should also be provided on services above 200°C in order to maintain the temperature of stuffing the box within the limits specified in accordance with the manufacturer's recommendations.

Air-operated diaphragms and springs should be selected to optimize a bench-setting range of 0.2–1 barg for the specified maximum upstream pressure with the downstream pressure of zero bar. The "Bench Setting Range" and the "In Service Stroking Range" should be specified on the control valve data plates. Air-operated control valves with an in-service stroking range other than 0.2–1 barg may be used if so dictated by availability of standard operators and the user's approval.

Manual loading–type hand operators should be considered in lieu of a side-mounted handwheel in relatively low pressure/pressure drop applications where block and bypass valves are not provided and a handwheel may cause a hazardous condition for automatic start-up or shutdown of the related equipment. These hand operators should consist of a three-way air switch and a handwheel-operated air regulator. The handle and ports should be clearly marked as, MAN-AUTO. The regulators may be common to other components.

For globe body control valves, the trim construction should be either single-seated with a heavy duty top guiding for the plug, double-seated with a top and bottom guiding for the plug, or cage type. For liquid services with a high pressure drop, i.e., boiler feed water and gas service (pressure let down), cage trims should be specified to have the plug supported at the critical area.

Balance type control valves in place of single seat valves in high-pressure services should be considered.

Control valves for steam-heated reboilers should be located in the steam lines and not in the condensate lines, unless otherwise agreed upon by the user.

Where control valves are liable to freezing due to operating or ambient conditions, they should be insulated or heat-traced.

16.3 CONTROL VALVE MATERIALS SELECTION

Since the majority of control valve applications are relatively noncorrosive at reasonable pressures and temperatures, cast iron and carbon steel bodies are the most common valve body materials used in the oil industries.

Most control valve materials can be placed in two categories:

The pressure containment materials for valve body, bonnet, bottom flange, and bolting.

The valve trim materials for valve plug, seat ring, cage, valve stem, guide bushing, and packing box parts.

For oxygen services, body and trim materials should be AISI-316 stainless steel. Body casting should be completely machined internally to a smooth surface to remove any casting imperfections.

For material selection of body, bolts, nuts, etc., the relevant piping class or any other information for the particular application should be adhered to.

Control valve material should be as specified in data sheets or should be selected from ANSI-B16.5 specifications and applicable sections of the codes and standards.

Supplier should comply with the pressure and temperature ratings of more common materials established by the ANSI-B16.5.

In the case where a corrosive condition would require very exotic materials, consideration may be given to a composite construction, such as an internal metallic lining of the body.

For very severe erosive services, the small fluid impact area inside the valve body should be covered with a hard facing.

The minimum requirement for the body material is that the valve should have a cast steel body, and the trim, consisting of plug, seat ring and stem, should have stainless steel 316, unless otherwise specified by the nature of the process fluid being handled and/or requested through relevant data sheets.

When valves are used for chlorine service or other fluids that become corrosive when in contact with a moist atmosphere, suitable valve stem material must be chosen or other precautions taken.

For chlorine services, neoprene diaphragm valves is recommended.

For extremely erosive and corrosive services, the hard-facing material made of two disks of tungsten carbide material in an angle pattern body can be used. This material is especially useful in oil production where severe sand erosion exists.

Hardened plug and seat rings should be selected for the following applications:

1. Erosive service.
2. Wet gas or wet steam service with a pressure drop above 5 bars, other services when the pressure drop is above 10 bar, at design condition.

Small-sized valves for erosive services should have their plug and seat rings made for solid satellite No. 6. For economical reasons, hardened stainless steel 440°C may be used as trim material if this is suitable for the particular process conditions.

When a tight shut off is required, a ball or plug valve, or a single-seated globe body valve should be selected. The seats should be of soft material, such as glass fiber–filled PTFE, and the selection should be based on suitability for the specified process conditions. The selected material should be suitable for temperatures of at least 50°C above the maximum process design conditions. The soft seat ring should be properly clamped between metal parts.

When valves are used for sour gas services, the trim and bolting material construction should comply with the recommendation of the National Association of Corrosion Engineer's MR-01-75 latest revision.

Packing glands should be equipped with flange–style gland followers with bolted constructions. A lubricator with a steel-isolating valve should be provided where packing lubrication is required.

Guide bushing should be a corrosion-resistant material. It is preferred that the guide bushing material be a minimum of 125 brinnel harder than the trim, i.e., 17-4 PH (precipitation hardened) stainless steel or better.

Stainless steel bellows seals may be considered for services with dangerous and or poisoning fluids such as TEL or TML (tetra ethyl lead, tetra methyl lead) but should be avoided wherever possible. A purge with suitable pressure should be used (monitored for purge) as an alternative method of sealing.

Butterfly valves material should be as specified in data sheet for the related service conditions or should be at manufacturer's option and in accordance with the applicable standard such as BS-5155.

Butterfly valves body material should be selected from those listed in Table 16.1, if they are not specified in the data sheet.

Butterfly valves trim material should be suitable for specified service conditions and compatible with the piping material.

Butterfly valves trim material including disks, shafts, bushings, body, and/or disk seating surfaces, internal keys, and pins and screws when in contact with the contained fluid should be selected from Table 16.1, if these are not specified in the data sheet.

Seats in the body and on the disk may be separate or integral. Seat facings may be applied to valve bodies and/or disks as deposited metal, integral metal, mechanically retained metal, or resilient materials.

16.4 CONTROL VALVE BODIES

A control valve consists of two major subassemblies, a valve body subassembly and an actuator. The valve body subassembly is the portion that actually controls the passing fluid. It consists of a housing, internal trim, bonnet, and sometimes a bottom flange (Fig. 16.1).

Body subassemblies occur in many shapes and working arrangements depending upon the individual service conditions and piping requirements. Each type has certain advantages and disadvantages for given service requirements and should, therefore, be selected with care.

Control valves operate by one of two primary motions: reciprocating (sliding stem) motion or rotary motion. The selection of a valve for a particular application is primarily a function of the process requirements. Some of the more common types of control valve bodies are discussed in the following sections.

TABLE 16.1 Basic Materials for Butterfly Valves

1	2	3
Component	**Material**	**BS Reference**
Body	Cast iron	1452
Body with integral seat	Austenitic cast iron	3468
Disk Handwheel	Spheroidal graphite iron	2789
Disk with integral seat	Carbon steel	1501.151
Rings fitted to body or disk for sealing, seating, or rataining purposes		1503.221
		1504.161
	Stainless steel	1501: Part 3
		1503
		1504, 3100
		1504
	Gunmetal	1400
	Aluminum bronze	1400
	Rings of deposited metal or resilient + material	
Shaft	Carbon steel	970: Part 1
	Stainless steel	970: Part 4
	Aluminum bronze	2672 or 2874
	Nickel copper alloy	3076
Shaft bearings seals (when fitted)	No requirement in this standard	
Internal fastenings	Carbon steel	
	Stainless steel	
	Phosphor bronze	2870, 2873
	Aluminum bronze	2872, 2874, 2875
	Nickel copper alloy	3076

+ When the resilient seal forms part of:
The body and the disk is of gray cast iron, spheroidal graphite iron, or carbon steel, it is recommended that the disk should be provided with a disk-facing ring deposit, on the edge, or coated all over.
The disk and the body is gray cast iron, spheroidal iron, or carbon steel; it is recommended that the body should be provided with a facing ring, a deposit on diameter in contact with the resilient seal, or coated all over.

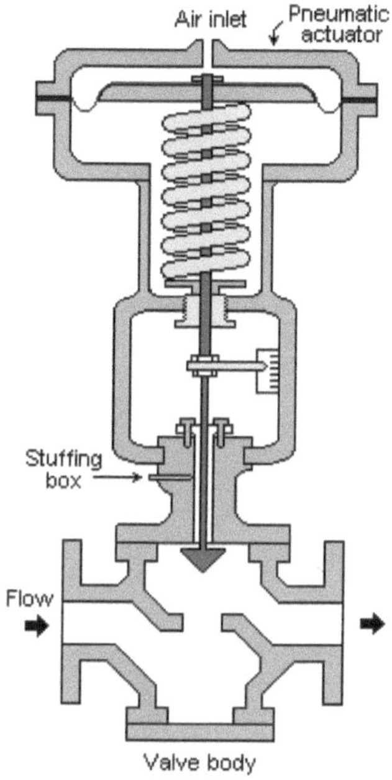

FIGURE 16.1 Typical control valve assembly.

16.4.1 Globe Body Control Valves

The most common control valve body style is in the form of a globe (Fig. 16.1), such a control valve body can be either single or double-seated:

A single-seat construction, for minimum leakage in the closed position should be employed.

A double-seat or balance construction, when requiring less actuator force but allowing for some leakage in the close position, should be used.

Single-seated valves should have a top-guided construction. The valve plug is guided within the lower portion of the valve bonnet (Fig. 16.2).

Double-seated valves should be top- and bottom-guided construction (Fig. 16.3).

Three-way valves are a design extension of a typical double-ported globe valve. They are used for diverting services and mixing or combining services (Fig. 16.4).

Control valve with a globe body should be considered for all applications (throttling or on–off control) except where adverse operating conditions such as high-pressure drops or high capacities make other types more suitable.

16.4.2 Angle Body Valves

Angle body valves should be considered for hydrocarbon services with a tendency for high-pressure drop or coking and erosive services such as slurries and applications where solid contaminants might settle in the valve body (Fig. 16.5).

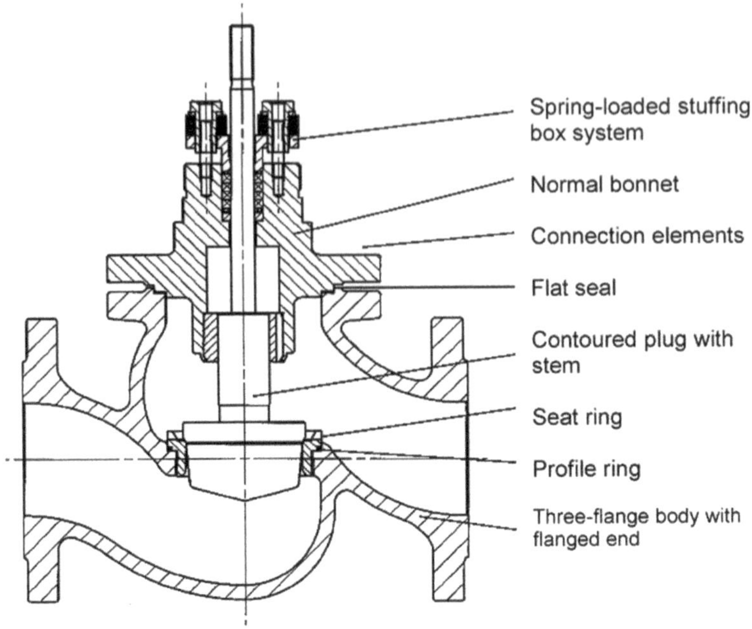

FIGURE 16.2 Globe body valve, top guided.

1	Body
12	Yoke Half Ring
20	Seat Ring
30	Seat Retainer
40	Bonnet (Standard)
50	Plug
55	Seat Ring Gasket
58	Bonnet Gasket
70	Bonnet Flange
76	Yoke Clamp
80	Gland Flange
82	Guide Liner Lower
83	Guide Retainer Lower
86	Guide Liner Upper
87	Guide Retainer Upper
88	Packing
93	Packing Spacer
108	Stud
114	Nut
201	Yoke
202	Cylinder
210	Adjusting Screw
211	Actuator Stem
213	Stroke Plate
225	Piston
227	Spring Button
228	Actuator Stem Spacer
229	Spring
248	Adjusting Screw Bracket
249	Stem Clamp
253	Yoke Bush
256	Retaining Ring
271	Piston O-Ring
272	Piston Stem O-Ring
274	Yoke O-Ring
275	Actuator Stem O-Ring
348	Actuator Stem Lock Nut

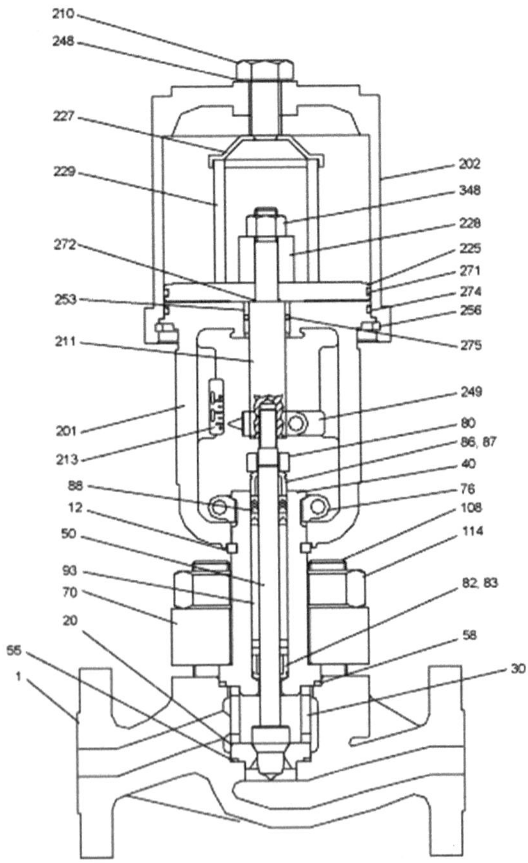

FIGURE 16.3 Globe body valve, top and bottom guided.

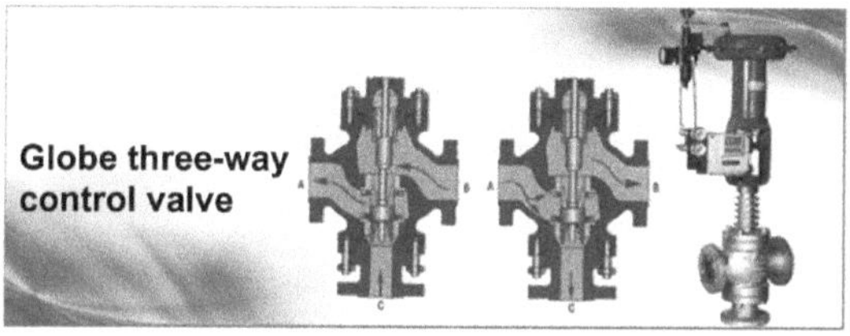

FIGURE 16.4 Three-way valve.

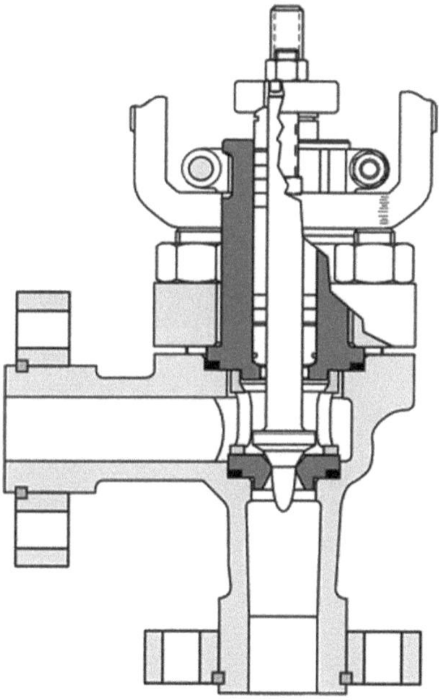

FIGURE 16.5 Angle body valve, low-noise trim.

16.4.3 Diaphragm Valves

Diaphragm valve may be considered for simple services and applications where the body lining in a standard valve becomes economically unattractive. When used for a throttling service, a characterized positioner may be required for obtaining the required valve characteristic (Fig. 16.6).

16.4.4 Cage-Guided Valves

Top entry or cage-guided valves have the advantages of easy trim removal. Valves of this type usually have streamlined body passages to permit increased flow capacity (Fig. 16.7).

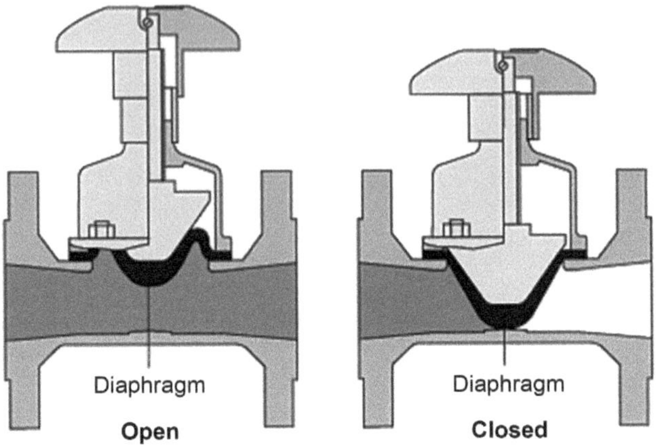

FIGURE 16.6 Diaphragm valve.

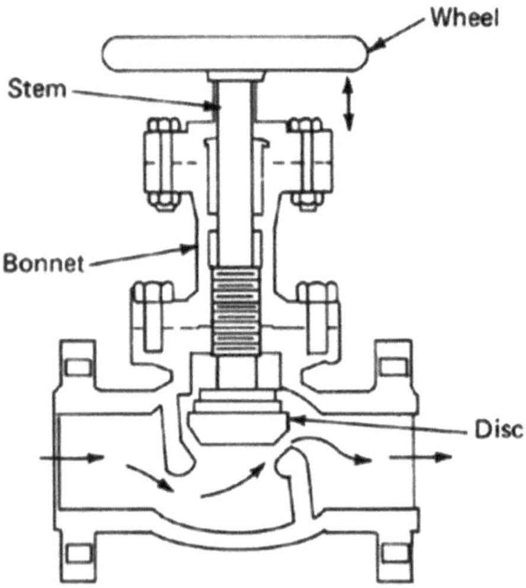

FIGURE 16.7 Globe body valve.

16.4.5 Rotary Type Control Valves

All types of rotary valves share certain basic advantages and disadvantages. Among the advantages are low weight, simplicity of design, high relative C_V, greater reliabilty, friction-free packing, and a generally low initial cost. They are generally not suitable in size, below 2 in., and pressure-drop ratings are limited.

16.4.5.1 Butterfly Valves

The most common type of rotary valve is the butterfly valve. Butterfly valves should be considered for high-capacity, low-pressure drops and where

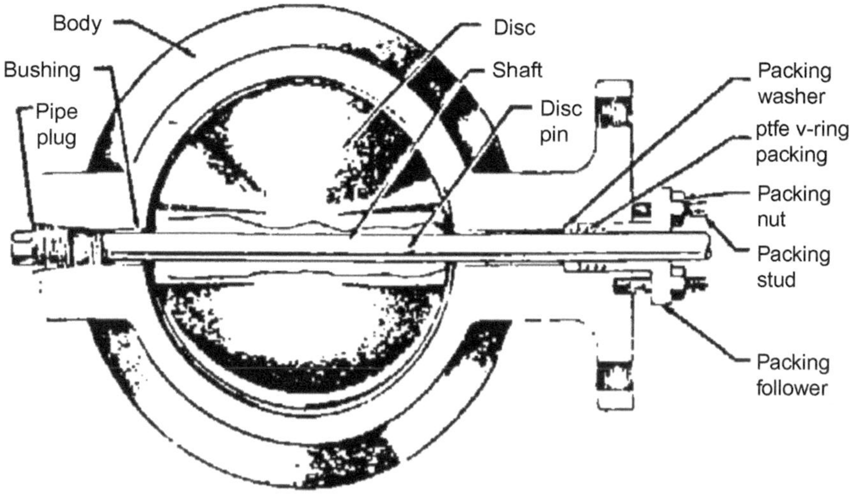

FIGURE 16.8 Typical construction details of 2–12 in. body sizes for butterfly valves.

no tight shutoff is required (Fig. 16.8). Although not normally used in minimum leakage applications, it is available with a piston ring, a pressurized seat, or various types of elastomer-seating surfaces, if minimum leakage is required.

Heavy-pattern butterfly valves should be used where they are practical and economical.

They should normally be furnished with diaphragm or piston actuators with positioners. Where a handwheel is required, the shaft-mounted declutchable type is preferred. Long-stroke position actuators should be used where practical.

16.4.5.2 Ball Valves

Ball valves may be considered for on–off and throttling services under moderate operating conditions.

Characterized ball valves may be used for fluids containing suspended solids or fluids likely to polymerize or crystallize (Fig. 16.9).

16.4.5.3 Emergency Shutoff Valves

For emergency shutoff valves on fuel services, a ball valve should be used for temperatures up to 150°C. Above this temperature single-seated, tight, shutoff globe valves should be used.

16.4.5.4 Plug Valves

Plug valves may be considered for special applications such as throttling control on slurry services in chemical plants (Fig. 16.10).

FIGURE 16.9 Characterized ball valve.

FIGURE 16.10 Eccentric plug valve.

16.4.5.5 Eccentric Rotating Plug Valves

Eccentric rotating plug valves are a general substitute for globe-body control valves provided that the application allows the use of long bolting (Fig. 16.10).

16.4.6 Special Type Control Valves

Special body types, such as angle, split-body (Fig. 16.11), low-noise (Fig. 16.12), low-flow valves should be considered where the process fluid may be erosive, viscous, or carrying suspended solids and/or high differential pressure is required. The flushing connection should be provided on slurry services.

16.4.6.1 Low-Noise Valves

For services at high-pressure drops, the application of a conventional valve trim often results in very high fluid velocities and unacceptable high noise levels.

Where this would be the case, the fluid velocity must be controlled by using a valve trim having specially designed multiple orifices in series and/or in parallel, or having a tortuous path forcing the fluid to change the direction continuously, causing high turbulence friction (Fig. 16.12).

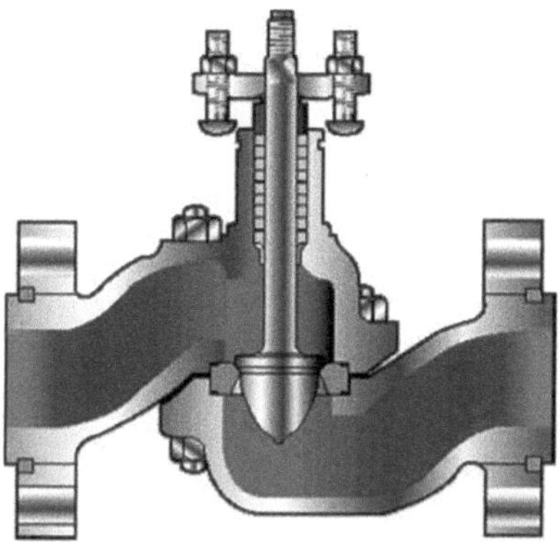

FIGURE 16.11 Split-body valve, stem guided.

FIGURE 16.12 Globe body valve, low-noise trim.

16.4.6.2 Low Flow or Miniature Valves

Where control valves with a very low capacity factor (C_V) are required, these may be of the miniature-valve—type with flanged or threaded connections and a needle trim.

16.5 CONTROL VALVE BODY SIZE AND FLANGE RATING

16.5.1 Globe Body Valves

16.5.1.1 Body Sizes

Nominal body sizes for the globe body should be selected from the following series:

(Inches)	1	1½	2	3	4	6	8	10	12	etc.

The use of odd sizes such as 1¼″, 2½″, 5″, 7″, 9″, etc., should be avoided. 1½″ and 3″ valves are less common in petroleum industries.

The minimum globe control valve body size to be used should be one inch screwed, unless flange type is specified, and the internal trim size should be in accordance to the requirements as specified in data sheet.

Body sizes smaller than 1 in. may be used for special applications, and pressure regulation services. For valve sizes smaller than 1 in., reduced trim in 1 in.-size bodies normally will be preferable.

Flange-rating-globe-control valves should normally have flanged ends, but flangeless bodies may be considered for special applications.

The flange rating should generally be in accordance with the piping class, but for carbon steel bodies the flange rating should be class 300 minimum.

For the pressure–temperature rating of globe-body control valves reference should be made to relevant standards.

All globe-body, control-valve manifolds, and bypass valves should follow the piping class and ratings. The dimensions, however, should be in accordance with the recognized standard such as ANSI/ISA RP-75.06 (Fig. 16.13).

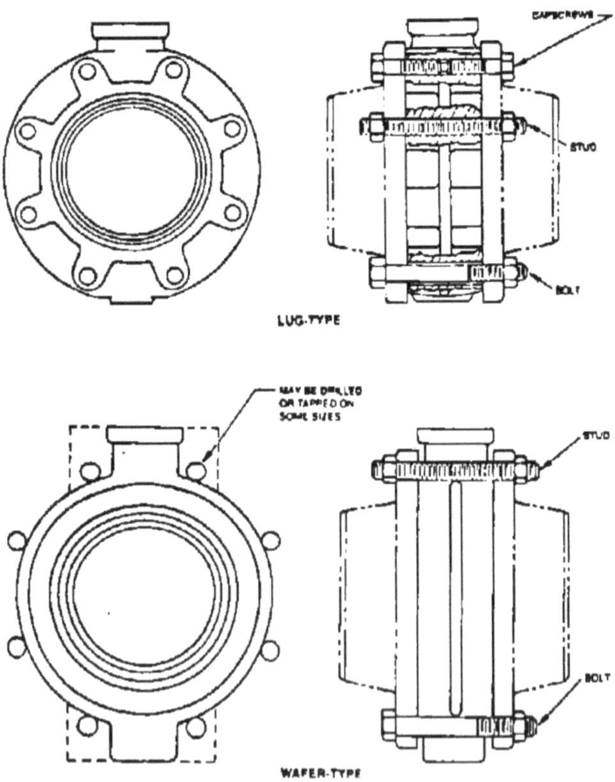

FIGURE 16.13 Illustrations of end types and external bolting options.

16.5.2 Butterfly Body Valves

Lug type and wafer type butterfly valves should have body pressure–temperature ratings for the selected American Society of Testing and Materials (ASTM) material specification in accordance with the applicable ANSI B-16 standard.

The wafer type butterfly valves, other than lugged type, should be provided with or without holes for the passage of bolts securing the connecting flanges dependent upon valve design.

Lugged type, wafer valves should be supplied with threaded or drilled holes the lugs to the size, nominal pressure rating, and type of connecting flange.

The end flanges of double-flanged steel butterfly control valves should be cast or forged integral with the body.

For other types of valves such as eccentric rotating plug valves or butterfly valves flanges should be a wafer type, i.e., suitable for installation between flanges.

Butterfly valves should be one of the following types shown in Figs. 16.14–16.17, with metal or resilient seating or linings:

1. *Double flanged*: A valve having flanged ends for connection to pipe flanges by individual bolting.
2. *Wafer*: A valve primarily intended for clamping between pipe flanges using through bolting:
   a. single flange;
   b. flangeless;
   c. U-section.

Note:

This type of valve when supplied with threaded holes may be suitable for terminal connections.

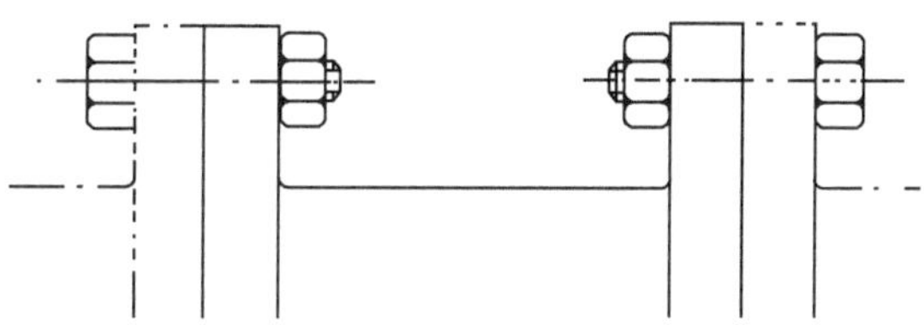

FIGURE 16.14 Butterfly valve: double-flanged type.

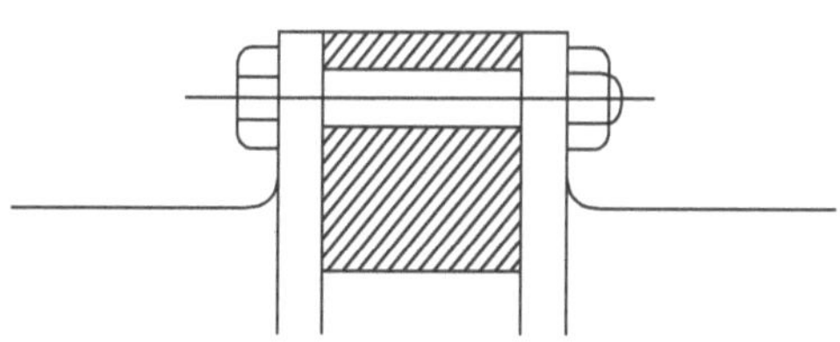

FIGURE 16.15 Butterfly valve: single-flange wafer type.

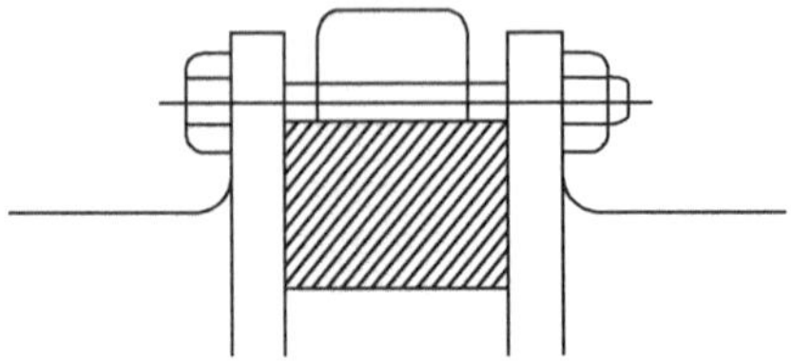

FIGURE 16.16 Butterfly valve: flangeless wafer type.

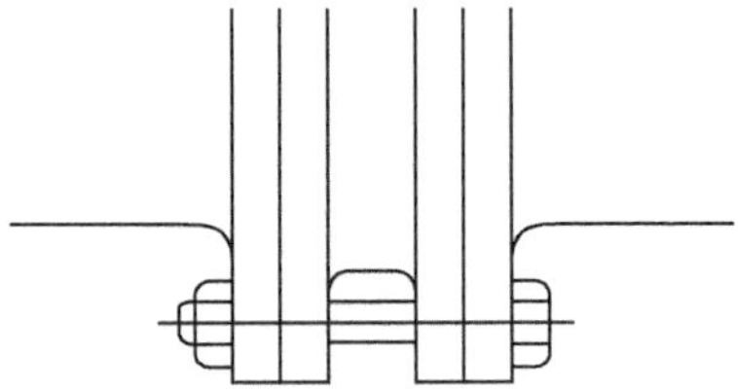

FIGURE 16.17 Butterfly valve: U-section wafer type.

Note:

This type of valve when supplied with threaded lugs may be suitable for terminal connections.

Notes:

1. This type of valve may be suitable for the individual bolting of each flange to the pipework, but this cannot be assumed.
2. This type of valve may be suitable for terminal connections.

16.5.3 Face-to-Face Dimensions

Face-to-face dimensions of flanged-bodied globe-style control valves should comply with the recognized standard such as ANSI/ISA-S 75.03 (Table 16.2).

Face-to-face dimensions of butterfly valves should be in accordance with the recognized standard such as BS-5155 (Table 16.3).

Tolerances on face-to-face dimensions for butterfly valves should be in accordance with BS-5155 standard (Table 16.4).

Flange dimensions of butterfly valves Class 125 should be in accordance with the recognized standard such as BS-5155 (Table 16.5, if a flange type is specified).

TABLE 16.2 Face-to-Face Dimensions for Flanged Globe-Style Control Valves

Nominal Valve Size	(ANSI Classes 150)		(ANSI Classes 300)		(ANSI Class 600)		Tolerance	
	Dimension "A"		Dimension "A"		Dimension "A"			
Inches	mm	Inches	mm	Inches	mm	Inches	mm	Inches
½	184	7.25	190	7.50	203	8.00	± 1.6	± 0.062
¾	184	7.25	194	7.62	206	8.12	± 1.6	± 0.062
1	184	7.25	197	7.75	210	8.25	± 1.6	± 0.062
1½	222	8.75	235	9.25	251	9.88	± 1.6	± 0.062
2	254	10.00	267	10.50	286	11.25	± 1.6	± 0.062
2½	276	10.88	292	11.50	311	12.25	± 1.6	± 0.062
3	298	11.75	318	12.50	337	13.25	± 1.6	± 0.062
4	352	13.88	368	14.50	394	15.50	± 1.6	± 0.062
6	451	17.75	473	18.62	508	20.00	± 1.6	± 0.062
8	543	21.38	568	22.38	610	24.00	± 1.6	± 0.062
10	673	26.50	708	27.88	752	29.62	± 1.6	± 0.062
12	737	29.00	775	30.50	819	32.25	± 3.2	± 0.125
14	889	35.00	927	36.50	972	38.25	± 3.2	± 0.125
16	1016	40.00	1057	41.62	1108	43.62	± 3.2	± 0.125

TABLE 16.3 Face-to-Face Dimensions of Butterfly Valves

1	2	3	4	5	6	7
Nominal Size (in.)	Double Flanged Short	Double Flanged Long	Wafer Short	Wafer Medium	Wafer Long	Wafer
	Face-To-Face Dimension For Nominal Pressures Not Exceeding		Face-To-Face Dimensions For Nominal Pressures Not Exceeding			Class 300
	Class 150	Class 300	Class 150	Class 150	Class 150	
	mm	mm	mm	mm	mm	mm
1½	106	140	33			
2	108	150	43			
2½	112	170	46			
3	114	180	46		64	49
4	127	190	52		64	56
5	140	200	56		70	64
6	140	210	56		76	70
8	152	230	60		89	71
10	165	250	68		114	76
12	178	270	78		114	83
14	190	290		92	127	127
16	216	310		102	140	140
18	222	330		114	152	160
20	229	350		127	152	170
24	267	390		154	178	200
28	292	430			229	
32	318	470			241	
36	330	510			241	
40	410	550			300	
48	470	630			350	
56	530	710			390	
64	600	790			440	
72	670	870			490	
80	760	950			540	

Note: Wafer type valves may not be available in all combinations of materials and face-to-face dimensions.

TABLE 16.4 Tolerances on Face-to-Face Dimensions of Butterfly Valves

Face-to-Face Dimension		Tolerance
	mm	mm
Up to and including	200	± 1
Above 200 up to and including	400	± 2
Above 400 up to and including	600	± 3
Above 600 up to and including	800	± 4
Above 800		± 5

TABLE 16.5 Dimensions of Class 125 (Cast Iron) Flanges of Butterfly Valves

1	2	3	4	5	6	7	
Nominal Size of Valve		Diameter of Flange	Minimum Thickness of Flange	Diameter of Bolt Circle	Number of Bolts	Diameter of Bolt Holes	
Inches		mm	mm	mm		mm	Inches
1½		127	14.3	98.4	4	15.9	5/8
2		152	15.9	120.6	4	19.0	¾
(2½)*		178	17.5	139.7	4	19.0	¾
3		190	19.0	152.4	4	19.0	¾
4		229	23.8	190.5	8	19.0	¾
5		254	23.8	215.9	8	22.2	7/8
6		279	25.4	241.3	8	22.2	7/8
8		343	28.6	298.4	8	22.2	7/8
10		406	30.2	362.0	12	25.4	1
12		483	31.8	431.8	12	25.4	1
14		533	34.9	476.2	12	28.6	1⅛
16		597	36.5	539.8	16	28.6	1⅛
18		635	39.7	577.8	16	31.8	1¼
20		698	42.9	635.0	20	31.8	1¼
24		813	47.6	749.3	20	34.9	1⅜

Note:
This size has been retained only for the purpose of replacing existing valves. Its use for new construction on piping systems using BS 1560: Part 2 flanges, should be avoided.
Class 125 (cast iron) valves should be used on special applications such as slurry and utility services.

Flange dimensions of butterfly body control valves should be in accordance with the recognized standard such as of BS-1560: Part 2.

For pressure–temperature rating of butterfly control valves reference should be made to relevant standards.

16.6 CONTROL VALVE SIZING AND CHARACTERISTICS

16.6.1 Control Valve Sizing

Control valve sizing is necessary to optimize operation, provide sufficient rangeability, and minimize cost. The key to correct control valve sizing is the proper determination of the required valve capacity coefficient (C_V).

By definition (C_V) is the number of gallon per minute of water at 15°C that will pass through a given flow restriction with a pressure drop of 1 psi. For example, a control valve that has a maximum flow coefficient (C_V) of 12 has an effective port area in the full open position such that it passes 12 gallons per minute of water with a pressure drop of 1 psi.

Determination of required (C_V) for a given application may be accomplished through formula or slide rule methods. For detailed information regarding control valve sizing equations refer to relevant standards.

Working equations are derived from the fundamental hydraulic equation and is converted to customary engineering units, then, equation becomes,

$$Q = C_V \frac{\Delta P}{G}$$

where C_V is the experimentally determined coefficient, ΔP is the differential pressure across the valve, and G is the specific gravity of the liquid, and Q is the quantity of liquid.

The equations of this standard are based on the use of experimentally determined capacity factors obtained by testing control valve specimen according to the procedures of ANSI/ISA S75.02 "Control Valve Capacity Test Procedure."

The equations are used to predict the flow rate of fluid through a valve when all factors, including those related to the fluid and its flowing condition, are known. When the equations are used to select a valve size it is often necessary to use capacity factors associated with the fully open or rated condition to predict an approximate required valve flow coefficient (C_V).

In using these methods, full knowledge of actual flowing conditions is essential. The primary factors that should be known for accurate sizing are:

1. The upstream and downstream pressures at the flow rates being considered.
2. The generic identity of process fluid.
3. The temperature of the fluid.
4. The fluid phase (gas, liquid, slurry, and so forth).
5. The density of the fluid (specific gravity, specific weight, molecular weight).

6. The viscosity (liquids).
7. The vapor pressure (liquids).

Valve sizing should be based on a maximum sizing capacity of 1.3× the normal maximum flow or 1.1× the absolute maximum flow, whichever is greater. The sizing pressure drop (ΔP sizing) should be sufficient to obtain good regulation at the normal maximum case as well as maintain maximum quantity and the normal minimum quantity within the rangeability of the selected valve.

If in a primary design stage the maximum flow is not available, then valves should be selected to have twice the C_V required for normal design flow at specified conditions.

Control valves with inherent high-pressure recovery characteristics can cause cavitation when fluid pressure and temperature conditions would indicate. Valves with low-pressure recovery, special trim should be used to minimize or prevent cavitation.

Flashing, like cavitation, can cause physical damage and decreased valve capacity. Manufacturers should be consulted for recommendations.

The pressure drop across the control valve at maximum process flow should be at least 20% of the pressure drop across the control valve at normal flow.

1. The control valve should be sized such that the C_V value of the control valve for maximum process flow with the pressure drop across the control valve at maximum process flow is approximately 80% of the maximum C_V value for that control valve. Furthermore, the control valve should never have less than 25% lift for minimum process flow at the specified pressure drop.

If neither a maximum nor a minimum process flow is stated, these flows should be assumed to be 120% and 80%, respectively, of the normal process flow.

Sizing calculations should be checked for at both extremes to assure controllability over the entire range of the flow rates and pressure drop.

Butterfly valves should be sized for maximum angle operating of 60 degrees. Proposals to use angles greater than 60 degrees should be submitted to the purchaser for approval.

Shafts of rotary actuated valves should be sized for pressure drop equal to maximum upstream pressure.

Control valve body size normally should not be less than half that of normal line size.

For more details, reference should be made to ISA Handbook of control valves (Chapter 6) or equivalent methods.

Cavitation effects on flow in control valves, due to the reduction of the calculated liquid flow coefficient of a valve C_V, will be observed if pressure drop is increased beyond a certain limit at a constant upstream pressure.

The cavitation index, K_c, defines the first point where reduction of C_V can be observed from experimental data.

16.6.2 Control Valve Characteristics

Control valve flow characteristics are determined principally by the design of the valve trim. The three inherent characteristics available are quick opening, linear, and equal percentage. These are shown in Fig. 16.18. A modified percentage characteristic generally falling between the linear and equal percentage characteristics is also available.

The three inherent characteristics can be described as follows:

16.6.2.1 Quick Opening

As the name implies, this characteristic provides a large opening as the plug is first lifted from the seat, with lesser flow increase as the plug opens further. This type is most commonly used where the valve will be either open or closed with no throttling of flow required.

16.6.2.2 Linear

Linear trim provides equal increases in C_V for equal increases in stem travel. Thus the C_V increase is linear with plug position throughout its travel.

16.6.2.3 Equal Percentage

Equal percentage trim provides equal percentage increases in C_V for equal increments of stem travel. This is accomplished by providing a very small opening for plug travel near the seat and very large increases toward the more open position. As a result, a wide rangeability of C_V is achieved.

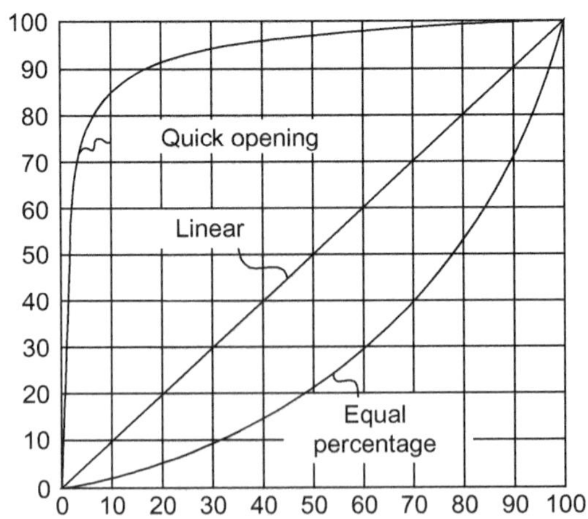

FIGURE 16.18 Percent of valve opening representative of inherent flow characteristic curves.

The pressure difference across the valve often varies with flow. This results in an "installed characteristic," which will differ from the inherent characteristic.

The characteristic of the inner valve should normally be an equal percentage except where system characteristics indicate otherwise. Linear and quick opening characteristics should be used where required. In general linear trim should be used only for split-range service or where the control valve pressure drop remains constant over the range of 10–100% of flow capacity.

Shutoff valves should normally have quick closing or equal percentage characteristic, but another characteristic (such as modified equal percentage) may be required for special cases, e.g., to avoid or reduce the consequence of hydraulic shock.

Characteristics of valves may change due to particular requirements.

Butterfly and angle valves and characterized ball valves ("V-Ball") should normally have equal percentage characteristics.

Three-way valves in control services should normally have linear characteristics.

Valves with shutoff function should be single seated.

The pressure drop across the control valve at maximum flow should be at least 25% of the pressure drop across the control valve at normal flow.

Three-way valves should be capable of operating against the maximum differential pressure that can exist across a single port. Each three-way valve should be specified as flow-mixing or flow splitting in accordance with the intended application.

The action of the valves on failure of the operating medium should be determined by process requirements with regard to safe operation and emergency shutdown.

Extension bonnets should be provided on services above 232°C and below −6.7°C or in accordance with the manufacturer's recommendation.

Pressure-balanced valves of the double diaphragm type should be considered for use on fuel gas to heaters in temperature control systems. When a single diaphragm type is used, a pneumatic ratio relay should be installed in the control air line with the input–output ratio as required.

Control valves installed in pipe lines should normally be at least one pipe size smaller than the computed line size. This is to allow margin for future expansion and a better controllability of the process.

Where it is necessary to reduce from a line size to a control valve size, swaged reducers should be used between the block valves and the control valve. Sufficient spacing between block valves should allow for installation of larger size control valves.

Oversized bodies with reduced trims should be used for valves in severe flashing or cavitating service. Angle type or multiple seat type valves may be considered for this service.

Valves used in pairs, as three-way valves, including rotary actuated valves, such as ball or butterfly types, should have linear characteristics. Characterized positioners may be used to meet this requirement. In this case calibration for the required characterization must be done by the valve manufacturer.

Gas compressor recycle control valves should have linear characteristics.

Valves in pressure-reducing services, where the pressure drop is constant, should have linear characteristics.

16.7 CONTROL VALVE MANIFOLD DESIGN

Control valves and bypass valves are mostly manifolded in piping systems to allow manual manipulation of the flow through the systems in those situations when the control valve is not in service. For more information, reference should be made to relevant standards.

For application, information and guidance reference should be made to ISA Handbook of Control Valves, API RP 550 or other relevant publications.

Dimensions for flanged globe control valves should be used as per ANSI/ISA-S75.03 standard.

Control valve body nominal size covered by the designs are 1, 1½, 2, 3, 4, and 6 in.

Reference should be made to ISA RP 75.06 "Control Valve Manifold Design" for additional information and dimensions for all ANSI classes.

16.8 CONTROL VALVE BLOCK AND BYPASS VALVES

Where significant future expansion is not anticipated, a less flexible but more economical approach that gives a minimum acceptable design is to make the block valves one size larger than the control valve (but not larger than line size).

The bypass line and valve should normally have a capacity at least equal to the calculated or required C_V of the control valve, but not greater than twice the selected C_V of the control valve.

Bypass valves in sizes of 4 in. or fewer are usually globe valves that allow throttling. For larger sizes, because of cost, gate valves are normally used.

Where block valves are provided, vent valves should be fitted between them so that pressure may be relieved and the control valve drained when the block valves are closed. Suitable drain lines should be provided where necessary.

Vent and drain connections should not be less than ¾ in. nominal bore.

A bypass connection and valve should be installed around each control valve unless other means are available for manual control when the control valve is out of service.

Consideration should be given to the elimination of bypass and block valves around control valves sizes 2 in. and over, but this should be by agreement with the user.

Block and bypass valve assemblies should be avoided in the following instances:

On hydrogen service.
Around three-way valves.
Around self-acting steam-pressure-reducing valves.
Around control valves forming part of a protective system.

Block and bypass valve assemblies should be provided in the following instances:

Where a valve controls a service common to a number of plants.
Where valves are in continuous operation and there is not sufficient assurance of reliability over the anticipated period between plant overhauls, e.g., on erosive or corrosive service or where the temperature is below 0°C. or above 180°C. The cost of a failure should also be taken into account.
Where failure of the control valve would necessitate continuous operator attention, e.g., on the fuel control to heaters.

Where bypass valves are not provided, a permanent side-mounted hand wheel should be fitted to the control valve. Where the cost of the hand wheel is greater than the cost of block and bypass valves, the latter should be provided except on hydrogen service and protective service.

Where block and bypass valves are not fitted initially adequate space should be allowed for possible future installation.

When control valves are placed in prestressed lines they should be in a bypass assembly to the main pipeline.

16.9 CONTROL VALVE PACKING AND SEALING

All valves should be drilled and tapped to accept a gland lubricator except when otherwise specified in data sheet.

The bottom flange or the bottom of the body of a control valve should not be drilled and tapped.

For special duties as specified in the data sheet, e.g., toxic control valve stems, they should be bellows sealed, with an independent gland seal, and the enclosed space should be monitored for bellows leakage.

When sealing by bellows is not possible, a purge should be used, monitored for flow failure. Bellows seals may also be required to prevent leakage of penetrating liquids.

On clean fluids, interlocking self-lubricating gland packings with spring followers may be used.

On higher temperature duties or where carbon or other deposits may settle on the stem, special packing should be used.

If dangerous fluids are encountered, horizontal lines should be fitted with suitable drains on the under side. This does not replace the vent.

Packing materials for butterfly valves should be suitable for the specified service conditions.

Where the controlled liquid contains particles or materials which would damage the valve guide, stem or packing, a purge system should be considered.

16.10 CONTROL VALVE NOISE AND VIBRATION CAUSED BY SONIC FLOW

Sonic flow occurs when the velocity of the fluid reaches the speed of sound in that medium.

At subsonic velocities the flow is characterized by turbulent mixing, and this is responsible for the noise produced. This noise is best described as a "hiss" for small jets or as a roar for larger jets but has no discrete dominating frequency. Its spectrum is continuous with a single, rather flat maximum.

As the pressure ratio increases past the critical ratio and the fluid reaches its sonic velocity, the sound emanating undergoes a fundamental change, while the roaring noise due to the turbulent mixing is still present, it may be almost completely dominated by a very powerful "whistle" or "serooch" of a completely different character. This noise is rather harsh and of a confused nature, becoming much more like a pure note.

At sonic flow, vibration can be caused in various frequency bands due to vertical/horizontal movement of control valve components (20–80 kHz), impingement of fluid on control valve internals at high velocities (400–1600 Hz), aerodynamic noise from shock waves by the sonic velocities (1200–4000 Hz), internal components vibrating at their natural frequencies (3000–6000 Hz), and high-pressure drop gas services (above 8000 Hz). For guidance in specifications the permissible noise exposure, is noted below:

Duration Hours per Day	Sound Level dB Slow Response
8	90
6	92
4	95
3	97
2	100
1½	102
1	105
½	110
¼ or less	115

16.10.1 Cause of Noise and Vibration

High-pressure drop gives rise to sonic flow. Sonic flow generates shock waves that in turn produce high frequency noise and vibration (1.2–4.8 kHz). The noise has a characteristic whistle or scream at its peak frequency, is directional in nature when discharged into the atmosphere, and is even more dependent upon fluid jet pressure than the turbulent mixing noise of subsonic flow.

The most dangerous vibration occurs in frequency band 3–7 kHz and is the result of resonance by the valve parts. This can lead to failure due to metal fatigue.

The most effective method of solution is to remove the cause, that is the high-pressure drop.

The absence of sonic flow means an absence of its effects of noise and vibration. In cases where the pressure drop must remain high, a special type of "low-noise" control valve is recommended.

If the calculated sound pressure level (SPL) value of a reducing valve under maximum load exceeds the stated limit by only 5–10 dB, then one of the following simple cures must be considered:-

1. Increase the pipe wall thickness downstream (doubling the wall thickness will decrease the SPL by 5 dB).
2. Use acoustical isolation downstream. This will reduce SPL by 0.2–0.5 dB/mm of insulation, depending on the density of the insulating material.

If the valve noise is 10 dB above the selected limit, then one must choose a different approach such as the use of downstream, in-line silencers. The silencers generally attenuate between 10 and 20 dB depending on the frequency range. The silencers must be installed directly adjacent to the valve body with the valve outlet velocity below sonic (i.e., a 1/3 match); otherwise, the silencer will act as a pressure-reducing device for which it is not suitable.

The use of expansion plates downstream of valves is recommended. The primary function of these plates is not to attenuate the valve noise, but to absorb some of the pressure reduction over the whole system. In this way the pressure across the control valve can be kept below critical. In a typical installation the expansion plate downstream flow area must be increased to compensate for the changes in the density due to pressure drop.

16.11 CONTROL VALVE ACTUATORS

Actuator: Any device designed for attachment to a general purpose industrial valve in order to provide the operation of valve.

The device is designed to operate by using motive energy that can be electrical, pneumatic, hydraulic, etc., or a combination of these. The movement is limited by travel, torque, or thrust.

Multiturn Actuator: An actuator that transmits to the valve a torque for at least one revolution and should be capable to withstand the thrust.

Torque: A turning moment transmitted through the mounting flanges and couplings.

Thrust: An axial force transmitted through the mounting flanges and couplings.

Control valve actuators should be selected, so that on failure of the operating medium, the valve will automatically take a position (open, closed, or locked) that will result in the safest configuration for the operating unit.

Actuator Travel Limit: In the opening direction, a stop should be engaged before the valve plug reaches its travel limit. The stop should have sufficient contact area to absorb any force transmitted to it. In the closing direction, the valve plug should seat before the actuator reaches its travel limit.

The valve connection to the actuator should be adjustable, with positive locking of the adjustment.

Rotary-actuated valves (such as butterfly, ball) should have shaft keyways that allow the action of the valve to be changed.

For Rotary-actuated valves in cryogenic services, the shaft thrust bearings should be provided.

Standard spring range should be 0.2–1 barg.

Handwheels, when specified, should be mounted and designed to operate in the following manner:

1. For globe valves, hand wheels should be mounted on the yoke, arranged so that the valve stem can be jacked in either direction.
2. Neutral position should be clearly indicated.
3. Handwheel operation should not add friction to the actuator.
4. Clutch/linkage mechanisms for handwheels on rotary valves should be designed such that control of valve position is not lost when engaging the handwheel.

Pneumatic actuators should have as low as an operating pressure as practicable in order to minimize the need for spare capacity in the instrument air system. In no case may the operating pressure exceed 4 bar.

For shutoff valves, the actuators should be capable of opening the valve against the full upstream pressure, with the downstream pressure assumed to be atmospheric.

For butterfly valves, the actuators should have sufficient force for coping with all operating conditions from the fully closed position to the fully open position, and for coping with all pressure drop and torque requirements.

The stroking time of control valves should be evaluated on the basis of the process control requirements. For critical analogue control systems, such as surge control of compressors, the stroking time should be less than 5 seconds. For other analogue control systems, longer stroking times may be acceptable.

For valves having spring-to-close action, the stroking time is determined by the spring force, the diameter of air exhaust ports in the actuator and solenoid valve, and the mechanical inertia of moving parts.

16.12 TYPES OF CONTROL VALVE ACTUATORS

There are many types of actuators for stroking control valves. These actuators may be classified into four general types:

1. Pneumatically operated diaphragm actuators.
2. Piston (cylinder) actuators.
3. Electrohydraulic actuators.
4. Electromechanical (motor) operated actuators.

16.12.1 Pneumatically Operated Diaphragm Actuators

Control valves should normally be operated by pneumatic diaphragm actuators. The actuator should normally be operated between 0.2–1.0 bar, and 0.4–2.0 bar may be used as specified for full stroke. Where the control signal is electric, the electropneumatic convertor should be used.

There are two types of diaphragm actuators:

- Direct acting
- Reverse acting

The actuator should be designed to provide dependable on–off or throttling operation of automatic control valve.

Reverse-acting diaphragm actuators using seals or glands are permitted only for those applications where the direct acting type of actuator is unsuitable (typical spring diaphragm actuators are shown in Figs. 16.19 and 16.20).

The air pressure required for stroking the valve may vary from the operating spring range. The maximum air pressure allowable on spring diaphragm actuators should not exceed 4 bar.

16.12.2 Pneumatic and Hydraulic Cylinder or Piston Actuators

Cylinder actuators should be used where a long stroke and high force is required, such as for dampers and louvers in large ducting for combustion air or flue gas services.

Piston actuators may be pneumatically or hydraulically operated.

The cylinders should be connected directly to the valve as an integral part. Actuator may also be purchased separately and mounted at site.

For throttling applications the cylinder actuator should be provided with a positioner mechanism and, where necessary, with a position transmitter. A pair of oil filter with isolating valves should be installed as close as possible to the positioner.

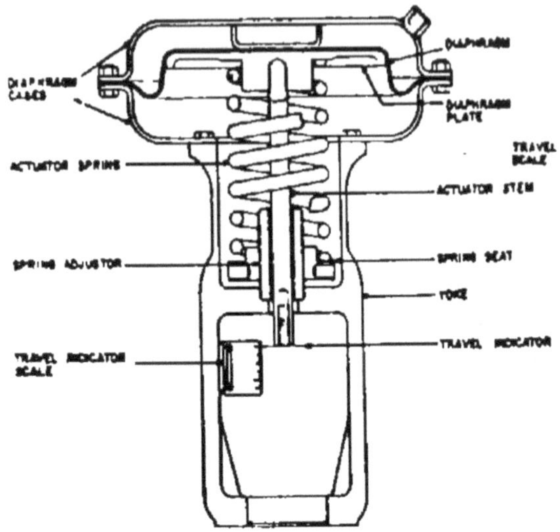

FIGURE 16.19 Direct-acting diaphragm actuator.

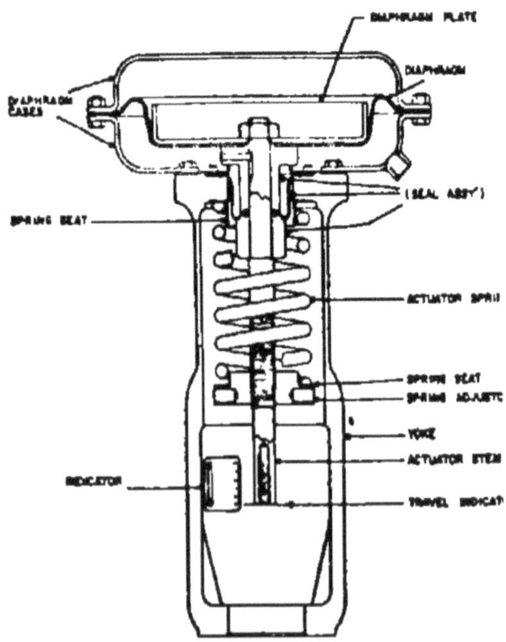

FIGURE 16.20 Reverse-acting diaphragm actuator.

Cylinder actuators on valves with provisions for manual (local) control should be provided with external bypass valves if these are not integral with the actuator. Four-way valves are often require in the piping of the cylinder to allow local operation.

For hydraulic cylinders, the following points should be considered:

1. If the hydraulic manifold is rigidly piped, it should be connected to the hydraulic fluid supply and return headers by a flexible metalic hose.
2. To assure a continuous supply of hydraulic to actuators, it is advisable to provide both an oil filter or strainer and a spare suitably valved and piped so that either unit may be removed and cleaned without shutting off the supply.
3. Vent valves should be provided at high points in the hydraulic fluid system.
4. Depending upon whether the valve served by the actuator will move if the hydraulic oil pressure is lost, it may be necessary to use automatic fluid trapping valves that lock the hydraulic fluid in the cylinders upon failure of the hydraulic system.

16.12.3 Electrohydraulic Actuators

Electrohydraulic actuators may be used to operate a large rotary and sliding stem valves. Electrical signal of 4–20 mA or 10–50 mA DC may be used.

Electrohydraulic actuators should be used for an electronic control loop, where fast stroking speeds, high thrust, and long strokes are required.

Electrohydraulic actuators may be used at locations where a suitable air supply is not available.

16.12.4 Electromechanical Actuators

The electromechanical valve actuator has essentially the same advantages as the electrohydraulic actuator with respect to field use. It is capable of being used over long distances with only inconsequential signal transmission delays. It is also immune from the pneumatic system problem of freeze-up in extremely cold ambient conditions. Electromechanical actuators are still, however, generally more expensive, although more efficient, than electrohydraulic units.

An Electromechanical valve actuator is composed of a motorized gear train and screw assembly which drives the valve stem or rotary shaft valves. A typical example is shown schematically in Fig. 16.21. The varying input signal, whose magnitude corresponds to the required position of the inner valve stem, is fed into the positioner (usually a differential amplifier) and produces a voltage to actuate the motorized gear train and screw.

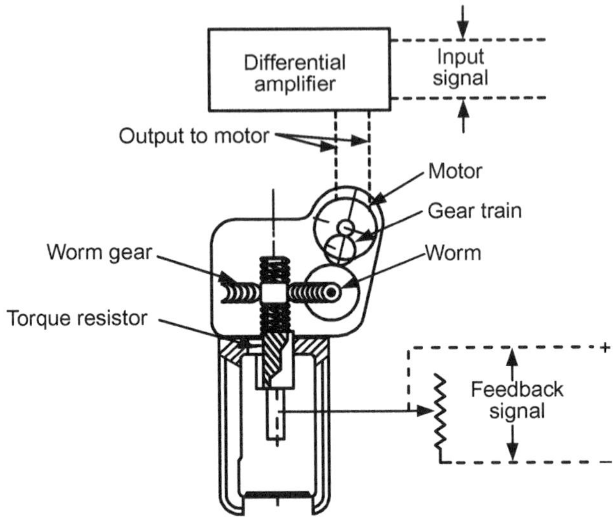

FIGURE 16.21 An electromechanical valve actuator.

The resultant movement of the stem and the take-off attached to it and to a potentiometer or linear differential transformer, produces a voltage that increases with stroke, and is sent into the positioner. When the input signal voltages and feed back voltages are equal, the output voltage to the motor goes to zero and the motor stops with the valve stem at the required position. Conversely, when the voltages are not equal, the motor is run in the direction to make them equal.

16.12.5 Motor-Operated Actuators

The actuator should consist of a motor-driven, reduction-gearing, thrust-bearing (where applicable) handwheel and local-position indicator, together with torque and limit switches, space heaters terminals, control power transformer, and integral motor starter controls, all furnished as a self-contained, totally enclosed unit.

Motor starter for remote mounting should be specified.

The motor should be sized for torque requirements according to the valve size, operating differential pressure and temperature, and speed operation.

The actuator terminal box should be double-sealed in such a way that when the actuator terminal box cover is removed for the connection of incoming cables, the remaining electrical components are still protected by the watertight enclosure.

The valve and actuator mounting bracket must be capable of withstanding the stall torque of the actuator, with torque and limit switches disconnected.

The actuator should provide both torque and position limitation in both directions. An automatic override should be provided to prevent the torque switches from tripping the motor on initial valve unseating.

The actuator should be capable of operating at any mounting angle.

The actuator should be designed so that there will be no release of high-stem thrust of torque-reaction spring forces when covers are removed from the actuator gear box, even when the valve is under full-line pressure conditions.

Failure of the motor, power, or motor gearing should not prevent manual operation of the valve through the use of the handwheel. When the motor drive is declutched, the handwheel drive should be engaged safely, with the motor running or stopped.

A means of locking the actuator in either the manual or motor condition should be provided. If not locked in either position, starting of the motor should automatically restore the power drive.

The actuator should be capable of functioning within the ambient temperature range as specified in the job specification, but in any case not less than −15°C to +80°C.

Clockwise rotation of the actuator handwheel should close the valve. The handwheel drive must be mechanically independent of the motor drive, and gearing should be such as to permit emergency manual operation in a reasonable time.

The motor should be of sufficient size to open and close the valve against maximum cold working pressure when voltage to motor terminals is within 10% of rated voltage. Motor nameplate rating should be 380 V, 3 phase, 50 Hz, unless otherwise specified.

Local push buttons should be provided for "Open-Stop-Close" control of valve, with a lockable selector switch providing positions "local control only," "local and remote," "remote plus local stop," and "off" position.

Provisions should have made for the addition of extra sets of limit switches in each actuator. Each set should be adjustable to any point of valve position. Each unit should be provided with auxiliary contacts, rated 5 A at 110 V AC for remote position indication. They should be adjustable continuously in the range of open-to-close positions.

Actuators should have an output speed of 0.4 rev/s unless otherwise specified on the data sheets (e.g., for high-speed emergency isolation). No valve should require more than 2 minutes for full operation, i.e., from closed to fully open or open to fully closed.

In the case of high-speed actuators a pulse timer should be included. This will give variable slow-down times for fast closing and fast opening, normally set to operate over the last 25% of the valve closure.

Provisions should also be made for remote operation through interposing relays supplied with DC power of 24 V unless otherwise specified.

Mechanical dial indication of the valve position should be incorporated in the actuator. Indication should be continuous if the valve is specified to be in a regulating device.

Position limit switches should be provided at each end of travel for remote indication and sequencing.

Torque and limit switches should be easily adjustable without special tools or removal of switch assembly from the actuator. Repeatability of switch actuation should be ±5% of the set point.

Control power transformers when used should have a fuse protection on the secondary. Fuses should be readily accessible for replacement or deactivation at the terminal board.

An electrically and mechanically interlocked motor starter should be provided in the actuator housing, unless the motor starter is specified for remote mounting.

All electrical components should be prewired by the actuator vendor to a legibly marked terminal strip. Power and control wiring should be segregated and insulated from each other. All wiring should be identified by the vendor, and access for maintenance provided.

Motor overload protection should be provided. One or more winding temperature detectors embedded in the motor winding, or three thermal overload relays in the motor controller, are acceptable. Either must be capable of being deactivated at the terminal board.

All motor operators should be explosion proof approved in accordance with the requirements of the National Electrical Code latest edition (NFPA No. 70) for use in the hazardous area classification, or unless otherwise specified in individual data sheets.

All electrical equipment and motors should be totally enclosed for outdoor services and should be water tight to IEC 34.5 and IEC 144-IP 67.

All electrical equipment used in hazardous areas should also meet the electrical area classification requirements as per relevant standards.

The vendor should supply the valve actuator compatible with the valve. All information required for sizing the actuator should be obtained from the purchaser and/or valve supplier.

Torque requirements of valve and torque characteristics of the actuator should be supplied to the purchaser for approval.

The actuating unit should include a three-phase electric motor, reduction-gearing geared-limit switches, torque switches, space heaters, and terminals, together with a handwheel for manual operation with declutching lever and valve-position indicator.

All gearing should be totally enclosed and continuously lubricated. All shafts should be mounted on ball or roller bearings. Limit switch drive should be stainless steel or bronze.

Power terminals should be of stud type, segregated by an insulating cover. Four conduit entrance taps should be provided as a minimum. Each tap will be provided with standard electrical connections in metric type such as M20.

The actuator terminal box should be double sealed in such a way that when the terminal box cover is removed for the connection of incoming cables the remaining electrical components are still protected by the watertight enclosure.

The actuator should have an integral motor starter, local controls, and lamp indication.

The starter should include a mechanically and electrically interlocked reversing contactor, with a control transformer having a grounded screen between primary and secondary windings. The common point of contactor coils and secondary winding should also be grounded, so that any ground fault will cause contactors to drop out. Terminals for remote controls should be provided.

The starter components should be readily accessible for inspection without disconnecting external cables. Internal wiring should be number-identified at both ends.

Lamp indication of "close" (green), intermediate (white), and "open" (Red) positions should be provided.

Open and close torque and/or position limit switches, plus two auxiliary limit switches at each end of travel should be provided. Switch ratings should be 5 A at 115 V AC or as specified in data sheet.

Internal control wiring of 5 A tropical-grade PVC cable should be provided, terminating in a separately sealed housing with stud terminals. The three phase leads of the motor should be brought to separately stainless studs.

The motor should be prelubricated and all bearings should be of antifriction type.

The motor should be sized for the torque requirements according to the valve size, operating pressure and temperature, and speed operation.

The motor should have class "B" insulation, short-time rated, with burn-out protection provided.

The motor should be of sufficient size to open and close the valve against maximum differential pressure when voltage to motor terminals is within 10% of rated voltage. Motor name plate rating should be 380 V, 3 phase, 50 Hz.

16.13 ACTUATOR CONSTRUCTION MATERIALS

Materials of construction should be manufacturer's standard for the specified environmental exposure.

The diaphragm housing material should be steel, unless otherwise specified. For piston type actuators, aluminum housing is acceptable except for valve on depressurizing or emergency shutoff services. In special cases, such as larger sizes of butterfly valves, consideration may be given to (long-stroke) cylinder actuators.

The enclosure housing the electrical components of a valve should be made of iron, steel, brass, bronze, aluminum, or an alloy containing not less than 85% aluminum. A metal such as zinc or magnesium or other alloys should not be used.

Copper should not be used for an enclosure for use in Class I group A locations. A copper alloy should not be used for an enclosure unless it is coated with tin nickel or another acceptable coating, or unless the copper content of the alloy is not more than 30%.

Construction material of actuators may be considered and selected according to the requirements. The following materials should be considered for different parts of actuators:

Diaphragm casing	: Steel, cast iron, or cast aluminum
Diaphragm	: Nitrile on nylon or nitrile on polyester
Diaphragm plate	: Cast iron, cast aluminum, or steel
Actuator spring	: Alloy steel
Spring adjuster	: Steel
Spring seat	: Steel or cast iron
Actuator stem	: Steel
Travel indicator	: Stainless steel
O-rings	: Nitrile
Seat bushing	: Brass
Stem connector	: Steel zinc plated
Yoke	: Iron or steel

16.14 SELF-ACTUATED REGULATORS

The self-actuated regulator is a variation of the diaphragm actuator and normally uses the process fluid as the operating medium. For pressure applications, some self-actuated regulators use bellows instead of diaphragms for the actuator. For temperature applications, bellows with a filled system and bulb should be used instead of diaphragms. Piping arrangements are shown in Fig. 16.22.

A regulator is a very simple control device in which all of the energy to operate it is derived from the controlled system. Consider using a regulator first whenever you have a requirement for:

- pressure control;
- level control;
- flow control.

All regulators, whether they are being used for pressure, level of flow control, fit into one of the following two basic categories:

1. Direct-operated
2. Pilot-operated

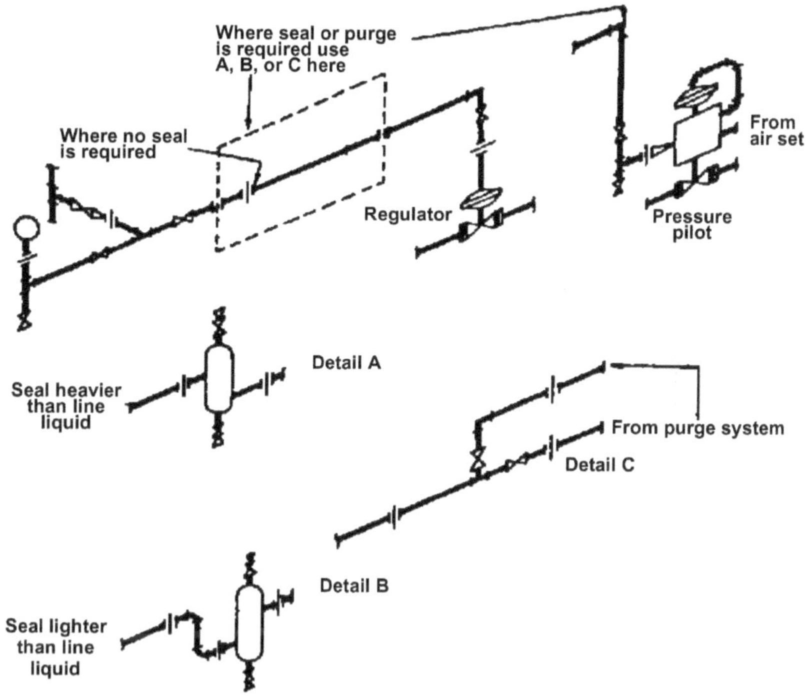

FIGURE 16.22 Piping at regulator valve or pressure pilots.

Characteristically, direct-operated regulators are adequate for narrow-range control, and where the allowable change in outlet pressure can be 10–20% of the outlet pressure setting.

Pilot-operated regulators are preferred for broad-range control, or where the allowable change in outlet pressure is required to be less than 10% of the outlet pressure setting. They are also commonly used when remote set point adjustment is required for a regulator application.

The globe-style pilot operated backpressure regulators or relief valves are used in gas or liquid service to maintain pressure on oil and gas separators and in pressure relief application in gas distribution systems.

With the pilot, pressure can be controlled, and set pressure is varied to individual requirements by the adjusting screw on the pilot. Pilot exhaust can be piped into the downstream line or vented to the atmosphere on gas service, but must always be piped downstream on liquid service.

Direct-operated regulators are used to provide constant reduced pressure to pneumatic instrumentation and other control equipments.

Pilot-operated service regulators are ideal for applications involving pressure factor measurement.

Construction material should be selected according to the process requirement, and indicated as per data sheet.

- Body material and spring case : Cast iron, steel, or stainless steel
- Major metal internal parts : Brass or stainless steel
- Valve plug seating surfaces and diaphragm : Neoprene or stainless steel

16.14.1 Self-Actuated Pressure Regulator Characteristic

All regulators should be installed in accordance with local and international standards and regulations.

Adequate overpressure protection should be installed to protect the regulator from over pressure, and also to protect all downstream equipment in the event of regulator failure.

Downstream pressures significantly higher than the regulator pressure setting may damage soft seats and other internal regulator parts.

The recommended selection for port diameter should be the smallest port diameter that will handle the flow.

Spring cases must be protected against the accumulation of water caused by condensation or other sources.

Control line connections (where required) should be made in a straight run of pipe 8 to 10 pipe diameters downstream of any area of turbulence such as elbows or block valves.

Regulator body size should never be larger than pipe size. In many cases, the regulator body should be one size smaller than the pipe size.

The self-operated regulators generally have a faster response to quick flow change than pilot-operated regulators.

Materials and temperature capabilities of the regulators must be checked to conform with process requirement. Stainless steel diaphragms and seats should be used for higher temperatures such as steam services.

Self-operated regulating valves may only be used for services where fixed gain control is acceptable. Where failure of the mechanism may give rise to a dangerous situation, e.g., heating of tanks the application of such valves should be discussed with the user.

16.14.2 Anti Freeze-Up Regulators

These regulators are self-operated, pressure-reducing regulators that resist hydrate formation and regulator freeze-up. These regulators are suitable for service with natural gas, air, propane, and other gases compatible with the internal parts. They are used on high-pressure lines from wellheads and separators.

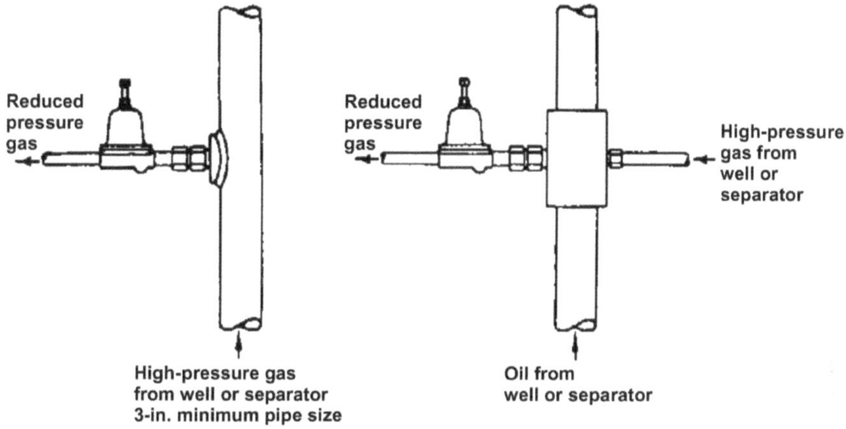

FIGURE 16.23 Typical regulator installations.

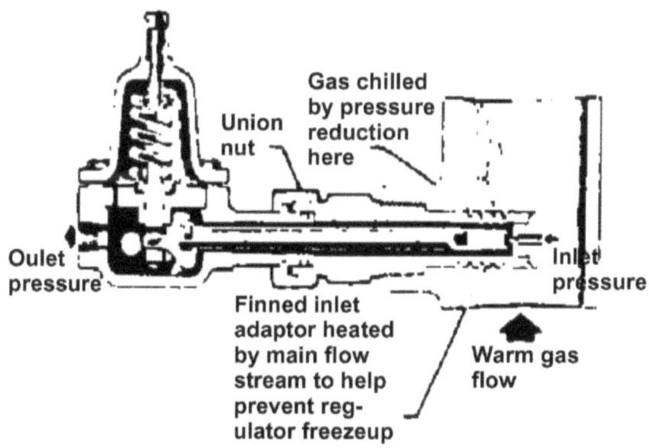

FIGURE 16.24 Type-1 regulator.

Regulator freeze-up resistance occurs as the pipe line gas warms the finned inlet adaptor and the seat ring area. As the gas cools within the inlet adaptor, due to pressure drop and volume expansion, the warm inlet adaptor helps keep the gas temperature above the freezing point of water and the hydrate formation temperature (see Figs. 16.23–16.25).

16.14.3 Self-Actuated Temperature Regulators

For temperature applications, bellows with a filled system and bulb are used.

The tube system assembly consist of the sensitive bulb, capillary tubing, the bellows assembly, an indicated dial thermometer, and the cap.

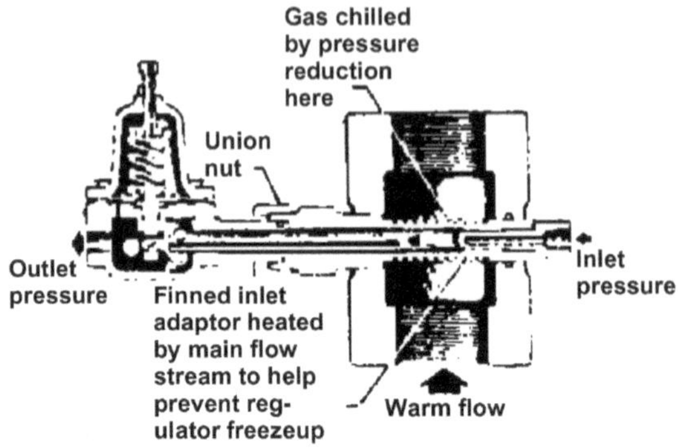

FIGURE 16.25 Type-2 regulator.

Self-operating temperature regulators are generally used on installations that require full pressure drop through the valve; the inlet pressure should not exceed the maximum allowable pressure drop.

The regulator should be suitable for installing in an accessible location on horizontal piping. Possible damage from moving parts, splashing of corrosive liquids, vibration, heat, etc., should be considered in deciding the location.

Similar consideration should also be given the capillary tubing and bulb. The capillary tubing on high-range instruments should be located where the temperature is at least (6.7°C) cooler than the control point.

Self-operating regulators are regularly furnished with the most sensitive temperature range unless otherwise specified. The most sensitive range span (approximately 10°C) may be changed to the corresponding wide span range (approx. 32°C) by replacing the sensitive range spring with the wide span range spring.

The maximum external pressure allowed on standard bulbs and sockets are 35 bar for copper or brass and 70 bar for steel or stainless steel, or unless otherwise specified by the data sheets.

Stainless steel trim valves are recommended on installation having pressure over 35 bar.

The tube system material assembly consist of the sensitive bulb, capillary tubing, and the bellows assembly, and an indicating dial thermometer; the cap should be specified in data sheets.

Packing gland for the regulators should be a teflon v ring (or graphite asbestos for high-temperature applications) packing sets with male and female adopters used as end rings, and a stainless steel compression spring. The spring loading of the packing should maintain proper compression of the rings and also to compensate for wear that occur at the seals (Fig. 16.26).

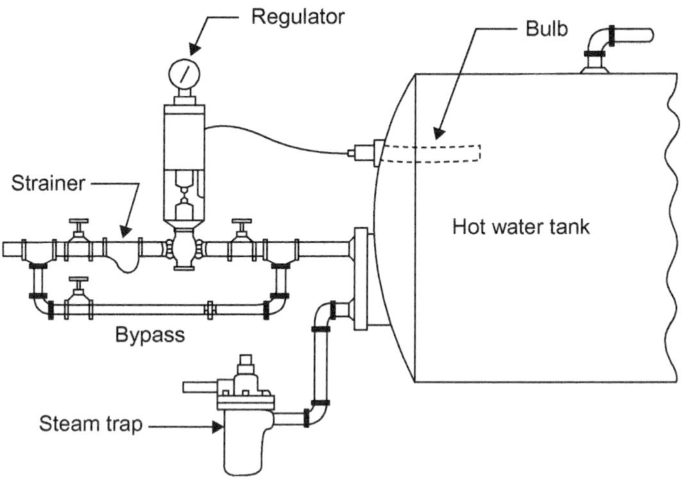

FIGURE 16.26 Typical application of self-actuated temperature regulator.

16.15 CONTROL VALVE ACCESSORIES

The most common types of pneumatic-control valve accessories that may be supplied with the control valve are solenoid valves, convertors, positioners, electropneumatic positioners, booster relays, extension bonnets, handwheels, air filter, limit switches, etc.

Since the positioner is often considered to be the most important of them, it is covered first.

16.15.1 Positioners

A pneumatic valve positioner is a device that precisely positions, by the use of air, the moving part (or parts) of a pneumatically operated valve in accordance with a pneumatic signal.

The valve positioner compares the valve stem position with the demand generated by the controller. If the valve stem is incorrectly positioned, the positioner either increases or decreases the air in the actuator until the correct valve stem position is obtained. The following is a list of six functions a positioner can accomplish:

1. Provide for split-range operation.
2. Improve transmission line speed of response to accommodate large actuator volumes at the end of signal-transmission lines.
3. Reverse the valve action without changing the "fail-safe" action of the spring in the actuator.

   (Note that this may also be done with a reversing-type relay.)

4. Increase the thrust in spring diaphragm actuators for use in high-pressure drop applications, and allow the same linearity in the installed characteristic as in the "bench setting" characteristic.
5. Change the control valve flow characteristic (cam type positioner).
6. Improve the resolution or sensitivity of the actuator where a high-precision valve control is required. Precision is enhanced by the availability of positioners with various gains, and by the fact that modern packings generally have equal static and dynamic coefficients of friction which eliminate the stick/slip behavior.

In the past a positioner was thought to reduce control loop stability for fast acting loops. Modern positioners with volume or pressure boosters, where required, can be made faster than any actuator without a positioner.

Pneumatic control valves may be operated over only part of the controller output range. This can be accomplished by either changing or adjusting the input spring of the positioner. A common arrangement is to have one valve and positioner operate over 0.2–0.6 barg of the controller output, while another valve and positioner operates over 0.6–1 barg of the controller output (split range operation). In the above, it is only necessary to modify the positioner springs.

Valve positioners should be provided with an integral pneumatic switch to bypass the positioner. The bypass may not be recommended where the valve will not operate without the positioner.

Valve positioners should be supplied with the requisite number of pressure gages, the controller output pressure, positioner output pressure, and supply pressure to the positioner.

Piston operators should be provided with positioners to ensure that the control valve position is always proportional to the control signal. The positioner should have a weather-proof enclosure.

16.15.2 Solenoid Valves

A solenoid valve is a combination of two basic functional units:

1. A solenoid (electromagnet) with its core.
2. A valve body containing one or more orifices.

Flow through an orifice is shut off or allowed by the movement of the core when the solenoid is energized or deener-gized.

A common application of a solenoid valve to a diaphragm control valve is illustrated in Fig. 16.27. In an emergency the solenoid valve can be switched, causing the control valve to go to the preselected position.

The solenoid valve is normally open and allows the positioner output to pass into the diaphragm case. Upon a power loss, the solenoid valve closes the port to the valve positioner and bleeds pressure from the diaphragm case of the control valve.

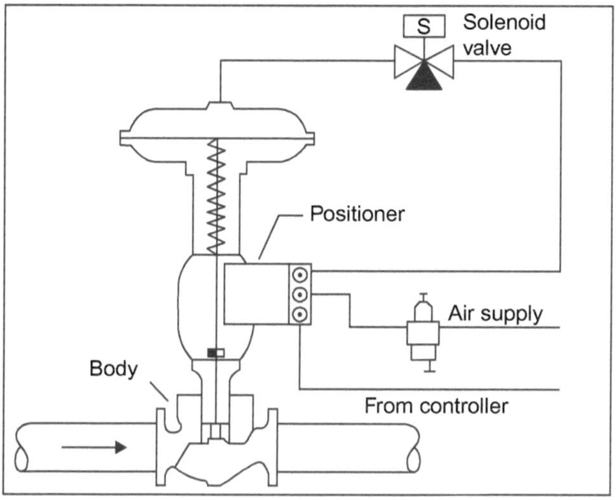

FIGURE 16.27 Typical application of a solenoid valve.

Where solenoid valves are installed in controlled air supplies to pneumatically operated valves to seal in diaphragm pressure, in the event of an electrical failure, the solenoid valves should incorporate a time delay and hand reset to prevent operation resulting from transient interruptions of the electrical supply.

For more information on solenoid valves, reference should be made to Section 18 of this Standard.

16.15.3 Convertors (Transducers)

A convertor is a device that converts one form of energy into another. This conversion may be pressure to movement, an electric current to a pressure, a liquid level to a twisting movement on a shaft, or any number of other combinations.

There are two basic types of sensors: one that produces an output proportional to a change in parameter is described as an analog device, one that produces an ON/OFF type of output is described as a digital device.

Electropneumatic transducers convert the electrical output signal from electronic controllers into pneumatic signal that may be used to operate diaphragm control valves.

Electropneumatic convertors should not be mounted on control valves. Sufficient capacity must be allowed in the pneumatic circuit to prevent interaction between convertors and valve positions. Where there is no possibility of local vibration ruining the valve positioner, consideration should be given to the use of valve-mounted electropneumatic valve positioner.

An I/P convertor operating a single valve should be mounted such that the length of tubing between the convertor and valve does not exceed 3 m. If a single I/P convertor is used to operate two or more valves, such as in split range service, the valves and convertor should be mounted such that the total length of tubing from the convertor to the valves does not exceed 3 m. Where this is not practical, a separate convertor should be supplied for each valve.

Electropneumatic convertors should be explosion proof or intrinsically safe and suitable for electrical area classification as indicated in data sheet.

Housing should be IP 54 (IEC-529) or, NEMA-3R weather proof.

Reference accuracy should be ±0.5% of full scale as detailed in ISA Standard S 51.1.

Input signals of: 4–20 mADC or unless otherwise specified.

Output signals: 0.2–1.0 bar, 0.4–2 bar, as required by data sheet.

Supply pressure: 1.4 bar (recommended). 3.5 bar maximum. The supply pressure medium must be clean, dry, and filtered.

An air filter regulator with an output pressure gage for the air supply to each convertor should be provided. Filter regulator should be mounted on an actuator.

16.15.3.1 Connection

- Supply pressure: ¼ in. NPT female
- Output pressure: ¼ in. NPT female
- Vents: ¼ in. NPT female
- Electrical: ¾ in. female conduit connector, or M 20 × 1.5.

16.15.3.2 Pressure Gage

On-filter regulator: 2½ in. diameter, dial with brass movement.

Construction material of housing and relay body should be die cast aluminum.

16.15.4 Booster Relays

Booster relays may be used to increase the speed of response of the control valve and are especially useful when the valve is remotely located from the controller. The function of the pressure booster is to amplify the signal from the controller to above 1.4 barg in certain applications (Fig. 16.28).

Volume boosters are used to increase the speed response of the control valve. An application with a booster relay on modern positioners with volume or pressure boosters, where required, can be made faster than actuator without a positioner (Fig. 16.29).

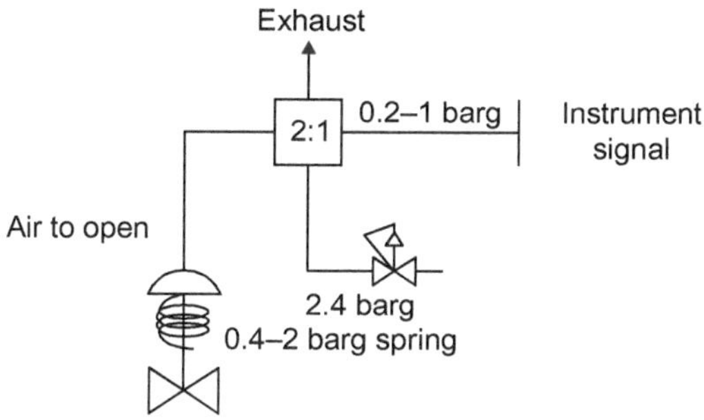

FIGURE 16.28 Pressure booster in a control valve loop.

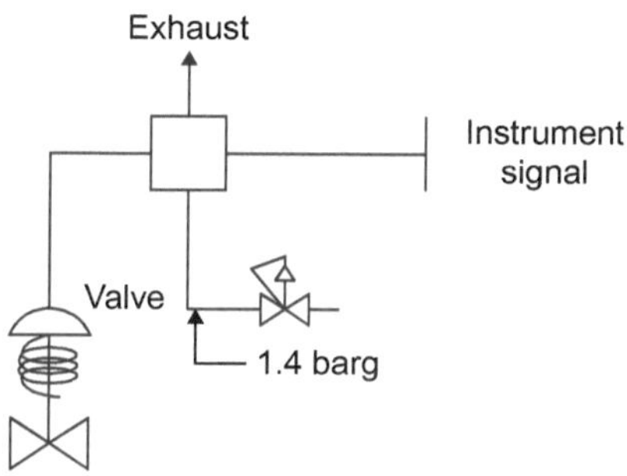

FIGURE 16.29 Volume booster in a control valve loop.

16.15.5 Extension Bonnets

The standard control valve bonnet, with the packing area relatively near the bonnet flange connection, is usually limited to temperatures not exceeding 232°C. For higher temperatures an extension bonnet containing sufficient area to provide radiating heat loss may be used. In no case should such a bonnet be covered with thermal insulating material (Fig. 16.30).

A similar bonnet design is employed on low-temperature applications (−29°C and below). This extension bonnet places the packing far enough away from the cold area of the valve to prevent freeze-up of the packing.

16.15.6 Handwheels

Handwheels can be supplied with most types of valves. They provide the operator with the means to override the control system and to operate the valve manually. Various designs are available, including those that can stroke the valve in either direction and those that stroke the valve in one direction, relying on the valve spring for the return stroke. Some handwheels are continuously connected. Others use a clutch, pin, or other means of

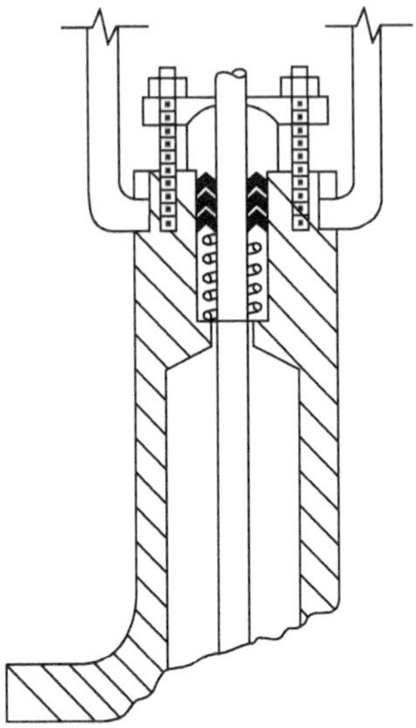

FIGURE 16.30 Extension bonnet.

engagement. These must be disengaged when not in use or damage may result.

Handwheels, when specified, should be mounted and designed to operate in the following manner.

1. For globe valves, handwheels should be mounted on the yoke, arranged so that the valve stem can be jacked in either direction.
2. Neutral position should be clearly indicated.
3. Handwheel operation should not add friction to the actuator.
4. Clutch/linkage mechanisms for handwheels on rotary valves should be designed such that control of the valve position is not lost when engaging the handwheel.

Top-mounted handwheels should be for the valves mounted relatively low (Fig. 16.31).

Side-mounted handwheels should be chosen for valves at higher elevations.

The side-mounted handwheels may also be operated by a chain fall plus chains to release and rest the locking levers.

Both of these operators can be used to facilitate the start-up of a control system (i.e., to preposition a valve to a given flow). They can also be used as devices to shut off the valve, eliminating the need in some systems for costly bypass valve arrangements.

The actuator should be equipped with a permanently attached handwheel of the automatic declutching type that precludes mechanical engagement of the handwheel while the drive is in operation.

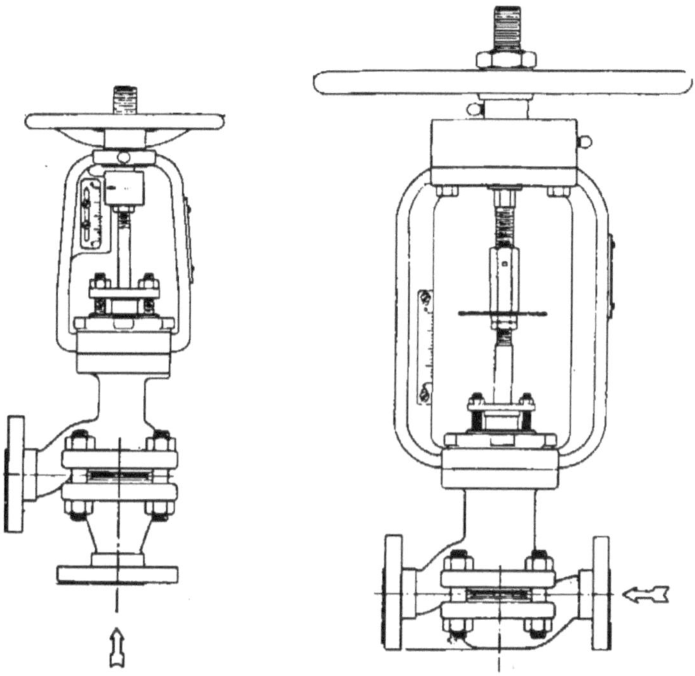

FIGURE 16.31 A top-mounted handwheel control valve.

The declutching device should:

1. Allow power-override of the handwheel operation at all times.
2. Permit manual handwheel operation of the valve in the event of a frozen or seized drive.

Handwheel drive should permit the valve to be stroked open or closed in 15 minutes or fewer.

Handwheel clockwise rotation should close the valve.

16.15.7 Air Locks

An air lock device is used for applications that require a control valve to hold its position in the event that the plant air supply pressure falls below a given level. One type is shown in Fig. 16.32. The plant air supply is fed into a chamber sealed by a spring-opposed diaphragm. In the event that the plant air decreases to a predetermined lower limit, the spring closes the connection to the actuator and locks the existing controller-signal pressure in the line connecting the valve operator.

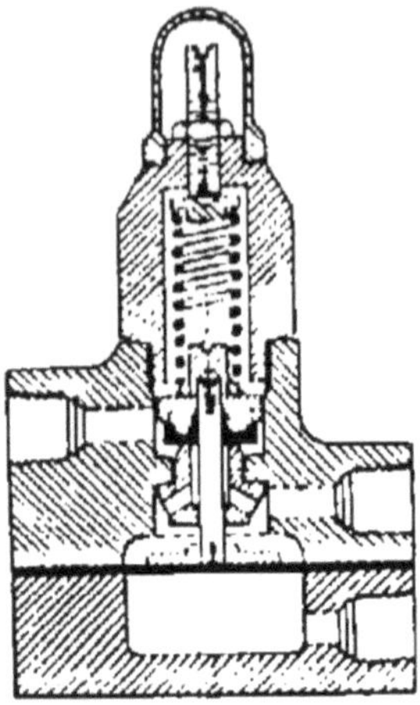

FIGURE 16.32 Air lock.

16.15.8 Pressure Sensing Trip Valves

Pressure-sensing trip valves are for control applications where a specific valve/actuator action is required when supply pressure fails below a specific point.

When the supply pressure fails below the trip point, the trip valve causes the actuator to fail in an up position, lock in the last position, or fail in the down position. When the supply pressure rises above the trip point, the trip valve automatically resets, allowing the system to return to normal operation.

The trip valve can be top-mounted on a manifold, yoke-mounted, or bracket-mounted to match the application requirements. Simplified sectional view of typical trip valve can be seen in Fig. 16.33.

16.15.8.1 Trip Valve Pressure Connections

Read the following information before making pressure connections:

1. Trip valve port A must receive the operating pressure that is intended for the top of the actuator cylinder. Depending on the actuator type and accessories being used, this operating pressure will be from a valve positioner or switching solenoid.
2. Trip valve port B must provide operating pressure to the top of the actuator cylinder. Depending on the actuator type and accessories being used, this port should be connected to the manifold assembly to the top of the cylinder, or to the cylinder connection on the hydraulic snubber (if one is used).
3. Trip valve port C must provide a fail-mode outlet for the operating pressure to or from the top of the actuator cylinder. For the fail-down mode, this port should be connected to the volume tank. For the fail-up mode, this port should vent to atmosphere. For the lock-in-last-position mode, this port should be plugged.

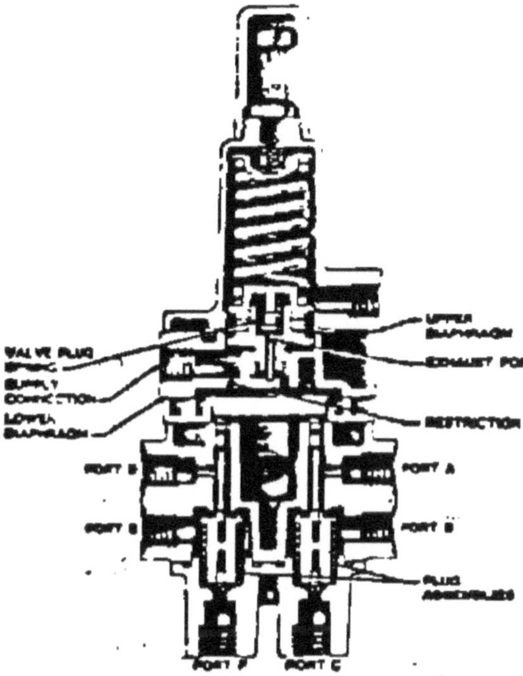

FIGURE 16.33 Simplified sectional view of trip valve.

4. Trip valve port D must receive the operating pressure that is intended for the bottom of the actuator cylinder. Depending on the actuator type and accessories being used, this operating pressure will be from a valve positioner or switching solenoid.
5. Trip valve port E must provide operating pressure to the bottom of the actuator cylinder. This port should always be connected to the bottom of the actuator cylinder.
6. Trip valve port F must provide a fail-mode outlet for the operating pressure to or from the bottom of the actuator cylinder. For the fail-down mode, this port should vent to the atmosphere. For the fail-up mode, this port should be connected to the volume tank. For the lock-in-last-position mode, this port should be plugged.

16.16 SOLENOID VALVES

The solenoid valve is basically a valve operated by a built-actuator in a form of an electrical coil (or solenoid) and a plunger. The valve is thus opened and closed by an electrical signal being returned to its original position (usually by a spring) when the signal is removed. Solenoid valves are produced in two modes: normally open or normally closed (referring to the state when the solenoid is not energized).

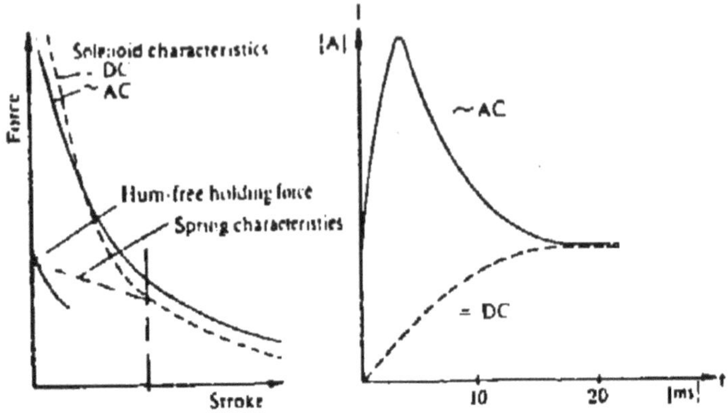

FIGURE 16.34 Force/stroke characteristics for a typical solenoid valve (left); and a time-based comparison of solenoid energization by AC and DC (right).

16.16.1 DC or AC Solenoids

The DC solenoids are generally preferred to AC because a DC operation is not subject to peak initial currents, which can cause overheating and coil damage with frequent cycling or accidental spool seizure. AC solenoids are preferred, however, where fast response is required, or where relay-type electric controls are used. Response time with AC solenoid-operated valves is of the order of 8–15 μs, compared with the 30–40 μs typical for DC solenoid operation.

There is an appreciable difference in the working characteristics of a solenoid supplied with DC and AC DC coils are slow in response time and can handle only low pressures. AC coils are quicker in response time and can handle higher pressures initially (see Fig. 16.34). They can thus be cycled at faster rates, if required. Electrical losses are higher, however, and proportional to AC frequency. (The power losses in a solenoid operated by AC of 60 Hz frequency, e.g., is higher than that of the same coil on a 50-Hz supply).

16.16.2 Return Spring Effect

With a two-way normally closed valve, both the spring force and the fluid inlet pressure act to close the valve. As a consequence the return spring can be made relatively weak, and in some designs eliminated entirely. The latter would require mounting the valve so that the solenoid was vertical, with a return action caused by gravity plus fluid pressure (Fig. 16.35A).

With a two-way normally open valve the spring holds the valve open, assisted by fluid pressure. The solenoid force must be sufficient to overcome both spring pressure and inlet pressure to close the valve.

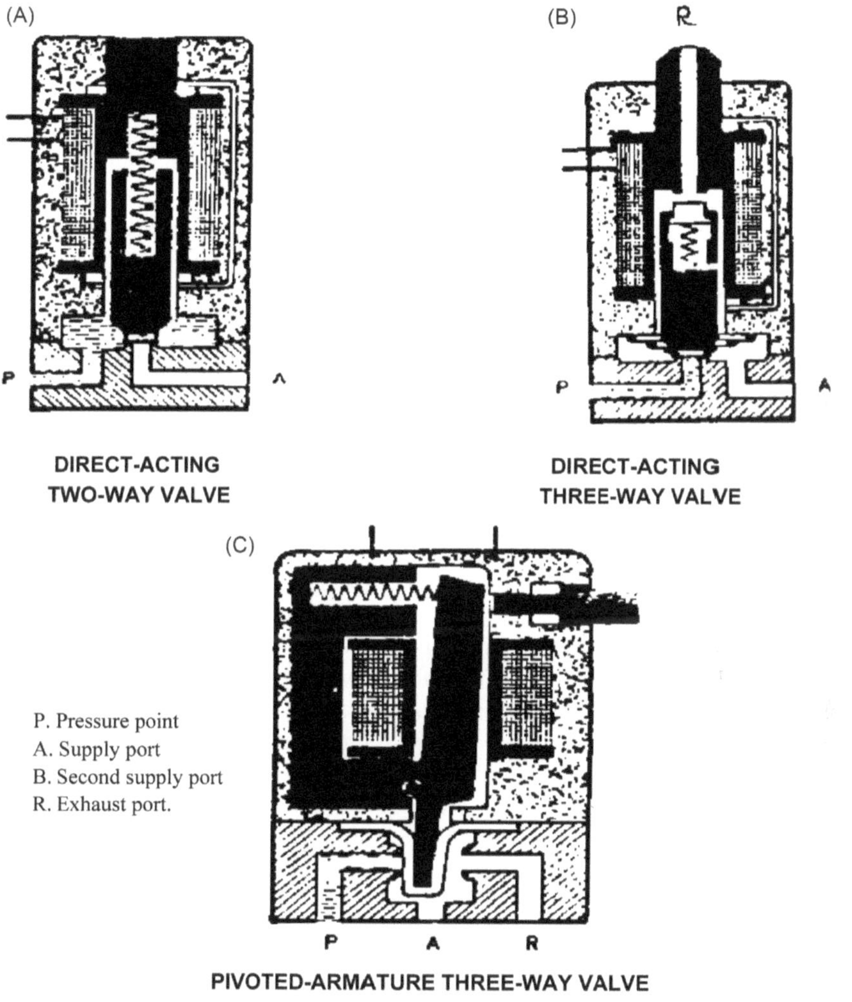

FIGURE 16.35 (A) Direct acting two-way valve, (B) direct acting three-way valve, and (C) pivoted-armature three-way valve.

Three-way valves require an upper and a lower spring. The lower spring presses the valve against its seal opened by inlet pressure. The upper spring acts in a direction to force the valve open (Fig. 16.35B and C). The following are the combination of spring strengths required :

	Lower Spring	Upper Spring
Three-way normally closed	Strong	Weak
Three-way normally open	Weak	Strong
Mixer valve	Medium	Medium
Divider valve	Strong	Weak

16.16.3 Special Types of Solenoid Valves

16.16.3.1 Pilot-Operated Valves

In general, a direct solenoid operation is restricted to smaller sizes of valves, i.e., up to about in bore size as a maximum. Larger solenoids needed for operating larger valves consume high levels of electricity and generate considerable heat. The pilot-operated valve offers a much more attractive proposition in such cases where a small solenoid is retained to operate a pilot valve, which in turn admits inlet pressure to an appropriate part of the valve to open the main valve. The pilot valve may be accommodated internally (internal pilot-operated) or externally (external pilot-operated).

For a pilot-operated valve to work there must be some differential pressure existing across the main valve for it to operate properly. It will then operate against higher pressures with low electrical input, with the pressure rating of the main valve the same as the pressure rating of the pilot valve, since both are subjected to the same line pressure. Pilot-operated valves have slower response time than direct-acting solenoid valves, although in many designs this is adjustable.

There is a further general class of solenoid-operated valves known as a semibalanced valve. This is a double-sealed valve with two plugs mounted on a common stem. The lower plug is slightly smaller than the upper plug. Line pressure is introduced below the lower plug and above the upper plug, creating a differential pressure to hold the valve on its seals, assisted by a spring, if necessary. The solenoid force required to open the valve is then only that which is due to the differential pressure (and the spring force, if present).

16.16.3.2 Solenoid-Operated Hydraulic Valves

Preference for the design of a solenoid-operated hydraulic valve is to use the solenoid for a "push" operation, utilizing spring action for "pull" motions. The solenoid must be powerful enough to override inertia and friction and also the spring and hydraulic forces. The latter may be extremely variable and not completely predictable, calling for a generous margin in the power of the solenoid and springs.

Solenoids may be of the "dry" or "wet" type. In general, "wet" solenoids can be smaller for the same duty because of their lower static and dynamic friction. They also have the advantage that all moving parts are enclosed and lubricated, and seals between the solenoid and valve body are eliminated. They are also described as glandless valves.

The size of directly operated solenoid valves is generally restricted to flow rates up to about 45 L/min, i.e., (⅛ in.) and (¼ in.) nominal valve sizes. Many of these valves can be switched directly from static systems, the outputs usually being 24 V DC and 20–65 W, depending on the system.

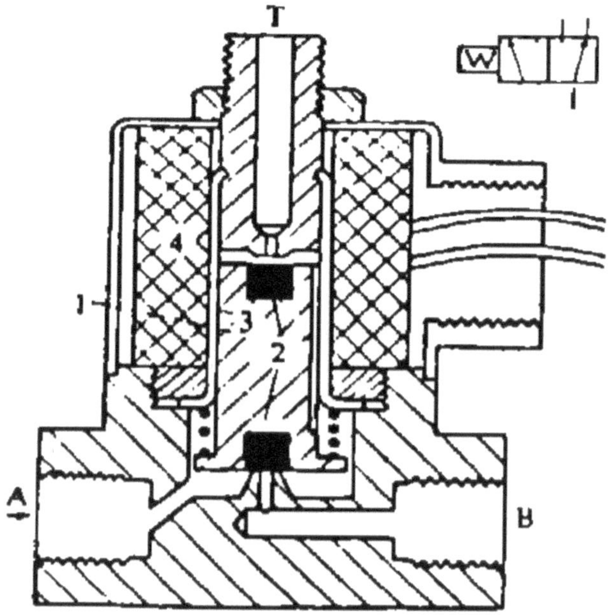

FIGURE 16.36 Glandless solenoid valve.

16.16.3.3 Glandless Solenoid Valves

By arranging the solenoid armature to work in a sealed tube with the solenoid coil enveloping it, the sealing glands can be dispensed with, thereby simplifying the construction and eliminating one possible point of leakage.

This principle has been applied extensively to the smaller valves. A typical type is shown in Fig. 16.36.

Glandless valves can be installed in any position and will withstand appreciable shock loads. Response time is extremely short, 5 μs on AC and 10–15 μs on DC and it is said that speeds of up to several hundred cycles per minute are possible.

For hazardous atmospheres, most manufacturers supply explosion-proof materials that are slightly heavier and bulkier than the standard type.

16.16.4 Solenoid Valves Characteristics

The valve body for solenoid valves should follow the instrument piping specifications when used in process lines. The manufacturer's standard bronze material should normally be used on air service.

Valve body connection sizes should be 1/8″, 1/4″, and 3/8″ or as required by the data sheet.

Coils for solenoid valves should be molded and encapsulated and specified continuous duty Class E, and F insulation at rated voltage and frequency. (Reference IEC-85, thermal insulation and classification of electrical insulation.)

The solenoid itself may be operated by DC or AC supply. Electrical rating of standard voltages should be 24 V AC or DC, 110 V AC 50 Hz, or as specified in the data sheet.

Solenoid coil should operate the valves by 10% of voltage variation, unless otherwise specified in the data sheet.

A variety of body materials are available to choose. Valve seat material should be selected to suit the requirement. Materials available are Buna N, and stainless steel discs, viton, Teflon, etc. Reference must be made to the specification detailed in data sheet for this selection.

External parts of solenoid construction in contact with fluid should be stainless steel.

Three-way and four-way packless solenoid valves that are direct acting and require no minimum operating pressure, may be installed on control valves. Both miniature and standard size solenoid valves are available along with both general purpose enclosure to protect from indirect splashing and dust, or explosion-proof and watertight enclosures. The requirement should be specified by user in the relevant data sheet.

The enclosure should be suitable for area classification as specified in data sheet.

16.17 INSPECTION, TEST, AND REPAIR OF DEFECTS

16.17.1 Inspection and Tests

When an inspection is specified in the purchase order by purchaser, then the inspection should be in accordance with API standard 598. If an inspection is not specified, control valves should meet the requirements for visual examination as described in API standard 598.

Each control valve should be hydrostatically tested by the manufacturer and the certified test reports should be provided confirming that the control valve have been tested in accordance with the test standard outlined in ANSI/ISA-S 750.1 or ANSI B16.37 test requirements.

The test requirement and the test procedure for obtaining the control valve various coefficients should be in accordance with ANSI/ISA-S 75.02 (control valve capacity test procedure).

After testing, each valve should be drained of test liquid, cleaned of any extra matter, and suitably protected in preparation for storage and transportation.

16.17.2 Repair of Defects

The user reserves the right to reject individual valves for bad workmanship or defects.

The repair of defects in cast iron or ductile iron castings, by welding, brazing, plugging, pinning, or impregnation is not permitted.

Defects in the body of carbon steel or alloy steel valve revealed by inspection or test may be repaired as permitted by the most nearly applicable ASTM material specification listed in Table 16.1 of ANSI/B16.34.

16.18 SPECIFIED MAXIMUM LEAKAGE

The maximum allowable leakage rate in terms of percent of valve maximum C_V value should be specified.

Another method of calculating the leakage rate is to specify the leakage rate per inch of valve seat orifice diameter per pound of differential pressure to compare with water or air leakage test specifications. These specifications are given in ISA recommended practice RP-39.6 (Fig. 16.37) lists leakage rate for various valve types.

Select a valve type that has a lower leakage rate than the maximum process leak rate allowed.

The given rates are based on factory tests of new valves, so an allowance must be made for leakage to increase with service usage.

Seat tests should be conducted as per ANSI/FCI 70-2 or ISA RP-39.6 (Fig. 16.37) metal-to-metal seated valve.

Note:

Tests conducted under factory test conditions with 50 psig air to atmosphere.

Body Tests: Test pressure for steel bodies should be per ANSI B16.34 cold pressure rating at 38°C ratings. For cast iron, brass, or bronze bodies, test pressure should be 2× primary pressure rating.

1 Example-0.46 cc/min for a 2-in. port orifice diameter in a globe, butterfly, or ball valve with 50 psi differential pressure air. Equivalent to three bubbles per minute from a ¼in. OD, 032-inch wall tube, and ¼ inch under water surface.

Note:

The terms bubble tight and drop tight are meaningless unless some leakage rate is specified. Lack of visible air bubbles using soap solution indicates leakage of less than 1×10^{-3}, 1×10^{-4} cc/s.

16.19 HYDROSTATIC TESTING OF CONTROL VALVES

Control valves having bodies, bonnets and cover plates made of carbon steel, low-alloy, and high-alloy (stainless) steel, cast iron and ductile iron should be hydrostatically tested as per recognized standard ANSI B16.37.

Pressure-measuring instruments used in testing should be of the indicating or recording type.

It is recommended that gages and recording instruments have a range of approximately double but not more than 4× the test pressure.

LEAKAGE CLASS ISA RP39.6	ALLOWABLE LEAKAGE RATE AIR OR WATER	VALVE TYPES	REMARKS
CLASS I	CATEGORY II, III OR IV, BUT NO TEST REQUIRED BY AGREEMENT BETWEEN USER AND SUPPLER.	VALVE TYPES LISTED IN CATEGORY II, III & IV	QUALITY OF mfg. IMPLIES THAT THESE VALVES DO NOT EXCEED LEAKAGE CLASSES II, III & IV, BUT NO GUARANTEE IS STIPULATED.
CLASS II	0.5% RATED VALVE CAPACITY. (MAXIMUM C_V)	GLOBE, DOUBLE SEATED. GLOBE, SINGLE SEATED, BALANCED WITH STEPPED METAL PISTON SEAT. BUTTERFLY, METAL LINED.	
CLASS III	0.1% OF RATED VVALVE CAPACITY	HIGH QUALITY GLOBE DOUBLE SEATED. GLOBE, SINGLE SEATED, BALANCED WITH CONTINUOUS METAL PISTON SEALS.	
CLASS IV	01% OF RATED VALVE CAPACITY	GLOBE, SINGLE SEATED. GLOBE, SINGLE SEATED, BALANCED WITH ELASTOMER PISTON SEALS. ROTARY ECCENTRIC CAM TYPE. BALL VALVES WITH METAL SEAT.	
CLASS V	5×10^{-4} cc.min. OF WATER PER INCH OF ORIFICE DIAMETER PER psi DIFFERENTIAL PRESSURE	GLOBE VALVES IN CLASS IV WITH HEAVY DUTY ACTUATORS TO INCREASE SEATING FORCE.	FEW VALVES CONTINUE TO REMAIN THIS TIGHT IN SERVICE UNLESS THE SEAT PLASTICALLY DEFORMS TO MAINTAIN CONTACT WITH THE PLUG.
CLASS VI	MAXIMUM PERMISSIBLE LEAKAGE ASSOCIATED WITH RESILLENT SEATING VALVES. EXPRESSED AS BUBBLES PER MIN AS PER RP39.6[1]	GLOBE WITH RESILIENT SEAT. BUTTERFLY, ELASTOMER LINED. ROTARY ECCENTRIC CAM WITH ELASTOMER SEAT. BALL WITH RESILIENT SEAT, SOLID BALL TYPE. DIAMPHRAGM, WEIR TYPE. PLUG VLAVES, ELASTOMER SEATED OR SEALANT INJECTION SEALING SYSTEM.	ELASTOMER SEALED VALVES REMAIN THIS TIGHT FOR MANY THOUSANDS OF CYCLES UNTILL THE SEAL IS WORN OR CUT.

FIGURE 16.37 Valve leakage rates.

16.20 PRESSURE TEST REQUIREMENTS FOR BUTTERFLY VALVES

All butterfly valves, completely assembled, should be pressure tested by the manufacturer before dispatch and in accordance with the recognized standard such as BS-5155 (pressure testing).

Testing should be carried out before valves are painted or otherwise externally coated with materials that are capable of sealing against leakage. Internal linings and external nonpressure sealing anticorrosion treatments should be permitted for the purposes of testing.

No valve undergoing pressure testing should be subject to shock loading.

Valves and connections should be purged of air prior to pressure testing.

The test fluid for all pressure tests should be either water with the addition of a suitable inhibitor, or another liquid whose viscosity at ambient temperature is equal to or less than that of water.

Note:

Attention is drawn to the need to control the chloride content of test water in contact with austenitic stainless steel components.

Test pressures should be determined by the following relationships:

1. Shell test: 1.5× maximum permissible, working pressure at 20°C.
2. Disk strength test (applicable to valves 14 in. and larger): 1.5× maximum permissible working pressure at 20°C.
3. Seat test: 1.1× maximum permissible working pressure at 20°C.

Test procedures for shell, disk, and seat of butterfly valves should be in accordance with the recognized standard such as BS-5155.

Test durations for butterfly valves should be conducted as per Table 10 of BS-5155.

After test carried out by Table 10 above, the maximum permissible leakage should be given in Table 11 of BS-5155 for each valve type.

16.21 SOLENOID VALVES TEST REQUIREMENT

Solenoid valves should be tested by the manufacturers before dispatch according to UL specification and approvals or other recognized testing organizations.

Test approvals by a third party is only necessary for equipment used in hazardous areas; the test should be certified in accordance to any approved bodies such as BASEFA, UL, FM, PTB, etc.

Solenoid valves are used in hazardous areas should meet the electrical area classification requirements.

16.22 MARKING

16.22.1 Nameplate

The nameplate should be 316 stainless steel and 70×50 mm and should be marked with the following information:

1. Tag number
2. Body material
3. Trim material
4. Body/trim sizes
5. Actuator air to open/close
6. Actuator operating pressure
7. Manufacturer's name and model number and serial number
8. Pressure rating
9. Maximum differential pressure
10. A tag indicating type of packing and lubricant
11. The arrow indicating direction of valve flow on the body
12. Bench set pressure
13. Valve characteristics

16.23 PACKING AND SHIPPING

Equipment must be carefully protected and packed to provide adequate protection during transit to destination and should be in accordance with any special provision contained in the specification or order. Special attention must be given to protection against corrosion during transit. All bright and machined parts must be painted with a rust preventative.

Ancillary items forming an integral part of the equipment should be packed preferably in a separate container if the equipment is normally cased or crated.

Alternatively the ancillary items should be fixed securely to the equipment and adequate precaution taken to ensure that the items do not come loose in transit or be otherwise damaged.

Unless export packaging is specified in the purchase order, valves should be shipped or packed in wooden boxes or crates, and fastened, so that shifting is prevented within the package.

Threaded openings of the valves should be plugged with suitable protective device to prevent entrance of dirt and to prevent damage to the threads.

Flanged faces should be coated with rust-ban or other suitable rust-preventive substance. Flanged faces should be protected by covers securely bolted to the flanges to prevent dirt from entering the valve interior.

Valves shipped with mounted actuators should be packed in a manner that will prevent damage while in transit.

Butterfly valves should be shipped with the shaft packing installed.

Butterfly valves should be shipped with the disk positioned so that the disk edges are within the body contact faces to prevent damage during normal handling.

After the receipt of the inspection report, the control valve should be prepared for shipment either to the plant area for installation, or to storage. The valve body's air or electrical connections should be plugged to keep out dirt. If the control valve is to be stored for any length of time, it should be packed for protection against environmental adverse effects.

16.24 DOCUMENTATION/LITERATURE

16.24.1 At Quotation Stage

Suppliers are to provide the following in the numbers requested at the time of quotation:

1. Comprehensive descriptive literature.
2. List of recommended commissioning spares with prices.
3. Details of any special tools required with prices.
4. Calculation, including C_V noise, etc.

16.24.2 At Ordering Stage

Suppliers are to provide the following in quantities and at times as detailed on the order:

1. List of recommended spares for two years of continuous operation.
2. Illustrated comprehensive spare parts manual with part numbers suitable for warehouse stocking.
3. Illustrated installation and operating instructions.
4. Maintenance manuals.

16.25 LINEAR MOTION CONTROL VALVE TYPES

Types of valves with a closure member that moves with a linear motion to modify the rate of flow through the valve.

16.25.1 Globe Valve

A valve with a linear motion closure member, one or more ports and a body distinguished by a globular shaped cavity around the port region. Typical globe valve types are illustrated below. Flow arrows shown indicate a commonly used flow direction.

16.25.1.1 Bottom Flange

A part that closes a valve body opening opposite the bonnet opening. It may include a guide bushing and/or serve to allow reversal of the valve action. In three-way valves it may provide the lower flow connection and its seat.

16.25.1.2 Globe Valve Trim

The internal parts of a valve that are in flowing contact with the controlled fluid. Examples are the plug, seat ring, cage, stem, and the parts used to attach the stem to the plug. The body, bonnet, bottom flange, guide means, and gaskets are not considered as part of the trim.

16.25.1.2.1 Antinoise Trim

A combination of plug and seat ring or plug and cage that by its geometry reduces the noise generated by fluid flowing through the valve.

16.25.1.2.2 Anticavitation Trim

A combination of plug and seat ring or plug and cage that by its geometry permits noncavitating operation or reduces the tendency to cavitate, thereby minimizing damage to the valve parts and the downstream piping.

16.25.1.2.3 Balanced Trim

An arrangement of ports and plug or combination of plug, cage, seals, and ports that tends to equalize the pressure above and below the valve plug to minimize the net static and dynamic fluid flow forces acting along the axis of the stem of a globe valve.

16.25.1.2.4 Erosion-Resistant Trim

Valve trim that has been faced with very hard material or manufactured from very hard material to resist the erosive effects of the controlled fluid flow.

16.25.1.2.5 Soft-Seated Trim

Globe valve trim with an elastomeric, plastic, or other readily deformable material used either in the valve plug or seat ring to provide tight shutoff with minimal actuator forces. See ANSI B16.104 for leakage classifications.

16.25.1.3 Globe Valve Plug Guides

The means by which the plug is aligned with the seat and held stable throughout its travel. The guide is held rigidly in the body or bonnet.

16.25.1.3.1 Stem Guide

A guide bushing closely fitted to the valve stem and aligned with the seat (Fig. 16.38B).

16.25.1.3.2 Post Guide

Guide bushing or bushings fitted to posts or extensions larger than the valve stem and aligned with the seat (Fig. 16.38D).

16.25.1.3.3 Cage Guide

A valve plug fitted to the inside diameter of the cage to align the plug with the seat (Fig. 16.38A).

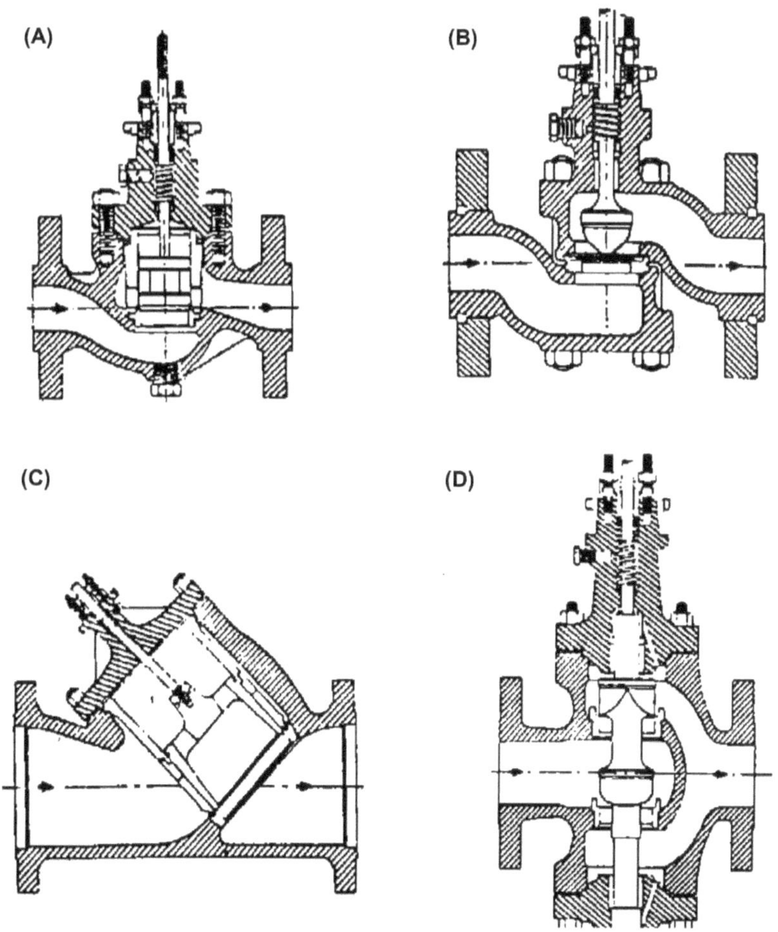

FIGURE 16.38 Two-way bodies. (A) Cage guided, (B) split body stem guided, (C) Y-type cage guided, (D) double-ported post or (top and bottom) guided, (E) angle body, (F) diverging, (G) converging, and (H) three position.

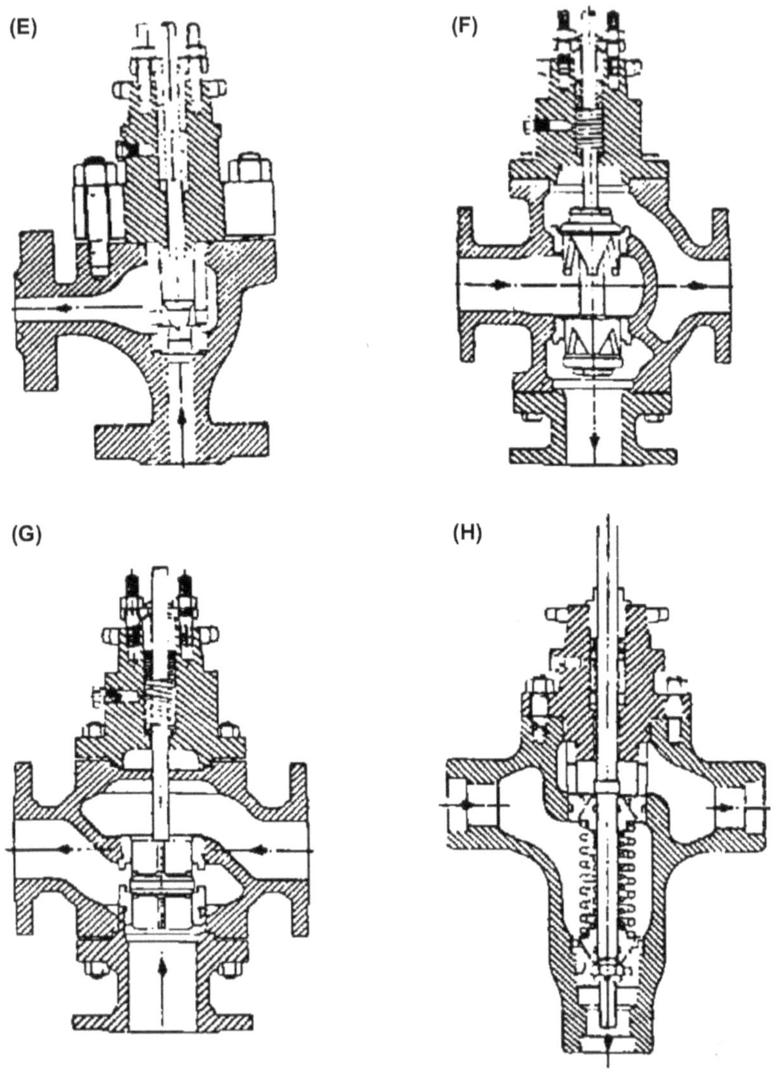

FIGURE 16.38 (Continued)

16.25.1.3.4 Port Guide

A valve plug with wings or a skirt fitted to the seat ring bore.

16.25.2 Diaphragm Valve

A valve with a flexible linear motion closure member that is forced into the internal flow passageway of the body by the actuator.

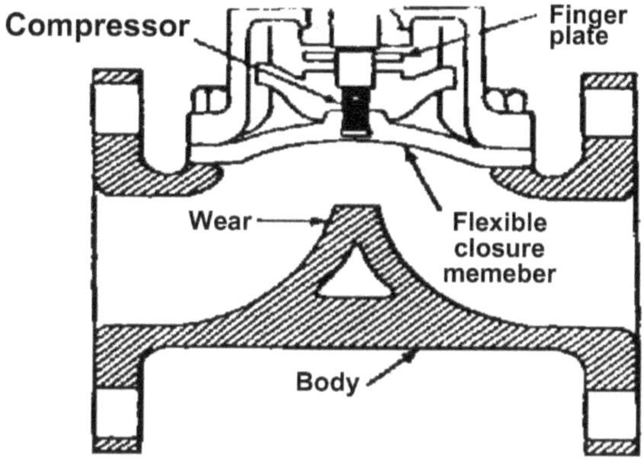

16.25.2.1 Valve Diaphragm

A flexible member that is moved into the fluid flow passageway of the body to modify the rate of flow through the valve.

16.25.2.2 Compressor

A device that the valve stem forces against the backside of the diaphragm to cause the diaphragm to move toward and seal against the internal flow passageway of the valve body.

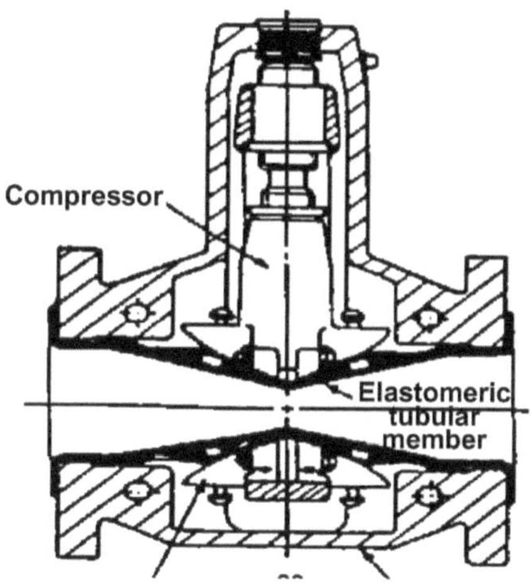

16.25.2.3 Finger Plate

A plate used to restrict the upward motion of the diaphragm and prevent diaphragm extrusion into the bonnet cavity in the full open position.

16.26 ROTARY MOTION CONTROL VALVE TYPES

Types of valves with a closure member that moves with a rotary motion to modify the rate of flow through the valve.

16.26.1 Ball Valve

A valve which modifies flow rates with rotary motion of the closure member, which is either a sphere with an internal passage or a segment of a spherical surface (Ref. 6.6.2).

16.26.2 Butterfly Valve

A valve with a circular body and a rotary motion disk closure member, pivotally supported by its stem.

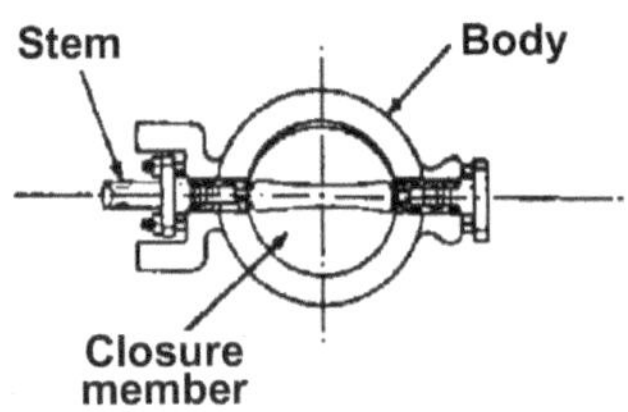

16.26.2.1 Body Types

16.26.2.1.1 Wafer Body

A body whose end surfaces mate with the pipeline flanges. It is located and clamped between the piping flanges by long bolts extending from flange to flange. A wafer body is also called a flangeless body.

16.26.2.1.2 Split Body

A body divided in half by a plane containing the longitudinal flow path axis.

16.26.2.1.3 Lined Body

A body having a lining that makes an interference fit with the disk in the closed position thus establishing a seal.

16.26.2.1.4 Unlined Body

A body without a lining.

16.26.2.2 Typical Disk Orientations

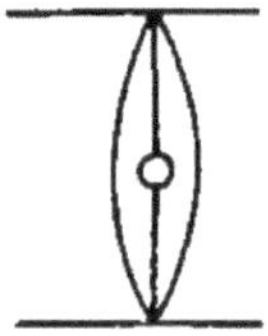

a) Aligned

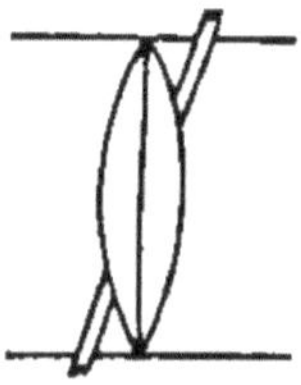

b) Aligned with canted stem

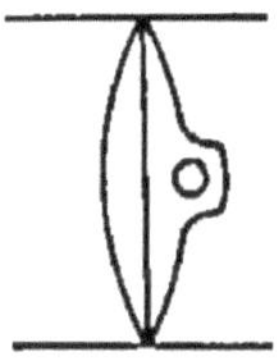

c) Offset

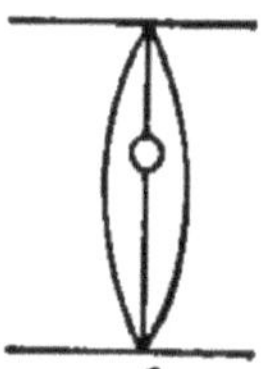

e) Canted disk

d) Cammed

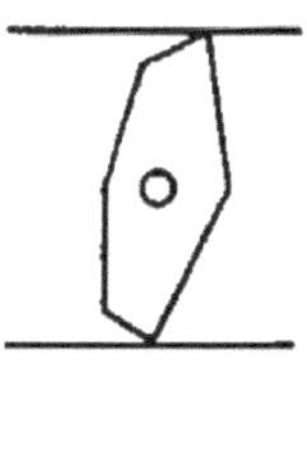

f) Angle seated

16.26.2.3 *Typical Disk Shapes*

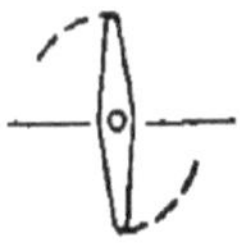

a) Flat

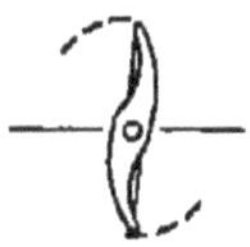

b) Cambered

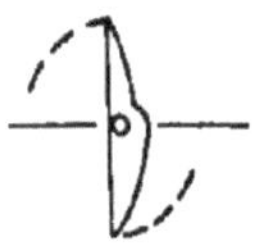

c) Nonsymmetrical edge

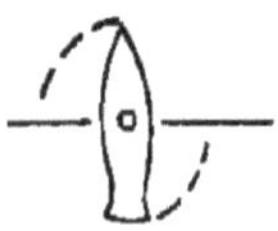

d) Contoured

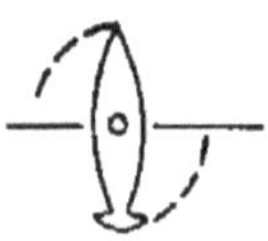

e) Knife

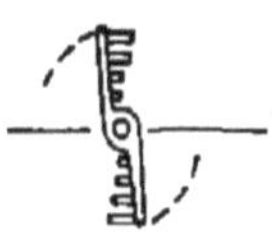

f) Fluted

16.26.2.4 *Seal on Disk*

A seal ring located in a groove in the disk circumference. The body is unlined in this case.

16.26.2.5 *Stem Bearings*

Butterfly stem bearings are referred to as either the outboard or the inboard type, depending on their location, outside, or inside of the stem seals.

16.27 SIZING THEORY AND APPLICATIONS

16.27.1 Sizing Theory and Applications[1] (L.R. Driskell)

16.27.1.1 Introduction

The problem of sizing a control valve may be broken into several more or less distinct parts. The person sizing a valve should carefully consider the effect of these factors on valve size.

1. **Flow Application Data:**
   Flow rate, maximum and minimum.
   Pressure upstream and downstream (at both maximum and minimum flows).
   Temperature of the stream.
2. **Fluid Data:**
   Name of fluid or its generic identity.
   Fluid phase-liquid, gas, slurry, etc.
   Density (specific gravity, specific weight, molecular weight, etc.).
   Viscosity (liquids).
   Vapor pressure (liquids).
3. **Piping Influence:**
   Pressure of reducers or other disturbances at the valve that will change the rated capacity.
4. **System Influence:**
   Control dynamics (is oversizing unimportant?).
   Economic factors (is downstream relief valve size affected?).
   Safety.
5. **Style of Valve-Selection Based on Application:**
   Capacity, order of magnitude of size.
   Rangeability.
   Corrosion or erosion resistance.
   Special requirements (tight shutoff, low noise, etc.).
6. **Sizing Calculations:**
   Manufacturer's sizing coefficients.
   Sizing formulas, slide rules, nomographs.
7. **Judgment:**

   Based on discriminating analysis of past successes and failures.

The first four items can be properly evaluated only by one who is knowledgeable about the whole process; therefore, this is usually the purchaser.

1. The sizing information and data contained in this chapter is based on the latest ISA Standards S-39.1 and S-39.3.

The principal responsibility of the valve manufacturer, in valve sizing, is to publish certified valve sizing coefficients. These values simply state the performance characteristics of his valves, based on standard test data and quality control of production. In some instances a valve manufacturer may assume additional responsibilities. This usually involves sizing the valve based on data provided by the purchaser and may include the selection of valve type and materials as well. This additional engineering on the part of the vendor involves added work and liability on his part and involves some degree of risk on the part of the purchaser. If there is divided responsibility, the application must fly without the last important ingredient—good judgment.

16.27.1.1.1 Theory

To be a good aircraft pilot, it is necessary to have the seat-of-the-pants feel of the ship. It is also important to know why the ship responds the way it does. For the same reasons, the art of valve sizing goes hand-in-hand with the science of fluid mechanics. It is helpful to know why valves work and what their limitations are, to properly apply the cook-book formulas.

16.27.1.1.2 Incompressible Fluids

A fluid flowing through control valves follows the same laws of conservation of mass and energy as expressed in the equations of fluid mechanics. First consider the flow of liquids, which essentially are incompressible fluids. When any fluid flowing inside a pipe, passes through a narrower passage or restriction, it must accelerate. The energy for this acceleration must be taken from the pressure of the fluid, or the static head.

After passing the restriction, the fluid slows down again and part of this head, or pressure, is recovered. The unrecovered part of the pressure has been converted into internal energy by friction. Fig. 16.39 shows the pressure gradient around a valve or an orifice.

Neglecting friction and other nonideal influences for a moment, Bernoulli's theorem tells us that

$$U_2^2 - U_1^2 = 2\,gh \tag{16.1}$$

where U_1 and U_2 are the mean axial speed of the fluid across areas a_1 and a_2, respectively; g is the acceleration of gravity; and h is difference in head (pressure) measured in feet of the fluid.[2]

2. In this part of the discussion of theory all units are in standard absolute units, such as pressures in pounds per square foot, areas in square feet, etc. The working equations discussed later use customary engineering units of pounds per square inches, etc. When metric units are applicable it will be noted.

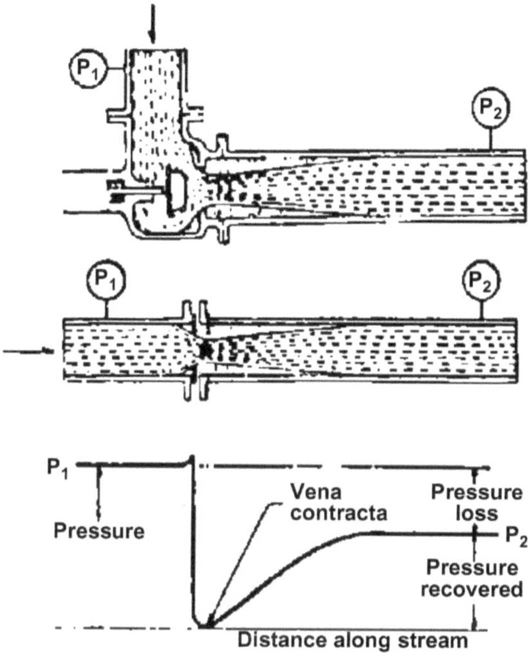

FIGURE 16.39 Control valve and orifice plate installations showing reduced area at vena contracta. The curve describes the pressure gradient around either device.

16.27.1.2 Compressible Fluids

Liquids are nearly incompressible and, as they pass through a valve, their specific weight remains constant. Gases and vapors, on the other hand, expand as the pressure drops; therefore, their specific weight decreases as they pass from the valve inlet to the vena contracta within the valve body. The effective specific weight, as far as flow is concerned, is approximately midway between the value upstream at p_1 and the value at the vena contracta.

The relationship of Y to x for any particular valve fortunately turns out to be an essentially linear curve. The slope of this curve is determined experimentally by methods prescribed by ISA Standard S39.4. With air as the test fluid, the slope of the curve is defined by the pressure drop ratio factor xT, which is the value of x when Y reaches its lower limit of 0.667. Mathematically, when air is the fluid Y has the value: $1 - 0.33\ (x/xT)$.

The relationships of Y versus x for several widely differing styles of valves are shown in Fig. 16.40. It will be noticed that all of the plots are straight lines and that they terminate at a value of $Y = 0.667$.

Although the true plots may deviate somewhat from a straight line, the test data to date indicate that the linear representation is within the tolerance

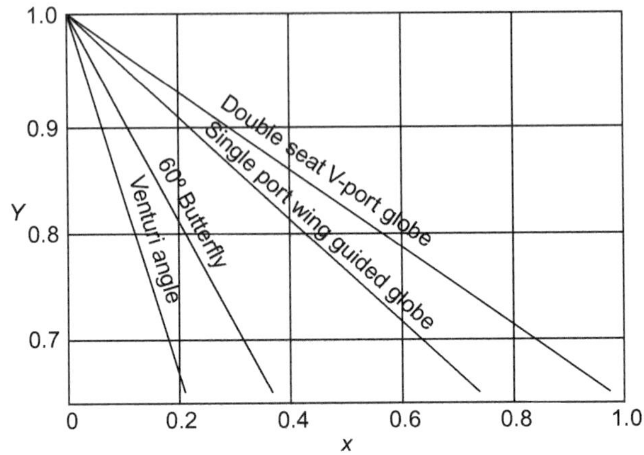

FIGURE 16.40 Expansion factor, Y as a function of pressure drop ratio, x, four widely differing styles of control valve.

of experimental error (about 2%). Termination at $Y = 0.667$ results from the fact that the ultimate flow through a restriction at any selected $p1$ must occur when the quantity $Y\sqrt{x}$ reaches a maximum. With the linear relationship of Y to x, it can be demonstrated by differential calculus that $Y\sqrt{x}$ reaches a maximum at $Y = 0.667$.

Nitrogen, oxygen, hydrogen, and other diatomic gases behave like air when passing through a valve, but some gases and vapors behave differently and require an adjustment to the slope of the Y curve. The adjustment is brought about by multiplying xT by the correction factor F_{k}, which is called the ratio of specific heats factor. Taking this effect into account, the resulting formula for Y for any gas or vapor becomes:

$$Y = 1 - 0.33\,(x/FK_{xT}) \tag{16.2}$$

and since the maximum flow is reached when Y attains its minimum value of 0.667, then the maximum useful value of x is FK_{xT}.

16.27.1.3 Working Equations

16.27.1.3.1 Slide Rules and Nomographs

Many control valve manufacturers provide slide rules and nomographs to be used for valve sizing. When properly designed and used, the results of these tools are as accurate as direct calculations. They, of course, have their advantages and disadvantages. Generally these methods are faster, because all constants in the equations are incorporated into the scales and require no additional settings. The tools must be used frequently in order to avoid rereading the procedural instructions. The units of measurement used must conform to those employed by the slide rule or nomograph.

The chances of error for the various methods depends upon the calculator. Usually no record is preserved when these devices are used; therefore, checking is difficult.

16.27.1.3.2 Low Reynolds Number

The nature of flow in pipes depends upon the combination of four variables (diameter, viscosity, density, velocity) arranged to produce a dimensionless number called the Reynolds number. The value of this number determines whether the flow is laminar or turbulent, and it also affects the coefficient of discharge of an orifice.

The flow coefficient, C_V, is determined at a reasonably high Reynolds number with the fluid in turbulent flow. The C_V changes very little as the Reynolds number is increased. However, as the Reynolds number is decreased, the effective C_V becomes smaller and a correction factor must be applied. At extremely low Reynolds numbers the flow becomes laminar rather than turbulent, and the flow rate is proportional to ΔP rather than $\sqrt{\Delta P}$. This change in flow mechanism is provided for by the correction factor, FR.

The Reynolds criterion is valid only when there is precise mechanical similarity between the devices. For example, in the determination of the friction factor for pipes the "length" dimension used to calculate Reynolds number is the pipe internal diameter, but the relative roughness of the pipe wall must also be the the same as that of the test specimen to validate the method.

16.28 CONTROL VALVE INSTALLATION (GENERAL)

This section represents the minimum and general technical requirements for the installation of different types of control valves and their accessories that are used in oil, gas and petrochemical industries. In any case, the manufacturer's installation instructions should be strictly followed.

Control valves should be installed so that they are readily accessible from grade or platforms. Wherever possible, they should be located at grade for ease of maintenance.

Wherever possible control valves should be installed with stems vertically above the body. Where line conditions prohibit this, suitable support must be provided for the valve top-works.

Where equipment must be observed, while on manual control, the control valve should be installed adjacent to the equipment.

Clearance should be provided above and below a control valve so that the bottom flange and plug or the top-works and plug may be removed with the valve body in the pipeline.

The valve body may swing around a selected bolt axis for maintenance access. However, clearance is still being provided to enable inspection of the valve plug without rotating the valve in the pipeline.

Control valves should have removable trims, and sufficient clearance should be allowed for access and removal.

Clearance also should be provided for hand-wheel operation and positioner maintenance.

Control valves at fired heaters should be located 15 m away from burners so that maintenance can be carried out without danger of flash-back.

Control valves for flammable and volatile fluids should not be installed adjacent to hot pumps, lines, or equipment.

Control valves should be located so that diaphragms and electric or electronic components are not damaged by heat radiated from vessels, heaters, and other equipment.

Where it is necessary to reduce from line size to control valve size, swaged reducers should be used between the block valves and the control valve. Sufficient spacing between block valves should allow for installation of larger size control valves.

The computed valve size should be at least one pipe size smaller than lines in which control valves are installed.

For toxic or other dangerous duties, control valve stems should be bellows sealed, with an independent gland seal, the enclosed space being monitored for bellows leakage.

When sealing is not possible, a purging system that monitors flow failure should be used.

Control valve vent and drain connections should not be less than ¾ in. nominal bore.

Butterfly valves should be installed with their shafts horizontal.

Where butterfly valves have to be installed in vertical lines, care should be taken that the diaphragm actuator stays clear from the piping.

Control valves should be installed in main lines but not in long straight runs. In the case of long straight runs, the control valves should be offset from the main line so that they will not be subjected to line stresses caused, e.g., by thermal expansion and weight of unsupported lines.

Extra clearance should be provided where extension bonnets or accessories are used. Clearance should also be provided on the side of the control valve for maintenance of positioners and other devices.

Long bolting used with flangeless valves can expand when exposed to fire and cause leakage. A fire deflection shield and/or insulation is recommended.

Certain rotary motion control valve types, which utilize low-friction, plastic-lined bearings and as a result are susceptible to static electricity, should be grounded. Manufacturer's recommendations should be followed.

16.29 ACTUATOR INSTALLATION

Electrically operated items such as motor actuators, solenoid valves, converters should be approved for installation under the applicable hazardous area classifications.

16.29.1 Pneumatic Diaphragm Actuators

Sliding stem spring loaded diaphragm actuators with air as the operating medium should be installed vertically above the body. If piping conditions prohibit this, the manufacturer recommendation should be considered.

16.29.2 Motor-Operated Valve Actuators

The M.O.V. actuator should be an integral unit suitable for direct mounting on the valve stem at any position. The unit should be self-centering and no special alignment should be required.

Electric-motor–driven actuators should be mounted so that the motor is above the gear box. This arrangement prevents gear oil from saturating the motor windings.

16.29.3 Electropneumatic Converters (Transducers)

Electropneumatic converters, where required, should be furnished and mounted independent of the control valve.

Standard mounting hardware should be provided for mounting electropneumatic converter on a pipestand, or a panel.

Electropneumatic converters should not be mounted on control valves. Sufficient capacity must be allowed in the pneumatic circuit to prevent interaction between converters and valve positions. Where there is no possibility of local vibration ruining the valve positioner, consideration should be given to the use of a valve-mounted, electropneumatic valve positioner.

Where two valves are used in the three-way service, one common controller and electropneumatic converter should be used for both valves. Air supply to this converter should be the same as the valve positioners and air locks.

16.29.4 Air Lock Devices

Air lock-up devices should be provided for all services requiring that the control valve remains in the position in which it was immediately before the air failure. On control valves with a positioner, the lock-up valve should be installed between the positioner output and the actuator.

If air lock valves are specified, they should be installed as close to the valve actuator as possible. However, any solenoid valves associated with protective system should be installed between the air lock valve and the actuator. The air supply for the air lock should be the same as for the valve positioner.

Valves with the air lock feature should, in addition, have the following:

1. A pressure gage indicating actual diaphragm pressure, for diaphragm actuators.
2. A pressure gage indicating air volume reserve tank pressure, for piston actuators.

16.29.5 Solenoid-Operated Valves

Solenoid-operated valves are extremely versatile and are frequently used with the control valves in a variety of on–off or switching applications such as equipment override, fail-safe interlock with two valves, and switching from one instrument or pressure line to another.

A typical solenoid valve installation on a pneumatically operated control valve is shown in Fig. 16.41. The solenoid valve is normally open and allows the positioner output into the diaphragm case. Upon a power loss, the solenoid valve closes the port to the valve positioner and bleeds pressure from the diaphragm case of the control valve.

Where solenoid valves are installed in control air supplies to pneumatically operated valves to seal-in diaphragm pressure in the event of an electrical failure, the solenoid valves should incorporate a time delay and hand reset to prevent inconsistent operation resulting from transient interruptions of the electrical supply.

Various solenoid/actuator mounting arrangements are shown in Fig. 16.42.

16.29.6 Pneumatic Cylinder Actuators

Pneumatic cylinder actuators that are not an integral part of the final control element should be located such that necessary linkage is short

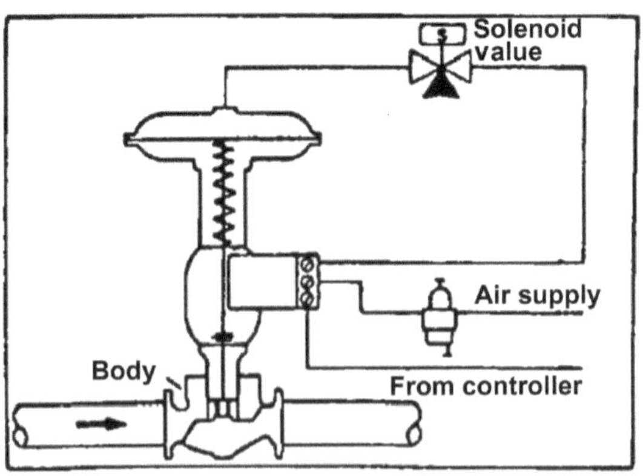

FIGURE 16.41 Typical installation of a solenoid valve.

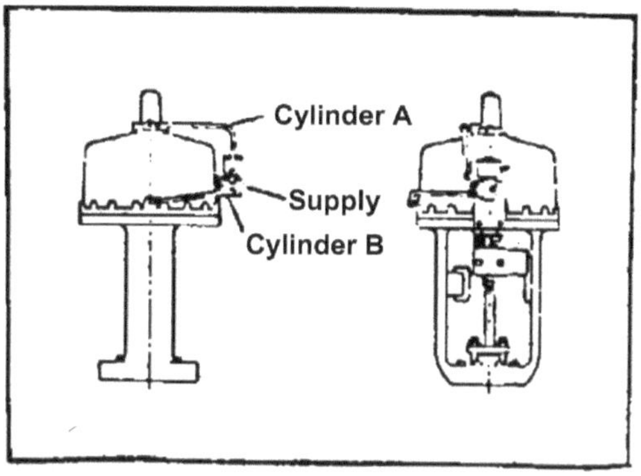

FIGURE 16.42 Typical four-way solenoid valve mounted on actuator.

and straight. They should be firmly anchored to floors or rigid structural members.

Where, however, the final control element (damper or louver) is installed in ducting, which changes position relative to fixed structures, e.g., due to thermal expansion, or where necessary to prevent long linkages, the actuator should be installed on the ducting. Ladders and walkways should then be provided for easy access.

Flexible hoses should be provided where the cylinders are not in a fixed and rigid position.

Oil misters should be installed in the line-up when the cylinder requires lubrication.

16.29.7 Hydraulic Cylinder Actuators

Hydraulic cylinder actuators are usually an integral part of the control valve. Care should be taken that the valve is installed with the cylinder in the position prescribed by the manufacturer.

Where necessary, access facilities should be available for maintenance and manual operation.

Supply and return piping should be of sufficient diameter to prevent excessive pressure drop, especially where fast response in emergencies is required.

Isolating valves should be installed close to the cylinder. Cylinder actuators on valves with provisions for manual (handwheel) control should be provided with external bypass valve. When manually controlled hydraulic operation near the actuator is required, four-way valves should be installed in the line-up.

Vent valves should be provided at all high points in the hydraulic piping.

When the valve is subject to vibration or installed in equipment subject to thermal expansion, flexible hoses should be provided. These hoses should be of corrugated stainless steel, or other suitable flexible material properly welded or fast-ended to steel unions.

For pilot-operated hydraulic actuators, dual filters should be installed immediately upstream of the actuator.

16.30 BLOCK AND BYPASS VALVES

A block and bypass valve system may not be necessary where the process can be shut down to repair the control valve without significant economic loss, or where the process cannot be feasibly operated on the bypass.

The consequences of shutting down a process unit to perform a simple task (such as replacing control valve packing) should always be considered.

In cases where the block and bypass valves are not used, some users require that the control valve be equipped with a handwheel or other operating device. A permanent side mounted hand wheel should be fitted to the control valve.

Where the cost of the hand wheel is greater than the cost of block and bypass valves, the latter should be provided except on hydrogen services and protective services.

Where the greatest flexibility is to be provided for future expansion, the block valves upstream and downstream of the control valve should be line size. In situations where the control valve is two or more sizes smaller than line size, the block valves may be one size smaller than line size.

A bypass connection and valve should be installed around each control valve unless other means are available for manual control when the control valve is out of service. The bypass valve should be one size larger than the control valve but may, by specific agreement, be the same size as the control valve.

If control valves are installed without a valved bypass, the piping layout should be such that block and bypass valves can easily be included later.

The layout of the manifold should be such that a line size control valve can be installed later (with sufficient clearances).

The block valves on either side of the control valve should be gate valves of a size equal to line size. When, however, the control valve is two or more nominal sizes smaller than the line size, the block valves should be of intermediate size.

Block and bypass valve assemblies should be provided in the following instances:

1. Where a valve controls a service common to a number of plants.
2. Where valves are in continuous operation and there is not sufficient assurance of reliability over the anticipated period between plant overhauls, e.g., on erosive or corrosive service or where the temperature is below 0°C or above 180°C.

   The cost of a failure should also be taken into account.
3. Where failure of the control valve would necessitate continuous operator attention, e.g., on the fuel control to heaters.

Block and bypass valve assemblies should be avoided in the following instances:

1. Around three-way valves.
2. Around self-acting steam pressure reducing valves.
3. Around control valves forming part of a protective system, unless agreed to otherwise by the Company.

The C_V factor of the bypass valve should be at least equal to and not more than twice the C_V factor of the control valve. The bypass valve should be a globe valve for sizes 4 in. and smaller, and a gate valve for sizes 6 in. and larger.

Provisions should be made for draining and/or depressurizing of the control valve.

At least one drain valve should be provided adjacent to the failed open control valve, either upstream or downstream, depending on the physical layout.

For failed closed (air to open) control valves, two drain valves are recommended.

For hazardous, corrosive, or toxic fluids, more extensive provisions may be required.

Control valves with handwheels are generally installed in the piping without block valves or valved bypass.

Block valves may, however, be required on longlines to prevent excessive loss of product or air pollution when the control valve is removed from the line.

For control valves without handwheels a manifold assembly comprising block valves and a valved bypass should be provided. However, no bypass should be provided for safety shutoff valves, depressurizing valves, and on some applications where solids suspended in the stream might collect and block the bypass valve.

The provision of handwheels, bypasses etc., is governed almost entirely by operational considerations; the *P* and *I* diagrams should therefore indicate the solution adopted for each application.

16.31 DIMENSIONS OF CONTROL VALVES

For the piping layout the dimensions of the complete control valve, including actuator diameter, and its distance to control valve body connections, should be taken into account.

All dimensions can vary from application to application except for the face-to-face dimensions of 300 lbs ANSI flanged globe-type control valve bodies up to and including 8 in. size (which are standardized).

It is stressed that manufacturer's certified drawings must be consulted for detailing the piping work. For instance, an otherwise normal type of control valve may have an oversize actuator; a top-mounted handwheel or extension bonnet will considerably increase total height; angle-type control valves may have outlet connections not equal in size to the inlet, etc.

16.32 MANIFOLD PIPING ARRANGEMENT

The manifold piping should be arranged to provide flexibility for removing control valves, particularly where ring type joints are used.

Flexibility of piping is also necessary to keep excessive stresses from being induced in the body of the control valve.

Manifold arrangements illustrative of various typical situation are presented in a series of diagrams on the following pages. These may be quite suitable for use as shown, or may be made suitable for specific requirements by minor modifications.

Six control valve manifold types are presented in this recommendation with space estimates for various sizes. Each of these six types consists of a straight-through globe control valve, isolating upstream block valves, and bypass piping with a manually operated valve.

For additional information and dimensions for all ANSI classes, reference should be made to ISA RP-75.06 "Control Valve Manifold Design." For ease of reference, control valve manifold dimensions for ANSI class 300 are presented here (Fig. 16.43)

Note:

Dimensions shown are suggested piping dimensions and may vary depending upon actual dimensions of components being used (Table 16.6, Fig. 16.44).

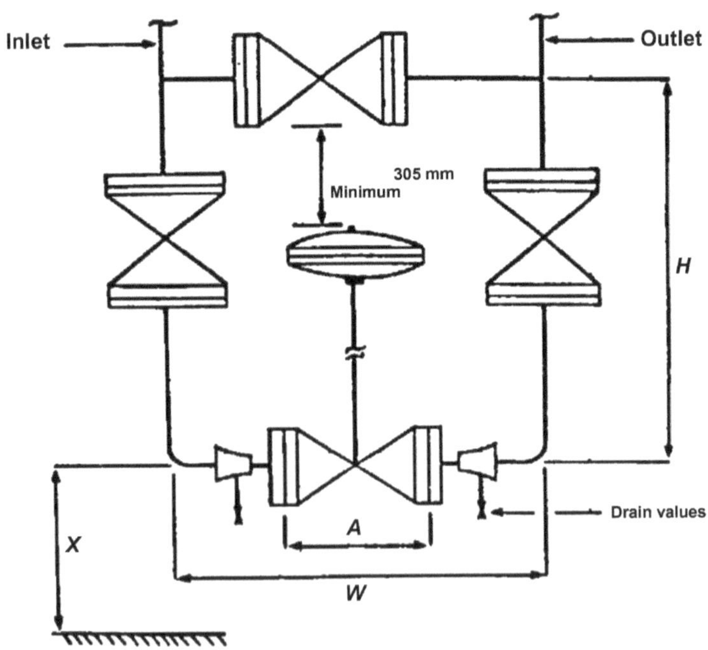

FIGURE 16.43 Elevation of type I control valve manifold.

TABLE 16.6 Type I Control Valve Manifold Dimensions for Ansi Class 300

Inches							Millimeters				
A^a	W^b	H^b	X^b	Actual Manifold Pipe Outside Diameter	Nominal Size		Actual Manifold Pipe Outside Diameter	A^a	W^b	H^b	X^b
					Manifold pipe	Control Valve					
7¾	27	39	23	1.315	1	1	33.4	197	690	990	580
9¼	27	39	23	1.900	1½	1½	48.3	235	690	990	580
7¾	27	39	23	1.900	1½	1	48.3	197	690	990	580
10½	27	39	23	2.375	2	2	60.3	267	690	990	580
9¼	27	39	23	2.375	2	1½	60.3	235	690	990	580
7¾	27	39	23	2.375	2	1	60.3	197	690	990	580
12½	30	42	27	3.500	3	3	88.9	317	760	1070	690
10½	30	42	27	3.500	3	2	88.9	267	760	1070	690
9¼	30	42	27	3.500	3	1½	88.9	235	769	1070	690
14½	35	43	30	4.500	4	4	114.3	368	890	1090	760
12½	35	43	30	4.500	4	3	114.3	317	890	1090	760
10½	35	43	30	4.500	4	2	114.3	267	890	1090	760
14½	45	54	39	6.625	6	4	168.3	368	1140	1370	990
12½	45	54	39	6.625	6	3	168.3	317	1140	1370	990
185/8	55	57	46	8.625	8	6	219.1	473	1400	1450	1170
14½	55	57	46	8.625	8	4	219.1	368	1400	1450	1170

[a] *Actual dimensions from manufacturers drawings.*
[b] *Suggested dimensions.*

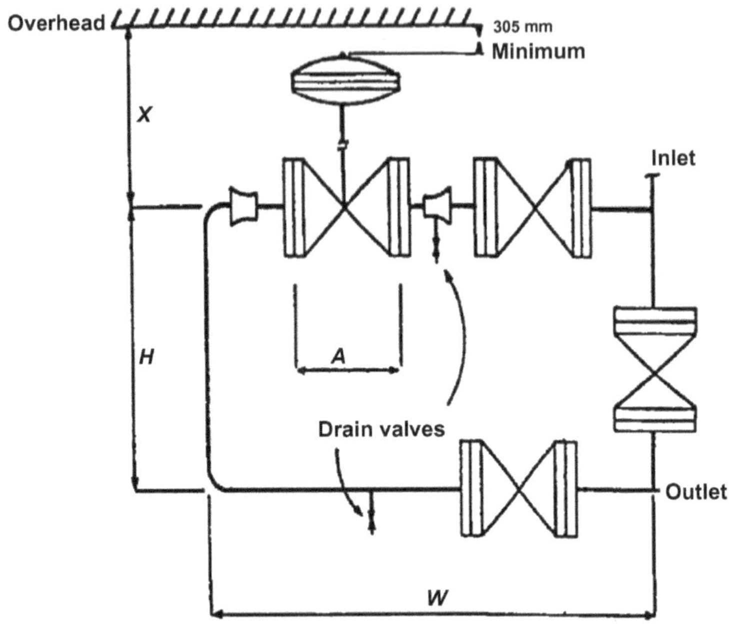

FIGURE 16.44 Elevation of type II control valve manifold.

Note:

Dimensions shown are suggested piping dimensions and may vary depending upon actual dimensions of components being used (Table 16.7, Fig. 16.45).

Note:

Dimensions shown are suggested piping dimensions and may vary depending upon actual dimensions of components being used (Table 16.8, Fig. 16.46).

Note:

Dimensions shown are suggested piping dimensions and may vary depending upon actual dimensions of compenents being used (Table 16.9, Fig. 16.47).

Note:

Dimensions shown are suggested piping dimensions and may vary depending upon actual dimensions of components being used (Table 16.10, Fig. 16.48).

Note:

Dimensions shown are suggested piping dimensions and may vary depending upon actual dimensions of components being used (Table 16.11).

TABLE 16.7 Type II Control Valve Manifold Dimensions for ANSI Class 300

Inches							Millimeters				
A^a	W^b	H^b	X^b	Actual Manifold Pipe Outside Diameter	Nominal Size		Actual Manifold Pipe Outside Diameter	A^a	W^b	H^b	X^b
					Manifold Pipe	Control Valve					
7¾	44	25	37	1.315	1	1	33.4	197	1120	640	940
9¼	44	25	37	1.900	1½	1½	48.3	235	1120	640	940
7¾	44	25	37	1.900	1½	1	48.3	197	1120	640	940
10½	44	25	37	2.375	2	2	60.3	267	1120	640	940
9¼	44	25	37	2.375	2	1½	60.3	235	1120	640	940
7¾	44	25	37	2.375	2	1	60.3	197	1120	640	940
12½	48	29	39	3.500	3	3	88.9	317	1220	740	990
10½	48	29	39	3.500	3	2	88.9	267	1220	740	990
9¼	48	29	39	3.500	3	1½	88.9	235	1220	740	990
14½	56	33	40	4.500	4	4	114.3	368	1430	840	1020
12½	56	33	40	4.500	4	3	114.3	317	1430	840	1020
10½	56	33	40	4.500	4	2	114.3	267	1430	840	1020
14½	70	43	50	6.625	6	4	168.3	368	1780	1090	1270
12½	70	43	50	6.625	6	3	168.3	317	1780	1090	1270
185/ 8	78	50	52	8.625	8	6	219.1	473	1990	1270	1320
14½	78	50	52	8.625	8	4	219.1	368	1990	1270	1320

[a] *Actual dimensions from manufacturers drawings.*
[b] *Suggested dimensions.*

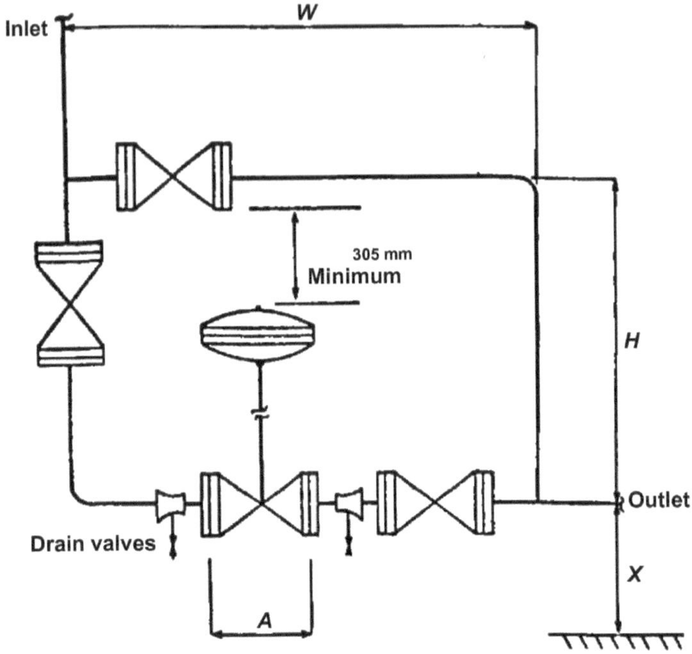

FIGURE 16.45 Elevation of type III control valve manifold.

16.33 SUMMARY OF INSTALLATION PRACTICES

The following detailed items are recommended for consideration in every installation:

The air supply used to operate pneumatic control valves should be free from oil and moisture.

Follow the manufacturer's recommendations for the hydraulic system to power hydraulically driven control valves.

Follow the manufacturer's recommendations for the electronic system for electrically or electrohydraulically-driven control valves.

Inlet piping to control valves with small passage ways, should be fitted with appropriate filters to eliminate internal valve damage from foreign matter in the piping system.

Do not put excessive stress on valve bodies when installing in the system. This is particularly important for split body valves.

Before initial start-up and after a maintenance shutdown, install screens ahead of the control valve to collect pipe scale, rust, and other debris.

TABLE 16.8 Type III Control Valve Manifold Dimensions for ANSI Class 300

Inches							Millimeters				
A^a	W^b	H^b	X^b	Actual Manifold Pipe Outside Diameter	Nominal Size		Actual Manifold Pipe Outside Diameter	A^a	W^b	H^b	X^b
					Manifold Pipe	Control Valve					
7¾	44	39	23	1.315	1	1	33.4	197	197	990	580
9¼	44	39	23	1.900	1½	1½	48.3	235	235	990	580
7¾	44	39	23	1.900	1½	1	48.3	197	197	990	580
10½	44	39	23	2.375	2	2	60.3	267	267	990	580
9¼	44	39	23	2.375	2	1½	60.3	235	235	990	580
7¾	44	39	23	2.375	2	1	60.3	197	197	990	580
12½	48	42	27	3.500	3	3	88.9	317	317	1070	690
10½	48	42	27	3.500	3	2	88.9	267	267	1070	690
9¼	48	42	27	3.500	3	1½	88.9	235	235	1070	690
14½	56	43	30	4.500	4	4	114.3	368	368	1090	760
12½	56	43	30	4.500	4	3	114.3	317	317	1090	760
10½	56	43	30	4.500	4	2	114.3	267	267	1090	760
14½	70	54	39	6.625	6	4	168.3	368	368	1370	990
12½	70	54	39	6.625	6	3	168.3	317	317	1370	990
185/8	78	57	46	8.625	8	6	219.1	473	473	1450	1170
14½	78	57	46	8.625	8	4	219.1	368	368	1450	1170

[a]*Actual dimensions from manufacturers drawings.*
[b]*Suggested dimensions.*

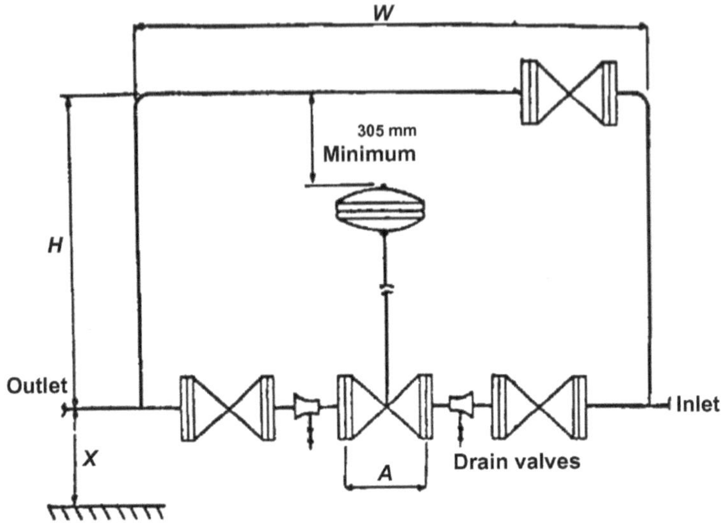

FIGURE 16.46 Elevation of type IV control valve manifold.

Whenever possible, the piping system should be fitted with a spool piece and flushed out prior to control valve installation.

If the valve is to operate in a dusty atmosphere, install a rubber or plastic boot around the stem to protect its polished surface from damage.

Be sure to follow all of the manufacturer's instructions for adjustments and switch positions for the accessories. For example, do not leave the valve positioner bypass switch in the bypass position.

If the control valve is to be removed from the system after installation, be sure that all block valves are closed and tagged. If the control valve contains damaging fluids or contaminants, it should be tagged accordingly, for proper cleaning, prior to disassembly.

Be sure the valve is installed with the flow direction arrow in the proper direction. Cases have been reported where the manufacturer furnished a valve with the arrow pointing in the wrong direction.

Review all of the control valve manufacturer's specific instructions prior to installation.

16.34 CONTROL VALVE TEST REQUIREMENTS (GENERAL)

For testing, all valves should be completely assembled with packing box fully packed and made up handtight. The valve stem may be lightly

TABLE 16.9 Type IV Control Valve Manifold Dimensions

Inches							Millimeters				
A^a	W^b	H^b	X^b	Actual Manifold Pipe Outside Diameter	Nominal Size		Actual Manifold pipe Outside Diameter	A^a	W^b	H^b	X^b
					Manifold Pipe	Control Valve					
7¾	58	39	23	1.315	1	1	33.4	197	1470	990	580
9¼	58	39	23	1.900	1½	1½	48.3	235	1470	990	580
7¾	58	39	23	1.900	1½	1	48.3	197	1470	990	580
10½	58	39	23	2.375	2	2	60.3	267	1470	990	580
9¼	58	39	23	2.375	2	1½	60.3	235	1470	990	580
7¾	58	39	23	2.375	2	1	60.3	197	1470	990	580
12½	67	42	27	3.500	3	3	88.9	317	1700	1070	690
10½	67	42	27	3.500	3	2	88.9	267	1700	1070	690
9¼	67	42	27	3.500	3	1½	88.9	235	1700	1070	690
14½	74	43	30	4.500	4	4	114.3	368	1880	1090	760
12½	74	43	30	4.500	4	3	114.3	317	1880	1090	760
10½	74	43	30	4.500	4	2	114.3	267	1880	1090	760
14½	97	54	39	6.625	6	4	168.3	368	2460	1370	990
12½	97	54	39	6.625	6	3	168.3	317	2460	1370	990
185/8	109	57	46	8.625	8	6	219.1	473	2770	1450	1170
14½	109	57	46	8.625	8	4	219.1	368	2770	1450	1170

[a]*Actual dimensions from manufacturers drawings.*
[b]*Suggested dimensions.*

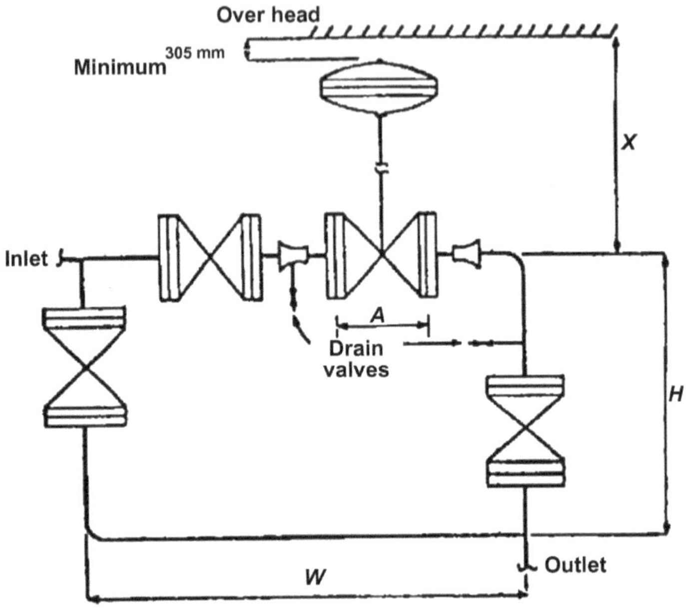

FIGURE 16.47 Elevation of type V control valve manifold.

lubricated. If the valve is equipped with a positioner, tests should be performed with the positioner bypassed.

Control valves should be checked for smooth stroking and correct input span. Attention should be given to spring action, split-range operation, reverse-action positioners, etc.

Packing boxes of all control valves should be inspected for presence of lantern ring and correct type of packing. Where packing is unsuitable for the intended service, or damaged during inspection, new packing should be applied.

Operational tests should consist of measuring the valve stem position for increasing and decreasing input signals with the positioner bypassed. Performance should meet the following standards:

1. Stem position error should not exceed $\pm 5\%$ of rated travel.
2. Hysteresis plus deadband should not exceed 5% of rated travel.

Actuators for variable pitch fans on air-cooled heat exchangers should be tested in situ. Final adjustment should be done when plant is in operation:

1. Minimum pitch (usually negative to compensate for heat conduction) should cause zero air flow.
2. Maximum pitch should coinside with maximum allowable motor current.

TABLE 16.10 Type V Control Valve Manifold Dimensions for ANSI Class 300

Inches							Millimeters				
A^a	W^b	H^b	X^b	Actual Manifold pipe Outside Diameter	Nominal Size		Actual Manifold Pipe Outside Diameter	A^a	W^b	H^b	X^b
					Manifold Pipe	Control Valve					
7¾	44	36	37	1.315	1	1	33.4	197	1120	910	940
9¼	44	36	37	1.900	1½	1½	48.3	235	1120	910	940
7¾	44	36	37	1.900	1½	1	48.3	197	1120	910	940
10½	44	36	37	2.375	2	2	60.3	267	1120	910	940
9¼	44	36	37	2.375	2	1½	60.3	235	1120	910	940
7¾	44	36	37	2.375	2	1	60.3	197	1120	910	940
12½	48	39	39	3.500	3	3	88.9	317	1220	990	990
10½	48	39	39	3.500	3	2	88.9	267	1220	990	990
9¼	48	39	39	3.500	3	1½	88.9	235	1220	990	990
14½	56	39	40	4.500	4	4	114.3	368	1430	990	1020
12½	56	39	40	4.500	4	3	114.3	317	1430	990	1020
10½	56	39	40	4.500	4	2	114.3	267	1430	990	1020
14½	70	46	50	6.625	6	4	168.3	368	1780	1170	1270
12½	70	46	50	6.625	6	3	168.3	317	1780	1170	1270
185/8	78	50	52	8.625	8	6	219.1	473	1990	1270	1320
14½	78	50	52	8.625	8	4	219.1	368	1990	1270	1320

[a]*Actual dimensions from manufacturers drawings.*
[b]*Suggested dimensions.*

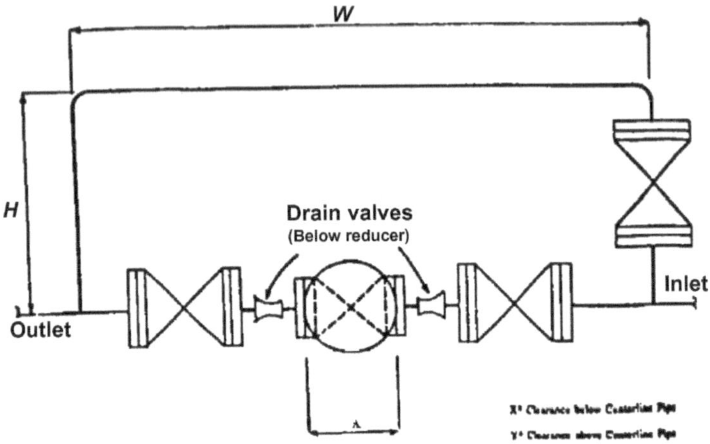

FIGURE 16.48 Plan view of type VI control valve manifold.

Cylinder actuators for dampers, etc., should be tested in situ for correct operation.

The following is considered the minimum inspection criteria for factory assembled control valves:

1. Visual examination, using the assembly drawing;
2. Hydrostatic test;
3. Leakage test;
4. Hysteresis check;
5. Valve-travel check by the operator;
6. Operational check of all accessories limit switches, valve positioners, etc.;
7. Electrical tests (Megger and multimeter) for electrical devices;
8. Packaging and shipping checks to specifications and procedures.

If these tests are properly performed at the factory, the receipt inspection will only require visual examination. If these tests were not performed at the factory, or if there is reason to suspect problems, these tests should be performed on site, as required.

During the start-up of any new facilities, care should be taken to keep scale, welding rods, and other foreign material from plugging or damaging control valves. One method is to remove the valve and substitute a spool piece during flashing operation.

TABLE 16.11 Type VI Control Valve Manifold Dimensions for ANSI Class 300

Inches								Millimeters					
A[a]	*W*[b]	*H*[b]	*X*[b]	*V*[b]	Actual Manifold Pipe Outside Diameter	Nominal Size		Actual Manifold Pipe Outside Diameter	*A*[a]	*W*[b]	*H*[b]	*X*[b]	*V*[b]
						Manifold Pipe	Control Valve						
7¾	58	21	23	37	1.315	1	1	33.4	197	1470	530	580	940
9¼	58	21	23	37	1.900	1½	1½	48.3	235	1470	530	580	940
7¾	58	21	23	37	1.900	1½	1	48.3	197	1470	530	580	940
10½	58	21	23	37	2.375	2	2	60.3	267	1470	530	580	940
9¼	58	21	23	37	2.375	2	1½	60.3	235	1470	530	580	940
7¾	58	21	23	37	2.375	2	1	60.3	197	1470	530	580	940
12½	67	26	27	39	3.500	3	3	88.9	317	1700	660	690	990
10½	67	26	27	39	3.500	3	2	88.9	267	1700	660	690	990
9¼	67	26	27	39	3.500	3	1½	88.9	235	1700	660	690	990
14½	74	29	30	40	4.500	4	4	114.3	368	1880	740	760	1020
12½	74	29	30	40	4.500	4	3	114.3	317	1880	740	760	1020
10½	74	29	30	40	4.500	4	2	114.3	267	1880	740	760	1020
14½	97	36	39	50	6.625	6	4	168.3	368	2460	910	990	1270
12½	97	36	39	50	6.625	6	3	168.3	317	2460	910	990	1270
185/8	109	41	46	52	8.625	8	6	219.1	473	2770	1040	1170	1320
14½	109	41	46	52	8.625	8	4	219.1	368	2770	1040	1170	1320

[a]*Actual dimensions from manufacturers drawings.*
[b]*Suggested dimensions.*

References

Bahadori, A., Vuthaluru, H.B., 2009. Prediction of bulk modulus and volumetric expansion coefficient of water for leak tightness test of pipelines. Int. J. Press. Vess. Pip. 86, 550–554.

Buijvoets, J., 2009. A new method of offshore pipe laying: an economic solution to exploit small offshore oil-and gasfields. Global Pipeline Monthly 5 (8).

De Waard, C., Lotz, U., Milliams, D.E., 1991. Predictive model for CO_2 corrosion engineering in wet natural gas pipelines. Paper 577. Corrosion 47 (12), 976.

De Waard, C., 1995. Influence of Liquids Flow Velocity on CO_2 Corrosion: A Semi-Emperical Model. Paper No.128. NACE, Houston, TX.

Duncan, R.N., 1980. Materials performance in Khuff Gas Service. Mater. Performance45.

Eriksrud, E., Sontvedt, T., 1984. Effect of flow on CO_2 corrosion rates in real and synthetic formation waters. Advances in CO_2 Corrosion. NACE, Houston, TX.

Halali, M.A., Azari, V., Arabloo, M., Mohammadi, A.H., Bahadori, A., 2015. Application of a radial basis function neural network to estimate pressure gradient in water–oil pipelines. J. Taiwan Inst. Chem. Eng. 58, 189–202.

Ikeda, A., Ueda, M., Mukai, S., 1985. Influence of environmental factors on corrosion in CO_2 source well. Advances in CO_2 Corrosion. NACE, Houston, TX.

Johnsen, R. A review of materials for pipelines transporting unprocessed well fluid. In: Int. Conf. on Pipeline Protection & Maintenance, Stavanger, Norway, 20–21 September 1988.

Kolts, J., Buck, E., Erickson, D.D., Achour, M., Stuever, W., 1990. Corrosion prediction and design considerations for internal corrosion in continuously inhibited wet gas pipelines. In: UK Corrosion 90, vol. 3, p. 217.

Liu, H., Hu, J., Yang, J., 2004. Existing state and developing trend of detection technology for long oil gas transmission pipelines in China. Nat. Gas Ind. 24 (11), 147–150.

Liu, X.-Y., Mao, Q.-J., Liu, L.-J., Li, X.-Q., Liu, J.-J., 2004. Affecting factors of the temperature drop of oil-gas-water three-phase buried pipeline. J. Eng. Thermophys. 8, 1343–1346.

Makarenko, V.D., Petrovs'kyi, V.A., Chernov, V.Yu., 2003. Mechanism of hydrogen delamination of pipe steels of oil and gas pipelines. Mater. Sci. 39 (6), 895–900.

Milliams, D.E., Kroese, C.J., 1979. Aqueous corrosion of steel by H_2S and H_2S/CO_2. In: III International Conference on Internal and External Pipe Protection, Paper HI. Minerals Management Service Database, Gulf of Mexico Region, New Orleans, LA.

NACE Standard MR-01-75, 1991. Sulphide Stress Cracking Resistant Metallic Materials for Oil Field Equipment. NACE, Houston, TX, p. 1984.

Okpobiri, G.A., Ajienka, J.A., 1989. Optimum energy cost design of oil and gas pipelines-part 1. J. Pipelines 7 (3), 235–241.

Palacios, C.A., Shadley, J.R., 1991. Performance of oilfield corrosion inhibitors under single and two-phase flow conditions. In: Corrosion 91, NACE, Houston, TX, 475.

RP-14E API, Recommended Practice for Design and Installation of Offshore Production Platform Piping Systems. API, Washington DC.

Satoh, K., Yamamoto, K., Kagawa, N., 1984. Prevention of CO2 Corrosion in Gas Gathering Systems. Advances in CO_2 Corrosion. NACE, Houston, TX.

Schmitt, G., 1984. CO_2 corrosion of steels: an attempt to range parameters and their effects. Advances in CO_2 Corrosion. NACE, Houston, TX.

Schmitt, G., Simon, T., Hausleor, R.H., 1990. CO_2 erosion corrosion and its inhibition under extreme shear stress-1. Development of Methodology, Corrosion 90, NACE, Houston, TX, .22.

Simon Thomas, M.J.J., De Waard, C., Smith, L.M., Controlling factors in the rate of CO_2 corrosion. In: UK Corrosion 87. Brighton, 26–28 October 1987, p. 147.

John, S., Smart, III, 1991. The meaning of API RP-14E formula for erosion corrosion in oil and gas production. In: Corrosion 1991, NACE, Houston, TX, Paper 468.

Videm, K., Dugstad, A., 1987. Effect of flowrate, pH, Fe^{2+} concentration and steel quality on the CO_2 corrosion of carbon steels. Corrosion 1987. NACE, Houston, TX, Paper 42.

Yang, F.-P., Zhang, L., Luo, J.-H., Zhao, X.-W., Zhang, G.-L., 2013. Study on the yield behavior of oil and gas pipe in typical stress states. Eng. Mech. 30 (11), 293–297.

Zhang, X.Y., Feng, J., 2013. Oil and gas pipeline residual strength research based on reliability. Appl. Mech. Mater. 318, 562–566.

ANSI (AMERICAN NATIONAL STANDARD INSTITUTION)

ANSI/ASME B31.3 "Chemical Plant and Petroleum Refinery Piping"
ANSI/ASME B31.5 "Refrigeration Piping"
ANSI/ASME B16.5 "Pipe Flanges and Flanged Fittings"

ASME (AMERICAN SOCIETY OF MECHANICAL ENGINEERS) BOILER AND PRESSURE VESSEL CODE

Section IX "Welding and Brazing Qualification"
Section V "Non-Destructive Examination"

ANSI/ASME (AMERICAN NATIONAL STANDARDS INSTITUTE/AMERICAN SOCIETY OF MECHANICAL ENGINEERS)

ANSI/SAME B31.1 "Power Piping" (1989)
ANSI/ASME B31.3 "Chemical Plant and Petroleum Refinery Piping" (1990)
ANSI/ASME B31.8 "Gas Transmission and Distribution Piping Systems"

AWS (AMERICAN WELDING SOCIETY)

ANSI/AWS A3.0 "Standard Welding Terms and Definitions"

ASNT (AMERICAN SOCIETY FOR NON-DESTRUCTIVE TESTING)

SNT-TC-1A "Recommended Practice for Non-Destructive Personnel"

BSI (BRITISH STANDARDS INSTITUTION)

BS 3923 "Methods for Ultrasonic Examination of Welds"

ANSI (AMERICAN NATIONAL STANDARD INSTITUTE)/ASME (AMERICAN SOCIETY OF MECHANICAL ENGINEERS)

B16.5 "Pipe Flanges and Flanged Fittings"
B31.3 "Chemical Plant and Petroleum Refinery Piping"
B36.10 "Welded and Seamless Wrought Steel Pipe"
B36.19 "Stainless Steel Pipe"
BVP Section VIII Div. I "Un-Fired Pressure Vessels"

API (AMERICAN PETROLEUM INSTITUTE)

Publication 941 "Steels for Hydrogen Service at Elevated Temperatures and Pressures in Petroleum Refineries and Petrochemical Plants"

NACE (NATIONAL ASSOCIATION OF CORROSION ENGINEERS)

Paper No. 67

API (AMERICAN PETROLEUM INSTITUTE)

5L-42nd. Edition, Jan. 2000 "Specification for Line Pipe"

ASTM (AMERICAN SOCIETY FOR TESTING AND MATERIALS)

7	"Definition of Terms Relating to Metallography"
E−7	"Definition of Terms Relating to Metallography"
E−112	"Standard Test Methods for Determining Average Grain Size"
E−165	"Practice for Liquid Peneterant Examination"
E−309	"Standard Practice for Eddy Current Examination of Steel Tubular Products Using Magnetic Saturation"
E−570	"Standard Practice for Flux Leakage Examination of Ferromagnetic Steel Tubular Products"
E−709	"Practice for Magnetic Particle Examination"
F−21	"Standard Test Method for Hydrophobic Surface Films by the Atomizer Test"

DIN (GERMAN STANDARD)

SEL−072	"Ultrasonically Tested Heavy Plates Technical Delivery Specifications"

ISO (INTERNATIONAL STANDARD)

1027	"Radiographic Image Quality Indicator for NonDestructive Testing"
2566-1	"Steel, Conversion of Elongation Values Part 1 Carbon and Low Alloy Steels"
9001	"Quality Systems-Model for Quality Assurance in Design/ Development, Production, Installation and Servicing"
9712	"Non-Destructive Testing, Qualification and Certification of Personnel"
10474	"Steel and steel products, Inspection Documents"

NACE (NATIONAL ASSOCIATION OF CORROSION ENGINEERS)

TM 0177-96	"Laboratory Testing of Metals for Resistance to Sulphide Stress Cracking in H2S Environments"
TM 0284-96	"Evaluation of Pipeline Steels for Resistance to Stepwise Cracking"

ASME (AMERICAN SOCIETY OF MECHANICAL ENGINEERS)

Section VIII	"Rules for Construction of Pressure Vessels"
Section V	"Non-Destructive Examination"
Section IX	"Qualification STD. for Welding and Brazing Procedures"
Section II	"Material Specification"

ANSI (AMERICAN NATIONAL STANDARD INSTITUTE)

B16.5	"Steel Pipe Flanges and Flanged Fittings"
B16.25	"Butt Welding Ends"
B31.8	"Gas Transmission & Distribution Systems"
B31.4	"Liquid Transportation Systems for Hydrocarbons"

ASTM **(AMERICAN SOCIETY FOR TESTING AND MATERIAL)**

A193/A193M "Specification for Alloy Steel and Stainless Steel Bolting Materials for High Temperature Services"

A194/A194M "Specification for Carbon and Alloy Steel Nuts for Bolts for High Pressure and High Temperature Service"

E92 "Test Method for Vickers Hardness of Metallic Materials"

API **(AMERICAN PETROLEUM INSTITUTE)**

5L "Specification for Line Pipe"

1104 "Welding of Pipelines and Related Facilities"

BSI **(BRITISH STANDARD INSTITUTION)**

BS 5500 "Specification for Unfired Fusion Welded Pressure Vessel"

BS 1133 "Packaging Code"

IP **(INSTITUTE OF PETROLEUM)**

Part 15 "Model Code of Safe Practice Electrical Part 1"

NACE **(NATIONAL ASSOCIATION OF CORROSION ENGINEERS)**

MR-0175 "Sulphide Stress Cracking Resistant Metallic Materials for Oil Field Equipment"

MSS **(MANUFACTURERS STANDARDIZATION SOCIETY)**

SP-58: 2002 "Pipe Hangers and Supports Materials, Design and Manufacturer"

SP-69: 2002 "Pipe Hangers and Supports Selection and Application"

SP-89: 1998 "Pipe Hangers and Supports Fabrication and Installation Practices

SP-90:2000 "Guidelines on Terminology for Pipe Hangers and Supports"

ASTM **(AMERICAN SOCIETY FOR TESTING AND MATERIALS)**

ASTM A6/A6M-05 "Standard Specification for General Requirements for Rolled Structural Steel Bars, Plates, Shapes, and Sheet Piling."

ASTM A123/A123M-02 "Standard Specification for Zinc (Hot Dip Galvanized) Coatings on Iron and Steel Products."

ASTM A153/A153M-05 "Standard Specification for Zinc Coating (Hot-Dip) on Iron and Steel Hardware."

ASTM A403/A403M-04 "Standard Specification for Wrought Austenitic Stainless Steel Piping Fittings."

ASTM B633-98e1 "Standard Specification for Electrodeposited Coatings of Zinc on Iron and Steel."

ASTM A653/A653M-04a "Standard Specification for Steel Sheet, Zinc Coated (Galvanized) or Zinc–Iron Alloy-Coated (Galvannealed) by the Hot-Dip Process."

ASTM B695-04 "Standard Specification for Coatings of Zinc Mechanically Deposited on Iron and Steel."

ASTM B696-00(2004)e1 "Standard Specification for Coatings of Cadmium Mechanically Deposited."

ASTM B766-86 (2003) "Standard Specification for Electrodeposited Coatings of Cadmium."

ASME (AMERICAN SOCIETY OF MECHANICAL ENGINEERS)

ASME B31.3-2004 "Process Piping"
ASME B31.4-2002 "Pipeline Transportation Systems for Hydrocarbons, and other Liquids"
ASME B31.8-2003 "Gas Transmission and Distribution Piping Systems"
ASME Section IX "Welding and Brazing Qualifications"

UL (UNDERWRITERS LABORATORIES)

UL 203:2005 "Pipe Hanger Equipment for Fire Protection Service"

ASTM (AMERICAN SOCIETY FOR TESTING AND MATERIALS)

D-887 "Practice for Sampling Water-Formed Deposits"

ASME/ANSI (AMERICAN SOCIETY MECHANICAL ENG.)

ASME/ANSI B 31.8 "Gas Transmission and Distribution Piping System"
ASME/ANSI B 31.4 "Liquid Transportation Systems for Hydrocarbons, Liquid Petroleum Gas, Anhydrous Ammonia and Alcohols"

API (AMERICAN PETROLEUM INSTITUTE)

API RP 1110 "Recommended Practice for Pressure Testing of Liquid Petroleum Pipeline"

ASME (AMERICAN SOCIETY OF MECHANICAL ENGINEERS)

B31.4 "Pipeline Transportation Systems for Liquid Hydrocarbon and other Liquids" 2002 Edition
B31.8 "Gas Transmission and Distribution Systems" 2001 Edition

API (AMERICAN PETROLEUM INSTITUTE)

6D "Specification for Pipeline Valves"
Spec. 5L "Specification for Line Pipe"
RP 1102 "Steel Pipelines Crossing Rail Roads and Highways"
1160 "Managing Systems Integrity for Hazardous Liquid Pipelines"

ASME (AMERICAN SOCIETY OF MECHANICAL ENGINEERS)

ASME B31.8 "Gas Transmission and Distribution Piping Systems Addenda B31.8A-1990 Addenda B31.8B-1990, Addenda B31.8C-1992."
ASME B31.4 "Liquid Transportation Systems for Hydrocarbons"

ASTM (AMERICAN SOCIETY FOR TESTING AND MATERIALS)

E92 "Standard Test Method for Vickers Hardness of Metallic Materials"

BSI (BRITISH STANDARDS INSTITUTION)

BS 2910 "Methods for Radiographic Examination of Fusion-Welded Circumferential Butt Joints in Steel Pipes"
BS 3971 "Specification for Image Quality Indicator for Industrial Radiography (Including Guidance on their Use)"
BS 5135 "Specification for Arc Welding of Car bon and Carbon Manganese Steels"

NACE (NATIONAL ASSOCIATION OF CORROSION ENGINEERS)

NACE MR-0175 "Sulfide Stress Cracking Resistant Metallic Materials for Oilfield Equipment"

ISO (INTERNATIONAL ORGANIZATION FOR STANDARDIZATION)

ISO 148 "Steel-Charpy Impact Test (V-notch)"

ASME (AMERICAN SOCIETY OF MECHANICAL ENGINEERS)

B-31.1:2004 "Power Piping"
B-31.3 :2004 "Process Piping"
B-36.10M :2004 "Welded and Seamless Wrought Steel Pipe"
B-16.5 :2003 "Pipe Flanges and Flanged Fittings"

API (AMERICAN PETROLEUM INSTITUTE)

RP-521 "Guide for Pressure-Relieving and Depressuring Systems"

BSI (BRITISH STANDARD INSTITUTION)

BS 3274:1960 "Tubular Heat Exchangers for General Purposes"
BS 3293:1960 "Specification for Carbon Steel Pipe Flanges (over 24" nominal size) for the Petroleum Industry"
BS 4485:1988 "Water Cooling Towers"
BS 6129-1:1981 "Selection and Application of Bellows Expansion Joints"

NACE (NATIONAL ASSOCIATION OF CORROSION ENGINEERS)

MR-01-75 "Standard Material Requirement Sulfide Stress Cracking Resistant Metallic Material for Oil Field Equipment"

NFPA (NATIONAL FIRE PROTECTION ASSOCIATION)

30:2003 "Flammable and Combustible Liquids Code"

MSS (MANUFACTURERS STANDARDIZATION SOCIETY)

MSS-SP-6:2001 "Standard Finishes for Contact Faces of Pipe Flanges and Connecting-End Flanges of Valves and Fittings"
MSS-SP-58:1993 "Pipe Hangers and Supports-Materials, Design and Manufacture"
MSS-SP-69:2002 "Pipe Hangers and Supports-Selection and Application"

ISO (INTERNATIONAL ORGANIZATION FOR STANDARDIZATION)

ISO 7-1.1982 "Pipe Threads Where Pressure-Tight Joints Are Made on the Threads"
ISO 148 "Charpy Impact Test (V Notch)"

ASTM (AMERICAN SOCIETY FOR TESTING AND MATERIALS)

ASTM A-370 "Test Method and Definitions for Mechanical Testing of Steel Products"
ASTM E-125-1985 "Reference Photographs for Magnetic Particle Indication on Ferrous Casting"
ASTM E-142-1986 "Method for Controlling Quality of Radiographic Testing"
ASTM E-165-1983 "Practice for Liquid Penetrant Inspection Method"
ASTM E-186-1984 "Reference Radiographs for Heavy Walled Steel Casting"
ASTM E-446-1984 "Referenced Radiographs for Steel Casting up to 51 mm (2") in Thickness"

ANSI (AMERICAN NATIONAL STANDARD INSTITUTE)

ANSI B-16.34-1988 "Valves-Flanged, Threaded, and Welding Ends"
ANSI B-31.3 "Chemical Plant and Petroleum Refinery Piping"

UL (UNDERWRITERS LABORATORIES)

ANSI/UL 331 (1977)	"Strainers for Flammable Fluids and Anhydrous Ammonia"
ANSI/UL 1105 (1977)	"Marine Use Filters, Strainers and Separators"
ANSI/UL 1193 (1985)	"Marine Filters and Strainers for Non-Flammable Liquids"

SIS (STANDARDISERINGS KOMISSIONEN I SVERIGE)

SIS 05 5900	"Swedish Standards Institution Practice Surface Preparation Standard for Painting Steel Surface"

API (AMERICAN PETROLEUM INSTITUTE)

Guide for Inspection of Refinery Equipment (Chapter II Conditions Causing Deterioration of or Failure)

API 5L	"Line Pipe"
API RP 14E "	Recommended Practice for Design and Installation of Offshore Production Platform Piping systems"
API Publication 941	"Recommended Practice for Calculation of Heater Tube Thickness in Petroleum Refinery"

ASME (AMERICAN SOCIETY OF MECHANICAL ENGINEERS)

ASME VIII	"Boiler and Pressure Vessel Codes"

ASTM (AMERICAN SOCIETY FOR TESTING AND MATERIALS)

ASTM	Corrosion Standard Test Method: G1 to G52
B 117	"Salt Spray (Fog) Testing"
B 287	"Acetic Acid-Salt Spray (Fog) Testing"
G 1	"Recommended Practice for Preparing, Cleaning, and Evaluating Corrosion Test Specimens"
G 2	"Recommended Practice for Aqueous Corrosion Testing of Samples of Zirconium and Zirconium Alloys"
G 3	"Recommended Practice for Conventions Applicable to Electrochemical Measurements in Corrosion Testing"
G 4	"Recommended Practice for Conducting Plant Corrosion Tests"
G 5	"Recommended Practice for Standard Reference Method for Making Potentiostatic and Potentiodynamic Anodic Polarization Measurements"
G 15	"Definition of Terms Relating to Corrosion and Corrosion Testing"
G 16	"Recommended Practice for Applying Statistics to Analysis of Corrosion Data"
G 28	"Detecting Susceptibility to Intergranular Attack in Wrought Nickel-Rich, Chromium-Bearing Alloys"
G 30	"Recommended Practice for Making and Using U-Bend Stress Corrosion Test Specimens"
G 31	"Recommended Practice for Laboratory Immersion Corrosion Testing of Metals"

ISA (INSTRUMENT SOCIETY OF AMERICA)

S75.01 ANSI/ISA	"Flow Equations for Sizing Control Valves" 1986
S70.02 ANSI/ISA	"Control Valve Capacity Test Procedure" 1988
S75.03 ANSI/ISA	"Face-to-Face Dimensions for Flanged Globe Style Control Valve Bodies (ANSI Class 125, 150, 300, 600)" 1985
S75.04 ANSI/ISA	"Face-to-Face Dimensions for Flangeless and Control Valve Bodies (ANSI Class 150, 300 and 600)" 1985
S75.05 ANSI/ISA	"Control Valves Terminology" 1986
S75.06 ANSI/ISA	"Control Valve Manifold Designs" 1981
RP-39.6 ISA	"Valve Leakage Rates"

API	**(AMERICAN PETROLEUM INSTITUTE)**	
	RP-550	"Process Instrumentation and Control Section 6-Control Valve and Accessories" 1985
	STD-598	"Valve Inspection and Tests" 1987
	STD-609	"Butterfly Valves Lug Type and Wafer Type" 1983
ANSI	**(AMERICAN NATIONAL STANDARDS INSTITUTE)**	
	B-2.1	"Pipe Threads" 1968
	B-16.5	"Pipe Flanges and Flanged Rating for Class 150, 300, 400, 600, 900, 1500 and 2500" 1988
	B-16.10	"Face-to-Face Dimensions" 1980
	B-16.34	"Steel Valve Flanged Buttwelding End" 1981
	B-16.37	"Hydrostatic Testing of Control Valves" 1980
	ANSI/FCI-70.2 (1976) (R 1982) Formerly ANSI B16.104 (1982) "Valve Leakage Classifications"	
BSI	**(BRITISH STANDARDS INSTITUTION)**	
	BS-5155	"Butterfly Valves" 1988
	BS-1560: Part 2	"Flange Dimensions for Butterfly Valves" Class 150 and Class 300
IEC	**(INTERNATIONAL ELECTROTECHNICAL COMMISSION)**	
	IEC-34.5	"Classification of Degree of Protection Provided by Enclosure for Rotating Machines" 1981
	IEC-144	"Degree of Protection of Enclosures" 1963
	IEC-534.1	"Control Valve Terminology" 1984
NACE	**(NATIONAL ASSOCIATION FOR CORROSION ENGINEERS)**	
	MR-01-75	"Material Recommendation in Corrosive Atmosphere" 1975
NFPA	**(NATIONAL FIRE PROTECTION ASSOCIATIONS)**	
	70	"Practical Safeguarding of Persons and Property from Hazards Arising from the Use of Electricity" 1989
UL	**(UNDERWRITERS LABORATORIES)**	
	1002	"Electrically Operated Valves for Use in Hazardous Locations" 1983
	G32	"Vibratory Cavitation Erosion Test"
	G33	"Recommended Practice for Recording Data from Atmospheric Corrosion Tests of Metallic-Coated Steel Specimens"
	G34	"Test for Exfoliation Corrosion Susceptibility in 7XXX Series Copper-Containing Aluminum Alloys (EXCO Test)"
	G35	"Recommended Practice for Determining the Susceptibility of Stainless Steels and Related Nickel-Chromium-Iron Alloys to Stress Corrosion Cracking in Polythionic Acids"
	G36	"Recommended Practice for Performing Stress Corrosion Cracking Tests in a Boiling Magnesium Chloride Solution"
	G37	"Recommended Practice for Use of Mattsson's Solution of pH 7.2 to Evaluate the Stress Corrosion Cracking Susceptibility of Copper-Zinc Alloys"

G38	"Recommended Practices for Making and Using the C-Ring Stress Corrosion Cracking Test Specimen"
G39	"Preparation and Use of Bent-Beam Stress Corrosion Specimens"
G40	"Definition of Terms Relating to Erosion by Cavitation and Impingement"
G41	"Recommended Practice for Determining Cracking Susceptibility of Titanium Alloys Exposed Under Stress to a Hot Salt Environment"
G43	"Acidified Synthetic Sea Water (Fog) Testing"
G44	"Recommended Practice for Alternate Immersion Stress Corrosion Testing in 3.5% Sodium Chloride Solution"
G46	"Recommended Practice for Examination and Evaluation of Pitting Corrosion"
G47	"Recommended Practice for Determining Susceptibility to Stress Corrosion Cracking of High Strength 7 × × × Aluminum"
G48	"Test for Pitting and Crevice Corrosion Resistance of Stainless Steels and Related Alloys by the Use of Ferric Chloride Solution"
G49	"Recommended Practice for Preparation and Use of Direct Tension Stress Corrosion Test Specimens"
G50	"Recommended Practice for Conducting Atmospheric Corrosion Tests on Metals"
G51	"Test for pH of Soil"
G52	"Recommended Practice for Conducting Surface Sea Water Exposure Tests on Metals and Alloys"

BS	**(BRITISH STANDARDS)**
BS 5500	"Unfired Fusion Welded Pressure Vessels"
BS 5903	"Corrosion Testing of Austenitic Stainless Steel"
BS 6682	"Bimetallic Corrosion Determination"
BS 7318 PP	"Materials Testing for Design"
BS 534	"Steel Pipes, Fittings and Specials for Water, Gas and Sewage"

ISO	**(INTERNATIONAL ORGANIZATION FOR STANDARDIZATION)**	
ISO 9000	"Quality Systems"	
ISO 8993	Corrosion	Anodized aluminum and aluminum alloys—Rating system for the evaluation of pitting #—Chart method
ISO 8994	Corrosion	Anodized aluminum and aluminum alloys—Rating system for the evaluation of pitting #–Grid method
ISO 3651-1	Corrosion	Austenitic stainless steels–Determination of resistance to intergranular #—Part 1: Corrosion test in nitric acid*
ISO 3651-2	Corrosion	Austenitic stainless steels-Determination of resistance to intergranular #—Part 2: Corrosion test in a sulphuric*
ISO 5952	Corrosion	Continuously hot-rolled steel sheet of structural quality with improved atmospheric # resistance
ISO 6957	Corrosion	Copper alloys-Ammonia test for stress # resistance
ISO 4539	Corrosion	Electrodeposited chromium coatings—Electrolytic # testing (EC test)
ISO 7441	Corrosion	Corrosion of metals and alloys-Determination of bimetallic # in outdoor exposure corrosion tests

ISO 8407	Corrosion	Corrosion of metals and alloys—Removal of # products from corrosion test specimens
ISO 7539-1	Corrosion	Corrosion of metals and alloys—Stress # testing—Part 1: General guidance on testing procedures
ISO 7539-2	Corrosion	Corrosion of metals and alloys—Stress # testing-Part 2: Preparation and use of bent-beam specimens
ISO 7539-3	Corrosion	Corrosion of metals and alloys—Stress # testing—Part 3: Preparation and use of U-bend specimens
ISO 7539-4	Corrosion	Corrosion of metals and alloys—Stress # testing—Part 4: Preparation and use of Uniaxially loaded tension specimens
ISO 7539-5	Corrosion	Corrosion of metals and alloys—Stress # testing—Part 5: Preparation and use of C-ring specimens
ISO 7539-6	Corrosion	Corrosion of metals and alloys—Stress # testing—Part 6: Preparation and use of pre-cracked specimens
ISO 7539-7	Corrosion	Corrosion of metals and alloys—Stress # testing—Part 7: Slow strain rate testing
ISO 10446	Corrosion	All-weld metal test assembly for the classification of # resisting chromium and chromium-nickel steel covered arc welding*
ISO 3651-1	Corrosion	Determination of resistance to intergranular corrosion—Part 1: # test in nitric acid medium by measurement of loss in*
ISO 3651-2	Corrosion	Determination of resistance to intergranular corrosion—Part 2: # test in a sulphuric acid/copper sulphate medium in the*
ISO 6743-8	Corrosion	Industrial oils and related products (class L)—Classification—Part 8: Family R (Temporary protection against #)
ISO 4536	Corrosion	Metallic and non-organic coatings on metallic substrates—Saline droplets # test (SD test)
ISO 4541	Corrosion	Metallic and other non-organic coatings—Corrodkote # test (CORR test)
ISO 4543	Corrosion	Metallic and other non-organic coatings—General rules for # tests applicable for storage conditions
ISO 8403	Corrosion	Metallic coatings—Coatings anodic to the substrate-Rating of test specimens subjected to # tests
ISO 4540	Corrosion	Metallic coatings—Coatings cathodic to the substrate—Rating of electroplated test specimens subjected to # tests
ISO 1462	Corrosion	Metallic coatings—Coatings other than those anodic to the basis metal —Accelerated # tests—Method for the evaluation*
ISO 1459	Corrosion	Metallic coatings—Protection against # by hot dip galvanizing—Guiding principles
ISO 4538	Corrosion	Metallic coatings—Thioacetamide #test (TAA test) ISO 8565
	Corrosion	Metals and alloys—Atmospheric # testing—General requirements for field tests
ISO 9400	Corrosion	Nickel-based alloys—Determination of resistance to intergranular # ISO 9591 Corrosion of aluminum alloys—Determination of resistance to stress corrosion cracking

ISO 9223	Corrosion of metals and alloys—Corrosivity of atmospheres—Classification	
ISO 9224	Corrosion of metals and alloys—Corrosivity of atmospheres—Guiding values for the corrosivity categories	
ISO 9225	Corrosion of metals and alloys—Corrosivity of atmospheres—Measurement of pollution	
ISO 7441	Corrosion of metals and alloys—Determination of bimetallic corrosion in outdoor exposure corrosion tests	
ISO 6509	Corrosion of metals and alloys—Determination of dezincification resistance of brass	
ISO 8407	Corrosion of metals and alloys—Removal of corrosion products from corrosion test specimens	
ISO 8044	Corrosion of metals and alloys—Vocabulary	
ISO 4623	Corrosion	Paints and varnishes—Filiform # test on steel
ISO/TR 10129	Corrosion	Plain bearings—Testing of bearing metals—Resistance to # by lubricants under static conditions
ISO 3506	Corrosion-resistant stainless steel fasteners—Specifications	
ISO 6315	Corrosion	Road vehicles—Brake linings—Seizure to ferrous mating surface due to # –Test procedure
ISO 6505	Corrosion	Rubber, vulcanized—Determination of adhesion to, and # of, metals ISO 9455-12 Corrosion • Soft soldering fluxes—Test methods—Part 12: Steel tube # test
ISO/TR 10217	Corrosion	Solar energy—Water heating systems—Guide to material selectionwith regard to internal#
ISO 4952	Corrosion	Structural steels with improved atmospheric # resistance
ISO 10062	Corrosion tests in artificial atmosphere at very low concentrations of polluting gas(es) ISO 7384 Corrosion tests in artificial atmosphere—General requirements	
ISO 9227	Corrosion tests in artificial atmospheres—salt spray tests	
ISO 3160-2	Corrosion	Watch cases and accessories—Gold alloy coverings—Part 2: Determination of fineness, thickness, # resistance and *
ISO 5662	Corrosive	Petroleum products—Electrical insulating oils—Detection of # sulphur
ISO 2746	Corrosive	Vitreous and porcelain enamels—Enameled articles for service under highly # conditions—High voltage test
ISO 6251	Corrosiveness	Liquefied petroleum gases—# to copper-copper Strip Test
ISO 2160	Corrosiveness	Petroleum products—# to copper—Copper strip test

NACE (NATIONAL ASSOCIATION OF CORROSION ENGINEERS)

Corrosion Test Methods

TM-01-69	"Laboratory Corrosion Testing of Metals for the Process Industries"
TM-02-70	"Methods of Conducting Controlled Velocity Laboratory Corrosion Tests"
TM-01-71	"Autoclave Corrosion Testing of Metals in High Temperature Water"
TM-01-77	"Testing of Metals for Resistance to Sulfide Stress Cracking at Ambient Temperatures"
MR-01-75	"Sulfide Stress Cracking Resistance Metallic Materials for Oil Field Equipment"
(1991 Rev)	
TM-01-70	"Visual Standard for Surface of New Steel Airblast Cleaned with Sand Abrasive"

TM-01-72 "Antirust Properties of Petroleum Products Pipeline Cargoes"

TM-01-73 "Methods for Determining Water Quality for Subsurface Injection Using Membrane Filters"

TM-01-74 "Laboratory Methods for the Evaluation of Protective Coatings Used as Lining Materials in Immersion Service"

TM-02-74 "Dynamic Corrosion testing of Metals in High Temperature Water"

TM-03-74 "Laboratory Screening Tests to Determine the Ability of Scale Inhibitors to Prevent the Precipitation of Calcium sulfate and Calcium Carbonate from Solution"

TM-01-75 "Visual Standard for Surfaces of New Steel Centrifugally Blast Cleaned with steel Grit and Shot"

TM-02-75 "Performance Testing of Sucker Rods by the Mixed String, Alternate Rod Method"

TM-02-75 "Performance Testing of Sucker Rods by the Mixed String, Alternate Rod Method"

TM-03-75 "Abrasion Resistance Testing of Thin Film Baked Coatings and Linings Using the Falling Sand Method"

Glossary of Terms

SOUR WATER SERVICE

All process streams containing either:

1. Liquid water and at least 10 ppm of H_2S or;
2. Liquid water and at least 10 ppm of H_2S, cyanides, and small amounts of water-soluble organic acid.

ENGINEER

Refers to person or party representing the Company for supervision of design, engineering services, and execution of project as required and specified by the Company.

MANUFACTURER

The party that manufactures or produces line pipe and piping components according to the requirements of relevant IPS Standards.

CONSULTANT

Is the party which carries out all or part of a pipeline design and engineering.

SPECIFIC TERMS

Design Factor

Ratio of the hoop stress developed in the pipeline by the design pressure and the specified minimum yield stress (SMYS) of the pipeline material.

Flammable Fluid

A fluid having a flash point lower than 100°C.

Flow Line

A pipeline (including valves and fittings) for transporting untreated hydrocarbons and other reservoir.

ENGINEER

The Engineer referred to in this standard is a person or a body appointed in writing by the Company to be project manager.

EXECUTOR

The Executor is the party which carries out all or part of construction for the pipeline project.

GAS GATHERING LINE

A pipeline (including valves, traps, and fittings) between the block valve on the wellhead separator (or wellhead separator cluster) gas outlet line and the block valve on the NGL plant or production unit gas inlet line.

GAS TRANSMISSION LINE (GAS TRUNK LINE)

A pipeline (including valves, traps, and fittings) between the block valve on the NGL plant or gas refinery or gas compressor station gas outlet line and the block valve at gas distribution terminal or consumers premises inlet line but excluding the piping, valves, fittings, etc. between the booster stations main inlet and outlet block valves.

INCIDENTAL PRESSURE

Pressure which occurs in a pipeline with limited frequency and within a limited period of time, such as surge pressures and thermal expansions, if not occurring most of the time.

MAIN OIL LINE (OIL TRUNK LINE)

A pipeline (including valves and fittings) between the main block valve on the production unit oil outlet line and the main block valve on crude oil terminal inlet line. But excluding the piping, valves, fittings, etc. between the booster stations main inlet and outlet block valves.

MAXIMUM ALLOWABLE INCIDENTAL PRESSURE (MAIP)

The maximum pressure that is allowed by ANSI/ASME B31.4 and B31.8 to occur in a pipeline with a limited frequency and during limited period of time.

MAXIMUM ALLOWABLE OPERATING PRESSURE (MAOP)

The maximum pressure at which a pipeline is allowed to be operated under steady state process conditions, in accordance with ANSI/ASME B31.4 and B31.8.

NGL LINE

A pipeline (including valves and fittings) between the block valve at the NGL plant liquid outlet line and the block valve at the NGL distribution terminal or LPG plant or consumers premises inlet line.

SPECIFIED MINIMUM YIELD STRESS (SMYS)

The level of stress which produces 0.5% total strain (API definition). This is specified by the Company and shall be guaranteed by the Manufacturers/Suppliers/Vendors.

STABLE FLUID

A fluid which has an NFPA reactivity grade number of zero. (Refer to IPS-E-SF-100).

TOXIC FLUID

Includes all fluids in the slightly toxic, toxic, and highly toxic categories.

CONSTRAINT

A device which restrains the pipe from displacement and/or rotation.

CRADLE OR PIPE SHOE

A device placed between the pipe and the supporting structure. Usually made from plate or cut from sections, welded or clamped to the pipe.

DUMMY SUPPORT

A support welded to an elbow of the prefabricated pipe. The pipe dummy will rest on the support structure and so support the pipe at the change of direction. A pipe dummy is made from pipe, usually of a smaller diameter than the pipe to which it is welded.

ENGINEER

A person who shall be appointed time to time by the Company to exercise the functions entrusted to him under the contract and whose appointment has been notified by the Company in writing to the Executor.

EXECUTOR

The Executor is the party which carries out all or part of construction and/or commissioning of the project.

PIPE STANCHION

A tubular supporting element with base plate, being part of the prefabricated pipe. When considerable site adjusting is required, the pipe stanchion may be of the 2-piece type, welded together after adjustment. In the latter case, the upper part shall be part of the prefabricated pipe.

SHOE

Normally a tee section attached to the pipe that transmits the load or forces to the adjacent structure.

STRUCTURAL ATTACHMENTS

Structural attachments include elements which are welded, bolted, or clamped to the pipe, such as clips, lugs, rings, clamps, clevises, straps, and skirts.

SUPPORTING ELEMENTS

The broad terms "Supporting Elements" or "Supports" shall encompass the entire range of the various methods of carrying the weight of pipelines, insulation, and the fluid carried. It therefore includes "Hangers" which are generally considered as those elements which carry the weight from above, with the supporting members being mainly in tension.

Likewise, it includes "Supports" which on occasion are delineated as those which carry the weight from below, with the supporting members being mainly in compression. In many cases a supporting element may be a combination of both of these.

SWAY STRUT

An adjustable device, usually applied for restraining movement of piping in one direction while providing for movement in another direction.

TRUNNION

A tubular supporting device, branching-off horizontally from a vertical line, and resting on, or suspended from, the supporting structure.

U-CLIP

A general description of the various types of U-shaped straps, bolted or welded to the supporting structure, clamping or guiding the line. A typical use of such a clip (welded as well as bolted) is on vibrating compressor lines. In some cases the U-clip may have one spring loaded bolt and, if necessary, a shim plate welded to the bottom of the pipe.

BATTERY LIMIT

The boundary of a process unit, enclosing all equipment and unit limit block valves.

COMPLEX

A group of units, the operation of which are interlinked. (On small plants, the term “complex” may refer to all the process units on the plant).

PIPING SYSTEM

It covers the overall systems of pipes and piping components, e.g. fittings, valves, nozzles and supports, that are employed to transfer liquid, gas, steam, etc. between the equipment such as tanks, pumps, vessels and so forth.

UNIT (AREA)

A main production component of a refinery plant or chemical complex, e.g. distillation unit, utility unit, etc.

UTILITIES

Air supply, water supply and treatment, steam generation, power generation and similar services.

DEPOT

A storage area with capacity less than 5000 tons with import and export facilities.

EXECUTOR

The party which carries out all, or part of the construction, installation and commissioning aspect for the projects.

STEAM SERVICES

Low-pressure service steam	up to 700 kPag (7 barg)
Medium-pressure service steam above	700 kPag (7 barg) up to 2400 kPag (24 barg)
High-pressure service steam	above 2400 kPag (24 barg), normally 4000 kPag (40 barg)

COLD SPRING

Cold spring is the intentional deformation of piping during assembly to produce a desired initial displacement and stress.

STRESS RELIEVING

Uniform heating of a structure or portion thereof to a sufficient temperature and maintaining for a specified period to relieve the major portion of the residual stresses followed by uniform and controlled cooling.

PIPING COMPONENT

Mechanical elements suitable for joining or assembly into pressure-tight fluid-containing piping systems. Components include pipe, tubing, fittings, flanges, gaskets, bolting, valves, and devices such as expansion joints, flexible joints, pressure hoses, traps, strainers, in-line portions of instruments, and separators.

ACCELERATION CORROSION TEST

Method designed to approximate, in a short-time, the deteriorating effect under normal long-term service condition.

AERATED

Solution containing more than 10 ppb (Parts Per billion) oxygen.

AGE HARDENING

Hardening by aging, usually after rapid cooling or cold working.

AGING

A change in the properties of certain metals and alloys that occurs at ambient or moderately elevated temperatures after hot working or a heat treatment (quench aging in ferrous alloys, natural or artificial aging in ferrous and non-ferrous alloys) or after a cold working operation (strain aging). The change in properties is often but not always, due to a phase change (precipitation), but never involves a change in chemical composition of the metal or alloy.

ALLOY STEEL

Is one that contains either silicon or manganese in amounts in excess of those quoted in plain carbon steel or that contains any other element, or elements, as the result of deliberately made alloying additions.

ANNEALING

A generic term denoting a treatment, consisting of heating to and holding at a suitable temperature, followed by cooling at a suitable rate, used primarily to soften metallic materials, but also to simultaneously produce microstructure. The purpose of such changes may be, but is not confined to improvement of machinability, facilitation of cold work, improvement of mechanical or electrical properties and/or increase in stability of dimensions. When the term is used by itself, full annealing is implied. When applied only for the relief of stress, the process is properly called stress relieving or stress-relief annealing.

AUSTENITE

A solid solution of one or more elements in face-centered cubic iron. Unless otherwise designated (such as nickel austenite) the solute is generally assumed to be carbon.

BAINITE

A metastable aggregate of ferrite and cementite resulting from the transformation of austenite at temperatures below the pearlite but above the martensite start temperature.

BRITTLE FRACTURE

Separation of a solid accompanied by little or no macroscopic plastic deformation. Typically, brittle fracture occurs by rapid crack propagation with less expenditure of energy than for ductile fracture.

CASE HARDENING

A generic term covering several processes applicable to steel that change the chemical composition of the surface layer by absorption of carbon, nitrogen or a mixture of the two. Also, by diffusion, create a concentration gradient. The outer portion, or case, is made substantially harder than the inner portion, or core.

CAUSTIC EMBRITTLEMENT

An obsolete historical term denoting a form of stress corrosion cracking most frequently encountered in carbon steels or iron-chromium-nickel alloys that are exposed to concentrated hydroxide solutions at temperatures of 200–250°C.

COLD WORKING

Deforming metal plastically under conditions of temperature and strain rate that induce strain hardening usually, but not necessarily, conducted at room temperature. Contrast with hot working.

HOT WORKING

Deforming metal plastically at such a temperature and strain rate that recrystallization takes place simultaneously with the deformation, thus avoiding any strain hardening. Contrast with cold working.

HYDROGEN SULFIDE

Without H_2S Environment containing less than 1 ppm H_2S.

With H_2S	Environment containing more than 1 ppm H_2S.
Low H_2S	Greater than 3 ppm but less than 6 ppm H_2S.
High H_2S	More than 6 ppm

KILLED STEEL

Thoroughly deoxidized steel, e.g., by addition of aluminum or silicon, in which the reduction between carbon and oxygen during solidification is suppressed.

NORMALIZING

Heating a ferrous alloy to a suitable temperature above the transformation range and then cooling in air to a temperature substantially below the transformation range.

PEARLITE

A metastable lamellar aggregate of ferrite and cementite resulting from the transformation of austenite at temperatures above the bainite range.

POWDER METALLURGY

The art of producing metal powders and utilizing metal powders for production of massive materials and shaped objects.

PRECIPITATION HARDENING

Hardening caused by the precipitation of a constituents from a supersaturated solution. See also age hardening and aging.

QUENCH AGING

Aging induced by rapid cooling after solution heat treatment.

QUENCH-AGE EMBRITTLEMENT

Embrittlement of low carbon steels resulting from precipitation of solute carbon of existing dislocations and from precipitation hardening of the steel caused by differences in the solid solubility of carbon in ferrite at different temperatures. Quench age embrittlement usually is caused by rapid cooling of the steel from temperature slightly below AC1 (The temperature at which austenite begins to form), and can be minimized by quenching from lower temperature.

QUENCH CRACKING

Fracture of a metal during quenching from elevated temperature. Most frequently observed in hardened carbon steel, alloy steel, or tool steel parts of high hardness and low toughness. Cracks often emanate from filets, holes, corners, or other stress raisers and result from high stresses due to the volume changes accompanying transformation to martensite.

QUENCH HARDENING

(1) In ferrous alloys, hardening by austenitizing and then cooling at a rate such that a substantial amount of austenite transforms to martensite. (2) In copper and titanium alloys hardening by solution treating and quenching to develop a martensite like structure.

QUENCHING

Rapid cooling of metals (often steels) from a suitably elevated temperature. This generally is accomplished by immersion in water, oil, polymer solution, or salt, although forced air is sometimes used.

SENSITIZATION HEAT TREATMENT

A heat treatment, whether accidental, intentional, or incidental (as during welding) that causes precipitation of constituents at grain boundaries, often causing the alloy to become susceptible to intergranullar corrosion, cracking or SCC (Stress Corrosion Cracking).

SENSITIZATION

In Austenitic stainless steels the precipitation of chromium carbide usually at grain boundaries, on exposure to temperatures of about 550–850°C.

STEEL

Ferrite

Ferrite is the name given to the body centered cubic allotropes of iron, á and ä iron, and to body centered cubic solid solutions.

Austenite

Austenite is the name given to the faced centered cubic, or ã, variety of iron, and to the face centered cubic solid solutions.

Cementite

Is the name given to the carbide of iron, Fe_3C. This is an extremely hard and brittle constituents.

Pearlite

Pearlite is the eutectoid mixture of ferrite and cementite, and is formed when Austenite decomposes during cooling. It consists of alternate thin layers, or lamellae, of ferrite and cementite.

Martensite

This is the name given to the very hard and brittle constituent that is formed when a steel is very rapidly cooled from the Austenitic state. It is a ferrite, highly super saturated with dissolved carbon.

Sorbite and Troostite

These are names given to the structures produced when martensite or bainite is tempered, that is, heated to same temperature not exceeding 700°C for the purpose of reducing brittleness and hardness.

Bainite

This is the term that is given to the decomposition product that is formed when austenite decomposes by either isotherm transformation, or at a cooling rate intermediate between the very rapid cooling necessary for martensite and the slower rate of cooling at which pearlite is formed.

Plain Carbon Steel

Is a steel containing up to 1.5% of carbon together with not more than 0.5% of silicon and not more than 1.5% of manganese, and only traces of other elements.

Alloy Steel

Is one that contains either silicon or manganese in amounts in excess of those quoted above, or that contains any other element, or elements, as the result of deliberately made alloying additions.

Tempering

To reheat hardened steel or hardened cast Iron to some temperature below the eutectoid temperature for the purpose of decreasing hardness and increasing toughness.

Transition Temperature

Temperature within a range in which the ductility changes rapidly with temperature.

Tensile Strength

The ratio of maximum load to original cross section area.

BODY

The part of the valve which is the main pressure boundary. The body also provides the pipe connecting ends, the fluid flow passageway, and may support the seating surfaces and the valve closure member.

BONNET

That portion of the valve pressure retaining boundary which may guide the stem and contains the packing box and stem seal. It may also provide the principal opening to the body cavity for assembly of internal parts or be an integral part of the valve body. It may also provide for the attachment of the actuator to the valve body.

BONNET TYPES

Typical bonnets are bolted, threaded, or welded to or integral with the body. Other types sometimes used are defined below.

EXTENSION BONNET

A bonnet with a packing box that is extended above the bonnet joint of the valve body so as to maintain the temperature of the packing above or below the temperature of the process fluid. The length of the extension bonnet is dependent upon the difference between the fluid temperature and the packing design temperature limit as well as upon the valve body design.

CLOSURE MEMBER

A movable part of the valve which is positioned in the flow path to modify the rate of flow through the valve.

FLOW CONTROL ORIFICE

The part of the flow passageway that, with the closure member, modifies the rate of flow through the valve. The orifice may be provided with a seating surface, to be contacted by or closely fitted to the closure member, to provide tight shutoff or limited leakage.

SEAT RING

A part that is assembled in the valve body and may provide part of the flow control orifice. The seat ring may have special material properties and may provide the contact surface for the closure member.

CAGE

A part in a globe valve surrounding the closure member to provide alignment and facilitate assembly of other parts of the valve trim. The cage may also provide flow characterization and/or a seating surface for globe valves and flow characterization for some plug valves.

INTEGRAL SEAT

A flow control orifice and seat that is an integral part of the body or cage material or may be constructed from material added to the body or cage.

STEM

The stem rod, shaft or spindle which connects the valve actuator with the closure member.

STEM SEALS

The part or parts needed to effect a pressure-tight seal around the stem while allowing movement of the stem.

PACKING

A sealing system consisting of deformable material of one or more mating and deformable elements contained in a packing box which may have an adjustable compression means to obtain or maintain an effective pressure seal.

PACKING BOX

The chamber, in the bonnet, surrounding the stem and containing packing and other stem sealing parts.

BUSHING

A fixed member which supports and/or guides the closure member, valve stem and/or actuator stem. The bushing supports the nonaxial loads on these parts and is subject to relative motion of the parts.

ABBREVIATIONS

Ap	action required by purchaser
CE	carbon equivalent
CTOD	crack tip opening displacement
DN	nominal diameter
DWTT	drop weight tear test
EMT	electromagnetic testing
ERW	electric resistance welding
ESD	emergency shut down
FBH	flat bottomed hole
FCAW	flux cored arc welding
GMAW	gas metal-arc welding
GRE	glass-reinforced epoxy
GRP	glass-reinforced plastics
HAZ	heat-affected zone
HFI	high-frequency induction
HFW	high-frequency electric welding
HIC	hydrogen-induced cracking
LPG	liquefied petroleum gas
MAIP	maximum allowable incidental pressure
MAOP	maximum allowable operating pressure
MT	magnetic particle testing
NDT	nondestructive testing
NGL	natural gas liquid
NPS	nominal pipe size (inch)
OD	outside diameter
PCM	material cracking parameter
PN	nominal pressure rating (class) designation
PT	liquid penetrant examination
RDH	radial drilled hole
Re	Reynolds No.
RF	raised face
ROW	right-of-way
RT	radiographic (or radiological) testing
RTJ	ring type joint
SAW	submerged arc welding
SI	International System of Units
SMLS	seamless
SMYS	specified minimum yield stress
SPW	spiral welding
SS	sour service
TMCP	thermomechanically controlled process
UT	ultrasonic testing

Index

Note: Page numbers followed by "*f*" and "*t*" refer to figures and tables, respectively.

B

C

E

F

G

N

O

P

S

T

U

V

X

Y

Z

FSC
www.fsc.org
MIX
Papier | Fördert
gute Waldnutzung
FSC® C083411

Kolibri 360 GmbH
Ferdinand-Jühlke-Straße 7
99095 Erfurt, Deutschland
produktsicherheit@kolibri360.de